HOMOGENEOUS TURBULENCE DYNAMICS

This book summarizes the most recent theoretical, computational, and experimental results dealing with homogeneous turbulence dynamics. A large class of flows is covered: flows governed by anisotropic production mechanisms (e.g., shear flows) and flows without production but dominated by waves (e.g., homogeneous rotating or stratified turbulence). Compressible turbulent flows are also considered. In each case, main trends are illustrated using computational and experimental results, and both linear and nonlinear theories and closures are discussed. Details about linear theories (e.g., Rapid Distortion Theory and variants) and nonlinear closures (e.g., EDQNM) are provided in dedicated chapters, following a fully unified approach. The emphasis is on homogeneous flows, including several interactions (rotation, stratification, shear, shock waves, acoustic waves, and more) that are pertinent to many applications fields – from aerospace engineering to astrophysics and Earth sciences.

After graduating from Université Pierre et Marie Curie–Paris 6 in 1995, Professor Pierre Sagaut worked as a research engineer at ONERA (French National Aerospace Research Center) until 2002. That year, he returned to accept a professorship in mechanics at the Université Pierre et Marie Curie and now teaches there, as well as at the Ecole Polytechnique. He remains a consultant at ONERA, with IFP and CERFACS (France). His primary research interests include fluid mechanics, aeroacoustics, numerical simulation of turbulent flows (both direct and large-eddy simulation), and numerical methods. He is also involved in uncertainty modeling for CFD. He authored and coauthored more than 90 papers in peer-reviewed international journals and 130 proceedings papers. He is the author of several books dealing with turbulence modeling and simulation. He is a member of several editorial boards: *Theoretical and Computational Fluid Dynamics*, *Journal of Scientific Computing*, and *Progress in CFD*. Three times he has received the ONERA award for the best publication and the John Green Prize (delivered by ICAS, 2002). He is the coauthor of *Large-Eddy Simulation for Acoustics*, also published by Cambridge University Press.

Dr. Claude Cambon has been a senior member of the CNRS (French National Center for Scientific Research) since 1979. He has consulted at ONERA since 1988. As chairman of the Pilot Center PEPIT, he recently rebuilt the network into the Centre Henri Bénard of the European Research Community on Flow, Turbulence & Combustion. Dr. Cambon is currently working in the Laboratoire de Mécanique des Fluides & d'Acoustique at Ecole Centrale de Lyon as a CNRS agent and is responsible for a team studying waves and turbulence. He lectures at the graduate level and in summer programs. During the summers he has visited at the Center for Turbulence Research, Stanford University/NASA Ames (1990 to 1999). Also in 1999, Dr. Cambon was invited for a two-month residency to the Turbulence Program at Isaac Newton Institute, Cambridge.

Homogeneous Turbulence Dynamics

PIERRE SAGAUT

Université Pierre et Marie Curie

CLAUDE CAMBON

Ecole Centrale de Lyon

CAMBRIDGE UNIVERSITY PRESS
Cambridge, New York, Melbourne, Madrid, Cape Town, Singapore, São Paulo, Delhi

Cambridge University Press
32 Avenue of the Americas, New York, NY 10013-2473, USA

www.cambridge.org
Information on this title: www.cambridge.org/9780521855488

© Pierre Sagaut and Claude Cambon 2008

This publication is in copyright. Subject to statutory exception
and to the provisions of relevant collective licensing agreements,
no reproduction of any part may take place without
the written permission of Cambridge University Press.

First published 2008

Printed in the United States of America

A catalog record for this publication is available from the British Library.

Library of Congress Cataloging in Publication Data

Sagaut, Pierre, 1967–
Homogeneous turbulence dynamics / Pierre Sagaut and Claude Cambon.
   p. cm.
Includes bibliographical references and index.
ISBN 978-0-521-85548-8 (hardback)
1. Turbulence – Mathematical models. 2. Anisotropy – Mathematical models.
3. Shear waves – Mathematical models. I. Cambon, Claude, 1952– II. Title.
TA357.5.T87S269 2008
620.1′064 – dc22   2008008774

ISBN   978-0-521-85548-8 hardback

Cambridge University Press has no responsibility for
the persistence or accuracy of URLs for external or
third-party Internet Web sites referred to in this publication
and does not guarantee that any content on such
Web sites is, or will remain, accurate or appropriate.

"Wir müssen wissen, wir werden wissen"
(We must know, we shall know)

David Hilbert

Contents

Abbreviations Used in This Book   *page* xvi

1 Introduction   1
    1.1   Scope of the Book   1
    1.2   Structure and Contents of the Book   3

      Bibliography   9

2 Statistical Analysis of Homogeneous Turbulent Flows: Reminders   10
    2.1   Background Deterministic Equations   10
          2.1.1   Mass Conservation   10
          2.1.2   The Navier–Stokes Momentum Equations   12
          2.1.3   Incompressible Turbulence   13
          2.1.4   First Insight into Compressibility Effects   14
          2.1.5   Reminder About Circulation and Vorticity   15
          2.1.6   Adding Body Forces or Mean Gradients   16
    2.2   Briefs About Statistical and Probabilistic Approaches   19
          2.2.1   Ensemble Averaging, Statistical Homogeneity   19
          2.2.2   Single-Point and Multipoint Moments   19
          2.2.3   Statistics for Velocity Increments   20
          2.2.4   Application of Reynolds Decomposition to Dynamical Equations   20
    2.3   Reynolds Stress Tensor and Related Equations   22
          2.3.1   RST Equations   22
          2.3.2   The Mean Flow Consistent With Homogeneity   24
          2.3.3   Homogeneous RST Equations. Briefs About Closure Methods   26
    2.4   Anisotropy in Physical Space. Single-Point and Two-Point Correlations   27
    2.5   Spectral Analysis, From Random Fields to Two-Point Correlations. Local Frame, Helical Modes   28
          2.5.1   Second-Order Statistics   28
          2.5.2   Poloidal–Toroidal Decomposition and Craya–Herring Frame of Reference   31
          2.5.3   Helical-Mode Decomposition   32
          2.5.4   Use of Projection Operators   33
          2.5.5   Nonlinear Dynamics   35
          2.5.6   Background Nonlinearity in Different Reference Frames   36

2.6	Anisotropy in Fourier Space		38
	2.6.1	Second-Order Velocity Statistics	38
	2.6.2	Some Comments About Higher-Order Statistics	43
2.7	A Synthetic Scheme of the Closure Problem: Nonlinearity and Nonlocality		43

Bibliography **47**

3 Incompressible Homogeneous Isotropic Turbulence 49

3.1	Observations and Measures in Forced and Freely Decaying Turbulence		49
	3.1.1	How to Generate Isotropic Turbulence?	49
	3.1.2	Main Observed Statistical Features of Developed Isotropic Turbulence	51
	3.1.3	Energy Decay Regimes	57
	3.1.4	Coherent Structures in Isotropic Turbulence	58
3.2	Self-Similar Decay Regimes, Symmetries, and Invariants		59
	3.2.1	Symmetries of Navier–Stokes Equations and Existence of Self-Similar Solutions	59
	3.2.2	Algebraic Decay Exponents Deduced From Symmetry Analysis	62
	3.2.3	Time-Variation Exponent and Inviscid Global Invariants	64
	3.2.4	Refined Analysis Without PLE Hypothesis	65
	3.2.5	Self-Similarity Breakdown	66
	3.2.6	Self-Similar Decay in the Final Region	67
3.3	Reynolds Stress Tensor and Analysis of Related Equations		68
3.4	Classical Statistical Analysis: Energy Cascade, Local Isotropy, Usual Characteristic Scales		70
	3.4.1	Double Correlations and Typical Scales	70
	3.4.2	(Very Brief) Reminder About Kolmogorov Legacy, Structure Functions, "Modern" Scaling Approach	71
	3.4.3	Turbulent Kinetic-Energy Cascade in Fourier Space	73
3.5	Advanced Analysis of Energy Transfers in Fourier Space		76
	3.5.1	The Background Triadic Interaction	76
	3.5.2	Nonlinear Energy Transfers and Triple Correlations	79
	3.5.3	Global and Detailed Conservation Properties	80
	3.5.4	Advanced Analysis of Triadic Transfers and Waleffe's Instability Assumption	81
	3.5.5	Further Discussions About the Instability Assumption	85
	3.5.6	Principle of Quasi-Normal Closures	86
	3.5.7	EDQNM for Isotropic Turbulence. Final Equations and Results	89
3.6	Topological Analysis, Coherent Events, and Related Dynamics		97

		3.6.1	Topological Analysis of Isotropic Turbulence	98
		3.6.2	Vortex Tube: Statistical Properties and Dynamics	102
		3.6.3	Bridging with Turbulence Dynamics and Intermittency	107
	3.7	Nonlinear Dynamics in the Physical Space		109
		3.7.1	On Vortices, Scales, Wavenumbers, and Wave Vectors – What are the Small Scales?	109
		3.7.2	Is There an Energy Cascade in the Physical Space?	111
		3.7.3	Self-Amplification of Velocity Gradients	112
		3.7.4	Non-Gaussianity and Depletion of Nonlinearity	116
	3.8	What are the Proper Features of Three-Dimensional Navier–Stokes Turbulence?		117
		3.8.1	Influence of the Space Dimension: Introduction to d-Dimensional Turbulence	117
		3.8.2	Pure 2D Turbulence and Dual Cascade	118
		3.8.3	Role of Pressure: A View of Burgers' Turbulence	120
		3.8.4	Sensitivity with Respect to Energy-Pumping Process: Turbulence with Hyperviscosity	122

Bibliography 123

4 Incompressible Homogeneous Anisotropic Turbulence: Pure Rotation 127

	4.1	Physical and Numerical Experiments		127
		4.1.1	Brief Review of Experiments, More or Less in the Configuration of Homogeneous Turbulence	129
	4.2	Governing Equations		131
		4.2.1	Generals	131
		4.2.2	Important Nondimensional Numbers. Particular Regimes	131
	4.3	Advanced Analysis of Energy Transfer by DNS		133
	4.4	Balance of RST Equations. A Case Without "Production." New Tensorial Modeling		135
	4.5	Inertial Waves. Linear Regime		139
		4.5.1	Analysis of Deterministic Solutions	139
		4.5.2	Analysis of Statistical Moments. Phase Mixing and Low-Dimensional Manifolds	143
	4.6	Nonlinear Theory and Modeling: Wave Turbulence and EDQNM		145
		4.6.1	Full Exact Nonlinear Equations. Wave Turbulence	145
		4.6.2	Second-Order Statistics: Identification of Relevant Spectral-Transfer Terms	148
		4.6.3	Toward a Rational Closure with an EDQNM Model	149
		4.6.4	Recovering the Asymptotic Theory of Inertial Wave Turbulence	150
	4.7	Fundamental Issues: Solved and Open Questions		153
		4.7.1	Eventual Two-Dimensionalization or Not	153

		4.7.2	Meaning of the Slow Manifold	155

- 4.7.2 Meaning of the Slow Manifold — 155
- 4.7.3 Are Present DNS and LES Useful for Theoretical Prediction? — 156
- 4.7.4 Is the Pure Linear Theory Relevant? — 157
- 4.7.5 Provisional Conclusions About Scaling Laws and Quantified Values of Key Descriptors — 158
- 4.8 Coherent Structures, Description, and Dynamics — 159

Bibliography — 164

5 Incompressible Homogeneous Anisotropic Turbulence: Strain — 167

- 5.1 Main Observations — 167
- 5.2 Experiments for Turbulence in the Presence of Mean Strain. Kinematics of the Mean Flow — 169
  - 5.2.1 Pure Irrotational Strain, Planar Distortion — 170
  - 5.2.2 Axisymmetric (Irrotational) Strain — 172
  - 5.2.3 The Most General Case for 3D Irrotational Case — 173
  - 5.2.4 More General Distortions. Kinematics of Rotational Mean Flows — 173
- 5.3 First Approach in Physical Space to Irrotational Mean Flows — 174
  - 5.3.1 Governing Equations, RST Balance, and Single-Point Modeling — 174
  - 5.3.2 General Assessment of RST Single-Point Closures — 177
  - 5.3.3 Linear Response of Turbulence to Irrotational Mean Strain — 178
- 5.4 The Fundamentals of Homogeneous RDT — 180
  - 5.4.1 Qualitative Trends Induced by the Green's Function — 183
  - 5.4.2 Results at Very Short Times. Relevance at Large Elapsed Times — 183
- 5.5 Final RDT Results for Mean Irrotational Strain — 184
  - 5.5.1 General RDT Solution — 184
  - 5.5.2 Linear Response of Turbulence to Axisymmetric Strain — 185
- 5.6 First Step Toward a Nonlinear Approach — 186
- 5.7 Nonhomogeneous Flow Cases. Coherent Structures in Strained Homogeneous Turbulence — 187

Bibliography — 189

6 Incompressible Homogeneous Anisotropic Turbulence: Pure Shear — 192

- 6.1 Physical and Numerical Experiments: Kinetic Energy, RST, Length Scales, Anisotropy — 192
  - 6.1.1 Experimental and Numerical Realizations — 193
  - 6.1.2 Main Observations — 193

	6.2	Reynolds Stress Tensor and Analysis of Related Equations	197
	6.3	Rapid Distortion Theory: Equations, Solutions, Algebraic Growth	199
		6.3.1 Some Properties of RDT Solutions	201
		6.3.2 Relevance of Homogeneous RDT	204
	6.4	Evidence and Uncertainties for Nonlinear Evolution: Kinetic-Energy Exponential Growth Using Spectral Theory	206
	6.5	Vortical–Structure Dynamics in Homogeneous Shear Turbulence	207
	6.6	Self-Sustaining Turbulent Cycle in Homogeneous Sheared Turbulence	209
	6.7	Self-Sustaining Processes in Nonhomogeneous Sheared Turbulence: Exact Coherent States and Traveling-Wave Solutions	210
	6.8	Local Isotropy in Homogeneous Shear Flows	214

Bibliography 217

7 Incompressible Homogeneous Anisotropic Turbulence: Buoyancy and Stable Stratification 219

	7.1	Observations, Propagating and Nonpropagating Motion. Collapse of Vertical Motion and Layering	219
	7.2	Simplified Equations, Using Navier–Stokes and Boussinesq Approximations, With Uniform Density Gradient	223
		7.2.1 Reynolds Stress Equations With Additional Scalar Variance and Flux	224
		7.2.2 First Look at Gravity Waves	225
	7.3	Eigenmode Decomposition. Physical Interpretation	226
	7.4	The Toroidal Cascade as a Strong Nonlinear Mechanism Explaining the Layering	229
	7.5	The Viewpoint of Modeling and Theory: RDT, Wave Turbulence, EDQNM	231
	7.6	Coherent Structures: Dynamics and Scaling of the Layered Flow, "Pancake" Dynamics, Instabilities	235
		7.6.1 Simplified Scaling Laws	235
		7.6.2 Pancake Structures, Zig-Zag, and Kelvin–Helmholtz Instabilities	237

Bibliography 241

8 Coupled Effects: Rotation, Stratification, Strain, and Shear 243

	8.1	Rotating Stratified Turbulence	243
		8.1.1 Basic Triadic Interaction for Quasi-Geostrophic Cascade	246
		8.1.2 About the Case With Small but Nonnegligible f/N Ratio	247

		8.1.3 The QG Model Revisited. Discussion	248
	8.2	Rotation or Stratification With Mean Shear	250
		8.2.1 The Rotating-Shear-Flow Case	253
		8.2.2 The Stratified-Shear-Flow Case	255
		8.2.3 Analogies and Differences Between the Two Cases	255
	8.3	Shear, Rotation, and Stratification. RDT Approach to Baroclinic Instability	256
		8.3.1 Physical Context, the Mean Flow	256
		8.3.2 RDT Equations	258
	8.4	Elliptical Flow Instability From "Homogeneous" RDT	259
	8.5	Axisymmetric Strain With Rotation	265
	8.6	Relevance of RDT and WKB RDT Variants for Analysis of Classical Instabilities	266

Bibliography 270

9 Compressible Homogeneous Isotropic Turbulence 273

9.1	Introduction to Modal Decomposition of Turbulent Fluctuations		273
	9.1.1	Statement of the Problem	273
	9.1.2	Kovasznay's Linear Decomposition	274
	9.1.3	Weakly Nonlinear Corrected Kovasznay Decomposition	278
	9.1.4	Helmholtz Decomposition and Its Extension	279
	9.1.5	Bridging Between Kovasznay and Helmholtz Decomposition	281
	9.1.6	On the Feasibility of a Fully General Modal Decomposition	281
9.2	Mean-Flow Equations, Reynolds Stress Tensor, and Energy Balance in Compressible Flows		281
	9.2.1	Arbitrary Flows	281
	9.2.2	Simplifications in the Isotropic Case	285
	9.2.3	Quasi-Isentropic Isotropic Turbulence: Physical and Spectral Descriptions	288
9.3	Different Regimes in Compressible Turbulence		291
	9.3.1	Quasi-Isentropic Turbulent Regime	292
	9.3.2	Weakly Compressible Thermal Regime	309
	9.3.3	Nonlinear Subsonic Regime	315
	9.3.4	Supersonic Regime	318
9.4	Structures in the Physical Space		319
	9.4.1	Turbulent Structures in Compressible Turbulence	320
	9.4.2	A Probabilistic Model for Shocklets	321

Bibliography 324

10 Compressible Homogeneous Anisotropic Turbulence — 327

- 10.1 Effects of Compressibility in Free-Shear Flows. Observations — 327
  - 10.1.1 RST Equations and Single-Point Modeling — 328
  - 10.1.2 Preliminary Linear Approach: Pressure-Released Limit and Irrotational Strain — 330
- 10.2 A General Quasi-Isentropic Approach to Homogeneous Compressible Shear Flows — 332
  - 10.2.1 Governing Equations and Admissible Mean Flows — 333
  - 10.2.2 Properties of Admissible Mean Flows — 335
  - 10.2.3 Linear Response in Fourier Space. Governing Equations — 336
- 10.3 Incompressible Turbulence With Compressible Mean-Flow Effects: Compressed Turbulence — 342
- 10.4 Compressible Turbulence in the Presence of Pure Plane Shear — 344
  - 10.4.1 Qualitative Results — 344
  - 10.4.2 Discussion of Results — 345
  - 10.4.3 Toward a Complete Linear Solution — 348
- 10.5 Perspectives and Open Issues — 349
  - 10.5.1 Homogeneous Shear Flows — 350
  - 10.5.2 Perspectives Toward Inhomogeneous Shear Flows — 350
- 10.6 Topological Analysis, Coherent Events and Related Dynamics — 351
  - 10.6.1 Nonlinear Dynamics in the Subsonic Regime — 352
  - 10.6.2 Topological Analysis of the Rate-of-Strain Tensor — 354
  - 10.6.3 Vortices, Shocklets, and Dynamics — 355

Bibliography — 356

11 Isotropic Turbulence–Shock Interaction — 358

- 11.1 Brief Survey of Existing Interaction Regimes — 358
  - 11.1.1 Destructive Interactions — 358
  - 11.1.2 Nondestructive Interactions — 359
- 11.2 Linear Nondestructive Interaction — 360
  - 11.2.1 Shock Modeling and Jump Relations — 360
  - 11.2.2 Introduction to the Linear Interaction Approximation Theory — 361
  - 11.2.3 Vortical Turbulence–Shock Interaction — 363
  - 11.2.4 Acoustic Turbulence–Shock Interaction — 370
  - 11.2.5 Mixed Turbulence–Shock Interaction — 373
  - 11.2.6 On the Use of RDT for Linear Nondestructive Interaction Modeling — 378
- 11.3 Nonlinear Nondestructive Interactions — 379
  - 11.3.1 Turbulent Jump Conditions for the Mean Field — 379
  - 11.3.2 Jump Conditions for an Incident Isotropic Turbulence — 381

Bibliography — 382

12 Linear Interaction Approximation for Shock–Perturbation Interaction ... 384

- 12.1 Shock Description and Emitted Fluctuating Field — 384
- 12.2 Calculation of Wave Vectors of Emitted Waves — 386
  - 12.2.1 General — 386
  - 12.2.2 Incident Entropy and Vorticity Waves — 386
  - 12.2.3 Incident Acoustic Waves — 389
- 12.3 Calculation of Amplitude of Emitted Waves — 391
  - 12.3.1 General Decompositions of the Perturbation Field — 391
  - 12.3.2 Calculation of Amplitudes of Emitted Waves — 393
- 12.4 Reconstruction of the Second-Order Moments — 395
  - 12.4.1 Case of a Single Incident Wave — 395
  - 12.4.2 Case of an Incident Turbulent Isotropic Field — 399
- 12.5 *A posteriori* Assessment of LIA — 403

Bibliography ... **405**

13 Linear Theories. From Rapid Distortion Theory to WKB Variants ... 406

- 13.1 Rapid Distortion Theory for Homogeneous Turbulence — 406
  - 13.1.1 Solutions for ODEs in Orthonormal Fixed Frames of Reference — 406
  - 13.1.2 Using Solenoidal Modes for a Green's Function with a Minimal Number of Components — 408
  - 13.1.3 Prediction of Statistical Quantities — 409
  - 13.1.4 RDT for Two-Time Correlations — 412
- 13.2 Zonal RDT and Short-Wave Stability Analysis — 412
  - 13.2.1 Irrotational Mean Flows — 413
  - 13.2.2 Zonal Stability Analysis With Disturbances Localized Around Base-Flow Trajectories — 413
  - 13.2.3 Using Characteristic Rays Related to Waves Instead of Trajectories — 415
- 13.3 Application to Statistical Modeling of Inhomogeneous Turbulence — 417
  - 13.3.1 Transport Models Along Mean Trajectories — 417
  - 13.3.2 Semiempirical Transport "Shell" Models — 418
- 13.4 Conclusions, Recent Perspectives Including Subgrid-Scale Dynamics Modeling — 419

Bibliography ... **421**

14 Anisotropic Nonlinear Triadic Closures ... 423

- 14.1 Canonical HIT, Dependence on the Eddy Damping for the Scaling of the Energy Spectrum in the Inertial Range — 423

		14.2	Solving the Linear Operator to Account for Strong Anisotropy	425
			14.2.1 Random and Averaged Nonlinear Green's Functions	425
			14.2.2 Homogeneous Anisotropic Turbulence with a Mean Flow	426
		14.3	A General EDQN Closure. Different Levels of Markovianization	428
			14.3.1 EDQNM2 Version	429
			14.3.2 A Simplified Version: EDQNM1	430
			14.3.3 The Most Sophisticated Version: EDQNM3	431
		14.4	Application of Three Versions to the Rotating Turbulence	433
		14.5	Other Cases of Flows With and Without Production	437
			14.5.1 Effects of the Distorting Mean Flow	437
			14.5.2 Flows Without Production Combining Strong and Weak Turbulence	438
			14.5.3 Role of the Nonlinear Decorrelation Time Scale	440
		14.6	Connection with Self-Consistent Theories: Single Time or Two Time?	441
		14.7	Applications to Weak Anisotropy	443
			14.7.1 A Self-Consistent Representation of the Spectral Tensor for Weak Anisotropy	443
			14.7.2 Brief Discussion of Concepts, Results, and Open Issues	445
		14.8	Open Numerical Problems	446
	Bibliography			**447**
15	**Conclusions and Perspectives**			**449**
		15.1	Homogenization of Turbulence. Local or Global Homogeneity? Physical Space or Fourier Space?	449
		15.2	Linear Theory, "Homogeneous" RDT, WKB Variants, and LIA	451
		15.3	Multipoint Closures for Weak and Strong Turbulence	453
			15.3.1 The Wave-Turbulence Limit	454
			15.3.2 Coexistence of Weak and Strong Turbulence, With Interactions	455
			15.3.3 Revisiting Basic Assumptions in MPC	455
		15.4	Structure Formation, Structuring Effects, and Individual Coherent Structures	456
		15.5	Anisotropy Including Dimensionality, a Main Theme	457
		15.6	Deriving Practical Models	459
	Bibliography			**460**
Index				461

Abbreviations Used in This Book

Abbreviation	Definition	Abbreviation	Definition
1D	one-dimensional	LRA	linear response analysis, Lagrangian renormalized approximation
2D	two-dimensional		
2D-2C	two-dimensional, two-component		
3D	three-dimensional	MPC	multipoint closure
4D	four-dimensional	NAM	normalized angular momentum
AG	ageostrophic		
AKA	anisotropic kinetic alpha	ODE	ordinary differential equation
APV	absolute potential vorticity		
AQNM	asymptotic quasi-normal Markovian	pdf	probability density function
		PIV	particle image velocimetry
DIA	direct interaction approximation	PLE	permanence of large eddies
DNS	direct numerical simulation	POD	proper orthogonal decomposition
ED	eddy-damping	PTV	particle-tracking velocimetry
EDQNM	eddy-damped quasi-normal Markovian		
ESS	extended self-similarity	QG	quasi-geostrophic
HAT	homogeneous anisotropic turbulence	QN	quasi-normal
		RANS	Reynolds averaged Navier–Stokes
HDG	horizontal density gradient	RDT	rapid distortion theory
HIT	homogeneous isotropic turbulence	rms	root-mean-square
		RSM	Reynolds stress model
KRR	Kassinos, Reynolds, and Rogers (2001)	RST	Reynolds stress tensor
		SPIV	stereoscopic particle image velocimetry
LES	large-eddy simulation	SSP	self-sustaining process
LHDIA	Lagrangian history direct interaction approximation	TCL	two-component limit
		TFM	test field model
LIA	linear interaction approximation	VSHF	vertically sheared horizontal flow

1 Introduction

1.1 Scope of the Book

Turbulence is well known to be one of the most complex and exciting fields of research that raises many theoretical issues and that is a key feature in a large number of application fields, ranging from engineering to geophysics and astrophysics. It is still a dominant research topic in fluid mechanics, and several conceptual tools developed within the framework of turbulence analysis have been applied in other fields dealing with nonlinear, chaotic phenomena (e.g., nonlinear optics, nonlinear acoustics, econophysics, etc.).

Despite more than a century of work and a number of important insights, a complete understanding of turbulence remains elusive, as witnessed by the lack of fully satisfactory theories of such basic aspects as transition and the Kolmogorov $k^{-5/3}$ spectrum. Nevertheless, quantitative predictions of turbulence have been developed. They are often based on theories and models that combine "true" dynamical equations and closure assumptions and are supported by physical and – more and more – numerical experiments.

Homogeneous turbulence remains a timely subject, even half a century after the publication of Batchelor's book in 1953, and this framework is pivotal in the present book. Homogeneous isotropic turbulence (HIT) is the best known canonical case; it is very well documented – even if not completely understood – from experiments and simple models to recent 4096^3 full direct numerical simulation (DNS). Of course, this case is addressed (in Chapter 3), but more generally emphasis is put on homogeneous anisotropic turbulence (HAT) in the presence of mean (velocity, temperature, etc.) gradients, body forces, or both. This context is illustrated by several physical and numerical experiments (the latter being easy to perform by slight modification of pseudo-spectral numerical methods designed for DNS of isotropic turbulence following the method introduced by Rogallo in the late 1970s), but its interest for developing fundamental understanding and improving theories and models is largely underestimated regarding the existing literature. Depending on the strength of the distortion (by mean gradients and/or body forces) and its time of application, it is possible to move from pure linear approaches, such as the *rapid distortion theory* (RDT), to fully nonlinear statistical theories, with the important intermediate step of "weak" turbulence theories, such as the *wave-turbulence theory*. As far as possible, it is proposed to pass from "weak" to "strong" turbulence by

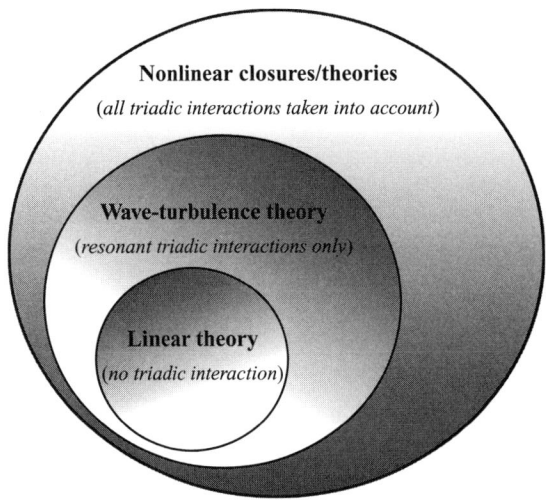

Figure 1.1. Sketch of the hierarchy of embedded turbulence theories and closures.

following a strict hierarchy of embedded models and theories, which is illustrated in Fig. 1.1.

This strategy was introduced by the second author in his contribution to the recent book *Theories of Turbulence* (Oberlack and Busse, 2002). Even if the most original part of the present book deals with two-point statistics, the Reynolds stress budget is very informative and therefore Reynolds stress equations are discussed before more complex approaches are addressed. Limits or failures of single-point closures are highlighted in each case.

A discussion of the physical relevance of the HAT cannot be avoided, and we show that homogeneous turbulence in the presence of space-uniform mean gradients is not so ideal and restrictive. In addition to physical and numerical experiments that are capable of reproducing HAT, some typical equations (e.g., Townsend or Craya equations) are shown to remain relevant for analyzing flows with nonuniform mean gradients [e.g. short-wave stability analyses, Wentzel–Kramers–Brillouin (WKB) RDT]. In some cases, pedagogical explanations for "pure" homogeneous turbulence can be extended toward inhomogeneous turbulence (e.g., near-wall turbulent shear flow). Another important point is that homogeneous sheared turbulence exhibits self-sustained cycles, which are key features of turbulence dynamics in near-wall regions.

A large number of books devoted to turbulence are available that put the emphasis on three aspects: statistical properties of isotropic, incompressible turbulence (e.g., Batchelor, 1953; Frisch, 1995; Tsinober, 2001, Davidson, 2004), descriptions of global dynamics and statistical properties of some academic flows (boundary layer, mixing layer, jet, wake, etc.; e.g., Townsend, 1976; Smits and Dussauge, 2006), and modeling of turbulent motion for engineering purpose (among others, Durbin and Petersson Reif, 2001; Wilcox, 2004). Only little information on the *dynamics* of turbulent scales is usually provided, and most authors put the emphasis on a particular feature. One should of course mention general-purpose textbooks (see Pope, 2000; Tennekes and Lumley, 1994; Bailly and Comte-Bellot, 2003), which provide the reader with a general survey of different issues related to turbulence research. Therefore, recent results dealing with dynamics of turbulent motion obtained from DNSs, advanced statistical models (linear theories and models, nonlinear triadic closures, etc.), and experiments are not available to the reader in a single book. Results are disseminated among a huge number of journal articles, technical reports, and conference papers that do not always use the same terminology.

The present book aims at providing a state-of-the-art sum of results and theories dealing with homogeneous turbulence, including anisotropic effects and compressiblity effects. The underlying idea is to gather the most recent results dealing with the dynamics of homogeneous turbulence when it interacts with external forcing (strain, rotation, etc.) and when compressibility effects are in play. Each chapter will be devoted to a given type of interaction and will present and compare experimental data, DNS/LES (large-eddy simulation) results, analysis of the Reynolds stress budget equations, and advanced linear and nonlinear theoretical models. The roles of both linear and nonlinear mechanisms are emphasized. The link between the statistical properties and the dynamics of coherent structures is also addressed. Despite its being restricted to homogeneous turbulence, this book will be of interest to all people involved in turbulence studies, as it will highlight basic physical mechanisms that are present in all turbulent flows.

Another interest of this book is the possiblity for the reader to find a unified presentation of the results and also a clear presentation of existing controversies and shortcomings in the theoretical background. Special attention is paid to bridging gaps among the results obtained in different research communities. This last point is developed concerning both results dealing with turbulence dynamics and the tools used to investigate it.

1.2 Structure and Contents of the Book

The presentation of the results is carried out in such a way that it allows for two levels of reading: a first level for readers interested in the results but who do not want to enter into the details of the tools (i.e., linear and nonlinear theoretical models) employed to get them, and a second level for readers interested in these details.

The book is organized in 15 chapters, with turbulent-flow cases ranging from HIT (without distortion, Chapter 3) to HAT subjected to various distorting processes (rotation, strain, shear, stratification) in Chapters 4–7. Flows subjected to coupled forcing effects are collected in Chapter 8, whereas compressible turbulence is addressed in Chapters 9–11. Chapter 2 presents the basis of dynamical (conservation equations) and statistical analyses of turbulence.

Technical details about theroretical tools and theories used in Chapters 3–11 are gathered in dedicated chapters whose reading is not mandatory. The linear-interaction theory for shock–turbulence interaction is presented in Chapter 12. Linear theories such as the RDT are detailed in Chapter 13, and two-point nonlinear closure theories [e.g., eddy-damped quasi-normal Markovian (EDQNM) theory] are addressed in Chapter 14. Some concluding comments are presented Chapter 15.

Constraints for ensuring consistency of statistical homogeneity – for turbulence – with the distorting processes are given in the most general way, for both incompressible (particularly in Chapter 2) and compressible (Chapter 10) flows. The physical relevance of this framework is also discussed.

Every typical flow case is revisited under different angles of attack, from observations and simulations, models, to theories, combining dynamical, statistical, and structural aspects, as follows:

1. observations, physical, and numerical experiments
2. analysis through *Reynolds stress tensor* (RST) equations, and balance and coupling terms
3. refined analysis using linear theory
4. refined analyses through full nonlinear theories and models for two-point statistics (if available)
5. phenomenological (and possibly dynamical) approach to structures, evolution, coupling.

It is worth noting that two classes of flows are discussed in this book. The first one is the class of flows without turbulence-production mechanisms (e.g., decaying isotropic turbulence, rotating homogeneous turbulence, stably stratified homogeneous turbulence, etc.) and flows with turbulent-kinetic-energy-production mechanisms (e.g. homogeneous sheared turbulence). In the former case, nonlinear dynamics and its modification by mean-flow effects are the sole features of the flow, whereas in the latter case linear mechanisms are the main dynamical characteristics. Therefore nonlinear models are the tools of choice in the first case (but eigenfunctions of the linear theories can provide an optimal basis to write them), whereas they are only briefly discussed in flows with productions for which linear theories are very powerful.

The most complete illustration of the hierarchy of models embedded in each other is the case of pure rotation (Chapter 4). Common models, such as RST closure models, are shown to present definite flaws in this case, and some limited attempts to improve single-point closure techniques are only briefly reviewed. As an important related point, linear theories such as the RDT were only briefly reviewed for irrotational mean flows only in other recent monographs about turbulence (e.g., Pope, 2000; Durbin and Petersson Reif, 2001), with the only exception of pure shear in the book by Townsend (1976), written a long time ago. In contrast, linear theory for HAT subjected to more general rotational mean flows is a very important part of the present book. In addition, our extended linear theory is a building block that may be useful for a wider community (e.g., elliptical-flow instability from the viewpoint of stability analysis, rotating and/or stratified shear flow, in Chapter 8).

The application domain of two-point nonlinear closures is even more restricted in existing monographs (e.g., Monin and Yaglom, 1975; Leslie, 1973; Lesieur, 1997; Frisch, 1995). Only isotropic turbulence is treated in a straightforward way, and only a few attempts to deal with small anisotropy are offered, whereas the linkage to linear models and wave turbulence is ignored.

1.2 Structure and Contents of the Book

The last item about "structures" deserves some clarification. On the one hand, it is recognized that typical structures can be evidenced by snapshots, or random realizations, of statistically homogeneous flows. The first example is the appearance of vortex tubes in isotropic turbulence. Other well-known structures are streaklike (in shear flows), cigar (in flows with dominant rotation), or pancake (flows with dominant stable stratification) structures. On the other hand, the relevance of low-order statistics to identify and quantify these structures is controversial. Second-order statistics, if they include fully anisotropic two-point correlations, can give real insight into these structures, with quantitative information (elongation parameters, aspect ratios). An objection can be made that phase coherence is lost in homogeneous statistics – at least for single-time second order – so that some aspects of coherent structures are not accounted for. Accordingly, we will speak of structures, or structuring effects, avoiding "coherent," when we identify them by using anisotropic statistics and not only using visualizations of snapshots.

The advanced models and theories selected here systematically incorporate dynamical operators that are really based on Navier–Stokes equations, even if they deal with "weak" turbulence only (e.g., linearized models, wave turbulence), not to mention exact triadic equations and conventional two-point closures based on them. Three-dimensional (3D) Fourier space is an unavoidable tool in HAT analysis; it is first considered here as a mathematical convenience to account for solenoidal properties (in isovolume turbulence) and to simplify related *modal decompositions*. Special use is made of decomposition of the fluctuating velocity in Fourier space, often referred to as the Craya–Herring decomposition, which amounts to a general Helmholtz decomposition, in terms of two solenoidal (toroidal–poloidal type), or vortical, modes and one dilatational (or divergent) mode. In incompressible turbulence, a Poisson equation is immediately recovered by projecting momentum equations onto the dilatational mode, the dilatational velocity mode being zero, so that dynamical equations deal with only the two solenoidal modes. This decomposition readily generates the helical-mode decomposition, and various "vortex-wave" decompositions when buoyancy fluctuation is accounted for (Chapters 7 and 8). The dilatational mode recovers its dynamic role, together with the pressure mode, when compressibility is introduced. The increase of the complexity of the system can be presented as follows:

1. Two-mode turbulence, in which the two independent unknowns are $u^{(1)}$, $u^{(2)}$ using the toroidal–poloidal decomposition, or $u^{(2)} \pm \iota u^{(1)}$ considering the helical-mode variant. The dilatational mode $u^{(3)}$ is strictly zero so that the pressure mode $u^{(4)}$ is completely solved in terms of the two solenoidal ones, and therefore removed from consideration (Chapters 2–6).
2. Three-mode turbulence. Same situation as before, but an additional buoyancy term is incorporated as a pseudo-dilatational mode. The physical problem with five components (three for velocity fluctuations, one for pressure fluctuations, one for buoyancy fluctuations) is turned into a three-mode one thanks to the Boussinesq approximation (divergence-free velocity field and the related

Poisson equation for pressure hold again, even if buoyancy exists), used in Chapters 7 and 8.

3. Four-mode turbulence $u^{(1)}, \ldots, u^{(4)}$, as in the quasi-isentropic flow cases addressed in Chapters 9 and 10. If the acoustic-equilibrium hypothesis holds, $u^{(3)}$ and $u^{(4)}$ can be combined as $[u^{(4)} \pm \iota u^{(3)}]$, where $u^{(3)}$ corresponds to the kinetic energy of acoustic waves and $u^{(4)}$ gives its potential counterpart.

4. Five-mode-turbulence, in which the last, fifth, entropy mode is added. In practice, decomposition in terms of five modes is possible, but not completely universal (discussed in Chapter 9). Introduction of a realistic entropy mode can be puzzling in homogeneous turbulence, not to mention the question of using density-weighted variables (velocity, momentum, or intermediate mixed quantity). Nevertheless, a decomposition very close to the $u^{(1)}$–$u^{(5)}$ one (toroidal–poloidal–dilatational–pressure–entropy) is used in Chapter 11 to describe upcoming perturbations passing through an idealized shock wave. Here, the Chu–Kovasznay decomposition is a preferential tool, as it makes it possible to split the incoming fluctuations into vortical, acoustic, and entropy modes. It is sufficient, however, to take the solenoidal (vortical) mode as one component only, so that four-mode turbulence is eventually used because upstream- and downstream-traveling acoustic perturbations must be treated in separate ways.

Isotropy is generally broken by the dynamical operators, so that a complete anisotropic description is needed, consistent with the symmetries of background equations, both in physical (two-point correlations) and in Fourier space (spectral tensors). It is worthwhile stressing that our detailed anisotropic description includes *dimensionality*, with a possibility of quantifying a 3D to two-dimensional (2D) [or to one-dimensional (1D)] transition. For instance, the structure-based modeling by Kassinos and Reynolds, which allows us to distinguish dimensionality and componentality, becomes a by-product of our general description, at least for homogeneous turbulence.

This viewpoint allows us to classify the theoretical approaches to turbulence as follows:

1. Theoretical "spectral-shell models" (as used by physicists to work on intermittency) are not considered in the present book, and empirical (spherically averaged) spectral models are only very briefly discussed (in Chapters 13 and 14), as solenoidal properties, and related exact pressure terms, cannot be preserved by spherically averaged transport equations in Fourier space. Consequently, equations that are exact and closed in the linear – rapid distortion – limit are no longer closed after spherical averaging.

2. "Modern" phenomenological theories about scaling and intermittency, from the legacy of Kolmogorov, are touched on, but in a minimal way, as they retain very little from Navier–Stokes equations. Only the Kolmogorov equation for the third-order structure function is partly based on Navier–Stokes, but it also relies on additional assumptions like local isotropy and quasi-steadiness. In

1.2 Structure and Contents of the Book

addition, a strong departure of the "anomalous exponents"[*] from the original Kolmogorov theory (which leads to $\zeta_n = n/3$) is interpreted as intermittency by physicists (see Bohr et al., 1998; Frisch, 1995). In contrast, wave turbulence based on (even weakly nonlinear) Navier–Stokes dynamics can radically question this viewpoint. Typical anomalous exponents can be found in the case of rapid rotation, even for low-order structure functions ($n = 2$ and $n = 3$), with no connection to intermittency (see Fig. 4.18 in Chapter 4 and the corresponding discussion). The "anomality" of exponents reflects the strong anisotropy linked to a partial transition from 3D to 2D structures and has probably nothing to do with intermittency in this example. Generally, the pure statistical description based on anomalous exponents, or extended self-similarity (ESS) laws, mixes anisotropy, inhomogeneity, and intermittency in an intricate way.

3. "Old-fashioned" statistical two-point "triadic" closures, the simplest one being EDQNM, are reconciled with linear models and wave-turbulence theory, and finally are shown to be still useful and relevant (especially with respect to the modern phenomenological theories quoted just before).
4. Low-order two-point (or more) moments are shown to be very informative: second-order moments for energy distribution, third-order moments for energy transfers (cascades), and fourth-order ones for typical closure, especially in connection with associated dynamical equations. Higher-order moments, by means of n-structure functions and full probability density functions (pdf's) are very briefly discussed.

Finally, Lagrangian statistics and passive scalar transport are not addressed, but it is worth noting that linear theories and two-point closures have relevant applications in these domains.

Let us go back to Chapters 9–11, dealing with dynamics of compressible turbulence. This issue is almost absent in most previous books dealing with turbulence fundamentals. Chapter 9 is devoted to presentation of state-of-the-art knowledge about the dynamics of compressible isotropic turbulence. The Chu–Kovazsnay modal decomposition of turbulent fluctuations is first introduced to provide the reader with a physical insight into coupling among acoustics, entropy, and vorticity. Then the different regimes observed in numerical simulations and theoretical analyses are described: the pseudo-acoustic regime, the subsonic regime (both pseudo-acoustic and thermal regimes are considered) and the supersonic regime. In each case, details of the interactions and transfers among scales and modes are discussed, and the link with the dynamics of coherent events (vortical structures, acoustic waves, shocklets, etc.) is made. Some low-Mach triadic-interaction-theory results are included, together with simplified models. Chapter 10 presents the coupling of compressible turbulence with mean-gradient effects. In this chapter, the emphasis is put on linear theory and DNS results because they are well suited to describe

---

[*] Often denoted ζ_n in the literature, n being the order of the structure function that is supposed to decay as $r^{-\zeta_n}$.

dominant dynamical mechanisms in such strongly anisotropic flows. The theory of compressible RDT is highlighted. Chapter 11 is dedicated to the shock–turbulence interaction, which has been proved to be very accurately predicted by the *linear interaction approximation* (LIA) for a large class of flows. The LIA is presented in Chapter 12 in its most achieved version, and it is used to illustrate the physics of the interaction of a shock with different kinds of fluctuations corresponding to the Chu–Kovazsnay modes. A comparison with DNS and experimental results is also made. Despite its being restricted to simple flow configurations, the basic physical mechanisms emphasized in this part are the building blocks for the interpretation and understanding of the properties of compressible turbulent flows in complex configurations.

Bibliography

BAILLY, C. AND COMTE-BELLOT, G. (2003). *Turbulence*, CNRS Editions.
BATCHELOR, G. K. (1953). *The Theory of Homogeneous Turbulence*. Cambridge University Press.
BOHR, T., JENSEN, M. H., PALADIN, G., AND VULPIANI, A., Dynamical systems approach to turbulence. Cambridge University Press.
DAVIDSON, P. A. (2004). *Turbulence*. Oxford University Press.
DURBIN, P. AND PETERSSON REIF, B. A. (2001). *Statistical Theory and Modeling for Turbulence Flow*. Wiley.
FRISCH, U. (1995). *Turbulence*. Cambridge University Press.
LESIEUR, M. (1997). *Turbulence in Fluids*, 3rd ed. Kluwer Academic.
LESLIE, D. C. (1973). *Developments in the Theory of Turbulence*. Clarendon.
MATHIEU, J. AND SCOTT, J. F. (2001). *Turbulent Flows: An Introduction*. Cambridge University Press.
MONIN, A. S. AND YAGLOM, A. M. (1971/1975). *Statistical Fluid Mechanics*. MIT Press, vol. I (1971); vol. II (1975).
OBERLACK, M. AND F. H. BUSSE, EDS. (2002). *Theories of Turbulence*. Springer, Chapter IV, pp. 197–251, from Claude Cambon.
PIQUET, J. (2001). *Turbulent Flows – Models and Physics*, 2nd revised printing. Springer.
POPE, S. B. (2000). *Turbulent Flows*. Cambridge University Press.
SMITS, A. J. AND DUSSAUGE, J. P. (2006). *Turbulent Shear Layers in Supersonic Flows*, 2nd ed. Springer.
TENNEKES, H. AND LUMLEY, J. L. (1994). *A First Course in Turbulence*. MIT Press.
TOWNSEND, A. A. (1976). *The Structure of Turbulent Shear Flow*, 2nd ed. Cambridge University Press.
TSINOBER, A. (2001). *An Informal Introduction to Turbulence*. Kluwer Academic.
WILCOX, D. C. (2004). *Turbulence Modeling for CFD*, 2nd ed. DCW Industries.

2 Statistical Analysis of Homogeneous Turbulent Flows: Reminders

2.1 Background Deterministic Equations

2.1.1 Mass Conservation

The equation of mass conservation is well known and does not need a long explanation to be derived. Both Eulerian and Lagrangian forms are subsequently given. The latter is less common in fluid dynamics but it deserves some attention, as it brings in some fundamental *Lagrangian* concepts and relationships.

Let us begin by addressing the Eulerian description. To this end, we consider a fixed arbitrary control volume $\mathcal{V}$, delineated by a surface S. The total mass of the fluid is governed by the following integral balance equation:

$$\underbrace{\frac{d}{dt} \iiint_{\mathcal{V}} \rho(\mathbf{x},t) d^3\mathbf{x}}_{\text{variation}} = \underbrace{-\iint_{S} \rho(\mathbf{x},t)\mathbf{u}(\mathbf{x},t) \cdot \mathbf{n} d\sigma}_{\text{flux}} + \underbrace{\iiint_{\mathcal{V}} m(\mathbf{x},t) d^3\mathbf{x}}_{\text{production}}, \quad (2.1)$$

in which ρ, $\mathbf{u}$, and m are the density, the velocity, and the rate of mass production, respectively. All these fields are assumed to be continuous fields in terms of time t and Eulerian and Cartesian coordinates $\mathbf{x}$. In this equation, $d^3\mathbf{x}$ is the elementary volume of a fluid particle, $d\sigma$ is the elementary surface with outward normal, and $\mathbf{n}$ is the unit vector. The classical Ostrogradsky formula yields $\iint_S \rho\mathbf{u} \cdot \mathbf{n} d\sigma = \iiint_{\mathcal{V}} \nabla \cdot (\rho\mathbf{u}) d^3\mathbf{x}$, so that the previous equation is rewritten as

$$\iiint_{\mathcal{V}} \left[\frac{\partial \rho}{\partial t} + \nabla \cdot (\rho\mathbf{u}) - m \right] d^3\mathbf{x}.$$

For the sake of clarity, the divergence of a vector $\mathbf{V}$ is denoted as $\nabla \cdot (\mathbf{V})$ or, alternatively, $\frac{\partial V_i}{\partial x_i}$ in the following. The classical local and instantaneous counterpart of the preceding equation is the continuity equation,

$$\frac{\partial \rho}{\partial t} + \nabla \cdot (\rho\mathbf{u}) = m. \quad (2.2)$$

In the Lagrangian description, fluid particles follow trajectories, which are given by the relationship

$$x_i = x_i^L(\mathbf{X}, t, t_0), \quad (2.3)$$

2.1 Background Deterministic Equations

which links the position of the fluid particle at time t to its initial position X at time t_0. The Lagrangian coordinates X characterize the initial position, and therefore label the trajectory. To avoid confusion, the trajectory equation is denoted as x_i^L, different from the Eulerian coordinates x_i. In the following, the superscript L is often omitted, but different notations are used for time derivatives.

On the one hand, $\frac{\partial}{\partial t}$ denotes an Eulerian time derivative, at constant x, as in Eq. (2.2). On the other hand, the overdot denotes the Lagrangian time derivative, at constant X. As a first example, the differential term in Eq. (2.3) can be expanded as

$$dx_i = u_i dt + F_{ij} dX_j \qquad (2.4)$$

(in which dx_i holds for dx_i^L), straightforwardly leading to

$$\dot{x}_i = \frac{\partial x_i^L}{\partial t} = u_i$$

and

$$F_{ij} = \frac{\partial x_i^L}{\partial X_j}.$$

The latter matrix, referred to as the Cauchy matrix, is denoted as $\partial x_i/\partial X_j$ from now on for the sake of brevity. It is the classical semi-Lagrangian displacement gradient in continuum mechanics (see Eringen, 1971, from whom notations are borrowed).

The preceding brief reminder is needed for deriving the continuity equation in the Lagrangian description. Now, one considers that the mass of an ensemble of fluid particles is conserved during its motion,

$$\frac{d}{dt} \iiint_\mathcal{V} \rho(X, t, t_0) d^3x = \underbrace{\iiint_\mathcal{V} m(X, t, t_0) d^3x}_{M},$$

but the moving domain $\mathcal{V}$ has to be considered as the mapping of an initial domain $\mathcal{V}_0$ following all individual trajectories with positions in this domain (m ought to be considered in Lagrangian coordinates, too, but a new specific notation is not introduced for the sake of simplicity). From the very definition of F_{ij}, its determinant (always nonzero positive) is the Jacobian of the x-to-X transformation, so that

$$d^3x = J d^3X,$$

where

$$J(X, t, t_0) = \text{Det}\, F \qquad (2.5)$$

is the local and instantaneous volumetric ratio following a trajectory. The conservation equation can be written as

$$\frac{d}{dt} \iiint_{\mathcal{V}_0} \rho J d^3X = M,$$

and the (Lagrangian) time derivative holds inside the integral, so that

$$\iiint_{V_0} (\dot{\rho} J + \rho \dot{J}) d^3 X = M.$$

From

$$\dot{F}_{ij} = \frac{\partial \dot{x}_i}{\partial X_j} = \frac{\partial u_i}{\partial x_n} F_{nj} \tag{2.6}$$

one derives

$$\dot{J} = \nabla \cdot (\boldsymbol{u}) J. \tag{2.7}$$

Finally, the continuity equation can be expressed as $(\rho \dot{J}) = Jm$, or

$$J(X, t, t_0) = \frac{\rho(X, t_0)}{\rho(X, t)} + \oint_{t_0}^{t} (Jm) dt, \tag{2.8}$$

using $J(X, t_0, t_0) = 1$ (where the time integral of m is computed along a trajectory), or, alternatively,

$$\dot{\rho} = -\rho \nabla \cdot (\boldsymbol{u}) + m. \tag{2.9}$$

Of course, the identity of the latter equation with Eq. (2.2) can be checked, using

$$\dot{\rho} = \frac{\partial \rho}{\partial t} + \dot{x}_j \frac{\partial \rho}{\partial x_j} = \frac{\partial \rho}{\partial t} + u_j \frac{\partial \rho}{\partial x_j}. \tag{2.10}$$

2.1.2 The Navier–Stokes Momentum Equations

In the same way as for the mass, the conservation of momentum yields

$$\underbrace{\frac{d}{dt} \iiint_V \rho u_i d^3 x}_{\text{variation}} = -\underbrace{\iint_S \rho u_i \boldsymbol{u} \cdot \boldsymbol{n} d\sigma}_{\text{flux}} + \underbrace{\iiint_V \rho f_i(\boldsymbol{x}, t) d^3 x}_{\text{production}}, \tag{2.11}$$

in which the "production" involves a body force per mass unit, denoted as $\boldsymbol{f}$, but the domain of fluid is not isolated: A *surfacic* strain tensor σ_{ij} is acting on it. Accordingly, a pure kinematic balance, as in the previous subsection for the mass, is no longer valid, and *dynamics* must be accounted for. Replacing $u_i \boldsymbol{u} \cdot \boldsymbol{n} = u_i u_j n_j$ with $(u_i u_j - \sigma_{ij}) n_j$, one obtains

$$\frac{\partial (\rho u_i)}{\partial t} + \frac{\partial (\rho u_i u_j - \sigma_{ij})}{\partial x_j} = \rho f_i, \tag{2.12}$$

or

$$\rho \dot{u}_i = \frac{\partial \sigma_{ij}}{\partial x_j} + \rho f_i, \tag{2.13}$$

using the continuity equation with $m = 0$.

2.1 Background Deterministic Equations

Finally, the classical Navier–Stokes equations* correspond to the following expression of the strain tensor in Eq. (2.13):

$$\sigma_{ij} = -p\delta_{ij} + \mu \underbrace{\left(\underbrace{\frac{\partial u_i}{\partial x_j} + \frac{\partial u_j}{\partial x_i} - \frac{2\delta_{ij}}{3}\frac{\partial u_n}{\partial x_n}}_{S_{ij}} \right) + \frac{3\lambda + 2\mu}{3}\frac{\partial u_n}{\partial x_n}\delta_{ij}}_{\sigma_{ij}^{\text{visc}}}, \quad (2.14)$$

in which the threefold decomposition includes a spherical term linked to pressure p, a shearing viscous term that involves the symmetric, trace-free velocity-gradient tensor, and a viscous "bulk" term. The fluid-dependent parameters μ and λ are the Lamé coefficients, which are often linked together by the Stokes relationship $3\lambda + 2\mu = 0$, which removes from consideration the pure volumic bulk dissipation process.

2.1.3 Incompressible Turbulence

Strict incompressibility is recovered assuming $\rho(x,t) = \rho_0$ in the continuity equation, so that the velocity field is divergence free or solenoidal. Ignoring the mass "production" term m for the sake of simplicity, mass conservation reduces to the divergence-free (solenoidal) condition, or to $J = 1$ from the Lagrangian viewpoint, and the momentum equation reduces to

$$\dot{u}_i = -\frac{1}{\rho_0}\frac{\partial p}{\partial x_i} - \nu\nabla^2 u_i + f_i, \quad (2.15)$$

in which the left-hand-side term is the acceleration, or

$$\dot{u}_i = \frac{\partial u_i}{\partial t} + \dot{x}_j\frac{\partial u_i}{\partial x_j} = \frac{\partial u_i}{\partial t} + u_j\frac{\partial u_i}{\partial x_j} \quad (2.16)$$

as for Eq. (2.10). $\nu = \mu/\rho_0$ is the kinematic viscosity, which is considered a constant parameter. The problem is self-consistent and well posed, with four dependent variables (u_1, u_2, u_3, p) and four equations (one for the divergence-free constraint, three for the preceding system of Navier–Stokes equations).

The pressure is no longer a thermodynamic, autonomous variable, but a simple Lagrange multiplier connected to the solenoidal constraint for velocity. Taking the divergence of Eq. (2.15), and accounting for the incompressibility constraint $\partial u_i/\partial x_i = 0$, one obtains

$$\frac{1}{\rho_0}\nabla^2 p = -\frac{\partial^2}{\partial x_i \partial x_j}(u_i u_j) - \frac{\partial f_i}{\partial x_i}. \quad (2.17)$$

---

* These equations were established by Claude Navier in 1823 and rediscovered or rederived at least four times: by Cauchy in 1823, by Poisson in 1829, by Saint-Venant in 1837, and by Stokes in 1847 (Darrigol, 2005). Navier had already distinguished two types of motion, "régulier" (mean) and "tumultueux" (turbulent), foreshadowing Osborne's Reynolds decomposition. The idea that two length scales are present was also considered by Saint-Venant.

This Poisson-type equation displays how the pressure is connected to the terms that are not divergence free in the Navier–Stokes equations[†]: the acceleration term itself (contributing to the first term on the right-hand side) and possibly the body-force term (second term on the right-hand side). In contrast, divergence-free terms, such as $\partial u_i/\partial t$ and $\nu \nabla^2 u_i$, are removed. This relationship between velocity and pressure is essential in many turbulent flows such as those discussed in Chapters 3–8. Two remarks can be made from the very beginning:

- This nonlocal and *instantaneous* relationship is not physical, because it implies that the speed of sound is infinite, so that a pressure disturbance in a remote position instantaneously responds to a velocity disturbance.
- However, this unphysical problem is very relevant to studying and to understanding low-Mach-number turbulence. It is now clear that the problem of turbulence is not only due to the nonlinearity of the acceleration term (2.16), as often advocated, and not only due to the lack of integrability of trajectories (2.3) (e.g., Lagrangian chaos). The "pressure-released" turbulence, illustrated by Burgers' equation in the 1D case and by the cosmological gas in three dimensions (e.g., Polyakov, 1995), is essentially solved! Hence, the role of pressure, or, identically, the restriction to solenoidal modes (projection onto a solenoidal subspace), is an essential point to understanding why turbulence is so complex. In addition, "solenoidal turbulence," as calculated using pseudo-spectral DNS at the highest resolution available [e.g., 4096^3, (Kaneda et al., 2003)], mimics all characteristics of – low-speed – "real" physical turbulence.

2.1.4 First Insight into Compressibility Effects

As soon as the solenoidal condition is relaxed, the coupling between pressure and velocity becomes very different. First, the problem with five components (ρ, u_1, u_2, u_3, p) is governed by only four equations, i.e., mass and momentum conservation, the latter with given σ_{ij}. The state law of the fluid provides a new equation, but also introduces a new variable, usually temperature or entropy. Consequently, a new conservation equation (for energy, enthalpy, entropy) is needed. As introduced in Chapter 9, the entropy term s is chosen in this book, so that the six-component $(\rho, u_1, u_2, u_3, p, s)$ compressible problem is addressed by use of a six-component system of equations: 1 (mass conservation) + 3 (momentum conservation) + 1 (state law) + 1 (entropy conservation).

As a first illustration, it is possible to derive an equation for ρ, combining mass and momentum equations, as follows:

$$\frac{\partial^2 \rho}{\partial t^2} - a_0^2 \nabla^2 \rho = \frac{\partial^2}{\partial x_i \partial x_j} \left[\underbrace{\rho u_i u_j - \sigma_{ij}^{\text{visc}} + \delta_{ij}(p - a_0^2 \rho)}_{T_{ij}} \right]. \qquad (2.18)$$

[†] It is important to note that such terms exist even if the velocity field is solenoidal.

2.1 Background Deterministic Equations

A characteristic speed of sound, a_0, is assumed to be space uniform in this equation, which is "exact" in this limit (only production terms m and f are ignored, for the sake of simplicity). For instance, with isentropic and low-Mach-number additional assumptions, T_{ij} reduces to its first term, and (2.18) can be used to support acoustic analogies (e.g., Lighthill and many followers). An equation similar to (2.18) can be found for the pressure, so that Eqs. (2.17) and (2.18) illustrate the different dynamics when compressibility is called into play.

2.1.5 Reminder About Circulation and Vorticity

The reader is referred to basic textbooks for the fundamentals about circulation, related Kelvin theorem, and vorticity, e.g., Saffman (1995). The vorticity is defined as the curl of velocity, i.e. $\boldsymbol{\omega} = \operatorname{curl} \boldsymbol{u}$, leading to

$$\omega_i = \epsilon_{ijn} \frac{\partial u_n}{\partial x_j}. \tag{2.19}$$

It may be noted that instead of using the tensorial expression with the third-order alternating pseudo-tensor ϵ_{ijn} (often referred to as the *Levi–Civita tensor*), one can use the symbolic notation $\nabla \times \boldsymbol{u}$, but with care because the symbolic operator ∇ is not a vector, and some permutation rules do not hold for it.

From momentum conservation law (2.13) it follows that

$$\dot{\omega}_i + \frac{\partial u_j}{\partial x_j} \omega_i - \frac{\partial u_i}{\partial x_j} \omega_j = \epsilon_{ijn} \frac{\partial}{\partial x_j} \left(\frac{1}{\rho} \frac{\partial \sigma_{nm}}{\partial x_m} + f_n \right).$$

The right-hand-side is exactly the curl of the acceleration $\dot{\boldsymbol{u}}$. Ignoring the viscous and body-force terms, one can rewrite this equation as

$$\frac{D}{Dt} \left(\frac{\omega_i}{\rho} \right) = \frac{\partial u_i}{\partial x_j} \frac{\omega_j}{\rho} = \frac{1}{\rho^3} \epsilon_{ijn} \frac{\partial \rho}{\partial x_j} \frac{\partial p}{\partial x_n}, \tag{2.20}$$

which makes the *baroclinic torque* appear on the right-hand side.

This term vanishes in the barotropic case $p = p(\rho)$, and not only in the pure incompressible case. In such a situation, Eq. (2.6) for F_{ij} is the same as the preceding equation for ω_i/ρ at fixed j. As a consequence, the evolution of the vorticity vector along trajectories is governed by the Cauchy matrix as

$$\omega_i(\boldsymbol{x}, t) = \frac{1}{\operatorname{Det} \mathbf{F}} F_{ij}(\boldsymbol{X}, t, t_0) \omega_j(\boldsymbol{X}, t_0). \tag{2.21}$$

A similar but less common equation for the velocity is the *Weber equation*:

$$u_i(\boldsymbol{x}, t) = F_{ji}^{-1}(\boldsymbol{X}, t, t_0) u_j(\boldsymbol{X}, t_0) + \frac{\partial \phi}{\partial x_i}, \tag{2.22}$$

where ϕ is a scalar potential. It is easy to derive the former from the latter by means of the curl operator, but the reciprocal is much more difficult to establish. An alternative way for proving Eq. (2.22) is to start from the Kelvin circulation theorem (Julian Hunt, private communication):

$$\oint_C \boldsymbol{u}(\boldsymbol{x},t) \cdot \delta\boldsymbol{x} = \oint_{C_0} \boldsymbol{u}(\boldsymbol{X},t_0) \cdot \delta\boldsymbol{X},$$

in which the closed chains of fluid particles, respectively C_0 at initial time and C at final time, are connected by trajectories. Because the result has to be independent of the form of the (e.g., initial) loop C_0, a differential formula can be derived from the integral one, as in thermodynamics for the first and second principles, yielding

$$\boldsymbol{u}(\boldsymbol{x},t) \cdot \delta\boldsymbol{x} - \boldsymbol{u}(\boldsymbol{X},t_0) \cdot \delta\boldsymbol{X} = d\phi,$$

in which $d\phi$ is a total differential. The Weber equation is recovered from the former differential equation by use of $\delta X_i = F_{ij}^{-1}\delta x_j$ and $d\phi = \frac{\partial \phi}{\partial x_i}\delta x_i$.

2.1.6 Adding Body Forces or Mean Gradients

In the absence of external forcing or turbulence production by means of interaction with a nonuniform mean-velocity field, the "incompressible" turbulence (only this case is discussed here) decays, and (but this is more controversial) returns toward isotropy. Therefore decaying isotropic turbulence, which is addressed in Chapter 3, is the best illustration for turbulence dynamics. Statistical homogeneity is a mandatory requirement to study such a turbulence, so that the concepts of homogeneity and isotropy are intimately connected in various fundamental approaches to *homogeneous isotropic turbulence* (HIT). A main theme of this book is to illustrate how the framework of *homogeneous anisotropic turbulence* (HAT) is informative and useful – up to careful definitions and some caveats.[‡] In this context, we focus on anisotropic forcings, which can render the turbulence anisotropic and inject energy, so that the ultimate decay is altered or even prevented.

How such a forcing can preserve homogeneity, as far as possible, while representing a *physical* process, is the first question. Our experience is that forcing processes often used in fundamental studies have nothing to do with actual flows. On the other hand, mean rotation, strain, shear, and density stratification are physically relevant effects, consistent or not with statistical homogeneity.

The case of turbulence in a rotating frame is helpful to introduce our concept of HAT. On the one hand, a solid-body motion

$$x_i^{(0)} = F_{ij}(t - t_0)X_j, \qquad (2.23)$$

in which **F** reduces to a space-uniform, orthogonal ($\mathbf{F}^{-1} = \tilde{\mathbf{F}}$) matrix can be superimposed to a disturbance motion that will be considered "turbulent." In the preceding equation, the motion can be considered as base one, X denotes the Lagrangian coordinates *associated with the base-flow motion*, and $F_{ij}(t)$ is also the base-flow counterpart of the general matrix introduced at the beginning of this chapter. In agreement

---

[‡] A precise definition of statistical homogeneity is premature here; this definition is given in Subsection 2.2.1.

2.1 Background Deterministic Equations

with the usual solid-body motion description, the velocity field is characterized by the angular-velocity vector Ωn with magnitude Ω and orientation n, so that

$$u^{(0)} = \Omega n \times x \quad \text{with} \quad \nabla \times u^{(0)} = 2\Omega n$$

up to a constant term, which can be set equal to zero by changing the frame of reference thanks to the Galilean invariance property. In other words, the base flow is characterized by a constant-velocity-gradient matrix $\mathbf{A}$:

$$\mathbf{A} = \left(\frac{\partial u_i^{(0)}}{\partial x_j}\right) = \begin{bmatrix} 0 & -\Omega & 0 \\ \Omega & 0 & 0 \\ 0 & 0 & 0 \end{bmatrix}, \quad (2.24)$$

choosing $n_i = \delta_{i3}$ without loss of generality. Accordingly, it is possible to replace u_i with $u_i^{(0)} + u_i'$ in the background equations and to study the turbulent flow (i.e., u') in the presence of a particular base flow $u^{(0)}$ with a constant, antisymmetric gradient matrix. In our simple example dealing with solid-body rotation, the base-displacement gradient matrix is

$$\left[F_{ij}^{(0)}\right] = \begin{bmatrix} \cos\Omega(t-t_0) & -\sin\Omega(t-t_0) & 0 \\ \sin\Omega(t-t_0) & \cos\Omega(t-t_0) & 0 \\ 0 & 0 & 1 \end{bmatrix}. \quad (2.25)$$

On the other hand, it is well known that it is easier to study turbulence in the rotating frame, projecting both position and velocity in this non-Galilean frame of reference. Replacing x_i with X_i and u_i with v_i, defined in the same way as in Eq. (2.23) by

$$u_i = F_{ij}^{(0)}(t-t_0)v_j,$$

one sees that v_i and X_i satisfy the same equations as u_i and x_i in the Galilean reference frame, up to additional, centrifugal, and Coriolis forces. This non-Galilean acceleration term is defined as

$$f = \underbrace{-2\Omega n \times v}_{\text{Coriolis}} - \Omega^2[n \times (n \times X)].$$

This simple example, addressed in more detail in Chapter 4, is used here to illustrate the relevance of adding a constant base velocity gradient or adding a body force. For solid-body rotation, energy is not directly injected into turbulence, because the Coriolis force produces no work (and the centrifugal force can be removed from consideration if it is incorporated in the pressure term), but the energy cascade is strongly altered and rendered highly anisotropic. Without anticipating the results dealing with statistical properties of this flow, which are presented in the relevant chapter, the presence of solid-body rotation can be shown to be consistent with statistical homogeneity of the turbulent flow: removing the base-flow motion for defining homogeneity (and hence restricting the analysis to disturbances) in the first case,

$u' = u - u^{(0)}$,§ and considering homogeneity for $u' \to v$ in the rotating frame in the second case.

To what extent can the pure antisymmetric base velocity-gradient matrix be replaced with a more general one, including both symmetric and antisymmetric parts? Craya introduced in 1958 a relevant formalism for this purpose, which has been completely revisited in Cambon's thesis (1982) and rediscovered later in the context of stability analysis (e.g., Craik and Criminale, 1986). In a large part of this book, a constant-mean-velocity-gradient matrix **A** is used for studying the turbulent velocity field:

$$u'_i = u_i - A_{ij}x_j,$$

The following comments can be made prior to analysis:

1. A special form of **A** is required for preserving statistical homogeneity of the turbulent field. In the incompressible case, however, these conditions are not very stringent, allowing hyperbolic, linear, and elliptical streamlines for the mean motion. As soon as **A** has a nonzero symmetric part, kinetic energy can be directly injected into the turbulent flow, i.e., some turbulence-production mechanisms can take place.
2. The strict analogy with the effect of body forces holds for solid-body motion only. The advection of turbulent motion by the mean flow can be removed by a convenient change of frame, even if **A** is not purely antisymmetric, but all other terms in the equations for u' are then rendered more complicated, as they will involve $F^{(0)}(t - t_0)$-dependent factors (Rogallo, 1981; Cambon, 1982).
3. Both linear [as in the *rapid distortion theory* (RDT) or in similar stability analyses] and *full* nonlinear approaches can be carried out with the additional constant-mean-velocity-gradient-matrix effect, keeping the context of homogeneous, but often highly anisotropic, turbulence.
4. What is the physical relevance of a mean flow without boundaries having the same **A** matrix over the whole space (often called *extensional base flow* in the community of hydrodynamic stability)? This question is essential. It received a clear answer, at least for linear theory, ranging from WKB RDT to short-wave disturbance analyses: One has to consider that the spatial homogeneity, and therefore the region in which the mean gradient is almost constant, is restricted to a domain that is large with respect to the size of relevant turbulent structures (turbulence in general), or large with respect to the wavelength of disturbances, from the linear-stability viewpoint.

This last point (4) suggests that HAT in the presence of constant **A** is not only a marginal domain in the field of turbulence research. We therefore propose to use it as one of the main threads in this book. Of course, this point has to be discussed with care, in order to delineate its relevance to understanding the dynamics of realistic shear flows of practical interest. In addition, it is shown at the end of this book that

---

§ It is worth noting that the base flow is not invariant by translation.

turbulence in the presence of a shock wave can be treated by use of the HAT formalism. Even if the effect of the shock is very far from a mean-gradient effect, it is consistent with the absence of typical length scale L for the distorting mechanism: L is considered infinite (or very large with respect to the size of turbulent structures) in the first case; it is zero (the thickness of the shock wave), in contrast, in the second case (or very small with respect to the size of turbulent or organized structures passing through the shock wave).

2.2 Briefs About Statistical and Probabilistic Approaches

A presentation of statistical tools is a "compulsory figure" in any book on turbulence. The reader is referred to, e.g., Tennekes and Lumley's monograph (1972) and to Chapter 3 of a more recent book (Mathieu and Scott, 2000) for a deep and comprehensive review. Because this aspect is well documented, we recall only unavoidable definitions and procedures in this section.

2.2.1 Ensemble Averaging, Statistical Homogeneity

The most fundamental statistical averaging deals with an ensemble of realizations of a random variable V and is denoted either by an overbar $\overline{V}$ or by angle brackets $\langle V \rangle$ in the following. Possible approximations using temporal or spatial averaging are not discussed here. We assume that the ensemble averaging has all the properties of commutation (with time and spatial derivatives), which are often referred to as the Reynolds axioms and can therefore be referred to as Reynolds averaging. Discussing ergodicity is also beyond the scope of this book. The *probability density function* (pdf) that underlies the calculation of any statistical moment, as $\overline{V^n}$ for a scalar or $\overline{\boldsymbol{V}^n}$ for a vector, is introduced only when it is used within a specific context.

In view of the importance of homogeneous turbulence in this book, one cannot ignore the excellent definition by Batchelor (1953), as follows:

> Given an infinite body of uniform fluid in which motions conform to the equations (1.2.1) and (1.2.2), and given that at some initial instant the velocity of the fluid is a random function of position described by certain probability laws which are independent of position, to determine the probability laws that describe the motions of the fluid at subsequent times.

2.2.2 Single-Point and Multipoint Moments

It is important to point out that the Reynolds decomposition, in terms of a mean-velocity field, $\overline{u}$, and a fluctuating-velocity field, u', remains useful in many applications. It is used as a mandatory requirement before the statistical tools are applied to u'_i, e.g., evaluating statistical moments of the turbulent field. In this sense, statistical modeling is restricted to a *centered* random variable $u' = u - \overline{u}$. In addition, Nth-order moments of u' can be taken at the same point in the spatial domain, or at different points, the number of sampling points ranging from 2 to N. Evolution equations can be derived for all these quantities, with a problem of closure revisited in the last section of this chapter. A given level of description can be labeled as $[N, P]$, where N is the order of correlations and P ($P \leq N$) is the number of undependent points. Emphasis in this book is put on low-order moments, with N ranging from 2 to 4, but possibly in a multipoint description.

2.2.3 Statistics for Velocity Increments

An alternative to the two-point description, for instance based on the velocity-correlation tensor

$$R_{ij}(x, r, t) = \overline{u_i(x, t) u_j(x + r, t)}, \qquad (2.26)$$

is to work with velocity increments

$$\delta u_i = u_i(x + r, t) - u_i(x, t)$$

and to consider their moments only. This analysis is discussed in Chapter 3. A local scaling (in terms of r) of related moments, or structure functions, is easier to justify, following Kolmogorov, because the velocity increments are naturally smaller and smaller as the distance $r = |r|$ decreases. This analysis is restricted to an inertial range of scales, with r significantly larger than the Kolmogorov scale (see definition in Table 3.4) and significantly smaller than a typical integral length scale. High-order moments, i.e., high-order structure functions, are investigated in order to characterize the internal intermittency, but the multipoint approach is always a two-point one.

2.2.4 Application of Reynolds Decomposition to Dynamical Equations

The velocity and pressure fields are first split into mean and fluctuating components, and equations for their time evolution are derived from the basic equations of motion of the fluid. Assuming incompressibility, as is done in this chapter unless explicitly stated, one obtains the following mean-flow equations,

$$\frac{\partial \overline{u}_i}{\partial t} + \overline{u}_j \frac{\partial \overline{u}_i}{\partial x_j} = -\frac{\partial \overline{p}}{\partial x_i} + \nu \frac{\partial^2 \overline{u}_i}{\partial x_j \partial x_j} - \underbrace{\frac{\partial \overline{u'_i u'_j}}{\partial x_j}}_{\text{Reynolds stress term}}, \qquad (2.27)$$

$$\frac{\partial \overline{u}_i}{\partial x_i} = 0, \qquad (2.28)$$

and the equations for the fluctuating component

$$\frac{\partial u'_i}{\partial t} + \overline{u}_j \frac{\partial u'_i}{\partial x_j} + u'_j \frac{\partial \overline{u}_i}{\partial x_j} + \underbrace{\frac{\partial}{\partial x_j}(u'_i u'_j - \overline{u'_i u'_j})}_{\text{nonlinear term}} = \underbrace{-\frac{\partial p'}{\partial x_i}}_{\text{pressure term}} + \underbrace{\nu \frac{\partial^2 u'_i}{\partial x_j \partial x_j}}_{\text{viscous term}} \quad (2.29)$$

2.2 Statistical and Probabilistic Approaches

and

$$\frac{\partial u'_i}{\partial x_i} = 0. \tag{2.30}$$

Here, $\bar{u}_i$ and $\bar{p}$ are the mean velocity and static pressure (divided by density), and u'_i and p' are the corresponding fluctuating quantities, usually interpreted as representing turbulence.

At various points, we will mention related works in the area of hydrodynamic stability. It is worth noting that in the inhomogeneous case Eqs. (2.29) and (2.30) for the fluctuating flow are essentially the same as those for a perturbation u'_i, about a basic flow $\bar{u}_i$, with an additional forcing term $\partial \overline{u'_i u'_j}/\partial x_j$. Although the aims of stability theory (to characterize the growth of perturbation) and of the theory of turbulence (to determine the statistics of u'_i) are different, we believe it is nonetheless valuable to draw parallels between these two fields. It is our hope that, in doing so, we will encourage specialists in both areas to become more conversant with each others' work.

Equation (2.29) is now used to derive equations for the time evolution of velocity moments, i.e., averages of products of u'_i with itself at one or more points in space. Setting up the equations for the nth-order velocity moments at n points, one discovers that there are two main difficulties. First, the term in (2.29) that is nonlinear with respect to the fluctuations leads to the appearance of $(n+1)$th-order moments in the evolution equations for nth-order moments. Second, the pressure term brings in pressure–velocity correlations.

The pressure field is intimately connected with the incompressibility constraint. Indeed, by taking the divergence of (2.29), one obtains a Poisson equation for the pressure fluctuations:

$$\nabla^2 p' = -\frac{\partial^2}{\partial x_i \partial x_j}(u'_i \bar{u}_j + \bar{u}_i u'_j + u'_i u'_j - \overline{u'_i u'_j}). \tag{2.31}$$

The solution of this equation based on the Green's functions expresses p' at any point in space in terms of an integral of the velocity field over the entire fluid domain, together with integrals over the boundaries, the details of whose expression in terms of velocity do not concern us here.

The solution of the Poisson equation $\nabla^2 p' = f'$ can be written as

$$p'(\boldsymbol{x}, t) = \int_{\mathbb{R}^3} \mathcal{G}(\boldsymbol{x}, \boldsymbol{x}') f'(\boldsymbol{x}', t) d^3 \boldsymbol{x}', \tag{2.32}$$

with the related Green's function given by

$$\mathcal{G}(\boldsymbol{x}, \boldsymbol{x}') = \frac{1}{4\pi} \frac{1}{|\boldsymbol{x}' - \boldsymbol{x}|} \tag{2.33}$$

in a 3D unbounded domain, getting rid of specific boundary conditions.

Replacing f' in Eq. (2.32) with the whole right-hand side of Eq. (2.31) yields both linear and nonlinear, nonlocal contributions from fluctuating velocity. Thus the pressure at a given point is nonlocally determined by the velocity field at all points of the flow, leading to integrodifferential equations for the velocity moments when the pressure–velocity moments are expressed in terms of the sole velocity. It must be observed that nonlocality is not specific to the use of statistical methods, but is an intrinsic feature of incompressible fluids, in which the pressure field responds instantaneously and nonlocally to changes in the flow to enforce incompressibility. The source term in Poisson equation (2.31) consists of parts that are linear and nonlinear with respect to the velocity fluctuation. Therefore the pressure can be decomposed as the sum of two components: a pressure term $p'^{(r)}$ associated with linear terms (and referred to as the *rapid pressure term*, as it responds immediatly to a change in the mean flow) and a second one, $p'^{(s)}$, which is associated with the nonlinear one (referred to as the *slow pressure term*, as it is not directly sensitive to a change in the mean flow):

$$\nabla^2 p'^{(r)} = -\frac{\partial^2}{\partial x_i \partial x_j}(u_i'\overline{u}_j + \overline{u}_i u_j'), \qquad (2.34)$$

$$\nabla^2 p'^{(s)} = -\frac{\partial^2}{\partial x_i \partial x_j}(u_i' u_j' - \overline{u_i' u_j'}). \qquad (2.35)$$

2.3 Reynolds Stress Tensor and Related Equations

2.3.1 RST Equations

In addition to simple closure models for the Reynolds-averaged Navier–Stokes equations, such as models of turbulent viscosity using a mixing-length assumption, second-order single-point [2,1] models offer both a dynamical and a statistical description of the turbulent field. The governing equations for the RST, turbulent kinetic energy, and for its dissipation rate can reflect the effects of convection, diffusion, distortion, pressure, and viscous stresses, which are present in the equations that govern the fluctuating field u_i'.

The exact evolution equation for the RST, $R_{ij} = \overline{u_i' u_j'}$ [with $r = 0$ in Eq. (2.26)], derived from Eq. (2.29), has the form

$$\frac{\partial R_{ij}}{\partial t} + \overline{u}_k \frac{\partial R_{ij}}{\partial x_k} = \mathcal{P}_{ij} + \Pi_{ij} - \varepsilon_{ij} - \frac{\partial \mathcal{D}_{ijk}}{\partial x_k}, \qquad (2.36)$$

where

$$\mathcal{P}_{ij} = -\frac{\partial \overline{u}_i}{\partial x_k} R_{kj} - \frac{\partial \overline{u}_j}{\partial x_k} R_{ki} \qquad (2.37)$$

is usually referred to as the production tensor and is the only term on the right-hand side of Eq. (2.36) that does not require modeling, as it is given in terms of the basic one-point variables $\overline{u}_i$ and R_{ij}. The remaining terms, which are not exactly expressible in terms of the basic one-point variables and heuristic approximations, forming the core of the model are introduced to close the equations.

2.3 Reynolds Stress Tensor and Related Equations

The second term on the right-hand side of Eq. (2.36) is associated with the fluctuating pressure and is given by

$$\Pi_{ij} = \overline{p'\left(\frac{\partial u'_i}{\partial x_j} + \frac{\partial u'_j}{\partial x_i}\right)}, \tag{2.38}$$

consisting of one-point correlations between the fluctuating pressure and rate-of-strain tensor. As discussed in the introduction, p' is nonlocally determined from the velocity field by Poisson equation (2.31), which, in principle, requires multipoint methods for its treatment. It is usual to decompose Π_{ij} into three parts,

$$\Pi_{ij} = \Pi_{ij}^{(r)} + \Pi_{ij}^{(s)} + \Pi_{ij}^{(w)}, \tag{2.39}$$

corresponding to the three components of the Green's function solution of (2.31). The first is known as the "rapid" pressure component and arises from the pressure component defined by Eq. (2.34). Being linear, this component is present in RDT, hence the term "rapid" component. The second term in (2.39) is the "slow" component and comes from (2.35). Finally, $\Pi_{ij}^{(w)}$ is the wall component and corresponds to a surface integral over the boundaries of the flow in the Green's function solution for p', which is additional to the volume integrals expressing the rapid and slow components. The three components of Π_{ij} have zero trace and are assumed to represent physically distinct mechanisms. Hence they are modeled separately. The pressure–strain tensor is traceless (because of the incompressibility constraint) and therefore corresponds to a mechanism of redistribution of energy between the different components of the RST. Linear and nonlinear mechanisms reflected in its rapid and slow parts, respectively, are discussed at the end of this subsection. In simple models, a mechanism of *isotropization of the production* is attributed to $\Pi_{ij}^{(r)}$, and a mechanism of *return to isotropy*, or *isotropization of the RST*, is attributed to $\Pi_{ij}^{(s)}$.

The dissipation tensor,

$$\varepsilon_{ij} = 2\nu \overline{\frac{\partial u'_i}{\partial x_k} \frac{\partial u'_j}{\partial x_k}}, \tag{2.40}$$

accounts for the destruction of kinetic energy by viscous effects. The usual scalar dissipation rate, denoted as ε, is defined as

$$\varepsilon \equiv \frac{1}{2}\varepsilon_{ii}. \tag{2.41}$$

The last term in Eq. (2.36) vanishes in homogeneous turbulence. This term is expressed as a flux of a third-order correlation tensor $\mathcal{D}_{ijk}$, which gathers triple-velocity correlations, pressure–velocity terms, and viscous-diffusion terms:

$$\mathcal{D}_{ijk} = \overline{u'_i u'_j u'_k} + \frac{1}{\rho}\left(\delta_{jk}\overline{p'u'_i} + \delta_{ik}\overline{p'u'_j}\right) + \nu\left(\overline{u'_i \frac{\partial u'_j}{\partial x_k}} + \overline{u'_j \frac{\partial u'_i}{\partial x_k}}\right). \tag{2.42}$$

Its role is essential to the spatial transfer of turbulent kinetic energy (and anisotropy), which is created near a wall, away from it. It is ignored, however, as far as strict statistical homogeneity is assumed in this book.

Given the importance of the $\mathcal{K} - \varepsilon$ model in engineering, together with the specific role of trace-free terms in the Reynolds stress equations, it is useful to introduce a trace-deviator decomposition for the RST,

$$R_{ij} = 2\mathcal{K}\left(\frac{\delta_{ij}}{3} + b_{ij}\right), \quad \mathcal{K} = \frac{1}{2}R_{ii}, \quad b_{ij} = \frac{R_{ij}}{2\mathcal{K}} - \frac{\delta_{ij}}{3} \qquad (2.43)$$

and to write the governing equations for both the kinetic energy $\mathcal{K}$ and the deviatoric, trace-free, and dimensionless anisotropy tensor b_{ij}.

The evolution equation for the kinetic energy derived from Eq. (2.36) is

$$\frac{\partial \mathcal{K}}{\partial t} + \bar{u}_k \frac{\partial \mathcal{K}}{\partial x_k} = \mathcal{P} - \varepsilon - \frac{\partial \mathcal{D}_k}{\partial x_k}, \qquad (2.44)$$

and a similar equation is derived for b_{ij}. All the terms present in the Reynolds stress equation contribute to the equation for the kinetic energy, except the – traceless – pressure–strain tensor. The scalar production term $\mathcal{P} = \mathcal{P}_{ii}/2$ can be rewritten as

$$\mathcal{P} = -\mathcal{K}\left[\frac{1}{3}\left(\frac{\partial \bar{u}_i}{\partial x_j} + \frac{\partial \bar{u}_j}{\partial x_i}\right) + \left(\frac{\partial \bar{u}_i}{\partial x_k}b_{kj} + \frac{\partial \bar{u}_j}{\partial x_k}b_{ki}\right)\right]. \qquad (2.45)$$

The deviatoric part of the dissipation tensor is either neglected or simply modeled similarly as the slow part of the pressure–strain tensor. Only the scalar dissipation rate ε is considered an independent variable, which is governed by its own equation. Because the exact evolution equation for ε is very complex, the model equation used in practice is obtained by deriving the equation for $\dot{\varepsilon}/\varepsilon$ from the one for $\dot{\mathcal{K}}/\mathcal{K}$, which is much easier to derive.

The terms that appear in the evolution equations for Reynolds stress models in homogeneous turbulence can be exactly expressed as integrals in Fourier space of contributions derived from the second-order spectral tensor $\hat{R}_{ij}$, which is the Fourier transform of double correlations at two points, and from the third-order "transfer" spectral tensor T_{ij}, which involves the Fourier transform of two-point triple-velocity correlations (see Section 2.5).

2.3.2 The Mean Flow Consistent With Homogeneity

Let us consider a mean flow, filling all the space, with space-uniform velocity gradients, which generalize the solid-body motion $\boldsymbol{u}^{(0)}$ introduced in Subsection 2.1.6:

$$\bar{u}_i(\boldsymbol{x}, t) = A_{ij}(t)x_j + u_i^0. \qquad (2.46)$$

Its presence can be consistent with statistical homogeneity for the fluctuating flow. This is a common background for homogeneous turbulence and recent linear stability analyses (see Craik and Criminale, 1986, among others).¶ Equations (2.27)

¶ It is important to stress that the feedback of the RST in (2.27) vanishes because of statistical homogeneity (zero gradient of any averaged quantity), so that the mean flow (2.46) has to be a particular solution of the Euler equations and can be considered a base flow for stability analysis. In turn, the form of (2.46) is consistent with the preservation of homogeneity of the fluctuating flow governed by (2.29) and (2.30), provided that homogeneity holds for the initial data. This explains why

2.3 Reynolds Stress Tensor and Related Equations

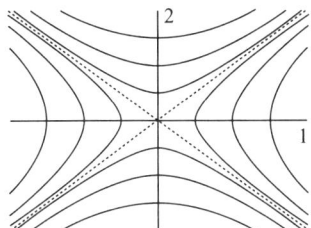

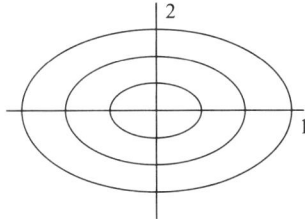

Figure 2.1. Sketch of isovalues of the stream function for the steady mean flow in homogeneous RDT: the three canonical cases, (a) elliptical $\Omega^2 > S^2$, (b) hyperbolical $\Omega^2 < S^2$, (c) linear $\Omega^2 = S^2$.

and (2.29) can be simplified by dropping the Reynolds stress term in both, so that (2.27) reduces to a particular Euler equation with a solution of type (2.46). As a consequence, the trace-free matrix **A** is subjected to the condition that $d\mathbf{A}/dt + \mathbf{A}^2$ must be symmetric, or, equivalently,

$$\epsilon_{ijk}\left(\frac{dA_{jk}}{dt} + A_{jn}A_{nk}\right) = 0, \quad A_{ii} = 0. \tag{2.47}$$

Irrotational mean flows, which are flows with a symmetric gradient matrix **A**, i.e., with $A_{ij} = A_{ji}$, are obvious solutions. Rotational mean flows yield more complicated linear solutions, and only the steady case has received much attention [Craik and co-workers and Bayly and co-workers performed recent developments in unsteady cases; see, e.g., Bayly, Holm, and Lifschitz (1996)]. Conditions (2.47) imply that **A** can be written as

$$\mathbf{A} = \begin{bmatrix} 0 & S-\Omega & 0 \\ S+\Omega & 0 & 0 \\ 0 & 0 & 0 \end{bmatrix} \tag{2.48}$$

in the steady, rotational case, when axes are chosen appropriately, where $S, \Omega \geq 0$. This corresponds to steady plane flows, combining vorticity 2Ω and irrotational straining S. The related stream function (sketched in Fig. 2.1) is

$$\psi = \frac{S}{2}(x_1^2 - x_2^2) + \frac{\Omega}{2}(x_1^2 + x_2^2), \tag{2.49}$$

homogeneous RDT can have the same starting point as a rigorous and complete linear-stability analysis in this case, before the random initialization of the fluctuating velocity field is considered.

with $\bar{u}_i = \epsilon_{i3j} \frac{\partial \psi}{\partial x_j}$. The problem with arbitrary S and Ω was analyzed in order to extend classical RDT results, which were restricted to pure strain and pure shear. For $S > \Omega$, the mean-flow streamlines are open and hyperbolic. For $S < \Omega$, the mean-flow streamlines are closed and elliptic about the stagnation point at the origin. The limiting case, $S = \Omega$, correspond to pure shearing of straight mean streamlines (see Chapter 5 for the fundamentals of RDT analysis).

2.3.3 Homogeneous RST Equations. Briefs About Closure Methods

Classical closure methods are now briefly addressed. The reader is referred to reference books for an exhaustive discussion about turbulence modeling, e.g., Piquet (2001). If we restrict our attention to homogeneous turbulence in the presence of mean-velocity gradients previously defined, the RST is *unsteady*: The steadiness of RST equations, often assumed in Reynolds averaged Navier–Stokes (RANS) methods, comes from the use of time averaging, and does not concern us here. Historically, basic concepts for deriving statistical closures were introduced in this unsteady homogeneous framework (Launder, Reece, and Rodi, 1975; Lumley, 1975). The most difficult term to close in homogeneous turbulence is the linear (rapid) contribution to the pressure–strain tensor in Eqs. (2.36) and (2.39), which one can write as

$$\Pi_{ij}^{(r)} = 2A_{mn}\left(M_{inmj} + M_{jnmi}\right), \tag{2.50}$$

with

$$M_{ijpq} = \frac{1}{4\pi} \frac{\partial^2}{\partial r_p \partial r_q} \iiint \frac{1}{|\boldsymbol{r} - \boldsymbol{r}'|} R_{ij}(\boldsymbol{r}', t) d^3\boldsymbol{r}', \tag{2.51}$$

using Eqs. (2.32) and (2.33). A slightly different form of M_{ijpq} can be found in Launder, Reece, and Rodi (1975) and Lumley (1978). The alternative relationship for M_{ijpq} in Fourier space, more tractable from our viewpoint, is given in Subsection 2.6.1.2, in exact agreement with, e.g., Kassinos, Reynolds, and Rogers (KRR) (2001). In general, there is no direct link between M_{ijpq} and the RST, even if the identity $M_{ijpp} = R_{ij}$ holds, and the problem of closure arises from the two-point structure in Eq. (2.51). In classical closures, the nondimensional tensor $M_{ijpq}/(2\mathcal{K})$ is sought as a tensorial function of the nondimensional deviatoric tensor b_{ij} defined in Eq. (2.43). Models range from linear (Launder, Reece, and Rodi, 1975) to cubic tensorial expansions.

Similarly, the slow pressure–strain tensor is assumed to be an isotropic tensorial function of b_{ij}. In the simplest version, $\Pi_{ij}^{(s)}$ is proportional to $-b_{ij}$, in agreement with an heuristic principle of return-to-isotropy.

Finally, the ε-equation is usually closed by pure analogy with the $\mathcal{K}$-equation. One can understand this by considering the following evolution equations for their logarithmic derivatives:

$$\frac{1}{\mathcal{K}} \frac{d\mathcal{K}}{dt} = \frac{\mathcal{P}}{\mathcal{K}} - \frac{\varepsilon}{\mathcal{K}} \tag{2.52}$$

2.4 Anisotropy in Physical Space

and

$$\frac{1}{\varepsilon}\frac{d\varepsilon}{dt} = C_{\epsilon 1}\frac{\mathcal{P}}{\mathcal{K}} - C_{\epsilon 2}\frac{\varepsilon}{\mathcal{K}}, \quad (2.53)$$

where $C_{\epsilon 1}$ and $C_{\epsilon 2}$ are two real arbitrary parameters, which are tuned to optimize the results on some very simple flows (e.g., decaying isotropic turbulence, homogeneous shear flow, turbulent flat-plate boundary layer, etc.). The first equation can be considered as exact in the homogeneous unsteady limit if the production term given by Eq. (2.45) is known (it derives from the RST, but it is evaluated from simplified b_{ij} models in linear and nonlinear $\mathcal{K} - \varepsilon$ models). On the other hand, the second equation is only a carbon copy of the first one, using two empirical constants, without linkage to the true enstrophy equation.

2.3.3.1 KRR's New Tensors

Even if the RST is recovered in contracting the last two indices of M_{ijpq}, the closure of this whole tensor in terms of the sole RST is a heuristic method, which was questioned, especially in the presence of a rotational mean flow. To capture more of the components of M_{ijpq}, KRR (2001) proposed introducing a *dimensionality tensor*,

$$D_{pq} = M_{iipq}, \quad (2.54)$$

along with a *circulicity tensor*, denoted as F_{ij} in KRR (2001), and a *stropholysis tensor*, Q_{ijn},

$$F_{ij} = \epsilon_{ipm}\epsilon_{jqn}M_{mnpq}, \quad Q_{ijn} = \epsilon_{ipq}M_{jqpn}, \quad (2.55)$$

with an alternative fully symmetrized version Q^*_{ijn}. In homogeneous turbulence, the circulicity tensor is not an independent one, in agreement with

$$F_{ij} = \mathcal{K}\delta_{ij} - D_{ij} - \overline{u'_i u'_j}. \quad (2.56)$$

Alternative definitions (also valid in inhomogeneous turbulence) are obtained with a vector potential, or *turbulence stream-function vector* ψ'_i, so that

$$u'_i = \epsilon_{imn}\psi'_{n,m}, \quad (2.57)$$

$$\overline{u'_i u'_j} = \epsilon_{imn}\epsilon_{ipq}\overline{\psi'_{n,m}\psi'_{q,p}}, \quad D_{ij} = \overline{\psi'_{n,i}\psi'_{n,j}}, \quad F_{ij} = \overline{\psi'_{i,n}\psi'_{j,n}}. \quad (2.58)$$

The reader is referred to KRR (2001) for the definition of a last tensor, denoted as C_{ij} and specifically inhomogeneous. As for M_{ijpq}, a new insight into this structure-based modeling will appear using spectral formalism.

2.4 Anisotropy in Physical Space. Single-Point and Two-Point Correlations

In single-point modeling used in RANS methods, the deviatoric part of the RST is used as the unique anisotropy indicator. An equation for b_{ij} is readily derived from (2.36), as the $\mathcal{K}$-equation (2.44).

Anisotropy for two-point correlations in physical space $\mathbf{R}(\mathbf{r}, t)$ is rarely investigated, because application of group-invariance properties (e.g., axisymmetry) must be made prior to application of the incompressibility constraint. Incompressibility yields implicit differential relationships between the different terms resulting from the symmetry-group analysis (Sreenivasan and Narasimha, 1978). In contrast, as we shall subsequently see, the incompressibility constraint *a priori* yields dramatic simplifications (reduction of the number of unknowns) for the spectral tensor, which is the 3D Fourier counterpart of $\mathbf{R}(\mathbf{r}, t)$, so that application of symmetry-group properties, as a second step, is much simpler. Similarly, anisotropy begins to be investigated for structure functions in the scaling–intermittency community, using the SO(3) symmetry group and spherical harmonic expansions (Arad, Lvov, and Proccacia, 1999). Nevertheless, we are not aware of any application to typically anisotropic flows, such as the ones addressed in Chapters 4–8.

The anisotropy tensor b_{ij} defined in Eq. (2.43) can be used to characterize the structure of the anisotropic flows. Following the Cayley–Hamilton therorem, one has

$$b_{ij}^3 + I_2 b_{ij} - I_3 \delta_{ij} = 0, \qquad (2.59)$$

where the second and third invariants of the anisotropy tensor are defined as $I_2 = -b_{ij}b_{ji}/2$ and $I_3 = \det(b)$, respectively. It was shown by Lumley and Newman (1977) that all physically admissible turbulent flows are contained within a finite region (often referred to as the *Lumley triangle*) in the space spanned by I_2 and I_3 (or, equivalently, in the space spanned by the two nonvanishing eigenvalues of the anisotropy tensor). Each admissible point in the anisotropy map corresponds to a specific shape of the ellipsoid generated by the three diagonal components of the RST. The classification of the main anisotropy states was recently clarified by Simonsen and Krogstadt (2005). The main elements of the classification are summarized in Table 2.1 and illustrated in Fig. 2.2.

This analysis can be applied to any deviatoric dimensionless and trace-free tensor derived from a definite-positive symmetrical tensor. The three positive eigenvalues of such a tensor, related to the orthogonal frame of eigenvectors (principal axes) can be used instead of the λ_i eigenvalues. Many instances are given throughout this book, including structure-based modeling and even spectral tensors.

2.5 Spectral Analysis, From Random Fields to Two-Point Correlations. Local Frame, Helical Modes

2.5.1 Second-Order Statistics

Regarding homogeneous turbulence, we aim to take into account the possible distorting effects of a mean flow defined by Eqs. (2.46) and (2.47), or effects of body forces, so that anisotropy is essential. Therefore, the emphasis is put on *homogeneous anisotropic turbulence* (HAT). The Fourier transform is a valuable tool to handle equations for velocity and pressure fluctuations, considered as random

2.5 Spectral Analysis

Table 2.1. *Characteristics of the RST and the anisotropy tensor*

State of turbulence	Invariants	Eigenvalues of b_{ij}	Shape of Reynolds stress ellipsoid
Isotropic	$I_2 = I_3 = 0$	$\lambda_1 = \lambda_2 = \lambda_3 = 0$	Sphere
Axisymmetric (one large λ_i)	$\frac{-I_2}{3} = \left(\frac{I_3}{2}\right)^{2/3}$	$0 < \lambda_1 < \frac{1}{3}$ $-\frac{1}{6} < \lambda_2 = \lambda_3 < 0$	Prolate spheroid
Axisymmetric (one small λ_i)	$\frac{-I_2}{3} = \left(-\frac{I_3}{2}\right)^{2/3}$	$0 < \lambda_1 < \frac{1}{3}$ $0 < \lambda_2 = \lambda_3 < \frac{1}{6}$	Oblate spheroid
One component	$I_3 = \frac{2}{27}\ I_2 = -\frac{1}{3}$	$\lambda_1 = \frac{2}{3}$ $\lambda_2 = \lambda_3 = -\frac{1}{3}$	Line
Two component (axisymmetric)	$I_3 = -\frac{1}{108}$ $I_2 = -\frac{1}{12}$	$\lambda_1 = -\frac{1}{3}$ $\lambda_2 = \lambda_3 = \frac{1}{6}$	Disk
Two component	$-\frac{I_2}{3} = \left(\frac{1}{27} + I_3\right)$	$\lambda_1 + \lambda_2 = \frac{1}{3}$ $\lambda_3 = -\frac{1}{3}$	Ellipsoid

Source: Adapted from Simonsen and Krogstad (2005).

variables, as well as their statistical multipoint correlations matrices. The relations between second-order tensors defined in both physical and Fourier spaces are displayed in Fig. 2.3.

The inverse Fourier transform that connects $\boldsymbol{u}'$ to $\hat{\boldsymbol{u}}$ is expressed as

$$u'_i(\boldsymbol{x}, t) = \iiint \hat{u}_i(\boldsymbol{k}, t) \exp(\imath \boldsymbol{k} \cdot \boldsymbol{x})\, d^3\boldsymbol{k}. \tag{2.60}$$

Applying it to the two-point correlation tensor, one obtains

$$\overline{u'_i(\boldsymbol{x}, t)u'_j(\boldsymbol{x}+\boldsymbol{r}, t)} = \iiint \hat{R}_{ij}(\boldsymbol{k}, t)\exp(\imath \boldsymbol{k}\cdot\boldsymbol{r})d^3\boldsymbol{k}. \tag{2.61}$$

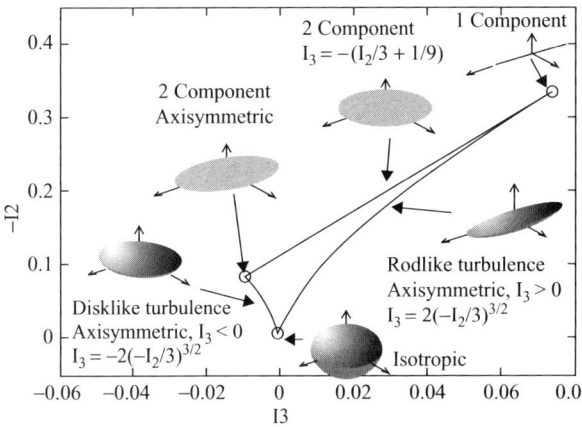

Figure 2.2. Lumley's anisotropy invariant map and related Reynolds stress ellipsoids. Admissible turbulent states are located inside the trianglelike subdomain. From Simonsen and Krogstad (2005), with permission of the American Institute of Physics.

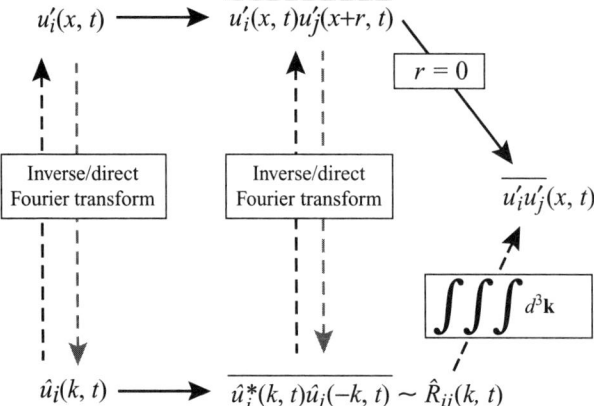

Figure 2.3. Schematic view of the relations that exist between second-order tensors defined in both physical and Fourier spaces.

One may recall here that the direct Fourier transform is written as

$$\hat{R}_{ij}(\mathbf{k}, t) = \frac{1}{(2\pi)^3} \iiint R_{ij}(\mathbf{r}, t) \exp(-\imath \mathbf{k} \cdot \mathbf{r}) d^3 \mathbf{r}. \tag{2.62}$$

It is worth noting that the prefactor $1/(2\pi)^3$ appears in Eq. (2.62), and not in Eq. (2.61). According to (2.61), the RST, which one obtains by setting $\mathbf{r} = \mathbf{0}$ in R_{ij}, derives from its spectral counterpart $\hat{\mathbf{R}}$ through a 3D integral:

$$\overline{u'_i(\mathbf{x}, t) u'_j(\mathbf{x}, t)} = \iiint \hat{R}_{ij}(\mathbf{k}, t) d^3 \mathbf{k}. \tag{2.63}$$

The final equation of note in this subsection is

$$\overline{\hat{u}_i^*(\mathbf{p}, t) \hat{u}_j(\mathbf{k}, t)} = \hat{R}_{ij}(\mathbf{k}, t) \delta^3(\mathbf{k} - \mathbf{p}). \tag{2.64}$$

Two alternative ways can be used to derive evolution equations for statistical quantities in spectral space. On the one hand, one can derive an equation for $\overline{u'_i(\mathbf{x}, t) u'_j(\mathbf{x} + \mathbf{r}, t)}$ by first using Eq. (2.29) and then obtaining the equation for $\hat{R}_{ij}$ by applying (2.62) (Oberlack, 2001). On the other hand, an equation for $\hat{u}_i(\mathbf{k}, t)$ can be directly obtained in Fourier space, from which one derives the equation for $\hat{R}_{ij}$ using Eq. (2.64). At least in homogeneous turbulence, the second way is simpler because the pressure term can be solved in the simplest way in the equation for $\hat{u}_i$. As a consequence, it is used in the following. The first way (Craya, 1958), even if more cumbersome, has the advantage of applying a Fourier transform on only statistical (smooth) quantities, without need for distribution theory. The reader is referred to Batchelor (1953) and to Chapter 6 of Mathieu and Scott (2000) for a detailed analysis of Fourier expansions of both random variables and their statistical moments and their limit in an infinite box.

2.5 Spectral Analysis

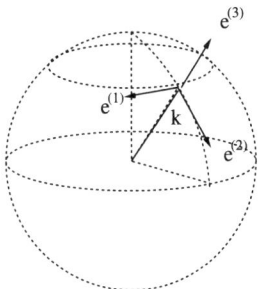

Figure 2.4. Polar-spherical system of coordinates for k and related Craya–Herring frame of reference.

2.5.2 Poloidal–Toroidal Decomposition and Craya–Herring Frame of Reference

The *poloidal–toroidal decomposition* (see, e.g., Chandrasekhar, 1981) is used to represent a three-component divergence-free velocity field in terms of two independent scalar terms, taking advantage of the presence of a privileged direction n:

$$u' = \underbrace{\nabla \times (s_{\text{tor}} n)}_{\text{toroidal part}} + \underbrace{\nabla \times [\nabla (s_{\text{pol}} n)]}_{\text{poloidal part}}, \quad (2.65)$$

in which s_{pol} and s_{tor} are scalar potentials. The axial vector n is chosen along the vertical direction, without loss of generality. As a caveat, some care is needed to represent *vertically sheared horizontal flows* (VSHFs) (as coined in Smith and Waleffe, 2002) that are defined as $u' = u'(x \cdot n, t)$ with $u' \cdot n = 0$.

In Fourier space, the preceding decomposition yields a pure geometrical representation,

$$\hat{u} = \underbrace{k \times n (\imath \hat{s}_{\text{tor}})}_{\text{toroidal mode}} - \underbrace{k \times (k \times n)(\hat{s}_{\text{pol}})}_{\text{poloidal mode}}, \quad (2.66)$$

and it appears immediately that the Fourier mode related to the vertical wave-vector direction, $k \parallel n$, has zero contribution. This gap in the spectral description precludes the capture of the VSHF mode in physical space. To solve this problem, one can define an orthonormal frame of reference, which is nothing but the local reference frame of a polar-spherical system of coordinates for k (see Fig. 2.4):

$$e^{(1)} = \frac{k \times n}{|k \times n|}, \quad e^{(2)} = e^{(3)} \times e^{(1)}, \quad e^{(3)} = \frac{k}{k}, \quad (2.67)$$

with $k \times n \neq 0$. Local frame vectors $e^{(1)}, e^{(2)}, e^{(3)}$ may coincide with the fixed frame of reference, with $e^{(3)} = n$ for $k \parallel n$. The only price to pay for matching toroidal–poloidal modes to the VSHF one is to accept a nonuniform definition for $k \times n = 0$ and $k \times n \neq 0$, but this is well known using any polar-spherical coordinate system. In the turbulence community, the local frame $(e^{(1)}, e^{(2)})$ of the plane normal to the wave vector is often referred to as the *Craya–Herring frame*. Accordingly the divergence-free velocity field in Fourier space has only two components in the Craya–Herring frame:

$$\hat{u}(k, t) = u^{(1)} e^{(1)} + u^{(2)} e^{(2)}. \quad (2.68)$$

For $\mathbf{k} \times \mathbf{n} \neq 0$, $u^{(1)}$ and $u^{(2)}$ are directly linked to the *toroidal mode* and the *poloidal mode*, respectively. For $\mathbf{k} \times \mathbf{n} = 0$, they correspond to the VSHF mode. For the vorticity fluctuation, a similar decomposition is found:

$$\hat{\omega}_i(\mathbf{k}, t) = \imath k \left[u^{(1)} e_i^{(2)} - u^{(2)} e_i^{(1)} \right], \qquad (2.69)$$

so that $u^{(1)}$ and $u^{(2)}$ are directly related to the spectral counterparts of Orr–Sommerfeld/Squires variables: $\hat{\omega}_3 = -k_\perp u^{(1)}$ for ω_3' and $-k^2 \hat{u}_3 = -kk_\perp u^{(2)}$ for $\nabla^2 u_3'$, with $k_\perp = |\mathbf{k} \times \mathbf{n}|$. A similar decomposition is used in Bayly, Holm, and Lifschitz (1996). Finally, the *wave-vortex decomposition* introduced in Riley, Metcalfe, and Weissman (1981) in the particular context of stably stratified turbulence (see Chapter 7) is also a particular case of Eq. (2.65).

RDT equations (and fully nonlinear ones, too) can be written in the Craya–Herring frame, resulting in a reduced Green's function with only four independent components (Cambon, 1982). Details are given in the following chapters, in which RDT solutions are discussed.

Finally, one may mention that the toroidal part of the flow is always a part of the "horizontal" velocity component $\mathbf{u}' \times \mathbf{n}$, the one that is divergence free only in terms of horizontal coordinates. This toroidal contribution does not reduce to a two-dimensional two-component (2D-2C) mode, because the toroidal potential s_{tor} also depends on the "vertical" coordinate $\mathbf{x} \cdot \mathbf{n}$ in general; it appears as a 3D stream function. Both the VSHF mode and the 2D-2C mode are low-dimension manifolds, which can be incorporated in the generalized poloidal–toroidal decomposition, *without need for specific additional terms*. In spite of the analogy of (2.65) with a "vortical-divergent" Helmholtz decomposition restricted to the horizontal flow (the vertical component being a part of the poloidal mode by means of $\mathbf{u}' \cdot \mathbf{n} = -\nabla^2 s_{\text{pol}}$), this decomposition is purely vortical in three dimensions, because it generates a 3D potential vector

$$\psi_i' = s_{\text{tor}} n_i + \epsilon_{ipq} \frac{\partial s_{\text{pol}}}{\partial x_p} n_q, \qquad (2.70)$$

as the one used by KRR [see Eq. (2.57)]. The detailed way to recover the two scalar functions s_{pol} and s_{tor} from the solenoidal velocity field is not recalled here for the sake of brevity.

2.5.3 Helical-Mode Decomposition

This decomposition is an alternative to the Craya–Herring decomposition and presents some advantages regarding frame-invariance properties, treatment of background nonlinearity, and rotating turbulence. The *helical modes* are defined from

$$N_i(\mathbf{k}) = e_i^{(2)}(\mathbf{k}) - \imath e_i^{(1)}(\mathbf{k}), \qquad (2.71)$$

2.5 Spectral Analysis

Table 2.2. *Local frame of reference in Fourier space*

Decomposition name	$V_3(k)$	$V_1(k)$	$V_2(k)$
Poloidal–toroidal	n	$\imath k \times n$	$-k \times (k \times n)$
Craya–Herring	$\dfrac{k}{k}$	$\dfrac{k \times n}{\|k \times n\|}$	$\dfrac{k}{k} \times \dfrac{k \times n}{\|k \times n\|}$
Helical	$\dfrac{k}{k}$	$\dfrac{k}{k} \times \dfrac{k \times n}{\|k \times n\|} - \imath \dfrac{k \times n}{\|k \times n\|}$	$\dfrac{k}{k} \times \dfrac{k \times n}{\|k \times n\|} + \imath \dfrac{k \times n}{\|k \times n\|}$

Note: The general form is $\hat{u}(k,t) = \chi_1(t) V_1(k) + \chi_2(t) V_2(k)$, where $[V_1(k), V_2(k)]$ is a local frame in the plane orthogonal to k. This local basis is supplemented by a third vector, V_3. The vector n is an arbitrary parameter in the three decompositions.

so that the solenoidal velocity field in Fourier space is decomposed as

$$\hat{u}(k,t) = \xi_+(k,t) N(k) + \xi_-(k,t) N(-k). \tag{2.72}$$

The preceding definition is the same as the one in Cambon's thesis (1982). It was used in all the subsequent papers (e.g., Cambon and Jacquin, 1989) from the same team. Particularly, this definition ensures Hermitian symmetry,

$$N(-k) = N^*(k),$$

where $*$ denotes a complex conjugate, because $e^{(1)}(-k) = -e^{(1)}(k)$ and $e^{(2)}(-k) = e^{(2)}(k)$. The most useful property is

$$\imath k \times N = k N, \tag{2.73}$$

which means that $N e^{\imath k \cdot x}$ and its complex conjugate are eigenmodes of the curl operator. Accordingly, the vorticity fluctuation in Fourier space is written as

$$\hat{\omega}_i(k,t) = k \left[\xi_+(k,t) N_i(k) - \xi_-(k,t) N_i(-k) \right]. \tag{2.74}$$

Helical modes (2.71) were also used by Waleffe (1992). If one looks at the literature from only the turbulence community, similar modes were introduced in the 1970s as *helicity waves* (Uriel Frish and Marcel Lesieur, private communication), but they were not used to get simplified dynamical equations.

Key elements of the three spectral decompositions just presented are summarized in Table 2.2.

2.5.4 Use of Projection Operators

Even when pure incompressible flows are considered, the solenoidal property for the velocity field is not satisfied by some terms in the governing equations, so that projection onto a solenoidal subspace is needed. The Helmholtz decomposition, which is addressed in Chapter 9, can be used in a simple way to define longitudinal and transverse projection operators.

A simple geometric decomposition is obvious for any vector V into a component along a given direction spanned by a unit vector a and a component contained in the plane normal to a,

$$V = V^{\parallel} + V^{\perp},$$

with $V^{\parallel} = (V \cdot a)a$ and, by difference, $V^{\perp} = V - (V \cdot a)a$. This decomposition brings in two projection matrices, $P_{ij}^{\parallel} = a_i a_j$, with $V_i^{\parallel} = P_{ij}^{\parallel} V_j$, and $P_{ij}^{\perp} = \delta_{ij} - a_i a_j$ with $V_i^{\perp} = P_{ij}^{\perp} V_j$.

If we now consider $V(k)$ as the Fourier transform of any term in the background equation that governs $\hat{u}$ and set a equal to the unit vector along k, i.e., $a_i = k_i/k$, the preceding geometric decomposition gives a simplified instance of the Helmholtz decomposition. Accordingly,

$$V_i^{\parallel}(k) = \frac{k_i k_j}{k^2} V_j(k)$$

corresponds to the projection onto the *dilatational mode*, and

$$V_i^{\perp}(k) = \left(\delta_{ij} - \frac{k_i k_j}{k^2} \right) V_j(k)$$

corresponds to the projection onto the *solenoidal mode*. This immediately suggests defining a longitudinal projection operator as

$$P_{ij}^{\parallel} = \frac{k_i k_j}{k^2} \qquad (2.75)$$

and a transverse projection operator as

$$P_{ij}^{\perp} = \delta_{ij} - \frac{k_i k_j}{k^2}. \qquad (2.76)$$

Let us consider the following generic model equation,

$$\frac{\partial u_i'}{\partial t} + S_i + \frac{\partial p'}{\partial x_i},$$

with a solenoidal vector field u' and an arbitrary term S_i. Its counterpart in Fourier space is

$$\frac{\partial \hat{u}_i}{\partial t} + \hat{S}_i + \imath k_i \hat{p} = 0,$$

and the solenoidal property is replaced with the condition that $\hat{u}$ and k are orthogonal. Application of the longitudinal projection operator yields

$$P_{ij}^{\parallel} \hat{u}_j = 0,$$

and

$$P_{ij}^{\parallel} \hat{S}_j + \imath k_i \hat{p} = 0,$$

which corresponds to the Poisson equation for the pressure term, whereas the transverse projection operator gives

$$P_{ij}^{\perp} \hat{u}_j = \hat{u}_i$$

2.5 Spectral Analysis

and

$$\frac{\partial \hat{u}_i}{\partial t} + P_{ij}^{\perp} \hat{S}_i = 0.$$

The latter equation is a pure solenoidal equation, which no longer includes the pressure term. When the emphasis is put on solenoidal turbulence, only the latter form of the dynamical equation is useful (as far as specific information on the pressure term is not needed), and only the tranverse projection operator $P^{\perp}(k)$ is needed. For the sake of simplicity the adjective "transverse" and the superscript $\perp$ will be omitted from now on. Of course, the projection operator has a simple expression in terms of Craya–Herring and helical modes:

$$P_{ij} = e_i^{(1)} e_j^{(1)} + e_i^{(2)} e_j^{(2)} = \Re\left(N_i N_j^*\right). \tag{2.77}$$

The decomposition of an arbitrary vector field $V(k)$, which is not *a priori* divergence free, requires the use of the three vectors $[e^{(1)}, e^{(2)}, e^{(3)}]$ of the Craya–Herring base, with $e_i^{(3)} = k_i/k$. One recovers here the fact that the third component is related to the dilatational mode (which is a 1D mode in the local reference frame) and that the first two components represent the solenoidal mode (which is *a priori* 2D in the local reference frame), in agreement with both the Helmholtz and toroidal–poloidal decompositions in physical space.

2.5.5 Nonlinear Dynamics

Considering a mean flow that preserves the statistical homogeneity of the fluctuating motion, one can recast nonlinear equation (2.29) as

$$\overset{\circ}{\hat{u}}_i + M_{ij} \widehat{u}_j = s_i - \nu k^2 \widehat{u}_i, \tag{2.78}$$

where

$$\overset{\circ}{\hat{u}}_i = \frac{\partial \hat{u}_i}{\partial t} + \frac{\partial \hat{u}_i}{\partial k_m} \frac{dk_m}{dt} = \frac{\partial \hat{u}_i}{\partial t} - A_{lm} k_l \frac{\partial \hat{u}_i}{\partial k_m} \tag{2.79}$$

is related to linear advection by the mean flow [see Eq. (2.46)], and $M_{ij} = A_{mj}(\delta_{im} - 2k_i k_m/k^2)$ gathers linear distortion and pressure terms (see Chapter 5). Once nonlinear and viscous terms have been summed, Eq. (2.78) generalizes the linear inviscid equation.

The nonlinear term s_i is given by

$$s_i(\mathbf{k}, t) = -\imath P_{ijk}(\mathbf{k}) \int_{\mathbf{p+q=k}} \widehat{u}_j(\mathbf{p}, t) \widehat{u}_k(\mathbf{q}, t) d^3 \mathbf{p}, \tag{2.80}$$

in which the third-order tensor $P_{ijk} = \frac{1}{2}(P_{ij}k_k + P_{ik}k_j)$ arises from the elimination of pressure using the incompressibility condition $k_i \widehat{u}_i(\mathbf{k}, t) = 0$, in agreement with the use of projection operator (2.76) as discussed in the previous subsection.

Equations (2.78)–(2.80) are completely generic and hold for other cases, including body forces and additional random variables. This is achieved in a straightforward manner, simply by changing the matrix **M** of the linear operator and/or the influence matrix P_{ijk} in the convolution product that reflects quadratic nonlinearity. The evolution equation for $\hat{R}_{ij}$ (Craya, 1958) derived from Eqs. (2.64) and (2.78) is

$$\overset{\circ}{\hat{R}}_{ij} + M_{ik}\hat{R}_{kj} + M_{jk}\hat{R}_{ik} = T_{ij} - 2\nu k^2 \hat{R}_{ij} \tag{2.81}$$

where the left-hand side arises from the linear inviscid part of Eq. (2.78). The term $\overset{\circ}{\hat{R}}_{ij}$ is a convective time derivative in k-space with distortion components. The second term on the right-hand side is the spectral counterpart of the dissipation tensor. The generalized transfer tensor T_{ij} is mediated by nonlinearity as

$$\langle \hat{u}_i^*(\boldsymbol{p},t) s_j(\boldsymbol{k},t) + s_i^*(\boldsymbol{p},t)\hat{u}_j(\boldsymbol{k},t) \rangle = T_{ij}(\boldsymbol{k},t)\delta(\boldsymbol{p}-\boldsymbol{k}). \tag{2.82}$$

This tensor involves triple-velocity correlations as shown by Eq. (2.80) for s_i. More details on it will be given in Chapter 3.

Although the purely linear theory closes the equations and simplifies mathematical analysis, its domain of applicability is rather limited because it neglects all interactions of turbulence with itself, including the physically important cascade process. Multipoint turbulence models that account for nonlinearity by means of closure lead to moment equations with a well-defined linear operator and nonlinear source terms. The view taken in this book is that, even when nonlinearity is significant, the behavior of the linear part of the model often has a significant influence. Thus it is important to first understand the properties of the linearized model. An additional interesting output of the linearized analysis is that it often allows for the definition of a simplified formulation of the nonlinear model by use of more appropriate variables.

2.5.6 Background Nonlinearity in Different Reference Frames

Equations given in Subsection 2.5.5 express background linear and nonlinear terms in 3D Fourier space. They can be rewritten in the different local reference frames previously introduced (results are summarized in Table 2.3).

In the Craya–Herring frame of reference, the $\hat{u}$ vector with three components is replaced with the $u^{(\alpha)}$ vector that has two components, and Eq. (2.78) becomes

$$\dot{u}^{(\alpha)} + m_{\alpha\beta} u^{(\beta)} = -\imath \int_{\boldsymbol{p}+\boldsymbol{q}=\boldsymbol{k}} P_{\alpha\beta\gamma} u^{(\beta)}(\boldsymbol{p},t) u^{(\gamma)}(\boldsymbol{q},t) d^3 \boldsymbol{p}, \tag{2.83}$$

with

$$m_{\alpha\beta}(\boldsymbol{k}) = e_i^{(\alpha)} M_{ij} e_j^{(\beta)} - \dot{e}_i^{(\alpha)} e_j^{(\beta)} \tag{2.84}$$

and

$$P_{\alpha\beta\gamma}(\boldsymbol{k},\boldsymbol{p}) = \frac{k}{2} \left\{ [e^{(\alpha)}(\boldsymbol{k}) \cdot e^{(\beta)}(\boldsymbol{p})][e^{(3)}(\boldsymbol{k}) \cdot e^{(\gamma)}(\boldsymbol{q})] \right.$$
$$\left. + [e^{(\alpha)}(\boldsymbol{k}) \cdot e^{(\gamma)}(\boldsymbol{q})][e^{(3)}(\boldsymbol{k}) \cdot e^{(\beta)}(\boldsymbol{p})] \right\}. \tag{2.85}$$

More details will be given in Chapters 3 and 7.

Table 2.3. *Expressions of the nonlinear momentum equation in both fixed Cartesian and local (Craya–Herring, helical) reference frames in Fourier space*

Reference frame	Convective term	Equation
Fixed Cartesian		$\dot{\widehat{u}}_i + M_{ij}\widehat{u}_j = s_i$
		$M_{ij}(\boldsymbol{k}) = A_{mj}(\delta_{im} - 2k_ik_m/k^2)$
	Conservative	$s_i(\boldsymbol{k}) = -\imath P_{ijk}(\boldsymbol{k}) \int_{\boldsymbol{p+q=k}} \widehat{u}_j(\boldsymbol{p},t)\widehat{u}_k(\boldsymbol{q},t) d^3\boldsymbol{p}$
Craya–Herring		$\dot{u}^{(\alpha)} + m_{\alpha\beta}u^{(\beta)} = -\imath \int_{\boldsymbol{p+q=k}} P_{\alpha\beta\gamma}u^{(\beta)}(\boldsymbol{p},t)u^{(\gamma)}(\boldsymbol{q},t) d^3\boldsymbol{p}$
		$m_{\alpha\beta}(\boldsymbol{k}) = e_i^{(\alpha)} M_{ij} e_j^{(\beta)} - \dot{e}_i^{(\alpha)} e_j^{(\beta)}$
	Conservative	$P_{\alpha\beta\gamma}(\boldsymbol{k},\boldsymbol{p}) = \frac{k}{2}\{[e^{(\alpha)}(\boldsymbol{k}) \cdot e^{(\beta)}(\boldsymbol{p})][e^{(3)}(\boldsymbol{k}) \cdot e^{(\gamma)}(\boldsymbol{q})] + [e^{(\alpha)}(\boldsymbol{k}) \cdot e^{(\gamma)}(\boldsymbol{q})][e^{(3)}(\boldsymbol{k}) \cdot e^{(\beta)}(\boldsymbol{p})]\}$
	Rotational	$P_{\alpha\beta\gamma} = \frac{1}{2}\epsilon_{\beta\delta 3} e^{(\alpha)}(\boldsymbol{k}) \cdot [qe^{(\delta)}(\boldsymbol{p}) \times e^{(\gamma)}(\boldsymbol{q}) + pe^{(\delta)}(\boldsymbol{q}) \times e^{(\gamma)}(\boldsymbol{p})]$
Helical modes		$\dot{\xi}_s + m_{ss'}\xi_{s'} = -\imath \int_{\boldsymbol{p+q=k}} M_{ss's''}\xi_{s'}(\boldsymbol{p},t)\xi_{s''}(\boldsymbol{q},t) d^3\boldsymbol{p}$
		$m_{ss'}(\boldsymbol{k}) = (1/2)N_i(-s\boldsymbol{k})M_{ij}N_j(s'\boldsymbol{k}) - (1/2)\dot{N}_i(-s\boldsymbol{k})N_j(s'\boldsymbol{k})$
	Conservative	$M_{ss's''} = \frac{k}{4}\{[N(-s\boldsymbol{k}) \cdot N(s'\boldsymbol{p})][e^{(3)}(\boldsymbol{k}) \cdot N(s''\boldsymbol{q})] + [N(-s\boldsymbol{k}) \cdot N(s''\boldsymbol{q})][e^{(3)}(\boldsymbol{k}) \cdot N(s'\boldsymbol{p})]\}$
	Rotational	$M_{ss's''} = \frac{1}{2}(s'p - s''q)N(-s\boldsymbol{k}) \cdot [N(s'\boldsymbol{p}) \times N(s''\boldsymbol{q})]$

Note: The tensor M_{ij} accounts for linear convective and pressure terms: $M_{ij} = A_{mj}(\delta_{im} - 2k_ik_m/k^2)$, where $A_{ij} = \partial \bar{u}_i/\partial x_j$.

When the helical-mode decomposition is used, with $\xi_s = (1/2)\hat{u}_i(\mathbf{k})N_i(-s\mathbf{k})$, background equation (2.83) becomes

$$\dot{\xi}_s + m_{ss'}\xi_{s'} = -\imath \int_{\mathbf{p}+\mathbf{q}=\mathbf{k}} M_{ss's''}\xi_{s'}(\mathbf{p},t)\xi_{s''}(\mathbf{q},t)d^3p, \qquad (2.86)$$

with

$$m_{ss'}(\mathbf{k}) = (1/2)N_i(-s\mathbf{k})M_{ij}N_j(s'\mathbf{k}) - (1/2)\dot{N}_i(-s\mathbf{k})N_j(s'\mathbf{k}) \qquad (2.87)$$

and

$$M_{ss's''} = \frac{k}{4}\Big\{[\mathbf{N}(-s\mathbf{k})\cdot\mathbf{N}(s'\mathbf{p})][\mathbf{e}^{(3)}(\mathbf{k})\cdot\mathbf{N}(s''\mathbf{q})]$$
$$+ [\mathbf{N}(-s\mathbf{k})\cdot\mathbf{N}(s''\mathbf{q})][\mathbf{e}^{(3)}(\mathbf{k})\cdot\mathbf{N}(s'\mathbf{p})]\Big\}. \qquad (2.88)$$

The signs s, s', s'' take only the values ± 1, and the Einstein convention on repeated indices is used. The last equation will be revisited in Chapters 3 and 7.

It is worth noting that Eqs. (2.85) and (2.88) directly use the expression of the basic nonlinearity as the solenoidal part of $\nabla(\mathbf{u}\otimes\mathbf{u})$. A very interesting variant is obtained starting from the solenoidal part of $\boldsymbol{\omega}\times\mathbf{u}$ (e.g., Waleffe, 1992). The counterpart of (2.85) is

$$P_{\alpha\beta\gamma} = \frac{1}{2}\epsilon_{\beta\delta 3}e^{(\alpha)}(\mathbf{k})\cdot\left[qe^{(\delta)}(\mathbf{p})\times e^{(\gamma)}(\mathbf{q}) + pe^{(\delta)}(\mathbf{q})\times e^{(\gamma)}(\mathbf{p})\right], \qquad (2.89)$$

whereas the counterpart of (2.88) is

$$M_{ss's''} = \frac{1}{2}(s'p - s''q)\mathbf{N}(-s\mathbf{k})\cdot[\mathbf{N}(s'\mathbf{p})\times\mathbf{N}(s''\mathbf{q})]. \qquad (2.90)$$

Applications of the first equation are given in Chapter 7, whereas applications of the second equations appear in Chapters 3 and 4.

2.6 Anisotropy in Fourier Space

2.6.1 Second-Order Velocity Statistics

Independent of closure, the spectral tensor $\hat{R}_{ij}$ is not a general complex matrix, but has a number of special properties, including the fact that it is Hermitian and positive-definite, as follows from Eq. (2.64). The incompressibility condition $k_j\hat{u}_j = 0$ and Eq. (2.64) also yield $\hat{R}_{ij}k_j = 0$. Taken together, these properties show that, instead of the 18 real degrees of freedom needed to describe a general complex tensor, $\hat{R}_{ij}$ can be represented by only four independent scalars. Indeed, by use of the spherical-polar coordinate system in $\mathbf{k}$-space defined by Eqs. (2.67) and (2.68), the tensor simplifies to

$$\hat{R} = \begin{bmatrix} \Phi^{11} & \Phi^{12} & 0 \\ \Phi^{12*} & \Phi^{22} & 0 \\ 0 & 0 & 0 \end{bmatrix}. \qquad (2.91)$$

2.6 Anisotropy in Fourier Space

Displaying the first two unit vectors of the Craya–Herring frame, one obtains

$$\hat{R}_{ij} = \Phi^{11} e_i^{(1)} e_j^{(1)} + \Phi^{12} e_i^{(1)} e_j^{(2)} + \Phi^{12*} e_i^{(2)} e_j^{(1)} + \Phi^{22} e_i^{(2)} e_j^{(2)}, \qquad (2.92)$$

or, in a more compact form,

$$\hat{R}_{ij} = \Phi^{\alpha\beta} e_i^{(\alpha)} e_j^{(\beta)},$$

in which the summation convention over repeated Greek indices, taking only the values 1 and 2, is used.

Similarly, the decomposition in terms of helical modes yields

$$\hat{R}_{ij} = \sum_{s=\pm 1} \sum_{s'=\pm 1} A^{ss'} N_i(-s\mathbf{k}) N_j(s'\mathbf{k}).$$

Even if these decompositions are essentially the same and rely on four independant real scalars, an optimal splitting can be found to identify the most "physical" and the most intrinsic (with respect to any change of the orthonormal frame of reference) quantities. Using

$$N_i N_j^* = P_{ij} + \iota \epsilon_{ijn} \frac{k_n}{k},$$

in which P_{ij} denotes the projection operator and ϵ_{ijn} the alternating third-order Ricci tensor, one can rewrite the latter equation as

$$\hat{R}_{ij} = e(\mathbf{k},t) P_{ij}(\mathbf{k}) + \Re[Z(\mathbf{k},t) N_i(\mathbf{k}) N_j(\mathbf{k})] + \iota \mathcal{H}(\mathbf{k},t) \epsilon_{ijn} \frac{k_n}{k}, \qquad (2.93)$$

where $e(\mathbf{k},t)$ and $\mathcal{H}(\mathbf{k},t)$ are real scalars and $Z(\mathbf{k},t) = Z_r + \iota Z_i$ is a complex scalar. The quantity

$$e(\mathbf{k},t) = \frac{1}{2}\hat{R}_{ii} = \frac{1}{2}\left(\Phi^{11} + \Phi^{22}\right) \qquad (2.94)$$

is the *energy density* in 3D $\mathbf{k}$-space, whereas

$$k\mathcal{H}(\mathbf{k},t) = \iota k_l \epsilon_{lij} \hat{R}_{ij} = -k\Im\Phi^{12} \qquad (2.95)$$

is the *helicity spectrum*. Global kinetic energy $\mathcal{K}$ and global helicity h are given by

$$\mathcal{K}(t) = \frac{1}{2}\overline{u_i' u_i'} = \int e(\mathbf{k},t) d^3\mathbf{k}, \quad h(t) = \frac{1}{2}\overline{\omega_i' u_i'} = \int k\mathcal{H}(\mathbf{k},t) d^3\mathbf{k}. \qquad (2.96)$$

The third term,

$$Z = \frac{1}{2}\hat{R}_{ij} N_i^* N_j^* = \frac{1}{2}\left(\Phi^{22} - \Phi^{11} + \iota\Re\Phi^{12}\right), \qquad (2.97)$$

characterizes a *polarization* anisotropy, as subsequently discussed.

Anisotropy is expressed through the variations of these scalars with respect to the direction of $\mathbf{k}$, as well as the departures of $\mathcal{H}$ and Z from zero at a given wavenumber. Whatever spectral closure is used, the number of real unknowns may be reduced to the preceding four scalar parameters when numerical simulations are carried out, and analysis of the results can be simplified by use of these variables, particularly when the turbulence is statistically axisymmetric.

In various homogeneous isotropic or anisotropic configurations, the *radial-energy spectrum* $E(k, t)$ is a key quantity, which is obtained from $e(\mathbf{k}, t)$ by averaging over spherical shells of radius $k = |\mathbf{k}|$:

$$E(k, t) = \int\!\!\int_{k=|\mathbf{k}|} e(\mathbf{k}, t) d^2\mathbf{k} = \int_0^\pi \int_0^{2\pi} e(k, \theta, \phi, t) k^2 \sin\theta \, d\theta \, d\phi, \quad (2.98)$$

the last integral specifying the use of conventional variables in polar-spherical coordinates. Other 1D energy spectra can be obtained by averaging over planes or cylinders. They are defined only when specific applications are addressed. For 3D isotropic turbulence (including mirror symmetry), the general set (e, Z, h) reduces to

$$e = \frac{E(k,t)}{4\pi k^2}, \quad Z = h = 0,$$

so that

$$\hat{R}_{ij}(\mathbf{k}, t) = \underbrace{\frac{E(k,t)}{4\pi k^2}}_{e} \underbrace{\left(\delta_{ij} - \frac{k_i k_j}{k^2}\right)}_{P_{ij}}. \quad (2.99)$$

2.6.1.1 Directional and Polarization Anisotropy – Intrinsic Form

Equation (2.93) can be written in any direct orthonormal system of Cartesian coordinates. It can be shown that e, $|Z|$, and $\mathcal{H}$ are invariants. If the fixed frame of reference is changed, or if the specific Craya–Herring frame is rotated around the wave vector $\mathbf{k}$, only the phase of Z will be modified. It is therefore possible to have access to the intrinsic (eigen)representation of the spectral tensor by specifying a unique angle directly related to the phase of Z. For physical convenience, let us discuss only the symmetric real part of the spectral tensor, ignoring the contribution from helicity. The real part of the spectral tensor can be represented in the orthonormal frame defined by its principal axes. The two nonzero eigenvalues $e + |Z|$ and $e - |Z|$ are associated with the two principal axes, which are orthogonal to $\mathbf{k}$ and to each other. The third eigenvalue, which is equal to 0, is related to the unit vector spanned by $\mathbf{k}$. Finally Z describes the anisotropic structure of the real part of the spectral tensor at a given $\mathbf{k}$: Its modulus is half the difference of the nonzero eigenvalues, whereas its phase is related to the angle for passing from the Craya–Herring frame to the eigenframe by rotation around $\mathbf{k}$ (see Fig. 2.5).

The anisotropic structure is then analyzed by isolating pure isotropic contribution (2.99) in Eq. (2.93), so that

$$\Re\left(\hat{R}_{ij}\right) = \underbrace{\frac{E(k)}{4\pi k^2} P_{ij}}_{\text{isotropic part}} + \underbrace{\left[e(\mathbf{k}) - \frac{E(k)}{4\pi k^2}\right] P_{ij}}_{\text{directional anisotropy}} + \underbrace{\Re\left[Z(\mathbf{k}, t) N_i N_j\right]}_{\text{polarization anisotropy}}. \quad (2.100)$$

It is now possible to distinguish the *directional anisotropy*, which means that all directions of $\mathbf{k}$ on a spherical shell do not have the same amount of energy, from the *polarization anisotropy*, which means that the orientations of the vector $\hat{\mathbf{u}}$, located

2.6 Anisotropy in Fourier Space

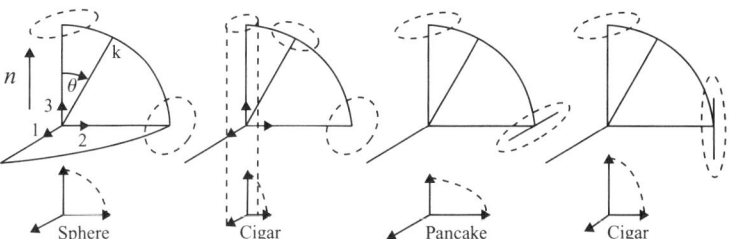

Figure 2.5. Schematic representation of anisotropy in spectral space. The top line displays the ellipse spanned by the four nonvanishing components of $\hat{R}$ that is the local Craya–Herring reference frame [see Eq. (2.91)], giving, for every k, $e + |Z|$ (length of the largest axis), $e - |Z|$ (length of the smallest axis), and their angle in the plane normal to k by means the phase of Z. The bottom line shows the associated ellipsoid generated by the corresponding Reynolds tensor in the physical space.

in the plane normal to a given wave vector k, are not statistically equivalent. The first kind of anisotropy is quantified by the angular distribution of $e - E/(4\pi k^2)$, whereas the second kind is measured by Z, whose modulus and phase are related to the intensity and the angle of polarization, respectively.

2.6.1.2 Induced Anisotropic Structure of Arbitrary Second-Order Statistical Quantities

The anisotropic decomposition previously introduced can be used to obtain a meaningful decomposition of any arbitrary second-order statistical tensor.

The following threefold splitting is obtained for the RST:

$$\overline{u_i u_j} = \iiint \hat{R}_{ij}(\boldsymbol{k}, t) d^3k = 2\mathcal{K} \left(\frac{\delta_{ij}}{3} + \underbrace{b_{ij}^{(e)} + b_{ij}^{(z)}}_{b_{ij}} \right), \qquad (2.101)$$

where

$$2\mathcal{K} b_{ij}^{(e)} = \iiint \left(e - \frac{E}{4\pi k^2} \right) P_{ij} d^3k, \quad 2\mathcal{K} b_{ij}^{(z)} = \iiint \Re\left(Z N_i N_j \right) d^3k. \qquad (2.102)$$

The fourth-order tensor M_{ijpq} can be expressed as

$$M_{ijpq} = \iiint \left[\frac{k_i k_p}{k^2} \hat{R}_{qj}(\boldsymbol{k}, t) + \frac{k_j k_p}{k^2} \hat{R}_{qi}(\boldsymbol{k}, t) \right] d^3k. \qquad (2.103)$$

2.6.1.3 Bridging with Dimensionality and Componentality

Similar decompositions can be found for the *dimensionality structure tensor* (KRR, 2001):

$$D_{ij} = \iiint \frac{k_i k_j}{k^2} 2e(\boldsymbol{k}, t) d^3k = 2\mathcal{K} \left[\frac{\delta_{ij}}{3} + 2b_{ij}^{(e)} + 0 \right],$$

and for the *vorticity correlation tensor***:

$$\overline{\omega'_i \omega'_j} = \omega^2 \left[\frac{\delta_{ij}}{3} + b_{ij}^{(k^2 e)} - b_{ij}^{(k^2 z)} \right], \qquad (2.104)$$

with

$$\omega^2 = \iiint k^2 e(k,t) d^3 k, \qquad (2.105)$$

$$\omega^2 b_{ij}^{(k^2 e)} = \iiint k^2 \left(e - \frac{E}{4\pi k^2} \right) P_{ij} d^3 k, \qquad (2.106)$$

$$\omega^2 b_{ij}^{(k^2 z)} = \iiint k^2 \Re (Z N_i N_j) d^3 k. \qquad (2.107)$$

Relations (2.106) and (2.107) show that the anisotropy tensor b_{ij} is the sum of two very different contributions: $b_{ij}^{(e)}$, which originates the directional (or dimensionality) anisotropy, and $b_{ij}^{(z)}$, which accounts for polarization anisotropy. Surprising RDT results in rotating flows are explained by this decomposition (see Chapter 4), and the formalism introduced in KRR (2001) appears as a by-product of Eq. (2.93) in homogeneous turbulence, the decomposition in terms of directional and polarization anisotropy lending support to componental and dimensional anisotropy.

The third tensor introduced by KRR, referred to as *stropholysis*, is also connected to the $e - Z$ decomposition, and its fully symmetrized form can be recast as

$$Q^*_{ijn} = \text{sym} \iiint \frac{k_n}{k} \Im(Z N_i N_j) d^3 k. \qquad (2.108)$$

In conclusion, it is worthwhile to point out that a fully anisotropic spectral (or two-point) description carries a very large amount of information, even if restricted to second-order statistics. In the inhomogeneous case, the *proper orthogonal decomposition* (POD) (see Lumley, 1967) has renewed interest in second-order two-point statistics, but this technique is applied to strongly inhomogeneous quasi-deterministic flows. It is said that POD spatial modes are only Fourier modes in homogeneous turbulence, without considering that a spectral tensor such as $\hat{R}$ ought to be diagonalized in order to exhibit its eigenmodes as POD modes in the *anisotropic* case.

** Rather than introducing the vorticity correlation tensor, KRR (2001) introduced a *circulicity tensor* F_{ij}, which involves larger scales. This tensor corresponds to

$$F_{ij} = 2\mathcal{K} \left[\frac{\delta_{ij}}{3} + b_{ij}^{(e)} - b_{ij}^{(z)} \right]$$

with our notation.

2.6.2 Some Comments About Higher-Order Statistics

N-order correlations at N points can be defined in homogeneous turbulence by means of spectral tensors, similar to the second-order case ($N = 2$). For instance,

$$\langle \hat{u}^*_{i_1}(\mathbf{k}_N)\hat{u}_{i_2}(\mathbf{k}_1)\ldots\hat{u}_{i_{N-1}}(\mathbf{k}_{N-2})\hat{u}_{i_N}(\mathbf{k}_{N-1})\rangle$$
$$= \hat{R}_{i_1 i_2 \ldots i_N}(\mathbf{k}_1, \mathbf{k}_2, \ldots, \mathbf{k}_{N-1})\delta\left(\mathbf{k}_N - \sum_{i=1}^{N-1}\mathbf{k}_i\right). \quad (2.109)$$

Because the N wave vectors form a closed polygon, only $N - 1$ of them, $\mathbf{k}_1$, $\mathbf{k}_2, \ldots, \mathbf{k}_{N-1}$, are independent, corresponding to the $N - 1$ independent separations vectors $\mathbf{r}_1, \ldots, \mathbf{r}_{N-1}$ in physical space. The interest of addressing the most complex configuration, with N independent points for representing Nth-order correlations, is discussed in the last section of this chapter.

As a first general result, for incompressible turbulence, it is possible to extend Eq. (2.81), which was derived for $N = 2$, to an arbitrary order, yielding an equation for $\hat{R}_{i_1 \ldots i_N}(\mathbf{k}_1, \ldots, \mathbf{k}_{N-1}, t)$. In this equation, all pressure effects can be exactly incorporated as functions of $\hat{R}_{i_1 \ldots i_N}$ itself, as all the linear effects, and of the $(N + 1)$th-order spectral tensor.

As a second result, it is possible to replace $\hat{u}(\mathbf{k}_n)$ with $u^{(\alpha)}(\mathbf{k}_n)$ using either the local Craya or poloidal–toroidal frame attached to $\mathbf{k}_n$. Accordingly, the Nth-order spectral tensor is shown to depend on $N + 1$ scalar components only, considering that $u^{(\alpha)}$ has two components and that the spectral tensor is left unchanged when permuting simultaneously the N wave vectors and the N indices. The latter result was found independently in the Ph.D. theses of Cambon (1982) and Lindborg (1996).

Of course, the most general case has very few applications, but the cases $N = 3$ and $N = 4$ are relevant in triadic and quasi-normal closure theories and models, which are addressed in the following chapters. Orders larger than $N = 4$ are commonly addressed in the "scaling-intermittency" community, for structure functions, but always restricted to two-point correlations.

2.7 A Synthetic Scheme of the Closure Problem: Nonlinearity and Nonlocality

Both the nonlinear pressure component and the nonlinear term appearing directly in Eq. (2.29) contribute to the closure problem: The equation for the nth-order velocity moments involves $(n + 1)$th-order moments. As a consequence, no finite subset of the infinite hierarchy of integrodifferential equations describing the velocity moments at all orders is closed, reflecting the fundamental difficulty of the turbulence problem, viewed through the classical statistical description in terms of statistical moments. The origin of the closure problem is *nonlinearity* of the Navier–Stokes equations, which is borne by the convective terms and the nonlinear part of the pressure fluctuations. *Nonlocality*, by itself, does not lead to problems, although the technical difficulties associated with integrodifferential are nontrivial.[††]

The nonlocal problem of closure is discarded only in models for *multipoint* statistical correlations, e.g., double correlations at two points or triple correlations at three points, so that in such models the problem of closure is determined by the sole nonlinearity.

The knowledge of the pdf of the velocity fluctuations is equivalent to the knowledge of all the statistical moments of arbitrary orders. Therefore the previously mentioned problem of the open hierarchy of the moment equations is precluded in a pdf-based approach. Consequently, the problem of closure induced by the nonlinearity is precluded by use of a pdf approach, but the nonlocal problem of closure remains, so that the equations for a local single-point velocity pdf involve a two-point velocity pdf, and equations for an n-point velocity pdf involve an $(n + 1)$-point velocity pdf (Lundgren, 1967). Therefore an open hierarchy of equations is recovered by a pdf approach but with respect to a multipoint spatial description!

To summarize all the consequences of the preceding discussion, a synthetic scheme using a triangle is shown in Fig. 2.6. The vertical axis bears the order of the statistical moments, from 1 (the mean velocity), 2 (second-order moments), up to arbitrarily high-order moments. For the sake of convenience, the moments of order greater than or equal to 2 are centered, so that they involve only the fluctuating-velocity field. Along the vertical axis, n corresponds to the number of possible different points for a multipoint description of the nth-order moment under consideration along the horizontal axis. Considering the nth-order moment, the number of points ranges from 1 (single-point correlation), 2 (two-point correlation), up to n. The possible solutions are observed to generate a triangle in this representation.

In other words, the vertical axis displays the open hierarchy that is due to nonlinearity, whereas the horizontal one deals with nonlocality. Each point in the triangle characterizes a level of description. As an example, the point [3, 2] is related to triple correlations at two points, which are associated with the spectral-energy transfer and the kinetic-energy cascade. In addition, one can state the problem of closure by looking at the adjacent points (if any) just above and just to the left. For instance, the main problem that concerns engineering, when solving Reynolds-averaged Navier–Stokes equations, is expressing the flux of the RST. This can be represented by an arrow from [2, 1] to [1, 1]. Then the equations that govern the RST [2, 1] need extra information (not given by [2, 1] itself, leading to the appearance of the closure problem) on second-order two-point correlations [2, 2] (involved in the "rapid" pressure–strain-rate term and the dissipation term), on triple-order single-point correlations [3, 1] and triple-order two-point correlations [3, 2] (involved in the "slow" pressure–strain-rate and diffusion terms). Of course, the RST [2, 1] is directly derived from second-order correlations at two points [2, 2], illustrating the simple

---

[††] Nonlocality ought to be only understood in *physical space* here. Of course, the operators related to pressure and dissipation will appear as local quantities in Fourier space, but this is only for the sake of mathematical convenience. Discussing the possible degree of locality of nonlinear interactions in Fourier space is briefly discussed in the next chapter.

2.7 A Synthetic Scheme of the Closure Problem: Nonlinearity and Nonlocality

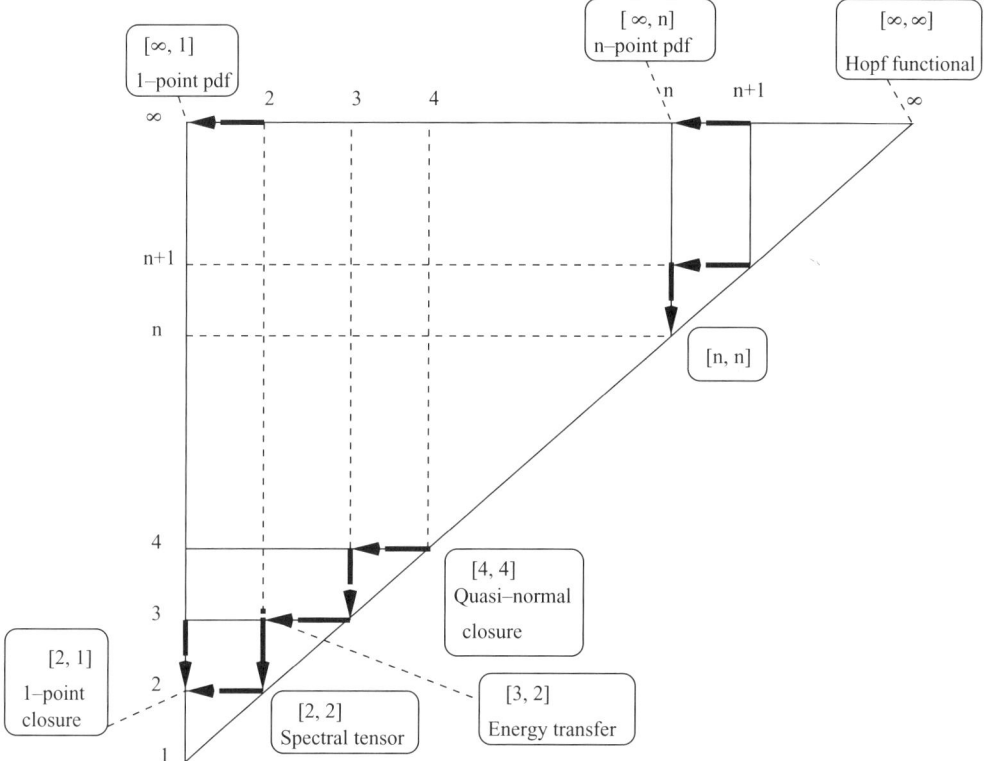

Figure 2.6. Synthetic scheme for statistical closures.

rule of *concentration of the information* when moving from the right to the left. The nonlocality issue, because of pressure and dissipation terms, is discarded if one looks at only $[n, n]$ correlations (located on the hypotenuse of the triangle in Fig. 2.6), leaving only the hierarchy that is due to nonlinearity. The governing equations for $[2, 2]$ need extra information on only $[3, 2]$. The equations that govern $[3, 3]$ require extra information on only $[4, 3]$. These two examples, which are directly involved in classical two-point closures, are discussed in Chapter 3.

The arrow from $[n + 1, n]$ to $[n, n]$ gives an obvious generalization of the optimal way to use multipoint closures and illustrates the open hierarchy of equations that is *due to the sole nonlinearity*. Often the closure relationship holds at the level $[n + 1, n + 1]$, from which is readily derived the level $[n + 1, n]$. For instance, the quasi-normal (QN) assumption, which is involved in all multipoint closures, as well as in wave-turbulence theories, calls into play the $[4, 4]$ level.

Regarding the pdf approach, we are concerned with the upper horizontal side of the triangle. It seems to be consistent to relate to the point $[\infty, 1]$ a description in terms of a local-velocity pdf. Accordingly, the arrow from $[\infty, 2]$ to $[\infty, 1]$ shows the need for extra information on the two-point pdf in the equations that govern single-point pdfs. In the same way, the arrow from $[\infty, n + 1]$ to $[\infty, n]$ shows the link between n-point and $(n + 1)$-point pdf's [Lundgren (1967)] and illustrates the open hierarchy of equations that is due to the sole nonlocality.

The last limit concerns the ultimate point $[\infty, \infty]$. It is consistent to consider that the limit of a joint pdf of velocity values at an infinite number of points is equivalent to the functional pdf description of Hopf (1952). In this case we reach the top right point of the triangle and there is no need for any extra information. The Hopf equation is closed, and it is possible to derive from it any multipoint pdf or statistical moment. It is interesting to point out that the bottom right point $[1, 1]$ gives the most crude information about the velocity field – its mean value – whereas the opposite point $[\infty, \infty]$ gives the most sophisticated.

As a last general comment, our synoptic scheme clearly shows that the problem of closure, which reflects a loss of information at a given level of statistical description, can be removed from consideration, at least partially, if additional degrees of freedom are introduced to enlarge the configuration space. For instance, introducing as a new dependent variable the vector that joins the two points in a two-point second-order description allows for the removal of the problem of closure that is due to nonlocality, which is present when a single-point second-order description is used. The introduction, as a new dependent variable, such as the test value Y_i of the random velocity field u'_i in a pdf approach,[††]

$$P(Y_i, \mathbf{x}, t) = \overline{\delta(u'_i(\mathbf{x}, t) - Y_i)},$$

allows for the removal of the problem of closure that is due to nonlinearity, which is present in any description in terms of statistical moments. Finally, any problem of closure is removed by use of the Hopf equation but the price to pay is an incredibly complicated configuration space! The probabilistic description, which is of practical interest regarding a *concentration scalar* field rather than a velocity field, is extensively addressed in the context of combustion modeling, and is no longer considered in this book.

[††] For the reader who is not acquainted with the definition using a "Dirac" function, this corresponds to the "fine-grained" pdf; which is well documented, for instance, in Pope's book.

Bibliography

ARAD, I., LVOV, V. S., AND PROCCACCIA, I. (1999). Correlations functions in isotropic and anisotropic turbulence: The role of the symmetry group, *Phys. Rev. E* **59**, 6753–6765.

BATCHELOR, G. K. (1953). *The Theory of Homogeneous Turbulence*, Cambridge University Press.

BAYLY, B. J., HOLM, D. D., AND LIFSCHITZ (1996). Three-dimensional stability of elliptical vortex columns in external strain flows, *Philos. Trans. R. Soc. London Ser. A* **354**, 895–926.

CAMBON, C. (1982). Etude spectrale d'un champ turbulent incompressible soumis à des effets couplés de déformation et rotation imposés extérieurement, *Thèse de Doctorat d'Etat, Université Lyon I, France*.

CAMBON, C. AND JACQUIN, L. (1989). Spectral approach to non-isotropic turbulence subjected to rotation, *J. Fluids Mech.* **202**, 295–317.

CAMBON, C., JACQUIN, L., AND LUBRANO, J.-L. (1992). Towards a new Reynolds stress model for rotating turbulent flows, *Phys. Fluids A* **4**, 812–824.

CAMBON, C. AND RUBINSTEIN, R. (2006). Anisotropic developments for homogeneous shear flows, *Phys. Fluids* **18**, 085106.

CAMBON, C. AND SCOTT, J. F. (1999). Linear and nonlinear models of anisotropic turbulence, *Annu. Rev. Fluid Mech.* **31**, 1–53.

CHANDRASEKHAR (1981). *Hydrodynamics and Hydrodynamic Stability*, Dover.

CRAIK, A. D. D. AND CRIMINALE, W. O. (1986). Evolution of wavelike disturbances in shear flows: A class of exact solutions of the Navier–Stokes equations, *Proc. R. Soc. London Ser. A* **406**, 13–26.

CRAYA, A. (1958). Contribution à l'analyse de la turbulence associée à des vitesses moyennes, P.S.T. n^0 345, Ministère de l'air, France.

DARRIGOL, O. (2005). *Worlds of Flow*, Oxford University Press.

ERINGEN, C. (1971). *Continuum Physics*, Academic Press, New York, Vol. 1.

HOPF, E. (1952). Statistical hydrodynamics and functional calculus, *J. Rat. Mech. Anal.* **1**, 87.

KANEDA, Y., ISHIHARA, T., YOKOKAWA, M., ITAKURA, K., AND UNO, A. (2003). Energy dissipation rate and energy spectrum in high resolution direct numerical simulations of turbulence in a periodic box, *Phys. Fluids* **15**, L21–L24.

KASSINOS, S. C., REYNOLDS, W. C., AND ROGERS, M. M. (2001). One-point turbulence structure tensors, *J. Fluid Mech.* **428**, 213–248.

LAUNDER, B. E., REECE, G. J., AND RODI, W. (1975). Progress in the development of a Reynolds stress turbulence closure, *J. Fluid Mech.* **68**, 537–566.

LINDBORG, E. (1996). Studies in classical turbulence theory, Ph.D. thesis, KTH (Stockholm, Sweden).

LUMLEY, J. L. (1967). The structure of inhomogeneous turbulent flows. In Yaglom, A. M. and Tatarski, V. I., eds., *Atmospheric Turbulence and Radio Waves Propagation*, Nauka, pp. 166–167.

LUMLEY, J. L. (1975). Pressure strain correlations, *Phys. Fluids* **18**, 750–751.

LUMLEY, J. L. (1978). Computation modelling of turbulent flows, *Adv. Appl. Mech.* **18**, 126–176.

LUMLEY, J. L. AND NEWMAN, G. (1977). The return to isotropy of homogeneous turbulence, *J. Fluid Mech.* **82**, 161–178.

LUNDGREN, T. S. (1967). Distribution function in the statistical theory of turbulence, *Phys. Fluids* **10**, 969–975.

MATHIEU, J. AND SCOTT, J. F. (2000). *Turbulent Flows: An Introduction*, Cambridge University Press.

OBERLACK, M. (2001). *In Theories of Turbulence*, Springer.

PIQUET, J. (2001). *Turbulent Flows. Models and Physics*, 2nd ed., Springer.

POLYAKOV, A. (1995). Turbulence without pressure, *Phys. Rev. E* **52**(6), 6183–6188.

POPE, S. B. (2000). *Turbulent Flows*, Cambridge University Press.

RILEY, J. J., METCALFE, R. W., AND WEISSMAN, M. A. (1981), in West, B. J., ed., *Proceedings of the AIP Conference on Nonlinear Properties of Internal Waves*, AIP, pp. 72–112.

ROGALLO, R. (1981). Numerical experiments in homogeneous turbulence, NASA Tech. Mem. No. 81315.

SAFFMAN, P. G. (1995). *Vortex Dynamics*, Cambridge University Press.

SIMONSEN, A. J. AND KROGSTAD, P. A. (2005). Turbulent stress invariant analysis: Clarification of existing terminology, *Phys. Fluids* **17**, 088103.

SMITH, L. M. AND WALEFFE, F. (2002). Generation of slow, large scales in forced rotating, stratified turbulence, *J. Fluid Mech.* **451**, 145–168.

SREENIVASAN, K. R. AND NARASIMHA, R. (1978). Rapid distortion theory of axisymmetric turbulence, *J. Fluid Mech.* **84**, 497–516.

TENNEKES, H. AND LUMLEY, J. L. (1972). *A First Course in Turbulence*, MIT Press.

WALEFFE, F. (1992). The nature of triad interactions in homogeneous turbulence, *Phys. Fluids* **4**, 350–363.

3 Incompressible Homogeneous Isotropic Turbulence

To caricature, it can be jokingly said that, once one has eliminated all features of a flow that one understands, what remains is turbulence. This sentence, taken from Mathieu and Scott (2000), is even more relevant in HIT, in which no interaction with a structuring effect (mean flow, body force, shock wave, wall, etc.) may occur. HIT, even if it can be described statistically with a few number of quantities, is really the core of the turbulence problem.

3.1 Observations and Measures in Forced and Freely Decaying Turbulence

3.1.1 How to Generate Isotropic Turbulence?

Isotropic turbulence can be investigated with both experimental and numerical approaches, despite the fact that it requires the existence of an unbounded domain from the theoretical point of view.

A quasi-isotropic fully developed turbulent state can be reached in wind tunnels with a grid to promote turbulence (see Fig. 3.1). In such a setup, boundary layers develop along solid walls, but an isotropic flow is recovered in the core of wind tunnel. The grid wake transforms a part of mean-flow kinetic energy into turbulent kinetic energy. Downstream of the grid, the mean flow is uniform and no more turbulence-production mechanism takes place. Therefore the turbulent-fluctuation dynamics is entirely governed by the advection that is due to the uniform mean flow, the nonlinear interactions, and the linear viscous effects, leading to a monotonic decay of the turbulent kinetic energy $\mathcal{K}$.

Several regions are usually identified downstream of the grid, which correspond to different dynamical regimes. These decay regimes are discussed in Subsection 3.1.3.

The spatial development of isotropic turbulence observed in wind tunnels cannot be exactly reproduced in numerical simulations because of the enormous computing power required. But it mimics switching from a spatially evolving flow to a time-developing flow. In this new configuration, periodic boundary conditions are imposed in all space directions, and a pseudo-turbulent initial condition is used. An isotropic time-decaying turbulent flow is then obtained. It can be made statistically steady in time by inserting an ad hoc forcing term. But it is worth noting that the use

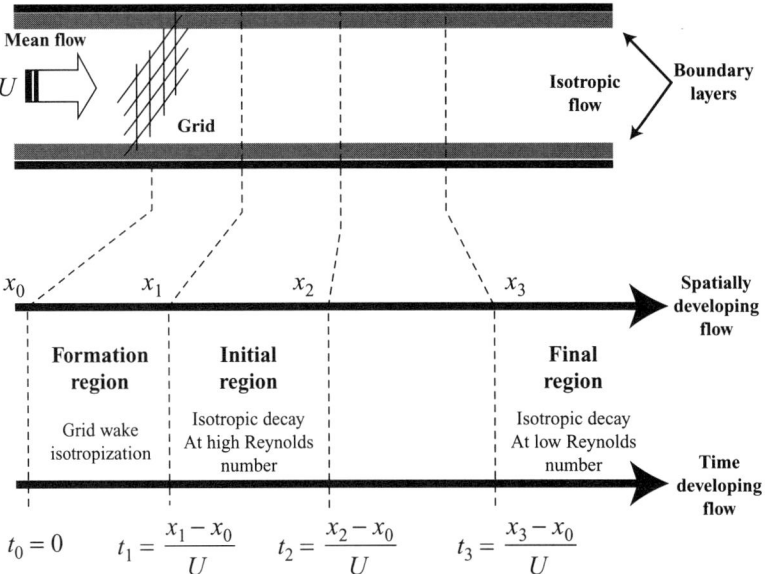

Figure 3.1. Schematic view of wind-tunnel setup for generating isotropic turbulence.

of periodic boundary conditions induces spurious couplings at scales of the order of the computational domain size and that the analysis of large-scale dynamics must be carried out with great care.

Spatially developing and time-evolving flows can be compared thanks to *Taylor's frozen-turbulence hypothesis*. In 1938, Taylor hypothesized that the turbulent velocity fluctuation $u(x, t)$ measured by a stationary probe can be interpreted as resulting from the advection of a frozen spatial structure by a uniform steady flow with velocity U, yielding

$$u(x, t) = u(x - Ut, 0). \tag{3.1}$$

This hypothesis can also be used to find an approximate relation between space and time derivatives. Let us consider consider a new reference frame advected at velocity U. Denoting quantities expressed in this new reference frame by a *tilde* ($\sim$) one has

$$\tilde{x} = x - Ut, \quad \tilde{t} = t, \quad \tilde{u}(\tilde{x}, \tilde{t}) = u(x, t) - U, \tag{3.2}$$

and

$$\frac{\partial u_i}{\partial t} = \frac{\partial \tilde{u}_i}{\partial x_j}\frac{\partial x_j}{\partial t} + \frac{\partial \tilde{u}_i}{\partial t} = \frac{\partial \tilde{u}_i}{\partial t} - U_j \frac{\partial \tilde{u}_i}{\partial x_j}. \tag{3.3}$$

If one now assumes that the signal is frozen in the advected frame, i.e., if $\partial \tilde{u}/\partial t \approx 0$, then the following relation holds:

$$\frac{\partial}{\partial t} \approx -U_j \frac{\partial}{\partial x_j}. \tag{3.4}$$

3.1 Forced and Freely Decaying Turbulence

It is important to note that the Taylor hypothesis does not hold in the following cases, at least from the theoretical viewpoint:

- A single advecting velocity cannot be defined. This is the case in compressible flows, in which hydrodynamic and acoustic perturbations do not have the same speed, and in flows in which the advection speed depends on the scale of the perturbation. This last case is met in some shear flows (e.g., mixing layers).
- The rate of change in the moving frame cannot be neglected. Let us consider a structure with characteristic size L and characteristic time T. The Taylor hypothesis is valid if

$$\frac{L}{U} \ll T. \tag{3.5}$$

Now using the relation $\sqrt{\mathcal{K}} \approx L/T$, the validity criterion can be recast as

$$\sqrt{\mathcal{K}} \ll U, \tag{3.6}$$

showing that the mean-flow speed must be large compared with the characteristic turbulent velocity scale.

One of the first experiments of decaying grid-generated turbulence, but perhaps one of the most documented, was carried out by Comte-Bellot and Corrsin (1966). To achieve a better isotropy, at least measured looking at the RST, a convergent duct was placed after the grid, in the "formation region." Without this additional device, the Reynolds stresses exhibit a mild axisymmetry with $\overline{u_1^2} > \overline{u_2^2} \sim \overline{u_3^2}$: The effect of the convergent duct is to diminish the Reynolds stress component in the axial direction (x_1 here) and to increase it in the radial directions, as shown by RDT (see Chapter 5). Unfortunately, such experiments cannot reproduce high-Reynolds-number flows; a typical value of the Reynolds number based on the Taylor microscale $Re_\lambda = u'\lambda/\nu$ is 70–80. Here, $u' = \sqrt{\frac{2}{3}\mathcal{K}}$ denotes the characteristic velocity scale of the large, energy-containing scales, and $\lambda \equiv \sqrt{15\nu u'^2/\varepsilon}$ is the Taylor microscale,* where ε is the kinetic-energy dissipation rate. Examples of available experimental data sets are listed in Table 3.1.

DNS began to reach higher Reynolds numbers in the early 1980s. A weakness of these simulations is that the large-scale forcing that is present in the simulation prevents recovering reliable information about the smallest wavenumbers.

Some examples are listed in Table 3.2.

3.1.2 Main Observed Statistical Features of Developed Isotropic Turbulence

The main results retrieved from laboratory experiments and numerical simulations are as follows:

- *Typical observed turbulent kinetic-energy spectrum shapes* are displayed in Fig. 3.2. A universal inertial range is observed in the turbulent kinetic-energy spectrum if

---

* It is recalled that the Taylor microscale is associated with scales at which the spectrum of kinetic-energy dissipation, or equivalently the enstrophy spectrum, exhibits its maximum.

Table 3.1. *A few experiments dealing with decaying incompressible isotropic turbulence*

Ref.	Re_{Grid}	Remarks
Batchelor and Townsend (1948)	5500, 11 000	
Wyatt (1955)	11 000, 22 000, 44 000	
Van Atta and Chen (1969)	25 600, 25 300	$Re_\lambda \simeq 35$–49
Comte-Bellot and Corrsin (1966)	34 000, 68 000	$Re_\lambda \simeq 37$–72
Sreenivasan et al. (1980)	7400	$Re_\lambda \simeq 34$
Uberoi (1963)	29 000	
Uberoi and Wallis (1967)	8750, 17 500	
Sirivat and Warhaft (1983)	5150, 9550	
Mohamed and LaRue (1990)	6000, 10 000, 12 000, 14 000	$Re_\lambda \simeq 28$–43
Kistler and Vrebalovich (1966)	10 500	$Re_\lambda \simeq 264$–669
Skrbek and Stalp (2000)	$\sim 10^5$	Superfluid helium
Midlarsky and Warhaft (1996)	Active grid	$Re_\lambda \simeq 400$–500
Midlarsky and Warhaft (1998)	Active grid	$Re_\lambda \simeq 730$
Kang et al. (2003)	Active grid	$Re_\lambda \simeq 630$–720

Note: For apparatus based on passive grids (i.e., turbulence is generated by the wake of rods), the Reynolds number is computed by use of the size of the mesh of the grid used to trigger turbulence and the mean velocity at the grid location. For active-grid-based experiments, the Taylor-scale-based Reynolds number at probe locations is indicated. It is worth noting that a nonnegligible residual anisotropic error is always observed in active-grid experiments.

Table 3.2. *A few DNSs dealing with incompressible isotropic turbulence*

Ref.	Re_λ	Forcing (Y/N)	Grid
Kerr (1990)	82	Y	128^3
Vincent and Meneguzzi (1991)	150	Y	240^3
Sanada (1992)	120	Y	256^3
Jimenez et al. (1993)	35	Y	64^3
Jimenez et al. (1993)	61	Y	128^3
Jimenez et al. (1993)	94	Y	256^3
Jimenez et al. (1993)	168	Y	512^3
Wang et al. (1996)	190	Y	512^3
Jimenez and Wray (1998)	37	Y	64^3
Jimenez and Wray (1998)	62	Y	128^3
Jimenez and Wray (1998)	95	Y	256^3
Jimenez and Wray (1998)	142	Y	384^3
Jimenez and Wray (1998)	163	Y	512^3
Jimenez and Wray (1998)	62	N	512^3
Kaneda et al. (2003)	167	Y	256^3
Kaneda et al. (2003)	257	Y	512^3
Kaneda et al. (2003)	471	Y	1024^3
Kaneda et al. (2003)	732	Y	2048^3
Kaneda et al. (2003)	1201	Y	4096^3

Note: Simulations without forcing are related to freely decaying turbulence. The Reynolds number is measured at the initial time of the simulation in the free-decay case and at statistical equilibrium in the forced case.

3.1 Forced and Freely Decaying Turbulence

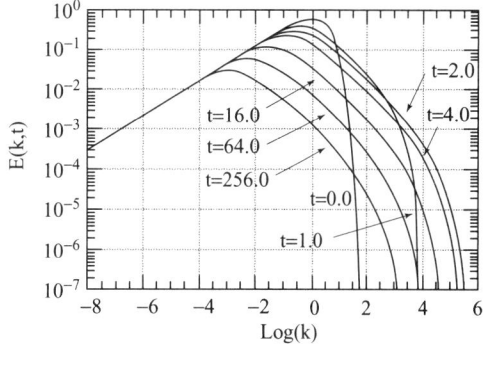

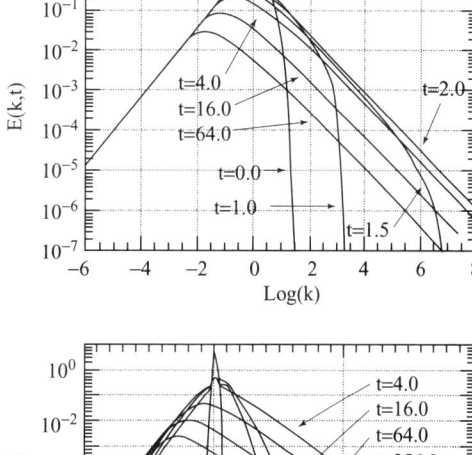

Figure 3.2. Evolution of the kinetic-energy spectrum in the initial stage decay computed with spectral closures. Top: emergence of self-similarity and validation of the PLE hypothesis for k^1 scaling at low wavenumbers. Middle: emergence of self-similarity and validation of the PLE hypothesis for k^2 scaling at low wavenumbers. Bottom: emergence of self-similarity with a k^4 behavior for initial Gaussian-shaped spectrum, and PLE hypothesis breakdown for the k^4 spectrum. From Clark and Zemach (1998) with permission of AIP.

the Reynolds number is high enough. At very high wavenumbers, viscous dissipation becomes dominant, and the energy spectrum falls very quickly. The physical assumption that the turbulent field is regular in the sense that the L_2 norm of all high-order spatial derivatives of the velocity field is finite suggests that the spectrum shape should exhibit an exponential decay at very high wavenumbers.

The spectrum shape at large scales (i.e. small wavenumbers) that do not belong to the inertial range is observed to be flow dependent.

The time evolution of the turbulent kinetic-energy spectrum is displayed in Figs. 3.2 and 3.3. Results dealing with both the free-decay case and the statistically steady case are presented. In the former case, no source of turbulent kinetic energy is present, and the turbulent kinetic energy is a monotonically decaying function of time, whereas in the latter a kinetic-energy source is used to reach a

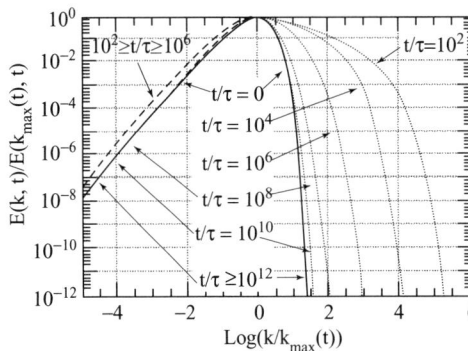

Figure 3.3. Evolution of the kinetic-energy spectrum with transition from the initial stage of decay to the final stage of decay. From Clark and Zemach (1998) with permission of AIP.

statistically steady state. In both cases, it is observed that the spectrum shape relaxes toward a universal shape at small scales (provided that the Reynolds number is high enough to allow for the existence of the inertial range). The change in the kinetic-energy spectrum shape is due to nonlinear interactions between modes. Two mechanisms are obviously at play: a direct kinetic-energy cascade from large to small scales (also referred to as the forward cascade) that is responsible for the existence of the inertial range, and an inverse kinetic-energy cascade from small to large scales (also named the backward cascade) that yields the growth of the energy spectrum at very small wavenumbers.

The celebrated hypotheses proposed by Kolmorogov in 1941 yield the following asymptotic spectrum shapes for small scales for which the local isotropy hypothesis holds:

$$E(k) = K_0 \varepsilon^{2/3} k^{-5/3} F(k\eta) \tag{3.7}$$

where K_0, ε, $\eta \equiv (\nu^3/\varepsilon)^{1/4}$, and F are the Kolmogorov constant, the dissipation rate, the Kolmogorov length scale, and a nondimensional function, respectively. The regularity of the derivatives of the velocity field is ensured by the function F, which must be a fast-decaying function, i.e.,

$$\int_0^{+\infty} x^n F(x) dx < \infty, \quad \forall n \geq 0. \tag{3.8}$$

Among the numerous proposals made for $F(x)$, a widely admitted one is

$$F(x) = C x^\alpha \exp(-\beta x^n), \tag{3.9}$$

where C, α, β, and n are real parameters. Not to mention values of n such as $n = 4/3$ (proposed by Pao, for pure mathematical convenience), $n = 2$ (suggested by Townsend, assuming linear response of small scales), $n = 1$ is consistently predicted by all "triadic" closures [EDQNM, direct interaction approximation (DIA), Lagrangian history direct interaction approximation (LHDIA), Lagrangian renormalized approximation (LRA)] (Kaneda, 1993) and supported by recent experimental and DNS results. The reader is referred to Ishihara et al. (2005) for a recent survey including new DNSs with the Taylor microscale Reynolds number Re_λ and resolution ranging up to about 1201 and 4096^3,

3.1 Forced and Freely Decaying Turbulence

respectively. In addition to $n = 1$, the values of α and β obtained by the latter DNS decrease monotonically with Re_λ and appear to converge toward asymptotic values as $Re_\lambda \to \infty$, but the convergence, especially that of β, is slow. A simple power-law fitting suggests the following asymptotic (infinite Re_λ) values:

$$\alpha = -2.9 + 7.2 Re_\lambda^{-0.47}, \quad \beta = 0.62 + 9.3 Re_\lambda^{-0.19}, \quad C = 0.038 + 23.5 Re_\lambda^{-0.42}.$$

Surprisingly, the previously mentioned closures predict $\alpha = 3$ (Kaneda, 1993). This positive value, however, does not yield an overshoot for the spectrum, between the end of the inertial range and the beginning of the dissipative range, because β is sufficiently large.

For small scales much larger than the Kolmogorov scale η, one recovers the inertial-range expression:

$$E(k) = K_0 \varepsilon^{2/3} k^{-5/3}. \tag{3.10}$$

The exact value of the Kolmogorov constant is not known. A large number of estimates are provided in the literature (Sreenivasan, 1995), coming from measures in the atmospheric boundary layer, from laboratory experiments, and from numerical simulations. This uncertainty comes from either the departure from isotropy in many flows or the absence of a large inertial range in the spectrum. A reliable estimate seems to be $K_0 = 1.5 \pm 0.1$.

It is important to note that the Kolmogorov scaling is valid in the asymptotic limit of very large Reynolds numbers only, and that the invariance of ε as a spectral flux comes from a simple dimensional analysis. Denoting L as the integral scale of turbulence, one finds that the usual estimates for the upper and lower bounds of the inertial range are

$$L_{\text{upper}} \simeq 5(Re_L)^{-1/2} L, \quad L_{\text{lower}} \simeq 50(Re_L)^{-3/4} L, \tag{3.11}$$

where $Re_L = Lu'/\nu$. The miminum Reynolds number for an inertial range to exist is an open issue, but there is evidence that the Taylor-scale-based Reynolds number Re_λ must be $O(100)$ for any natural inertial range to exist and that $Re_\lambda = O(1000)$ for a decade of inertial range.

- *Turbulent velocity fluctuations are not Gaussian random variables.*

A first manifestation of non-Gaussianity of the turbulent velocity field is that its odd-order statistical moments are not zero, whereas they are identically zero for a random Gaussian field. The skewness factor for a single-point velocity distribution can be almost zero (result given by either isotropy or Gaussianity), whereas the skewness of velocity gradients has a very significant negative value.

A measure of this difference is therefore gained by looking at the skewness and the flatness parameters[†] based on velocity increments (or equivalently

---

[†] Let us recall that the flatness factor $F(a)$ and the skewness factor $S(a)$ of the random field $\boldsymbol{a}$ are defined as

$$F(a) \equiv \frac{\langle a^4 \rangle}{\langle a^2 \rangle^2}, \quad S(a) \equiv \frac{\langle a^3 \rangle}{\langle a^2 \rangle^{3/2}}. \tag{3.12}$$

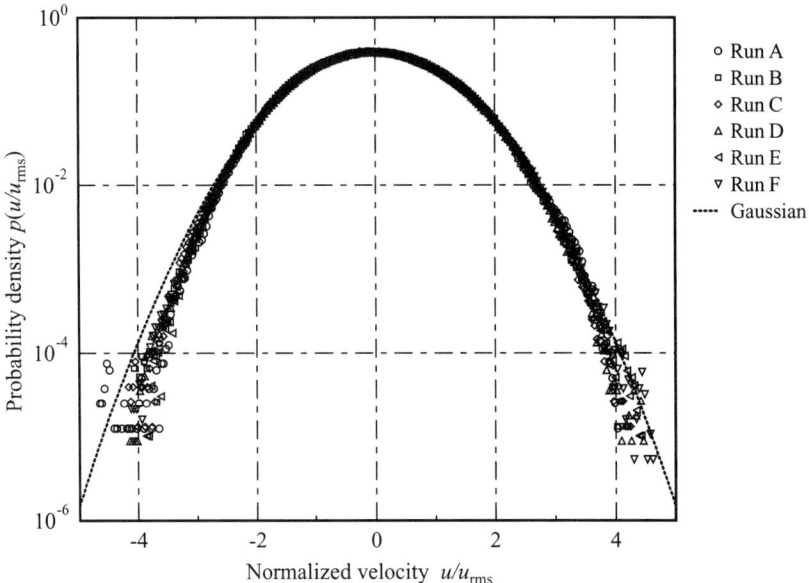

Figure 3.4. Probability density function of normalized velocity fluctuation in isotropic turbulence. From Noullez et al. (1997) with permission of CUP.

the velocity gradients). Common reported values of the skewness factor are $S_0 = -0.4 \pm 0.1$ (instead of $S_0 = 0$ for a Gaussian field), whereas the *flatness factor*, F_0, ranges from 4 to 40, depending on the Reynolds number (instead of $F_0 = 3$ for a Gaussian field).

It is worth noting that single-point even moments of velocity fluctuations exhibit a quasi-normal distribution (see Fig. 3.4), whereas velocity increments are not Gaussian random variables. Therefore the one-point analysis of the turbulent velocity field is not sufficient to analyze the lack of Gaussianity of turbulence: Two-point quantities must be considered. Extreme velocity events, which correspond to the very end of the tails of the pdf plots, are observed to escape the normal distribution. A possible explanation is that these extreme events are (at least partially) governed by the physical mechanisms responsible for the production of turbulent kinetic energy.[‡] Therefore they are flow dependent and do not

---

If a is a Gaussian field, then
$$F(a) = 3, \quad S(a) = 0. \tag{3.13}$$

Still assuming that a is a random Gaussian field, and defining $\boldsymbol{\omega}_a = \text{curl}\,\boldsymbol{a}$ and $\mathbf{S}_a = \frac{1}{2}(\nabla \boldsymbol{a} + \nabla^T \boldsymbol{a})$, one has
$$F(\omega_a) = 5/3, \quad F(S_a^2) = 7/5. \tag{3.14}$$

Another important point is that almost all nonlinear functions of a will exhibit non-Gaussian behavior.

[‡] It can be shown (Falkovich and Lebedev, 1997) that Gaussian random forcing having a correlation scale l_F and a time scale τ_F yields velocity pdf tails of the form
$$\ln P(u) \propto -u^4 \text{ for } u \gg \max(u_{\text{rms}}, l_F/\tau_F),$$

3.1 Forced and Freely Decaying Turbulence

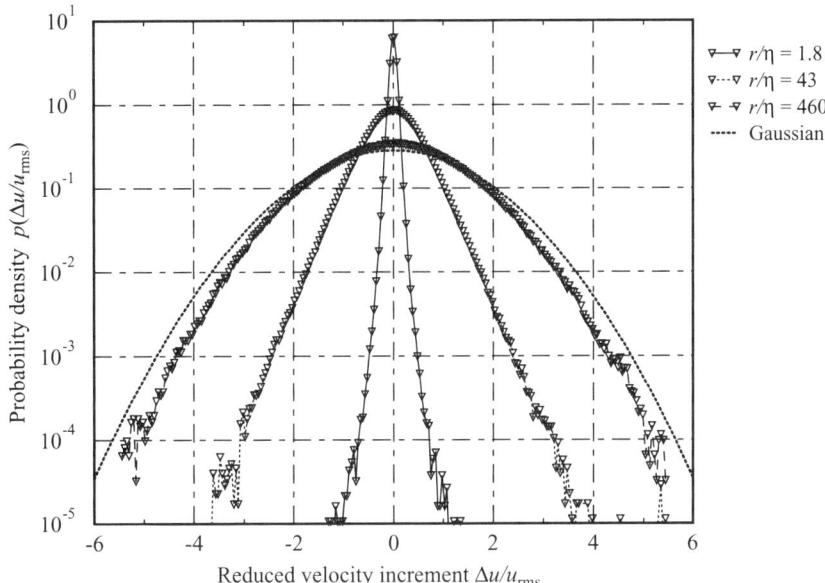

Figure 3.5. Probability density function of normalized velocity increment in isotropic turbulence. From Noullez et al. (1997) with permission of CUP.

exhibit a universal behavior, because they are sensitive to the characteristic time scale of turbulence production at large scales.

The analysis of the pdf's of the longitudinal velocity increments shows that the lack of Gaussianity is scale dependent (see Fig. 3.5), in the sense that velocity increments at small scales exhibit larger differences with the normal distribution than velocity increment at larger scales.

The lack of Gaussianity is an intrinsic feature of turbulence because of the nonlinearity of the Navier–Stokes equations. This point is addressed in Subsection 3.7.4.

3.1.3 Energy Decay Regimes

Turbulent kinetic energy $\mathcal{K}$ is observed to follow different regimes, depending on the position in the wake of the turbulence-generating grid. Three regions are usually identified, which are subsequently presented. They have an universal character, as they are observed in almost all clean experimental data sets.

1. The **formation region**, in which the wakes of the rods of the grid interact and merge. These interactions lead to a loss of memory of turbulent fluctuations and

where $P(u)$ is the pdf of the velocity fluctuation u. For short-correlated forcing such that $\tau_F \ll l_F/u_{\rm rms}$, one obtains

$$\ln P(u) \propto -u^3 \text{ for } l_F/\tau_F \gg u \gg u_{\rm rms}.$$

Therefore it is seen that the interplay between the external forcing and the turbulence nonlinearity leads to an automatic breakdown of Gaussianity for very intense events.

to the rise of an quasi-isotropic state.[§] It is important to note that this return to isotropy is not observed if the initial Reynolds number is too low.

2. The **initial region**, in which the flow can be considered isotropic and is strongly energetic. In this region, the Taylor-scale-based Reynolds number $Re_\lambda = \lambda u'/\nu$ is high, meaning that the nonlinear effects are dominant. In this region, turbulent kinetic energy $\mathcal{K}$ is observed to decay approximately as t^{-n} with $n \approx 1$, whereas the Taylor scale grows as $t^{0.35-0.4}$. Experimental data yield $6/5 \leq n \leq 4/3$. It is important to notice that experimental uncertainties dealing with the measure of the decay exponent are high, as this measure relies on several strong assumptions (Mohamed and Larue, 1990; Skrbek and Stalp, 2000).

   Theoretical analyses based on two-point closures, like EDQNM (see Subsection 3.5.6) reveal that the decay exponent n is sensitive to many parameters related to the initial condition, such as the slope of the turbulent kinetic-energy spectrum at very small wavenumbers at initial time, but also to possible saturation effects that are due to the finite size of both experimental facilities and computational domains (Skrbek and Stalp, 2000). The turbulence is observed to exhibit self-similar states during this decay stage. The analysis of these states is presented in Subsection 3.2.

3. The **final region**, which is defined as the region in which the Taylor-based Reynolds number is so low that the viscous linear effects are dominant. The criterion $Re_\lambda \leq 100$ is sometimes used to define the final region. The turbulent kinetic energy now decays more quickly, leading to $\mathcal{K} \sim t^{-n}$ with $n \approx 2$, whereas the Taylor microscale grows as $\sqrt{t}$. It is important to note that, at such a low Reynolds number, isotropy is very difficult to achieve, either in laboratory experiments or in numerical simulations, because of couplings between large and small scales. As in the previous case, the decay rate is expected to be sensitive to the slope of the spectrum at very low wavenumbers and various parameters of the experimental apparatus. Experimental realizations of the final region are very rare, and it seems that the transition between the initial and the final regions has never been observed experimentally, as it would require very long wind tunnels (Skrbek and Stalp, 2000). Details about this decay regime are given in Subsection 3.2.6.

3.1.4 Coherent Structures in Isotropic Turbulence

Statistical isotropy does not imply that isotropic turbulence fluctuations are uncoherent. Since the pioneering simulations of Siggia (1981), it has been observed that vortical coherent events are present in isotropic turbulence. One usually distinguishes elongated vortices, referred to as *worms* or *vortex tubes*, and flat *vortex sheets*. These structures, their dynamics, and their role in the turbulence dynamics are discussed in Section 3.6.

[§] The term quasi-isotropic refers here to a state in which at least second-order statistical moments are isotropic. But some anisotropic effects that are due to turbulence memory may remain on higher-order moments.

3.2 Self-Similar Decay Regimes, Symmetries, and Invariants

3.2.1 Symmetries of Navier–Stokes Equations and Existence of Self-Similar Solutions

Let us first recall that a physical law $F(x, t; u_1, \ldots, u_N)$ (where x and t denote the space and time, respectively, and $u_i, i = 1, N$ are physical quantities) is said to be invariant under the transformation $F \longrightarrow F^*, x \longrightarrow x^*, t \longrightarrow t^*, u_i \longrightarrow u_i^* (i = 1, N)$ if and only if

$$F(x, t; u_1, \ldots, u_N) = F(x^*, t^*; u_1^*, \ldots, u_N^*), \tag{3.15}$$

i.e., the physical law is not modified by the change of variables. The Navier–Stokes equations for an incompressible fluid in an unbounded domain (i.e., without boundary conditions) are known to admit the following one-parameter set of symmetries¶ (which has the mathematical structure of a Lie group):

- Time translation:
$$(t, \boldsymbol{x}, \boldsymbol{u}, p) \longrightarrow (t + t_0, \boldsymbol{x}, \boldsymbol{u}, p). \tag{3.16}$$

- Pressure translation:
$$(t, \boldsymbol{x}, \boldsymbol{u}, p) \longrightarrow [t, \boldsymbol{x}, \boldsymbol{u}, p + \zeta(t)]. \tag{3.17}$$

- Rotation (with $\mathbf{Q}$ as a constant-rotation matrix):
$$(t, \boldsymbol{x}, \boldsymbol{u}, p) \longrightarrow (t, \mathbf{Q}\boldsymbol{x}, \mathbf{Q}\boldsymbol{u}, p). \tag{3.18}$$

- Generalized Galilean transformation:
$$(t, \boldsymbol{x}, \boldsymbol{u}, p) \longrightarrow [t, \boldsymbol{x} + v(t), \boldsymbol{u} + \dot{v}(t), p - \rho \boldsymbol{x} \cdot \ddot{v}(t)]. \tag{3.19}$$

- Scaling I:
$$(t, \boldsymbol{x}, \boldsymbol{u}, p) \longrightarrow (\lambda^2 t, \lambda \boldsymbol{x}, \lambda^{-1} \boldsymbol{u}, \lambda^{-2} p). \tag{3.20}$$

- Scaling II:
$$(t, \boldsymbol{x}, \boldsymbol{u}, p, \nu) \longrightarrow (t, \lambda \boldsymbol{x}, \lambda \boldsymbol{u}, \lambda^2 p, \lambda^2 \nu), \tag{3.21}$$

where λ is an arbitrary strictly positive real parameter.

It is important to note that these symmetries are identified by conducting an exact mathematical analysis of the incompressible Navier–Stokes equations without introducing any hypothesis or modeling assumptions.**

¶ Other symmetries, such as mirror symmetry, exist but are not one-parameter symmetries.

** This analysis is performed considering the following one-parameter (Lie group) transformation:

$$T_a : y \to \hat{y} = \hat{y}(y, a), \quad y = (t, \boldsymbol{x}, \boldsymbol{u}, p, \nu), \tag{3.22}$$

which depends continuously on the real parameter a. Let us write formally the Navier–Stokes equations as $\mathcal{NS}(y) = 0$. T_a is said to be a symmetry of the Navier–Stokes equations if

$$\mathcal{NS}(y) = 0 \iff \mathcal{NS}(\hat{y}) = 0. \tag{3.23}$$

Boundary conditions may eventually decrease the number of symmetries, but cannot introduce new symmetries. It is worth noting that scalings I and II are particular forms (taking $h = -1$ and $h = 1$) of the even more general rescaling:

$$(t, x, u, p, v) \rightarrow (\lambda^{1-h}t, \lambda x, \lambda^h u, \lambda^{2h}, \lambda^{1+h}v). \quad (3.25)$$

We now focus on isotropic turbulence. In this case, symmetries such as rotation invariance, Galilean invariance, pressure, and time translation are implicitly met. Therefore the emphasis is to be put on the scaling symmetries and looking at the statistical moments of the turbulent velocity field. Let r and f be the correlation distance and the normalized two-point double-velocity correlation, respectively (see Subsection 3.4.1 for a detailed description). In the limit of very large Reynolds numbers (i.e., vanishing molecular viscosity), these quantities are transformed as follows (Oberlack, 2002):

$$t^* = \lambda_2 t, \quad r^* = \lambda_1 r, \quad \overline{u'^2}^* = (\lambda_1/\lambda_2)^2 \overline{u'^2}, \quad f^* = f. \quad (3.26)$$

In the case of a finite Reynolds number, the only possible solution is $\lambda_2 = \lambda_1^2$. A set of invariants $\check{r}, \check{f}, \check{u}, \check{p}$ can be defined:

$$\check{r} = \frac{r}{t^{\frac{2}{\sigma+3}}}, \quad \check{f} = \frac{\overline{u'^2}f}{t^{-2\frac{\sigma+1}{\sigma+3}}}, \quad \check{u} = \frac{u}{t^{-\frac{\sigma+1}{\sigma+3}}}, \quad \check{p} = \frac{p}{t^{-2\frac{\sigma+1}{\sigma+3}}}, \quad (3.27)$$

where

$$\sigma = 2\frac{\ln \lambda_2}{\ln \lambda_1} - 3. \quad (3.28)$$

It is worth noting that, in the finite-Reynolds-number case, $\sigma = 1$ is the only possible value. It can be shown, still considering the high-Reynolds-number limit, that the parameter σ is related to the spatial decay of the two-point correlations and the shape of the kinetic-energy spectrum at a low wavenumber:

$$\lim_{r \to +\infty} f(r) \sim r^{-\sigma}, \quad \lim_{k \to 0} E(k) \sim k^\sigma. \quad (3.29)$$

It is important to note that the constants involved in these scaling laws are assumed to be independent of time, corresponding to the so-called *permanence of large eddies (PLE) hypothesis*.

We now show that the existence of self-similar solutions for the isotropic decay problem can be deduced from symmetry analysis (and not assumed *a priori*). To this end, let us consider the following one-parameter subgroup of transformation (Clark and Zemach, 1998):

$$t^* = \lambda(t + t_0) - t_0, \quad \mathbf{x}^* = \lambda^\gamma \mathbf{x}, \quad (3.30)$$

The set of symmetries constitutes a local one-parameter Lie group, referred to as a symmetry group of the Navier–Stokes equations. Assuming that the neutral element of this group (i.e., the identity transformation) corresponds to $a = 0$, the group is characterized by the variation of y under T_a around $a = 0$, which is represented by the *infinitesimal generator X*:

$$X \equiv \left.\frac{\partial \hat{y}}{\partial a}\right|_{a=0} = \sum_i \xi_i \frac{\partial}{\partial y_i}, \quad \xi_i \equiv \left.\frac{\partial \hat{y}_i}{\partial a}\right|_{a=0}. \quad (3.24)$$

Once X is known, all elements of the symmetry group T_a can be calculated.

3.2 Self-Similar Decay Regimes, Symmetries, and Invariants

where λ is an arbitrary real parameter. This subgroup is labeled by γ and t_0, which are two real parameters. We are now looking for turbulent flows such that the shape of the kinetic-energy spectrum $E(k, t)$ is invariant under transformation (3.30). Simple dimensional analysis yields

$$\lambda^{3\gamma-2} E(k, t) = E[\lambda^{-k} k, \lambda(t + t_0) - t_0]. \tag{3.31}$$

The preceding property holds for all group elements if it holds for the *infinitesimal element*, i.e., for the group element $\lambda = 1 + \delta\lambda$, with $\delta\lambda \ll 1$. To this end, one differentiates (3.31) with respect to λ and then takes $\lambda = 1$. The result is the following *determining equation*:

$$(3\gamma - 2) E(k, t) = -\gamma k \frac{\partial E}{\partial k} + (t + t_0) \frac{\partial E}{\partial t}, \tag{3.32}$$

which can be solved by the method of characteristics in spectral space. The right-hand side of Eq. (3.32) leads to the following characteristic line equation:

$$\frac{d}{dt} k(t) = -\frac{\gamma}{t + t_0} k(t), \tag{3.33}$$

and therefore the wavenumber k evolves as

$$k(t) = \left(1 + \frac{t}{t_0}\right)^{-\gamma} k(0) \tag{3.34}$$

along the characteristic line spanned by $k(0)$. Along this line, the kinetic-energy spectrum evolution is given by

$$\frac{d}{dt} E(k(t), t) = \frac{\partial E}{\partial k} \frac{dk}{dt} + \frac{\partial E}{\partial t} = \frac{3\gamma - 2}{t + t_0} E[k(t), t], \tag{3.35}$$

leading to the following solution:

$$E(k(t), t) = \left(1 + \frac{t}{t_0}\right)^{3\gamma - 2} E_0[k(0)]$$

$$= \left(1 + \frac{t}{t_0}\right)^{3\gamma - 2} E_0\left[k(t)\left(1 + \frac{t}{t_0}\right)^{\gamma}\right]. \tag{3.36}$$

Introducing the length scale $L(t)$ and the energy scale $K(t)$ such that

$$L(t) = L_0 \left(1 + \frac{t}{t_0}\right)^{\gamma}, \quad K(t) = K_0 \left(1 + \frac{t}{t_0}\right)^{2\gamma - 2}, \tag{3.37}$$

one obtains

$$E(k, t) = K(t) L(t) F(k L(t)), \tag{3.38}$$

in which the nondimensional shape function F is such that

$$F(\xi) = \frac{E_0(\xi/L_0)}{K_0 L_0}. \tag{3.39}$$

Therefore, *the solution obeys a self-similar decay regime if there exist a single length scale $L(t)$, a single velocity scale[††] $\sqrt{K(t)}$, and a single nondimensional shape*

[††] An adequate choice for K_0 yields $K(t) = \mathcal{K}(t)$.

Table 3.3. *Time-evolution exponents in self-similar decay of isotropic turbulence deduced from symmetry analysis, assuming the PLE hypothesis holds*

	$\sigma = 1$	$\sigma = 2$	$\sigma = 4$	$\sigma = +\infty$
$L(t) \sim$	$t^{1/2}$	$t^{2/5}$	$t^{2/7}$	Const.
$\mathcal{K}(t) \sim$	t^{-1}	$t^{-6/5}$	$t^{-10/7}$	t^{-2}
$Re_t(t) \sim$	Const.	$t^{-1/5}$	$t^{-3/7}$	t^{-1}
Invariant name	Re_t	Birkhoff	Loitsyansky	$L(t)$
Invariant definition	$L\sqrt{\mathcal{K}}/\nu$	Eq. (3.46)	Eq. (3.44)	$\int_0^{+\infty} f(r)dr$
Associated spectrum		Saffman	Batchelor	

function F such that relation (3.38) is satisfied at all times and all wavenumbers. The important conclusion is that (3.38) is not postulated as in early studies like those of Karman and Howarth in the late 1930s, but deduced as being a consequence of the symmetries of the governing equations in the limit of very high Reynolds numbers.

3.2.2 Algebraic Decay Exponents Deduced From Symmetry Analysis

The symmetry analysis introduced in the previous subsection can also be used to recover some information about the time evolution of the solution. One does this by finding the values of σ in Eq. (3.28) or γ in Eq. (3.30).

Time-scaling laws for turbulent kinetic energy $\mathcal{K}(t)$, turbulent dissipation $\varepsilon(t)$, integral length scale $L(t)$, and turbulent Reynolds number Re_t are deduced from relation (3.27) in a straightforward way:

$$L(t) = \int_0^{+\infty} f(r,t)dr = t^{\frac{2}{\sigma+3}} \int_0^{+\infty} f(r^*)dr^* \sim t^{\frac{2}{\sigma+3}}, \tag{3.40}$$

$$\mathcal{K}(t) \sim t^{-2\frac{\sigma+1}{\sigma+3}}, \tag{3.41}$$

$$Re_t(t) = \frac{L(t)\sqrt{\mathcal{K}(t)}}{\nu} \sim t^{-\frac{\sigma-1}{\sigma+3}}, \tag{3.42}$$

$$\varepsilon(t) \sim \frac{d}{dt}\mathcal{K}(t) \sim t^{-\frac{3\sigma+5}{\sigma+3}}. \tag{3.43}$$

It is seen that the time-evolution exponents of these global turbulent parameters are explicit functions of σ. Because σ is also related to the shape of the kinetic-energy spectrum at very large scales [see Eq. (3.29)], this leads to the conclusion that the self-similar decay regime is governed by the very large scales of turbulence.

Different values for σ have been proposed during the past few decades, which are now briefly surveyed (corresponding time-evolution exponents are displayed in Table 3.3). A value of σ is associated with the existence of an invariant quantity that will remain constant during the decay (see subsequent discussion). In some cases the existence and the physical meaning of this invariant quantity are easily handled,

3.2 Self-Similar Decay Regimes, Symmetries, and Invariants

whereas some controversies exist in other cases. The most popular values for the parameter σ are as follows:

- σ = 4. According to the Loitsyansky–Landau theory, it was hypothesized by Loitsyansky in 1939 that the following integral quantity (referred to as the *Loitsyansky integral* or the *Loitsyansky invariant*),

$$\mathfrak{I} = -\int r^2 \overline{\boldsymbol{u}(\boldsymbol{x}) \cdot \boldsymbol{u}(\boldsymbol{x}+\boldsymbol{r})} d^3\boldsymbol{r} = 8\pi \overline{u'^2} \int_0^{+\infty} r^4 f(r) dr, \quad (3.44)$$

[with $f(r)$ given by (3.76)], is invariant in time during the decay phase. The corresponding time-evolution exponents were derived by Kolmogorov in 1941. The associated form of the kinetic-energy spectrum is referred to as the *Batchelor spectrum*:

$$E(k) = \frac{\mathfrak{I}}{24\pi^2} k^4 + \cdots (kL \ll 1), \quad (3.45)$$

where L is the integral length scale. The time invariance of $\mathfrak{I}$ is a controversial issue because it depends on the decay rate of a velocity two-point correlation at long range. It is constant if velocity long-range interactions decay fast enough, which is not obvious because the pressure fluctuations may induce strong long-range interactions.[‡‡] The controversy was initiated by Proudman and Reid in 1954, followed by Batchelor and Proudman in 1956, who advocated that long-range interactions are strong enough to render $\mathfrak{I}$ time dependent. Since that time, $\mathfrak{I}$ has been observed to be time dependent in many numerical simulations, in agreement with predictions of many two-point closures like EDQNM. This issue was very recently revisited by Davidson (2004) and Ishida, Davidson, and Kaneda (2006), who observed that $\mathfrak{I}$ becomes time independent after a transient phase in high-resolution DNS, provided that the domain size is much larger that of the turbulent integral scale (they considered a ratio up to 80) and that the Reynolds number is larger than 100. Therefore time dependency observed in previous simulations was an artefact that was due to spurious long-range correlations induced by the insufficient domain size and periodic boundary conditions. The fact that pressure fluctuations do not lead to a strong long-range coupling may be attributed to a *screening effect* in fully developed turbulence: Long-range correlations are weakened by opposite canceling effects of the very intricate turbulent vorticity field.

- σ = 2. This second value was proposed in 1954 by Birkhoff, who made the hypothesis that the following integral quantity is invariant (referred to as the *Birkhoff integral* but also as the *Saffman integral*):

$$\mathfrak{S} = \int \overline{\boldsymbol{u}(\boldsymbol{x}) \cdot \boldsymbol{u}(\boldsymbol{x}+\boldsymbol{r})} d^3\boldsymbol{r} = 4\pi \overline{u'^2} \int_0^{+\infty} r^2 \left(3f + r\frac{\partial f}{\partial r}\right) dr, \quad (3.46)$$

[‡‡] This point is easily understood if one looks at the Green's function solution given by Eqs. (2.32) and (2.33), which show that the pressure fluctuations caused by an eddy at a distance r from this eddy have an intensity $p' \sim r^{-3}$ for large r.

with $f(r)$ given by (3.76). The corresponding time behavior of the solution was derived by Saffman in 1967, after he argued that the Loitsyansky integral is diverging in isotropic turbulence. For that purpose, Saffman revised the approach introduced by Comte-Bellot and Corrsin in 1966 to investigate the connection between the energy spectrum and the energy decay. The associated spectrum shape (the *Saffman spectrum*) at large scales is

$$E(k) = \frac{\mathfrak{S}}{4\pi^2} k^2 \ldots (kL \ll 1). \tag{3.47}$$

- $\sigma = 1$. This value was proposed by Oberlack (2002), who emphasizes that this is the only value of σ that allows for the full similarity of the Kármán–Howarth equation [see Eq. (3.78) and the corresponding subsection] at finite Reynolds number. A noticeable feature of this solution is that the decay occurs at constant turbulent Reynolds number.
- $\sigma = +\infty$. This solution was also proposed by Oberlack in 2002. It corresponds to a decay with a constant integral scale.

3.2.3 Time-Variation Exponent and Inviscid Global Invariants

The direct physical interpretation of the value of the decay parameter σ is unclear in some cases. One can get a deeper insight into the related physics by looking at the links that exist between the choice of a value for σ and the conservation of exact invariants of inviscid flows.

Following Oberlack (2002), let us first recall that, for an inviscid flow in an unbounded domain V, the following nonlocal conservation laws are exact:

$$\frac{d}{dt} \int_V \boldsymbol{u} \cdot \boldsymbol{u} d^3 x = 0 \quad \text{(kinetic-energy conservation)}, \tag{3.48}$$

$$\frac{d}{dt} \int_V \boldsymbol{x} \times \boldsymbol{u} d^3 x = 0 \quad \text{(angular-momentum conservation)}, \tag{3.49}$$

$$\frac{d}{dt} \int_V \boldsymbol{u} \cdot (\nabla \times \boldsymbol{u}) d^3 x = 0 \quad \text{(linear-impulse or helicity conservation)}. \tag{3.50}$$

Now, using the change of variable based on the invariants introduced in Eq. (3.27), we can rewrite the three conservation laws as follows:

$$(\sigma - 2) \int_V \boldsymbol{\check{u}} \cdot \boldsymbol{\check{u}} d^3 \check{x} = 0 \quad \text{(kinetic-energy conservation)}, \tag{3.51}$$

$$(\sigma - 7) \int_V \boldsymbol{\check{x}} \times \boldsymbol{\check{u}} d^3 \check{x} = 0 \quad \text{(angular-momentum conservation)}, \tag{3.52}$$

$$(\sigma - 1) \int_V \boldsymbol{\check{u}} \cdot (\check{\nabla} \times \boldsymbol{\check{u}}) d^3 \check{x} = 0 \quad \text{(linear-impulse or helicity conservation)}. \tag{3.53}$$

Kinetic energy is strictly positive in a turbulent flow, whereas the sign and the absolute value of angular momentum and helicity are not *a priori* known. Because a choice for σ can enforce only one of the three conservation laws just given, it

3.2 Self-Similar Decay Regimes, Symmetries, and Invariants

makes sense to assume that the total linear momentum and helicity are identically null, whereas the kinetic energy is preserved, yielding $\sigma = 2$. Therefore the Birkhoff–Saffman theory is coherent with the preservation of kinetic energy at infinite Reynolds numbers.

Another interpretation is possible (Davidson, 2004) because both Loitsyansky and Birkhoff integral quantities are related to exact dynamical invariants of inviscid motion in an unbounded domain. The first one is the *linear impulse*, $\mathfrak{I}_{LI}$, and the second is the *angular momentum*, $\mathfrak{I}_{AM}$, with

$$\mathfrak{I}_{LI} = \frac{1}{2} \int_V (\boldsymbol{x} \times \operatorname{curl} \boldsymbol{u}) dV, \tag{3.54}$$

$$\mathfrak{I}_{AM} = \int_V (\boldsymbol{x} \times \boldsymbol{u}) dV. \tag{3.55}$$

Considering a volume V filled by isotropic turbulence with a characteristic length much larger that the integral length scale of the turbulent motion, the following relations hold:

$$\frac{\langle \mathfrak{I}_{LI}^2 \rangle}{V} \simeq \mathfrak{I}, \tag{3.56}$$

$$\frac{\langle \mathfrak{I}_{AM}^2 \rangle}{V} \simeq \mathfrak{S}. \tag{3.57}$$

If turbulent eddies have a finite, nonnegligible linear momentum, then $\mathfrak{S} \neq 0$ and therefore the spectrum will be of the Saffman type and $\sigma = 2$. If their linear momentum is very small but their angular momentum is finite, then $\mathfrak{S} \simeq 0$ and $\mathfrak{I} \neq 0$, yielding a Batchelor-like spectrum and $\sigma = 4$.

Let us just note that, if the two-point correlations fall sufficiently rapidly to ensure that all integrals are convergent, the following Taylor series expansion holds at small wavenumbers:

$$E(k) = \frac{\mathfrak{S}}{4\pi^2} k^2 + \frac{\mathfrak{I}}{24\pi^2} k^4 + \cdots (kL \ll 1). \tag{3.58}$$

3.2.4 Refined Analysis Without PLE Hypothesis

The analysis previously conducted relies on the PLE hypothesis, which is observed to fail in some cases. Let us now consider a self-similar state whose low-wavenumber kinetic-energy spectrum is

$$E(k, t) = C_\sigma(t) k^\sigma \quad (kL \ll 1), \tag{3.59}$$

where C_σ is *a priori* a time-dependent function. It is recalled that the PLE hypothesis states that C_σ is time independent. Simple dimensional analysis and solution self-similarity lead to the following scaling laws,

$$\mathcal{K} \propto t^{-n}, \quad L \propto t^{-b}, \quad \varepsilon \propto t^{-(n+1)}, \tag{3.60}$$

from which it follows that

$$Re_L = \frac{Lu'}{\nu} \propto t^{(2\gamma_\sigma+1-\sigma)/(\sigma+3)}, \quad \frac{\sqrt{\mathcal{K}}}{L} = \frac{\varepsilon}{\mathcal{K}} \propto t^{-1}, \qquad (3.61)$$

where the coefficient γ_σ is defined as

$$\gamma_\sigma = \frac{1}{C_\sigma} \frac{dC_\sigma}{d \log t} \qquad (3.62)$$

and where the coefficients are tied by the following relation:

$$-n + (\sigma + 1)b = \gamma_\sigma. \qquad (3.63)$$

The theoretical analysis based on the EDQNM closure shows that C_σ is constant in time and $\gamma_\sigma = 0$ for $\sigma < 4$, whereas $\gamma_4 = 0.16$. Therefore one has $n = -6/5$ for $\sigma = 2$ and $n = -1.38$ for $\sigma = 4$. Neglecting the time variation of C_4, one recovers the Kolmogorov value of $n = -10/7 \simeq -1.43$ for $\sigma = 4$.[§§]

Typical numerical results (obtained by use of simplified numerical models, as these data cannot be obtained by experimental means and are out of range of available supercomputing facilities) are displayed in Fig. 3.2. Turbulent flows with an initial small-wavenumber slope higher than 4 are observed to relax toward the $\sigma = 4$ self-similar state.

3.2.5 Self-Similarity Breakdown

According to Eq. (3.38), which is a definition, self-similarity implies that the spectrum can be described with a single length scale, a single velocity (or energy) scale, and a single shape function. All previous developments are compatible with the existence of a self-similar solution. The issue of the possible occurence of breakdown of self-similarity for solutions with $4 - p < \sigma < 4$, where p is the coefficient such that

$$C_\sigma(t) \propto L^p(t), \qquad (3.64)$$

was raised by Eyink and Thomson (2000) on the grounds of theoretical arguments, where L is the integral length scale. This self-similarity breakdown was observed in 1D Burgers' turbulence simulations by Noullez et al. (2005) (see Subsection 3.8.3 for more details about Burgers' turbulence). The value $p = 0.55$ can be inferred from EDQNM results. In this non-self-similar regime, the small-wavenumber part of the spectrum should be divided into two parts, one with $E(k, t) \propto Ak^\sigma$ for $k \leq 1/l^*(t)$

[§§] The value $n = -1.38$ for $\sigma = 4$ is associated with a time-varying Loitsyanski integral: $\mathfrak{J} \sim t^{0.16}$. This results conflicts with the most recent DNS results (Ishida, Davidson, and Kaneda, 2006). This can be understood looking at the expansion of the nonlinear transfer term mediated by strongly nonlocal triadic interactions in the limit of very small wavenumbers retrieved from two-point closures (e.g. EDQNM):

$$T(k \to 0) = \partial E/\partial t \sim Ak^4 - 2\nu_{\text{turb}} k^2 E,$$

where ν_{turb} is an eddy viscosity. An error on the constant A may yield an error on the energy balance at very small wavenumbers, inducing a spurious time evolution of $\mathfrak{J}$.

3.2 Self-Similar Decay Regimes, Symmetries, and Invariants

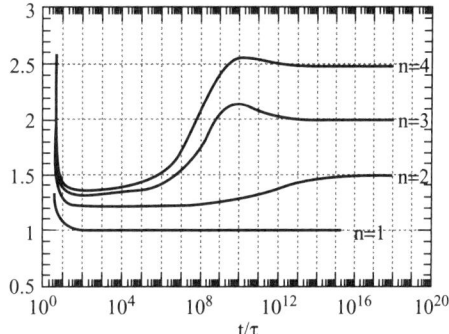

Figure 3.6. Evolution of the effective decay exponent σ_K for different values of the low-wavenumber spectrum exponent, n. The time scale τ is the same as in Fig. 3.3. From Clark and Zemach (1998) with permission of AIP.

and one with $E(k,t) \propto C_4(t)k^4$ for $1/l^*(t) \leq k \leq 1/L(t)$. Here the new length scale $l^*(t)$ evolves like $l^*(t) \propto L^{p/(4-\sigma)}(t)$. Self-similarity is broken if the two length scales are necessary to describe the spectrum evolution over long times, i.e., if $l^*(t) \gg L(t)$ for $t \gg 1$. This is the case if $4 - p < \sigma < 4$.

3.2.6 Self-Similar Decay in the Final Region

All results previously displayed in this section are related to the inital stage of decay, which is governed by nonlinear interactions. We now discuss the possibility of self-similar behavior in the final stage of decay. Simple calculations show that, if the PLE hypothesis holds, the kinetic energy evolves as

$$\mathcal{K}(t) \sim t^{-(\sigma+1)/2} \quad \text{if} \quad E(k,0) \sim k^\sigma \quad (kL \ll 1). \tag{3.65}$$

A long time evolution of the kinetic-energy spectrum computed with an EDQNM closure is displayed in Fig. 3.3. It is observed that a self-similar final stage of decay is reached after a very long time. Here τ denotes the eddy turnover time scale associated with the peak of the spectrum at the initial time: $\tau = [k_{\max}^3(0)E(k_{\max},0)]^{-1/2}$. It is also noticed that, in the final viscous-decay stage, the PLE holds, even in the present case in which $E(k,0) \sim k^4$ at very large scales.

An interesting problem is the time needed to reached the final stage of decay starting from a high-Reynolds-number solution initially governed by a nonlinear initial decay stage. The time evolution of the effective time-decay exponent σ_K defined as

$$\sigma_K^{-1}(t) = -\frac{\partial}{\partial t}\left[\frac{\mathcal{K}(t)}{\partial \mathcal{K}(t)/\partial t}\right] \tag{3.66}$$

is displayed in Fig. 3.6 for different values of the spectrum low-wavenumber power-law exponent. A trivial calculation shows that $\sigma_K = (\sigma + 1)/2$ if the turbulent kinetic energy obeys algebraic decay law (3.65).

It is observed that in all cases a very long time is needed before the solution reaches the final stage of decay, i.e., $\sigma_K = $ const. The turbulence quickly reaches a cascade-dissipation equilibrium for the k^1 spectrum.[¶¶] For other spectrum shapes, the transition is too long to be observed (Clark and Zemach, 1998). Considering a wind tunnel with air and a mean-flow velocity equal to $20\,\text{m}\,\text{s}^{-1}$, a grid-generated turbulence such that $\mathcal{K}(0) = 20\,\text{m}^2\,\text{s}^{-2}$ and the initial turbulent Reynolds number is equal to $Re_L = 3000$, the wind-tunnel length required to reach the final stage is of the order of $10^{16}\,\text{m}$ (about 1 light year, or about one-third the distance to the nearest star!) for k^2-shaped spectrum, and $5 \times 10^6\,\text{m}$ (almost an Earth radius!) for k^4 spectra.

3.3 Reynolds Stress Tensor and Analysis of Related Equations

For decaying HIT, the RST reduces to a spherical form, as the dissipation tensor, so that

$$\overline{u'_i u'_j} = 2\mathcal{K}\frac{\delta_{ij}}{3}, \quad \varepsilon_{ij} = 2\varepsilon\frac{\delta_{ij}}{3},$$

and Eqs. (2.52) and (2.53) simplify to

$$\frac{d\mathcal{K}}{dt} = -\varepsilon \tag{3.67}$$

and

$$\frac{d\varepsilon}{dt} = -C_{\epsilon 2}\frac{\varepsilon^2}{\mathcal{K}}. \tag{3.68}$$

In the absence of production because of the uniformity of the mean flow, the first equation is exact. Using the logarithmic derivatives, one can simply solve the preceding system of equations. It admits power-law solutions of the form

$$\mathcal{K}(t) = \mathcal{K}(0)\left(1 + \frac{t}{t_0}\right)^{-n}; \quad \varepsilon(t) = n\frac{\mathcal{K}(0)}{t_0}\left(1 + \frac{t}{t_0}\right)^{-n-1}, \tag{3.69}$$

with

$$t_0 = n\frac{\mathcal{K}(0)}{\varepsilon(0)}, \quad C_{\epsilon 2} = -\frac{d(\log \varepsilon)}{d(\log \mathcal{K})},$$

yielding

$$C_{\epsilon 2} = \frac{n+1}{n}. \tag{3.70}$$

Accordingly, a direct link of $C_{\epsilon 2}$ to the exponent of the decay law is given. Following the results given in Subsection 3.1.3, one obtains $m = 2, n = 6/5$, and $C_{\epsilon 2} = 11/6$ for a Saffman spectrum and $m = 4, n = 1.38$ and $C_{\epsilon 2} = 1.72$ for a Batchelor spectrum. The analysis of the self-similar initial decay stage given in the previous section emphasized that the decay exponent is directly tied to the power-law behavior of the kinetic-energy spectrum at low wavenumbers. Therefore $C_{\epsilon 2}$ can also be

---

[¶¶] This is consistent with the fact that it is the only solution that is fully consistent with the symmetry analysis at a finite Reynolds number.

3.3 Reynolds Stress Tensor and Analysis of Related Equations

recast as a function of the spectrum shape at very large scales:

$$C_{\epsilon 2} = 1 + \frac{\sigma + 3}{2(\sigma + 1)}, \quad \text{with} \quad E(k) \sim k^{\sigma}, \quad (kL \ll 1). \tag{3.71}$$

A direct consequence is that there is no really universal value for $C_{\epsilon 2}$ and that a $\mathcal{K} - \varepsilon$ model with fixed parameters is not able to capture the subtle changes in the decay rate of $\mathcal{K}$ that may occur.

All these developments hold for large values of the Reynolds number only. At a low Reynolds number, i.e., during the final decay period, more complex expressions for $C_{\epsilon 2}$ must be found. Because the high-Reynolds-number asymptotic analysis can no longer be used, only empirical expressions are available. Most of them rely on an exponential interpolation between asymptotic values. As an example, let us mention the model proposed by Coleman and Mansour (1991):

$$C_{\epsilon 2}(Re_T) = 1. - 0.222 \exp(-0.1677\sqrt{Re_T}), \tag{3.72}$$

where the Reynolds number Re_T is defined as $Re_T = \mathcal{K}^2/\nu\varepsilon$. A limitation of this model, which is shared by almost all other models, is that it does not take into account other parameters, like the initial condition. Considering a fully linear evolution, the turbulent kinetic-energy spectrum evolves as

$$E(k, t) = E(k, 0) e^{-2\nu k^2 t}. \tag{3.73}$$

For small wavenumbers, one obtains

$$E(k, t) = k^m e^{-2\nu k^2 t}, \tag{3.74}$$

which leads to $\mathcal{K}(t) \propto t^{-(m+1)/2}$. Available experimental data, in which nonlinear effects are small but not identically zero, lead to $m \simeq 3$. In the strictly linear limit, one expects to recover either the Batchelor solution ($m = 2, C_{\epsilon 2} = 1.67$) or the Saffman solution ($m = 4, C_{\epsilon 2} = 1.4$).

The analysis can be further extended to account for the influence of the skewness of velocity gradients. This point is not discussed here (see Piquet, 2001, for a detailed discussion of the modeling issues related to the free-decay case).

It is clear that the main trends of high-Reynolds dynamics of decaying HIT can be predicted by the simplest $\mathcal{K} - \varepsilon$ model if the initial conditions are taken into account, including the initial spectrum shape. But the discussion just presented also shows that, even for a very simple turbulent flow such as HIT, several physical mechanisms escape the formalism of the $\mathcal{K} - \varepsilon$ model defined by Eqs. (3.67) and (3.68). The very reason why is that the turbulent decay depends on both the large and the small scales and that most turbulence models written in the physical space are not able to account for spectral features of turbulence.

It is also worth emphasizing that *prediction is not explanation* and that our knowledge of HIT remains elusive. Internal intermittency that is reflected in the scaling of high-order moments is an open problem; the formation of microstructures like worms is shown in physical and numerical experiments but not really explained from the analysis of Navier–Stokes equations.

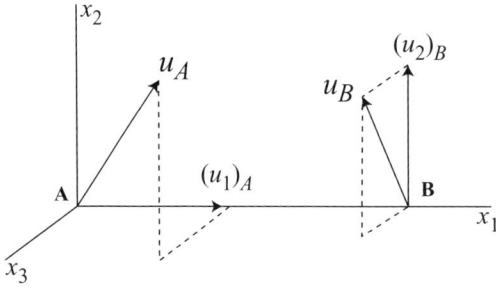
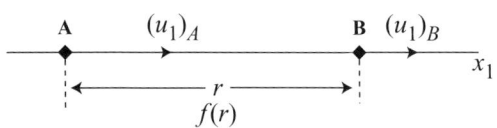
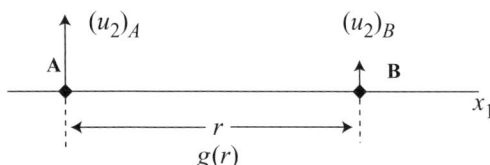

Figure 3.7. Schematic view of multipoint correlations. Top: general sketch of the correlation between two velocity components taken at two differents points A and B. Bottom: illustration of the physical meaning of the longitudinal correlation function $f(r)$ and its transverse counterpart $g(r)$.

3.4 Classical Statistical Analysis: Energy Cascade, Local Isotropy, Usual Characteristic Scales

3.4.1 Double Correlations and Typical Scales

Isotropy implies that the two-point second-order correlation tensor

$$R_{ij}(r) = \langle u'_i(x) u'_j(x+r) \rangle$$

(time is omitted for the sake of brevity) can be expressed as $R_{ij} = A(r)\delta_{ij} + B(r)r_i r_j$, or

$$R_{ij}(r) = u'^2 \left\{ g(r)\delta_{ij} + [f(r) - g(r)] \frac{r_i r_j}{r^2} \right\}, \qquad (3.75)$$

introducing the scaling factor $u'^2 = \frac{2}{3}\mathcal{K}$ and using the longitudinal correlation function

$$u'^2 f(r) = R_{ij}(r) \frac{r_i r_j}{r^2}, \qquad (3.76)$$

and its tranverse counterpart

$$u'^2 g(r) = R_{ij}(r) n_i n_j, \qquad (3.77)$$

in which n is a unit vector normal to r (see Fig. 3.7).

The scalar correlation functions f and g are linked by the incompressibility constraint. Using $\frac{\partial R_{ij}}{\partial r_j} = 0$ one obtains

$$g(r) = f(r) + \frac{r}{2} \frac{\partial f}{\partial r}.$$

3.4 Classical Statistical Analysis

Finally, reintroducing the time dependency, the evolution equation for the two-point second-order tensor amounts to the single scalar equation, e.g., for f, as follows:

$$\frac{\partial}{\partial t}(u'^2 f) = \left(\frac{\partial}{\partial r} + \frac{4}{r}\right)\left[R_{LL,L}(r,t) + 2\nu\frac{\partial}{\partial r}(u'^2 f)\right], \quad (3.78)$$

which is referred to as the *Karman–Howarth equation*. The term $R_{LL,L}$ represents the longitudinal two-point third-order correlation function, which is involved by means of the quadratic nonlinearity. It is defined as

$$R_{LL,L}(r,t) = \overline{u'_i(x,t)u'_i(x,t)u'_m(x+r,t)}\frac{r_m}{r}. \quad (3.79)$$

A slightly different form can be found in Mathieu and Scott (2000).

Typical spatial length scales of turbulence can be defined by functions $f(r)$ and $g(r)$. The *longitudinal and transverse integral length scales*, denoted by L_f and L_g, respectively, are defined as

$$L_f = \int_0^\infty f(r)dr, \quad L_g = \int_0^\infty g(r)dr, \quad (3.80)$$

and the *longitudinal and transverse Taylor microscales*, λ_f and λ_g, are computed as

$$\lambda_f = \sqrt{-\frac{2}{\frac{\partial^2 f}{\partial r^2}}}, \quad \lambda_g = \sqrt{-\frac{2}{\frac{\partial^2 g}{\partial r^2}}}, \quad (3.81)$$

respectively. The definitions of the latter come from the Taylor series expansions of f and g at small r, such as $f(r) = 1 - \frac{r^2}{\lambda_f^2} + \cdots$. Isotropy implies

$$L_g = \frac{1}{2}L_f; \quad \lambda_g^2 = \frac{1}{2}\lambda_f^2. \quad (3.82)$$

Finally, the dissipation rate is usually expressed as

$$\varepsilon = 30\nu\frac{u'^2}{\lambda_f^2} = -15\nu u'^2\frac{\partial^2 f}{\partial r^2}. \quad (3.83)$$

Assuming that the dissipation rate ε is constant, one can derive many scaling laws, which are summarized in Table 3.4. The scaling laws for ε are further discussed in Subsection 3.5.7.6.

3.4.2 (Very Brief) Reminder About Kolmogorov Legacy, Structure Functions, "Modern" Scaling Approach

Structure functions are interesting alternatives to velocity correlations at two points, using equivalent r (two-point) separation vectors, but velocity increments $\delta u' = u'(x+r) - u'(x)$ instead of $u'(x)$ or $u'(x+r)$. The counterpart of the longitudinal correlation f is the (longitudinal) second-order structure function

$$\langle \delta u_\parallel^2 \rangle = \left\langle \left\{[u'(x+r) - u'(x)] \cdot \frac{r}{r}\right\}^2 \right\rangle.$$

Table 3.4. *Usual scaling laws derived from Kolmogorov's 1941 framework*

Name	Symbol	Definition	Scaling with η	Remark
Integral scale	L		$L/\eta \propto Re_L^{3/4} \propto Re_\lambda^{3/2}$	Large energy-containing scales
Taylor microscale	λ	$\sqrt{15\nu u'^2/\varepsilon}$	$\lambda/\eta \propto Re_L^{1/4} \propto \sqrt{Re_\lambda}$	Small scales, maximum of dissipation/enstrophy spectrum
Kolmogorov scale	η	$(\nu^3/\varepsilon)^{1/4}$		Local Reynolds number equal to one

Note: The dissipation rate is assumed to be scale independent. Reynolds numbers are defined as $Re_L = u'L/\nu$ and $Re_\lambda = u'\lambda/\nu$. The local Reynolds number at scale l is defined as $Re_l = u(l)l/\nu$, where $u(l)$ is the characteristic velocity scale associated with l.

Definition of the nth-order *longitudinal structure function* $\langle \delta u_\parallel^n \rangle$ is readily obtained from the preceding equation. In homogeneous turbulence, the second-order longitudinal structure function, for instance, is given by

$$\langle \delta u_\parallel^2 \rangle = \frac{2}{3}\mathcal{K}\left[1 - f(r)\right]. \tag{3.84}$$

More generally, one can keep in mind that structures functions give information on two-point statistics for $r \neq 0$ and tend to zero with vanishing r.

Kolmogorov originally proposed scaling the structure functions in terms of r and the dissipation rate ε only, the first and simplest version (denoted K41 since the seminal paper of Kolmogorov was published in 1941) reducing to

$$\langle \delta u_\parallel^n \rangle \sim (\varepsilon r)^{n/3}. \tag{3.85}$$

The scaling results only from dimensional analysis, once the physical parameters have been chosen. Of course, this choice relies on nontrivial phenomenological aspects. The scaling holds for an inertial range, i.e., for $L \gg r \gg \eta$, delineated by a large-scale L, comparable with L_f in Eqs. (3.80) and the Kolmogorov scale η.

It is important to notice that the classical Taylor series expansion $u_i(x+r) = u_i(x) + \frac{\partial u_i}{\partial r_l} r_l + \cdots$ would yield a different scaling law: $\langle \delta u_\parallel^n \rangle \sim r^n$. This result, which holds for a smooth velocity field, may be valid for the smallest scales, i.e., $r < \eta$. The simple fact that the K41 exponent is fractional ($n/3$) means that the velocity field appears not to be differentiable in the inertial range and that self-similar dynamics is expected.

Modern phenomenological theories continue in search of a more general scaling, replacing the $n/3$ exponents with new ones, ζ_n, called "anomalous exponents," because the former are in question in the case of internal intermittency. The background argument for introducing such new scaling is to consider a local dissipation rate (or energy flux) that is no longer independent of the size r. The reader is referred to the following books for more details: Monin and Yaglom (1975), Frisch (1995), and Mathieu and Scott (2000).

3.4 Classical Statistical Analysis

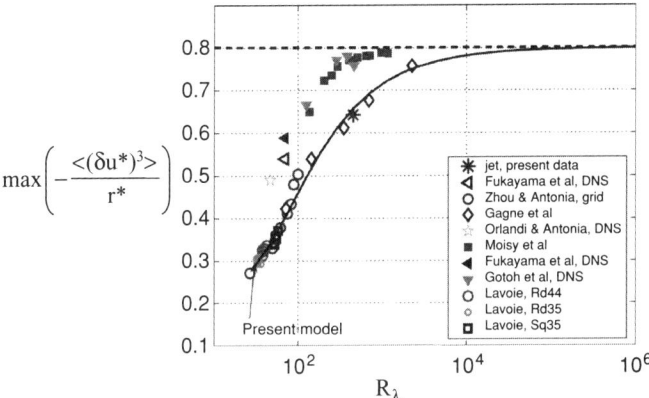

Figure 3.8. Value of the constant in inviscid asymptotic law (3.87) vs. the Taylor-microscale-based Reynolds number. From Antonia (2007), courtesy of R.A. Antonia.

Finally, let us just mention the famous *Kolmogorov's four-fifths law*,

$$\langle \delta u_\parallel^3 \rangle = -\frac{4}{5}\epsilon r + 6\nu \frac{\partial}{\partial r}\langle \delta u_\parallel^2 \rangle, \tag{3.86}$$

which appears as a simplified form of the Kármán–Howarth equation, and is one of the few "exact" equation in the theory of turbulence. Accordingly, the K41 scaling remains unquestioned (at least in HIT at very high Reynolds number) for $n = 3$.

But it is important noting that the further simplified inviscid form,

$$\langle \delta u_\parallel^3 \rangle = -\frac{4}{5}\epsilon r, \tag{3.87}$$

is nothing but an asymptotic limit, whose range of validity is still an open question. Recent experimental and numerical results show that it holds for very large Reynolds numbers only, as shown in Fig. 3.8: $Re_\lambda = O(10^3)$ may be considered a lower bound for flows with turbulence production, whereas much higher values such as $Re_\lambda = O(10^5) - O(10^6)$ are certainly more realistic in decaying turbulence.

3.4.3 Turbulent Kinetic-Energy Cascade in Fourier Space

It is often easier to investigate two-point statistics by use of 3D Fourier space. The counterpart of Eq. (3.75) in the Fourier space is Eq. (2.99), repeated here:

$$\hat{R}_{ij}(\boldsymbol{k}, t) = \underbrace{\frac{E(k, t)}{4\pi k^2}}_{e(k,t)} \underbrace{\left(\delta_{ij} - \frac{k_i k_j}{k^2}\right)}_{P_{ij}}.$$

It should be borne in mind that isotropy yields a very special form of the spectral tensor. The involved parameters are the following: $E(k, t)$, with $k = |\boldsymbol{k}|$, is the usual energy spectrum, representing the distribution of turbulent kinetic energy over different scales and the quantity in parentheses will be recognized as the projection matrix, $P_{ij}(\boldsymbol{k})$. Thus $\hat{R}_{ij}$ is determined by a single real scalar quantity, E, which is a function of the sole magnitude of $\boldsymbol{k}$. Therefore both the form of $\hat{R}_{ij}$ at a single point and its distribution over $\boldsymbol{k}$-space are strongly constrained by isotropy.

The evolution of the energy spectrum is governed by the *Lin equation*,*

$$\frac{\partial E(k,t)}{\partial t} + 2\nu k^2 E(k,t) = T(k,t) \qquad (3.88)$$

in which the third-order correlations are involved in the scalar spectral transfer term $T(k,t)$.† This equation can be seen as a spectral counterpart of Kármán–Howarth equation (3.78). The exact relationship among $E(k)$, $T(k)$, and all the correlations defined in physical space can be found in Mathieu and Scott (2000).

This equation derives from Craya's equation (2.81) by canceling mean-gradient terms and by assuming isotropy, so that

$$E(k,t) = 2\pi k^2 \hat{R}_{ii}(k,t), \quad T(k,t) = 2\pi k^2 T_{ii}(k,t). \qquad (3.89)$$

Integrating the equation over k yields

$$\mathcal{K}(t) = \int_0^\infty E(k,t)dk \quad \varepsilon = 2\nu \int_0^\infty k^2 E(k,t)dk \qquad (3.90)$$

and

$$\int_0^\infty T(k,t)dk = 0. \qquad (3.91)$$

This allows us to recover basic equation (3.67) and shows that $T(k,t)$ is a pure redistribution term in the Fourier space. The last relation accounts for the fact that the convection term conserves the total kinetic energy, leading to the well-known result that global kinetic energy is an invariant in inviscid incompressible flows (without boundary conditions).

One can notice that the counterpart of inviscid Kolmogorov 4/5 law (3.87) is

$$F(k) = \int_k^\infty T(k)dk = \varepsilon, \qquad (3.92)$$

where $F(k)$ is the spectral flux across a wavenumber k, is "exact" in very similar conditions, i.e., at very high Reynolds numbers only. This is observed in Fig. 3.9, in which a constant $F(k)$ is observed on hardly one decade at $Re_\lambda = 1132$, whereas it is not observed at all for $Re_\lambda = 732$.

Typical shapes of $E(k)$, $2\pi k^2 E(k)$, and $T(k)$ are displayed in Fig. 3.12 in Section 3.5. It is observed that the peak of the energy spectrum, $E(k)$, is significantly separated from the one of the dissipation spectra $2\nu k^2 E(k)$ at large Reynolds numbers. The transfer term is almost zero in a small zone within the inertial range, negative for smallest k and positive for largest k, the areas of both positive and negative values being exactly balanced. The physical meaning is that small-wavenumber

---

* The name *Lin equation* is used in agreement with a paper published by von Kármán and Lin in 1949, in which seminal talks by C. C. Lin are quoted.
† Let us emphasize here the physical meaning of the sign of $T(k)$. The net effect of nonlinearity on modes k such that $T(k) > 0$ is a kinetic-energy gain [which must be balanced by viscous effects in the statistically steady case $\partial E(k,t)/\partial t = 0$], whereas modes such that $T(k) < 0$ lose more kinetic energy than they gain through nonlinear interactions (these scales must be fed by a forcing term to obtain a statistically steady state). Last, scales such that $T(k) = 0$ are in equilibrium, in the sense that they do not lose or gain kinetic energy on the mean.

3.4 Classical Statistical Analysis

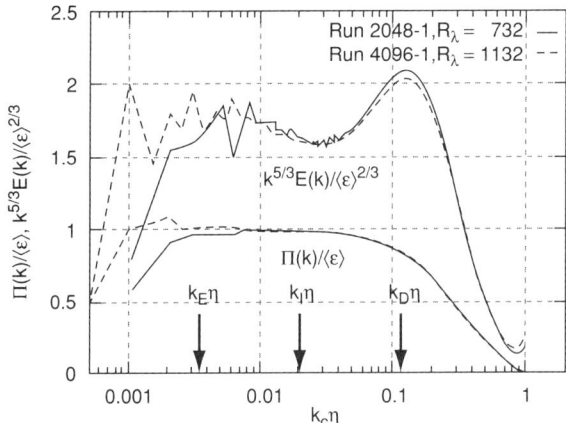

Figure 3.9. Value of the spectral flux $F(k)$ [denoted here as $\Pi(k)$] vs. the wave number k in high-resolution DNS. From Aoyama et al. (2005) with permission of Physical Society of Japan.

modes lose kinetic energy on the mean because of the nonlinear interactions, whereas large-wavenumber modes gain kinetic energy. The scale located within the inertial range has a zero net transfer. The associated dynamic picture is the celebrated *forward energy cascade* process[‡]: Turbulent kinetic energy is injected into the system (by external forcing, hydrodynamic instabilities, etc.) at small-wavenumber modes. The energy is then pumped toward higher-wavenumber modes by the nonlinear interactions, "streaming" in some sense toward modes at which it will be transformed into heat by viscous mechanisms. The inertial range is defined as the zone in which the net transfer $T(k)$ is zero [or, equivalently, the spectral flux $F(k)$ is constant]. In the inertial zone, the classical Kolmogorov scaling[§]

$$E(k) = K_0 \varepsilon^{2/3} k^{-5/3} \tag{3.93}$$

is observed in both experimental and numerical data sets.

The evolution equation for the dissipation ε is recovered from Eq. (3.88) by integrating it over k after multiplication by the factor $2\nu k^2$, yielding

$$\frac{d\varepsilon}{dt} = \int_0^\infty 2\nu k^2 T(k) dk - \int_0^\infty (2\nu k^2)^2 E(k) dk. \tag{3.94}$$

From this equation, it is clear that the second term on the right-hand side is negative and corresponds to a viscous-destruction mechanism. The first term is essentially positive. A part of $T(k)$ (at large k) is privileged by the k^2 weighting factor and can be interpreted as a *production* of ε by nonlinear interactions. This point will be further discussed in Section 3.7.

[‡] The term *cascade* was coined by Onsager in the late 1940s.
[§] This scaling is consistent with the content of the papers published by Kolmogorov in 1941. But it is worth noting that Kolmogorov never worked in Fourier space. The expression of the turbulent kinetic spectrum was given by his Ph.D. student, A. Obhukov, and almost independently rendered popular by Heisenberg.

The fact that the evolution of ε results from the imbalance between two very different terms, whose sum can be efficiently modeled by the purely negative term $-\frac{n}{n+1}\frac{\varepsilon^2}{\mathcal{K}}$ in Eq. (3.68), is certainly true in HIT at high Reynolds numbers, but remains not completely understood.

A brief introduction to 2D turbulence is given in Subsection 3.8.2.

3.5 Advanced Analysis of Energy Transfers in Fourier Space

3.5.1 The Background Triadic Interaction

The equation introduced in the previous chapter,

$$\frac{\partial \hat{u}_i}{\partial t}(\boldsymbol{k},t) = \imath P_{imn}(\boldsymbol{k}) \underbrace{\sum_{\Delta} \hat{u}_m^*(\boldsymbol{p},t)\hat{u}_n^*(\boldsymbol{q},t)}_{s_i}, \qquad (3.95)$$

with

$$P_{imn} = \frac{1}{2}\left[k_m P_{in}(\boldsymbol{k}) + k_n P_{im}(\boldsymbol{k})\right], \qquad (3.96)$$

is now detailed. Viscous effects are omitted and the symbol $\sum_\Delta$ for summation over triads is used in a generic way to avoid distinguishing between the discrete and the continuous formulations from the beginning. The use of complex conjugates for the Fourier coefficients in the sum (or integral) is consistent with a fully symmetric relationship for the triad, i.e.,

$$\boldsymbol{k} + \boldsymbol{p} + \boldsymbol{q} = 0 \qquad (3.97)$$

instead of $\boldsymbol{p} + \boldsymbol{q} = \boldsymbol{k}$ coming from the convolution product.

A slightly different form of the nonlinear coupling term is found to replace the term $\frac{\partial u_i u_j}{\partial x_j}$ in physical space with $\epsilon_{ijn}\omega_j u_n$. The corresponding form in Fourier space is $\imath s_i = P_{im}\epsilon_{mjn}\sum_\Delta \hat{\omega}_j(\boldsymbol{p},t)\hat{u}_n(\boldsymbol{q},t)$, which can be shown to be the same as the previous one, using the Ricci relationship and a symmetric form with respect to $\boldsymbol{p}$ and $\boldsymbol{q}$. This formulation is more convenient when the helical-mode basis is used.

In terms of the helical modes, Eq. (3.95) has the generic form

$$\frac{\partial \xi_s(\boldsymbol{k})}{\partial t} = \imath \sum_\Delta \underbrace{M_{ss's''}(\boldsymbol{k},\boldsymbol{p})}_{\text{I}} \underbrace{\xi_{s'}^*(\boldsymbol{p},t)\xi_{s''}^*(\boldsymbol{q},t)}_{\text{II}}, \qquad (3.98)$$

with $\xi_s(\boldsymbol{k}) = (1/2)\hat{\boldsymbol{u}} \cdot \boldsymbol{N}(-s\boldsymbol{k})$ and $\hat{\boldsymbol{u}}(\boldsymbol{p}) = \sum_{s'} \xi_{s'}\boldsymbol{N}(s'\boldsymbol{p})$. The signs s, s', s'', or *polarities*, take values of ± 1 only. It is worth noting that, in Eq. (3.98), term I is related to only the topology of the triad (i.e., is a purely geometric factor), whereas term II depends on only the amplitude of the modes, i.e., on the turbulent field itself. From Eq. (3.95) it is found that (e.g., Cambon and Jacquin, 1989)

$$M_{ss's''}(\boldsymbol{k},\boldsymbol{p}) = \frac{1}{2}\left\{[\boldsymbol{N}(-s\boldsymbol{k})\cdot\boldsymbol{N}(-s'\boldsymbol{p})][\boldsymbol{k}\cdot\boldsymbol{N}(-s''\boldsymbol{q})]\right.$$
$$\left. + [\boldsymbol{N}(-s\boldsymbol{k})\cdot\boldsymbol{N}(-s'\boldsymbol{q})][\boldsymbol{k}\cdot\boldsymbol{N}(-s''\boldsymbol{p})]\right\}. \qquad (3.99)$$

3.5 Advanced Analysis of Energy Transfers in Fourier Space

The second formulation, using $\boldsymbol{\omega} \times \boldsymbol{u}$ as the basic nonlinearity (Waleffe, 1992, 1993), yields

$$M_{ss's''}(\boldsymbol{k}, \boldsymbol{p}) = \frac{1}{2}(s'p - s''q)N(-s\boldsymbol{k}) \cdot [N(s'\boldsymbol{p}) \times N(s''\boldsymbol{q})], \quad (3.100)$$

using the aditional relationship

$$\hat{\boldsymbol{\omega}}(\boldsymbol{p}) = p \sum_{s'} s'\xi_{s'}(\boldsymbol{p}, t)N(s'\boldsymbol{p}) \quad (3.101)$$

and the antisymmetry of the triple scalar product.

The use of helical modes allows for an optimal factorization of the coupling terms in terms of the moduli k, p, q, and the angular variables: The former depend on only the geometry of the triangle whereas the latter also depend on the orientation of its plane. For further analysis, it is better to start from Eq. (3.100) because it appears more symmetric than (3.99) in terms of the three vectors of the triads, involving a *triple scalar product*, without need for additional calculations.

For instance, Eqs. (3.98) and (3.100) can be rewritten as

$$\frac{\partial \xi_s(\boldsymbol{k})}{\partial t} = \sum_{s's''} \sum_{\Delta} (s'p - s''q)K(s\boldsymbol{k}, s'\boldsymbol{p}, s'\boldsymbol{q})\xi_{s'}^*(\boldsymbol{p}, t)\xi_{s''}^*(\boldsymbol{q}, t), \quad (3.102)$$

with

$$K(s\boldsymbol{k}, s'\boldsymbol{p}, s''\boldsymbol{q}) = \frac{6}{4}N(-s\boldsymbol{k}) \cdot [N(-s'\boldsymbol{p}) \times N(-s''\boldsymbol{q})]. \quad (3.103)$$

The principle of triad instability stated by Waleffe (see Subsection 3.5.4) takes advantage of the full symmetry of the coupling coefficient K with respect to any simultaneous permutation of vectors and polarities within a given triad.

A last set of equations allows us to express K (and other related coefficients in statistical closures) in terms of the parameters of the triad. The idea is to turn from local reference frames (or helical modes) defined with respect to a fixed polar axis to their counterparts defined with respect to the normal unit vector of the triad (or almost equivalently with respect to a fixed $\boldsymbol{k}$, if $\boldsymbol{p}$ and $\boldsymbol{q}$ are under consideration). The unit normal vector is defined as

$$\boldsymbol{\gamma} = \frac{\boldsymbol{k} \times \boldsymbol{p}}{|\boldsymbol{k} \times \boldsymbol{p}|}, \quad (3.104)$$

and unit vectors in the plane spanned by the triad, normal to $\boldsymbol{k}, \boldsymbol{p}, \boldsymbol{q}$, respectively, are

$$\boldsymbol{\beta} = \frac{\boldsymbol{k}}{k} \times \boldsymbol{\gamma}, \quad \boldsymbol{\beta}' = \frac{\boldsymbol{p}}{p} \times \boldsymbol{\gamma}, \quad \boldsymbol{\beta}'' = \frac{\boldsymbol{q}}{q} \times \boldsymbol{\gamma}. \quad (3.105)$$

"Triadic" helical modes are defined by

$$W(s) = \boldsymbol{\beta} + \imath s\boldsymbol{\gamma}, \quad W(s') = \boldsymbol{\beta}' + \imath s'\boldsymbol{\gamma}, \quad W(s'') = \boldsymbol{\beta}'' + \imath s''\boldsymbol{\gamma}, \quad (3.106)$$

and they are related to the original ones by

$$N(s\boldsymbol{k}) = e^{\imath s\lambda}W(s), \quad N(s'\boldsymbol{p}) = e^{\imath s'\lambda'}W'(s'), \quad N(s''\boldsymbol{q}) = e^{\imath s''\lambda''}W''(s''), \quad (3.107)$$

where λ, λ', and λ'' are angles that characterize the rotation of the plane of the triad around $\mathbf{k}$, $\mathbf{p}$, and $\mathbf{q}$, respectively (see Chapter 14).

The advantage of $\mathbf{W}(s)$, $\mathbf{W}'(s')$, $\mathbf{W}''(s'')$ with respect to $\mathbf{N}(s\mathbf{k})$, $\mathbf{N}(s'\mathbf{p})$, $\mathbf{N}(s''\mathbf{q})$ is that any invariant combination (double or triple scalar product) of the former will rely on only the geometry of the triad, and therefore can be expressed in terms of the moduli k, p, q only. As a first useful application, the coefficient K can be expressed as

$$K = \frac{i}{4} e^{-\iota(s\lambda + s'\lambda' + s''\lambda'')} \mathbf{W}(s) \cdot [\mathbf{W}'(s') \times \mathbf{W}''(s'')].$$

The triple scalar product involves the sines of the internal angles of the triad:

$$\mathbf{W}(s) \cdot [\mathbf{W}'(s') \times \mathbf{W}''(s'')] = \iota(s's'' \sin\alpha + ss'' \sin\beta + ss' \sin\gamma).$$

These sines are connected to the lengths of the triangle through

$$\frac{\sin\alpha}{k} = \frac{\sin\beta}{p} = \frac{\sin\gamma}{q} = C_{kpq}, \quad (3.108)$$

so that

$$K(s\mathbf{k}, s'\mathbf{p}, s''\mathbf{q}) = e^{-\iota(s\lambda + s'\lambda' + s''\lambda'')} \frac{ss's''}{4} (sk + s'p + s''q) C_{kpq}, \quad (3.109)$$

with

$$C_{kpq} = \frac{\sqrt{2k^2 p^2 + 2p^2 q^2 + 2q^2 k^2 - k^4 - p^4 - q^4}}{2kpq}, \quad (3.110)$$

which appears in EDQNM models (see Subsection 3.5.7).

As a second application, the system of dependent variables $\mathbf{k}, \mathbf{p}, \mathbf{q}, \lambda$ is well suited for representing the triadic interactions. If the symbolic operator $\sum_\Delta$ is replaced with the integral, the relevant term is written as

$$\iiint S(\mathbf{k}, \mathbf{p}, t) d^3 \mathbf{p},$$

where $S(\mathbf{k}, \mathbf{p}, t)$ originates from $T = (1/2) T_{ii}$ using Eqs. (2.80) and (2.82). Its general expression in terms of triple-velocity correlations [see also Eq. (3.112)] is not important here, because only the change of dependent variables at fixed $\mathbf{k}$ [switching from (p_1, p_2, p_3) to (p, q, λ)] is considered, for any integrand S.

If $\mathbf{q}$ is expressed as $-\mathbf{k} - \mathbf{p}$ in S, then the factors p, q, and λ can replace p_1, p_2, p_3, yielding

$$\iiint S(\mathbf{k}, \mathbf{p}, t) d^3 \mathbf{p} = \iint_{\Delta_k} \frac{pq}{k} dp\, dq \int_0^{2\pi} S(k, p, q, \lambda) d\lambda. \quad (3.111)$$

The coefficient pq/k is the Jacobian of the change of integration variables, and Δ_k is the domain of p, q, so that k (fixed), p, q are the lengths of the sides of a triangle.

Finally, the other angular variables in Eq. (3.109), λ' and λ'', also can be expressed as functions of k, p, q, and λ.

3.5 Advanced Analysis of Energy Transfers in Fourier Space

3.5.2 Nonlinear Energy Transfers and Triple Correlations

The transfer term $T(k)$ in Eq. (3.88) involves triple-velocity correlations under summation on triads. We do not discuss here simplified closure models relying on a spectral-flux term $F(k)$ such that $T(k) = \partial F/\partial k$, where $F(k)$ is a more or less local (in Fourier space) function of k and $E(k)$. These models do not take into account the detailed triadic structure of the background equations, but can in some cases yield interesting results on the evolution of $E(k)$. In some models, a quasi-local flux in wave space can reflect very distant interactions, which are mediated by flat triads, so that the term "local" is misleading: For instance, the cusp in eddy viscosity corresponds to $k \sim p, q \sim 0$, whereas the plateau at small k corresponds to $p \sim q \gg k$ (Kraichnan, 1976). One could speak of "dyadic" models, instead of "local" models, in this case.

The techniques addressed in this subsection offer a closure for triple correlations at three points. They are developed in Fourier space for the sake of mathematical convenience. A third-order spectral tensor can be defined as

$$\langle \hat{u}_i(\boldsymbol{k})\hat{u}_j(\boldsymbol{p})\hat{u}_n(\boldsymbol{q})\rangle = \imath S_{ijn}(\boldsymbol{k},\boldsymbol{p},t)\delta(\boldsymbol{k}+\boldsymbol{p}+\boldsymbol{q}), \tag{3.112}$$

which corresponds to the general definition given in Chapter 2, up to a factor $\imath$. The transfer tensor that incorporates their contribution in the equation for the second-order spectral tensor is given by

$$T_{ij}(\boldsymbol{k})\delta(\boldsymbol{k}+\boldsymbol{p}) = \langle s_i(\boldsymbol{p})\hat{u}_j(\boldsymbol{k})\rangle + \langle \hat{u}_i(\boldsymbol{p})s_j(\boldsymbol{k})\rangle,$$

or

$$T_{ij}(\boldsymbol{k}) = \tau_{ij}(\boldsymbol{k}) + \tau_{ji}^*(\boldsymbol{k}), \tag{3.113}$$

with

$$\tau_{ij}(\boldsymbol{k}) = P_{imn} \int S_{jmn}(\boldsymbol{k},\boldsymbol{p}) d^3\boldsymbol{p}. \tag{3.114}$$

Two contributions can be distinguished in T_{ij}. The first one is given by

$$\frac{1}{2}\left[k_n \int (S_{jin} + S_{jin}^*) d^3\boldsymbol{p} + k_m \int (S_{jmi} + S_{imj}^*) d^3\boldsymbol{p}\right]$$

and corresponds to a true transfer tensor with zero integral. The complementary contribution,

$$\frac{1}{2}\frac{k_m k_n}{k^2}\left(k_i \int S_{jmn} d^3\boldsymbol{p} + k_j \int S_{imn}^* d^3\boldsymbol{p}\right),$$

gives by integration the "slow" pressure–strain tensor Π_{ij}^s introduced in Subsection 2.3.1.

Of course, we are interested in only

$$T(k) = 2\pi k^2 T_{ii} = 2\pi k^2(\tau_{ii} + \tau_{ji}^*)$$

in HIT, but it is necessary to address the equation for S_{ijn} to derive a consistent closure.

Similar to the equation for the second-order spectral tensor, the equation that governs S_{ijn} is found as

$$\left[\frac{\partial}{\partial t} + \nu(k^2 + p^2 + q^2)\right] S_{ijn}(k, p) = T_{ijn}(k, p) + T_{jni}(p, q) + T_{nij}(q, k).$$

The first term (the other ones are derived by circular permutations) on the right-hand side is exactly expressed as

$$\delta(k + p + q)T_{ijn} = \iota\langle s_i(k)\hat{u}_j(p)\hat{u}_n(q)\rangle$$
$$= \int_{k=r+s} P_{irs}(k)\langle \hat{u}_r(r)\hat{u}_s(s)\hat{u}_j(p)\hat{u}_n(q)\rangle d^3r, \quad (3.115)$$

and involves fourth-order correlations.

3.5.3 Global and Detailed Conservation Properties

Some global conservation properties of the Navier–Stokes equations in the limit of vanishing molecular viscosity can be easily recast in the Fourier space, providing some useful constraints on the triadic nonlinear transfer term.

We consider here the conservation of the global kinetic energy and the global helicity (in an unbounded domain and in the absence of external forcing):

$$\frac{\partial}{\partial t}\int u(x) \cdot u(x) d^3x = 0, \quad (3.116)$$

$$\frac{\partial}{\partial t}\int u(x) \cdot \omega(x) d^3x = 0, \quad \omega = \mathrm{curl}(u). \quad (3.117)$$

These two relations illustrate the fact that the nonlinear term redistributes energy and helicity among the different modes. As shown by Kraichnan, these global conservation properties can be supplemented by other ones, which hold at the level of each triad, leading to *detailed conservation properties*.

Let us consider a triad (k, p, q) that satisfies constraint (3.97). Using the helical-mode decomposition and rewriting relation (3.102) for the single triad under consideration, one obtains

$$\frac{\partial \xi_s(k)}{\partial t} = (s'p - s''q)K(sk, s'p, s'q)\xi_{s'}^*(p, t)\xi_{s''}^*(q, t), \quad (3.118)$$

$$\frac{\partial \xi_{s'}(p)}{\partial t} = (s''q - sk)K(sk, s'p, s'q)\xi_s^*(k, t)\xi_{s''}^*(q, t), \quad (3.119)$$

$$\frac{\partial \xi_{s''}(q)}{\partial t} = (sk - s'p)K(sk, s'p, s'q)\xi_s^*(k, t)\xi_{s'}^*(p, t). \quad (3.120)$$

It is obvious from these equations that

$$\dot{\xi}_s(k)\xi_s^*(k) + \dot{\xi}_{s'}(p)\xi_{s'}^*(p) + \dot{\xi}_{s''}(q)\xi_{s''}^*(q) = 0,$$

3.5 Advanced Analysis of Energy Transfers in Fourier Space

because $(s'p - s''q) + (s''q - sk) + (sk - s'p) = 0$, all other terms being perfectly symmetric in terms of $(s\mathbf{k}, s'\mathbf{p}, s''\mathbf{q})$, as the factor K is. Here,

$$e(k) = \frac{1}{2}\xi_s(\mathbf{k})\xi_s^*(\mathbf{k}) = \frac{1}{2}\hat{u}(\mathbf{k}) \cdot \hat{u}^*(\mathbf{k}) \tag{3.121}$$

denotes the spectral density of energy. In other words, examination of the very simplified form for $M_{ss's''}$ given in Eqs. (3.118)–(3.120) immediately shows that

$$M_{ss's''}(\mathbf{k},\mathbf{p}) + M_{s's''s}(\mathbf{p},\mathbf{q}) + M_{s''ss'}(\mathbf{q},\mathbf{k}) = 0, \tag{3.122}$$

using the same nomenclature as for the nonlinear terms in Subsection 3.5.1, so that the detailed conservation of energy is found in an optimal way.

Detailed conservation of helicity is an even more striking result because of the optimal modal decomposition, with

$$sk\dot{\xi}_s(\mathbf{k})\xi_s^*(\mathbf{k}) + s'p\dot{\xi}_{s'}(\mathbf{p})\xi_{s'}^*(\mathbf{p}) + s''q\dot{\xi}_{s''}(\mathbf{q})\xi_{s''}^*(\mathbf{q}) = 0, \tag{3.123}$$

resulting from $sk(s'p - s''q) + s'p(s''q - sk) + s''q(sk - s'p) = 0$, which implies that

$$skM_{ss's''}(\mathbf{k},\mathbf{p}) + s'pM_{s's''s}(\mathbf{p},\mathbf{q}) + s''qM_{s''ss'}(\mathbf{q},\mathbf{k}) = 0. \tag{3.124}$$

The spectral density of helicity is given by

$$h(k) \sim \sum_{s=\pm 1} \iota sk\xi_s(\mathbf{k})\xi_s^*(\mathbf{k}) = \left(\frac{1}{2}\right)\hat{u}^*(\mathbf{k}) \cdot \hat{\omega}(\mathbf{k}). \tag{3.125}$$

The related interesting result is

$$\frac{M_{ss's''}(\mathbf{k},\mathbf{p})}{s''q - s'p} = \frac{M_{s's''s}(\mathbf{p},\mathbf{q})}{sk - s''q} = \frac{M_{s''ss'}(\mathbf{q},\mathbf{k})}{s'p - sk} = -\iota K(s\mathbf{k}, s'\mathbf{p}, s''\mathbf{q}). \tag{3.126}$$

Equations (3.122) and (3.124) show that the nonlinear interactions among modes within a given triad conserve both kinetic energy and helicity. A look at Eqs. (3.118)–(3.120) also shows that two modes with the same wavenumber and the same polarity do not force the third one in the triad. In the previous example, one has $\partial \xi_{s''}(\mathbf{q})/\partial t = 0$ if $k = p$ and $s = s'$.

3.5.4 Advanced Analysis of Triadic Transfers and Waleffe's Instability Assumption

The analysis of triadic interactions can be further refined, distinguishing among the three following types of interactions:

- **local interactions**, which correspond to triads $(\mathbf{k}, \mathbf{p}, \mathbf{q})$ such that $k \simeq p \simeq q$. A usual definition is that $\max(k, p, q)/\min(k, p, q) \leq 2 - 3$.
- **distant interactions**, which are such that $\max(k, p, q)/\min(k, p, q) \geq 7\text{--}10$.
- **nonlocal interactions**, which correspond to all other cases.

It is important to stress that the detailed conservation of energy can be shown in terms of primitive variables, $\hat{u}$, with some consequences on the triadic transfers, subsequently discussed first, below, but new properties of these transfers are displayed by use of helical modes and taking advantage of the formal analogy of Eqs. (3.118)–(3.120) with the Euler problem for the angular momentum of a solid body (see Subsection 3.5.5).

Brasseur and co-workers addressed the question of the relative intensity of the transfers associated with each type of triadic interaction (Brasseur and Wei, 1994) at large wavenumbers contained in the inertial range of the energy spectrum. Considering the triad (k, p, q), the nonlinear term that appears in the evolution equation of $e(k) = \hat{u}^*(k) \cdot \hat{u}(k)$ associated with this single triad is

$$\dot{e}(k)_{\text{NL}} = -\imath \left\{ [\hat{u}(k) \cdot \hat{u}(p)] [k \cdot \hat{u}(q)] + [\hat{u}(k) \cdot \hat{u}(q)] [k \cdot \hat{u}(p)] \right\} + \text{c.c.} \quad (3.127)$$

in terms of primitive variables, instead of

$$\dot{e}(k)_{\text{NL}} = \imath M_{ss's''}(p, q) \xi_s^*(k) \xi_{s'}^*(p) \xi_{s''}^*(q),$$

using helical modes. In any case, detailed energy conservation implies that

$$\dot{e}(k)_{\text{NL}} + \dot{e}(p)_{\text{NL}} + \dot{e}(q)_{\text{NL}} = 0. \quad (3.128)$$

Numerical simulations have shown that distant interactions play a very important role in the dynamics of small scales. This observation can be explained as follows.

First, let us consider a distant triad that couples a low wavenumber k to two high wavenumbers p and q, and let us introduce the small parameter $\delta = k/p \simeq k/q$. We obtain from Eq. (3.127) the following scaling laws:

$$\dot{e}(k)_{\text{NL}} = O(\delta), \quad (3.129)$$

$$\dot{e}(p)_{\text{NL}} = -\dot{e}(q)_{\text{NL}} = -\imath \left\{ [\hat{u}(p) \cdot \hat{u}(q)] [p \cdot \hat{u}(k)] \right\} + \text{c.c.} + O(\delta), \quad (3.130)$$

which show that energy transfers take place between the two high-wavenumber modes, leading to the existence of a local energy transfer associated with a distant interaction. In the asymptotic limit $\delta \longrightarrow 0$, one can see that no energy is exchanged between large and small scales: The low-wavenumber mode acts only as a catalyst. But it is important to note that small- and larger-wavenumber modes are coupled through the distant interactions, even if no energy is exchanged between them, because distant interactions can propagate low-wavenumber (i.e., large-scale) anisotropy at small scales.

The magnitude of the rate of energy exchange of a high-wavenumber mode k ($k \gg 1$) that is due to distant interactions can be evaluated as

$$\dot{e}(k)_{\text{NL}} \propto e(k) k \sqrt{e(p)} \quad \text{(distant interactions)}, \quad (3.131)$$

where p is the energy-containing mode (i.e., the low-wavenumber mode of the distant triad), whereas, for the local interactions, one obtains

$$\dot{e}(k)_{\text{NL}} \propto e(k) k \sqrt{e(k)} \quad \text{(local interactions)}. \quad (3.132)$$

These evaluations show that the distant interactions induce a much larger energy transfer than the local ones, because the energy of the low-wavenumber mode

3.5 Advanced Analysis of Energy Transfers in Fourier Space

in the distant triad, $e(p)$, is much higher than $e(k)$. As a consequence, the effect of distant interactions is important at large wavenumbers. The relative importance of transfers associated with distant triads with respect to those associated to local triads is an increasing function of the ratio of the energy contained in the small- and high-wavenumber modes.

Direct numerical simulations have also shown that

- the energy transfer from large to small scales (i.e., the kinetic-energy cascade) is local across the spectrum,
- for energetic scales (i.e., wavenumbers located near the peak of the spectrum), the kinetic-energy transfer toward the smaller scales is mainly due to local interactions, and
- for small scales (i.e., high wavenumber located within the inertial range), the energy-transfer is governed by distant interactions involving one mode in the energy-containing range.

A finer analysis of numerical databases also reveals that all distant interactions do not contribute in same way to the energy transfer toward smaller scales, i.e., that distant triads do not redistribute kinetic energy in the same way among the three interacting modes. To explain this and to provide a detailed analysis of all possible transfers within a single distant triad, Waleffe (1992, 1993) developed a theory based on the *instability assumption*.

The first step in Waleffe's analysis is to consider the stability of system (3.118)–(3.120) around its steady solutions. There are three steady solutions. Considering the steady solution given by (the two others can be deduced by simple permutations)

$$\frac{\partial \xi_s(k)}{\partial t} = A, \quad \frac{\partial \xi_{s'}(p)}{\partial t} = \frac{\partial \xi_{s''}(q)}{\partial t} = 0, \tag{3.133}$$

one obtains

$$\frac{\partial^2 \xi_{s'}(p)}{\partial t^2} = (s''q - sk)(sk - s'p)|K(sk, s'p, s'q)|^2 |A|^2 \xi_{s'}^*(p), \tag{3.134}$$

where the modulus of the complex parameters is defined as follows:

$$|K(sk, s'p, s'q)|^2 \equiv K(sk, s'p, s'q)K^*(sk, s'p, s'q), \quad |A|^2 \equiv AA^*. \tag{3.135}$$

The disturbance in $\xi_{s'}^*(p)$ will grow exponentially if $(s''q - sk)(sk - s'p) > 0$. This happens if sk is intermediate between $s''q$ and $s'p$, leading to a stability criterion based on the intermediate mode. Now combining energy-detailed conservation relation (3.122) and the trivial geometric relation

$$(s''q - sk) + (sk - s'p) + (s'p - s''q) = 0, \tag{3.136}$$

one can see that the unstable mode is the mode whose coefficient $M_{ss's''}$ has a sign opposite to the two others, with the highest absolute value. From this observation it follows that

- the mode associated with the largest wavenumber can never be unstable,

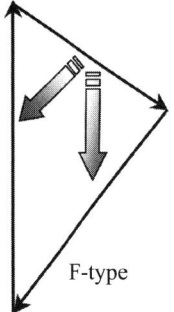
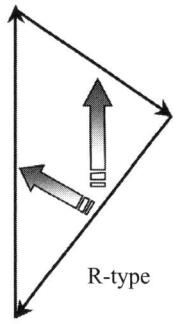

Figure 3.10. Schematic view of kinetic-energy transfers according to Waleffe's instability assumption among modes within a single triad. The two types of interactions are represented. Thick arrows denote the energy transfers.

F-type   R-type

- the mode associated with the smallest wavenumber is unstable if the two larger-wavenumber modes have opposite polarities (i.e., helicities of opposite sign),
- the mode associated with the intermediate wavenumber is unstable if it has the same polarity as the mode asssociated with the largest wavenumber.

The *instability assumption* advocated by Waleffe is that *the mode that releases energy toward the two others within a single triad is the instable mode.*

The combination of the two possible polarities for the three wave vectors leads to the existence of eight possible triadic interactions, which can be grouped in two classes according to the resulting kinetic-energy transfer (see Fig. 3.10):

- The *forward interactions* (F-type in the parlance of Waleffe), for which the two smallest wavenumbers have opposite polarities. In this case, the preceding analysis shows that the energy is released by the smallest wavenumber (i.e., the largest scale), leading to a forward energy cascade.
- The *reverse interactions* (R-type), for which the smallest wave vectors have the same polarity. In this case, the intermediate mode can be the unstable one, leading to a transfer of energy toward both a larger (backward energy cascade) and a smaller scale (forward energy cascade).

Let us now focus on the distant interactions, which are of particular importance in the large-wavenumber mode dynamics. Let (k, p, q) be a distant triad with $|q - p| < k \ll p \simeq q$. The triad-related geometrical factor that appears in the definition of $M_{ss's''}$ scales as $(sk + s'p + s''q)$. As a consequence, the energy transfer scales as $\pm k \pm (p - q)$ for distant F-type interactions and as $\pm k \pm (p + q)$ for distant R-type interactions. Therefore, on average, distant triads mostly induce energy transfers of the R-type, yielding a local energy transfer between the largest wave vectors.

Among the four possible R-type triadic interactions, two contribute in the mean to the backward energy cascade from the large-wavenumber modes toward the small-wavenumber modes. The direction of the cascade associated with the two others depends on the value of the ratio between the smallest and the intermediate wavenumbers.

On average, within the inertial range, the net effect of the R-type interactions is a backward energy transfer toward the small-wavenumber modes, the direct

3.5 Advanced Analysis of Energy Transfers in Fourier Space

energy cascade being due to F-type interactions. Consequently, the net energy transfer within the inertial range is a direct energy cascade that is due to the large local energy transfer associated with distant interactions.

Quantitative evaluations of the different energy fluxes were performed by Waleffe (1992, 1993), who used additional statistical assumptions about self-similarity, or a statistical closure as EDQNM [or test field model (TFM), which is almost identical to EDQNM for 3D HIT]. The EDQNM model allows us to reach much higher Reynolds numbers than DNS, and it may be more accurate in terms of spectral discretization, avoiding errors of cancellation. *In addition, it can be developed in terms of helical modes too, separating the eight different kinds of triads in exact agreement with detailed conservation of energy and helicity.*

3.5.5 Further Discussions About the Instability Assumption

We now discuss some analogies that exist between the instability principle and other problems.

As stated by Waleffe in his seminal paper (1992), the instability principle presented in the previous section is formally similar to the problem of the instability of a rigid body rotating around one of its principal axes of inertia. Let first note that system (3.118)–(3.120) can be recast in the following compact form:

$$\frac{d\boldsymbol{\xi}}{dt} = K(s\boldsymbol{k}, s'\boldsymbol{p}, s''\boldsymbol{q})(\mathbf{D}\boldsymbol{\xi}^*) \times \boldsymbol{\xi}^*, \quad (3.137)$$

where $\boldsymbol{\xi} = [\xi_s(\boldsymbol{k}), \xi_{s'}(\boldsymbol{p}), \xi_{s''}(\boldsymbol{q})]^T$ and

$$\mathbf{D} = \begin{bmatrix} sk & 0 & 0 \\ 0 & s'p & 0 \\ 0 & 0 & s''q \end{bmatrix}. \quad (3.138)$$

Detailed conservation laws of energy and helicity within the triad yield

$$\frac{d}{dt}(\boldsymbol{\xi} \cdot \boldsymbol{\xi}^*) = \frac{d}{dt}(\boldsymbol{\xi} \cdot \mathbf{D}\boldsymbol{\xi}^*) = 0. \quad (3.139)$$

Let us consider a solid body in rotation, where $\boldsymbol{L}$ and $\boldsymbol{\omega}$ are its angular-momentum and angular-velocity vectors, respectively. The Euler equations that describe this motion are

$$\frac{d\boldsymbol{L}}{dt} = \boldsymbol{L} \times \boldsymbol{\omega}. \quad (3.140)$$

Now, introducing the tensor of inertia of the solid, denoted as $\mathbf{I}$, one can write the angular momentum as the product of $\mathbf{I}$ with the rotation vector $\mathbf{I}\boldsymbol{\omega}$. Problem (3.140) can be rewritten in the principal axes of the inertia matrix as follows:

$$I_1 \dot{\omega}_1 = (I_2 - I_3)\omega_2 \omega_3, \quad (3.141)$$

$$I_2 \dot{\omega}_2 = (I_3 - I_1)\omega_3 \omega_1, \quad (3.142)$$

$$I_3 \dot{\omega}_3 = (I_1 - I_2)\omega_1 \omega_2. \quad (3.143)$$

Therefore the first conservation law is for the rotational kinetic energy $I_1\omega_1^2 + I_2\omega_2^2 + I_3\omega_3^2$ [equivalent to the triadic kinetic-energy conservation law – (3.122)], and the second one for the norm of the angular momentum $(I_1\omega_1)^2 + (I_2\omega_2)^2 + (I_3\omega_3)^2$ [equivalent to triadic helicity conservation law – (3.124)]. Systems (3.137) and (3.140) are mathematically similar, with **D** and $\boldsymbol{\xi}$ playing the role of **I** and **L**, respectively. It is known that there exist three steady-state solutions for the problem of the rotating solid, which correspond to rotation around any one of the principal axes of inertia. Rotation around the axis of middle inertia is unstable, whereas the two other cases are stable solutions. This implies that the smallest wavenumber is unstable if the two largest wavenumbers have helicities of opposite sign and that the medium wavenumber is unstable otherwise. Therefore it is seen that the analogy enables us to recover the results of the previous section. But it is worthwhile remarking that components of **D** can exhibit negative values.

The second point discussed by Waleffe deals with the link between the F-type interactions and the elliptical instability. Let us first recall that elliptical instability is the 3D instability of flows with locally elliptical streamlines. The instable modes are resonant inertial waves associated with the uniform background rotation (see Section 4.5). These waves are helical modes of opposite polarities and eigenfrequencies, say f^+ and f^-. A detailed analysis (see Waleffe, 1992, for technical details) shows that the elliptical instability corresponds to an F-interaction: The two modes with eigenfrequencies f^+ and f^- have opposite polarities and are coupled with the mean flow, which is associated with a zero frequency. It can also be shown that there exists a low-wavenumber cutoff: The wavenumber of the perturbation must remain higher than the effective wavenumber of the elliptic background flow for the instability to develop. Therefore elliptical instability originates in an interaction that leads to the instability of the smallest-wavenumber mode in a triad through interactions with two larger-wavenumber modes of opposite polarities.

3.5.6 Principle of Quasi-Normal Closures

The previous equations for $\hat{u}_i$, $\hat{R}_{ij}$, and S_{inj} illustrate the infinite hierarchy of open equations, which is usually formally written as

$$\frac{\partial}{\partial t} u = uu,$$

$$\frac{\partial}{\partial t} \langle uu \rangle = \langle uuu \rangle,$$

$$\frac{\partial}{\partial t} \langle uuu \rangle = \langle uuuu \rangle,$$

$$\ldots = \ldots$$

A common feature of triadic closures, from the EDQNM model (Orszag, 1970) to the most sophisticated Kraichnan's theories, is a quasi-normal (QN) relationship. Any technique that aims at expressing high-order moments as products of low-order ones is a good candidate for closing the previously mentioned infinite hierarchy of open equations. Instead of moments, cumulants directly express the difference of

3.5 Advanced Analysis of Energy Transfers in Fourier Space

moments with respect to their factorized expression in terms of lower-order ones, so that classical closures rely on small estimates of cumulants. Historically, the assumption of a vanishing fourth-order cumulant for the turbulent velocity fluctuations, i.e.,

$$\langle u^a u^b u^c u^d \rangle - \langle u^a u^b \rangle \langle u^c u^d \rangle - \langle u^a u^c \rangle \langle u^b u^d \rangle$$
$$- \langle u^a u^d \rangle \langle u^b u^c \rangle = 0, \quad (3.144)$$

was first proposed by Milionschikov (1941) and then by Proudman and Reid (1954). In the preceding equation, different superscripts are used to distinguish different velocity modes, possibly in physical space with four different positions and for different components, and finally in Fourier space for mathematical convenience. The assumption of a vanishing fourth-order cumulant is usually referred to as the *quasinormal* (QN) *approximation*, but not as a normal (or Gaussian) approximation because nothing is said about third-order cumulants (or third-order moments because there is no contribution from $\langle u \rangle \langle uu \rangle$). Of course, an estimate for third-order moments is sought, so that a pure Gaussian relationship, which removes them, is meaningless (except in some rapid distortion limit, which will be addressed in a subsequent chapter). In addition, a QN assumption can be supported mathematically and physically in the weak-turbulence theory of wave turbulence, as illustrated by Benney and Newell (1969) and Zakharov, Lvov, and Falkowitch (1991) (this approach will be revisited in Chapter 4).

Starting from the exact definition for T_{ijn}, the QN assumption yields

$$\delta(\boldsymbol{k}+\boldsymbol{p}+\boldsymbol{q})T_{ijn} = P_{irs}(\boldsymbol{k}) \int_{-\boldsymbol{k}+\boldsymbol{r}+\boldsymbol{s}=0} d^3p \times [\langle \hat{u}_r(\boldsymbol{r})\hat{u}_s(\boldsymbol{s})\rangle\langle \hat{u}_j(\boldsymbol{p})\hat{u}_n(\boldsymbol{q})\rangle$$
$$+ \langle \hat{u}_r(\boldsymbol{r})\hat{u}_j(\boldsymbol{p})\rangle\langle \hat{u}_s(\boldsymbol{s})\hat{u}_n(\boldsymbol{q})\rangle$$
$$+ \langle \hat{u}_r(\boldsymbol{r})\hat{u}_n(\boldsymbol{q})\rangle\langle \hat{u}_j(\boldsymbol{p})\hat{u}_s(\boldsymbol{s})\rangle]. \quad (3.145)$$

Using $\langle \hat{u}_r(\boldsymbol{r})\hat{u}_s(\boldsymbol{s})\rangle = \hat{R}_{rs}(\boldsymbol{s})\delta(\boldsymbol{r}+\boldsymbol{s})$, one finds the contribution from the first term to be zero because $\hat{R}_{rs}(\boldsymbol{k}=0) = 0$, so that

$$T_{ijn}^{\rm QN}(\boldsymbol{k},\boldsymbol{p}) = P_{ir}k_s \left[\hat{R}_{rj}(\boldsymbol{p})\hat{R}_{sn}(\boldsymbol{q}) + \hat{R}_{rn}(\boldsymbol{q})\hat{R}_{sj}(\boldsymbol{p}) \right] \quad (3.146)$$

or, equivalently,

$$T_{ijn}^{\rm QN}(\boldsymbol{k},\boldsymbol{p}) = P_{irs}\hat{R}_{rj}(\boldsymbol{p})\hat{R}_{sn}(\boldsymbol{q}). \quad (3.147)$$

Finally, one obtains the following *QN closure*:

$$\left[\frac{\partial}{\partial t} + \nu(k^2+p^2+q^2)\right]S_{ijn}(\boldsymbol{k},\boldsymbol{p}) = T_{ijn}^{\rm QN}(\boldsymbol{k},\boldsymbol{p}) + T_{jni}^{\rm QN}(\boldsymbol{p},\boldsymbol{q}) + T_{nij}^{\rm QN}(\boldsymbol{q},\boldsymbol{k}). \quad (3.148)$$

Even though the QN closure was proposed a long time ago, the resolution of the corresponding Lin equation requires significant numerical resources. First, numerical solutions obtained in the late 1960s (Ogura, 1963; O'Brien and Francis, 1963) exhibited incorrect behavior for a long time evolution. A negative zone appeared at small k in the energy spectrum, because of a too strong energy transfer from the largest structures. This lack of realizability was shown to result from a too high estimate of the right-hand side of the preceding equation. To solve this problem, Orszag (1970) proposed adding an *eddy-damping* (ED) *term*, so that

$$T_{ijn}(\mathbf{k}, \mathbf{p}, t) - T_{ijn}^{QN}(\mathbf{k}, \mathbf{p}, t) = -\underbrace{\eta(k,t) S_{ijn}(\mathbf{k}, \mathbf{p}, t)}_{\text{damping term}}.$$

Similar relationships are obtained for other wave-vector pairs by permuting the wave vectors of the triad. The special form of the linear relationship between fourth-order and third-order cumulants was partly suggested by Kraichnan's DIA theory. The left-hand side represents the contribution from fourth-order *cumulants* and the right-hand side deals with third-order cumulants. The ED coefficient plays the role of an extra dissipation, reinforcing the dissipative laminar effect, which is not sufficient to ensure realizability in the primitive QN closure. Gathering the dissipative terms into a single one,

$$\mu_{kpq} = \theta_{kpq}^{-1} = \nu(k^2 + p^2 + q^2) + \eta(k,t) + \eta(p,t) + \eta(q,t), \quad (3.149)$$

the EDQN counterpart of Eq. (3.148) is easily obtained from it, replacing $\nu(k^2 + p^2 + q^2)$ with θ_{kpq}^{-1}. The solution of the latter equation is found as

$$S_{ijn}(\mathbf{k}, \mathbf{p}, t) = \exp\left[-\mu_{kpq}(t - t_0)\right] S_{ijn}(\mathbf{k}, \mathbf{p}, t_0)$$

$$+ \int_{t_0}^{t} \exp\left[-\int_{t'}^{t} \mu_{kpq}(t'') dt''\right] \left[T_{ijn}^{QN}(\mathbf{k}, \mathbf{p}, t') + \cdots\right] dt'. \quad (3.150)$$

Conventionaly, the last procedure, called *Markovianization*, yields neglect of the intrinsic history of T_{ijn}^{QN}, or, equivalently, that of $\hat{R}_{ij}$, in the time integral. In other words $\hat{\mathbf{R}}$ and T^{QN} are considered as slowly varying quantities, so that one can take $t' = t$ in them, whereas the exponential term is considered as rapidly varying. Ignoring the initial data for triple correlations, consistently with large $t - t_0$, the simplest *EDQNM closure* (in the absence of complex additional linear terms) is

$$S_{ijn}(\mathbf{k}, \mathbf{p}, t) = \theta_{kpq} \left[T_{ijn}^{QN}(\mathbf{k}, \mathbf{p}, t) + T_{jni}^{QN}(\mathbf{p}, \mathbf{q}, t) + T_{nij}^{QN}(\mathbf{q}, \mathbf{k}, t)\right]. \quad (3.151)$$

The latter equation illustrates an instantaneous relationship between third- and second-order correlations, but nonlocality in spectral space and triadic structure is preserved.

The tensor τ_{ij} defined in Eq. (3.114) is then expressed as

$$\tau_{ij} = \int \theta_{kpq} P_{jnm}(\mathbf{k}) \left[P_{irs}(\mathbf{k}) \hat{R}_{rn}(\mathbf{p}) \hat{R}_{sm}(\mathbf{q}) + P_{nrs}(\mathbf{p}) \hat{R}_{rm}(\mathbf{q}) \hat{R}_{si}(\mathbf{k})\right.$$

$$\left. + P_{mrs}(\mathbf{q}) \hat{R}_{rm}(\mathbf{k}) \hat{R}_{sn}(\mathbf{p})\right] d^3 p, \quad (3.152)$$

in which the characteristic time θ_{kpq} is given by relation (3.149). Permuting $\mathbf{p}$ and $\mathbf{q}$ in the last term, the simplified form

$$\tau_{ij} = P_{jnm}(\mathbf{k}) \int \theta_{kpq} \hat{R}_{sm}(\mathbf{q}) \left[P_{irs}(\mathbf{k}) \hat{R}_{rn}(\mathbf{p}) + 2 P_{nrs}(\mathbf{p}) \hat{R}_{ri}(\mathbf{k})\right] d^3 p \quad (3.153)$$

is finally obtained.

3.5 Advanced Analysis of Energy Transfers in Fourier Space

3.5.7 EDQNM for Isotropic Turbulence. Final Equations and Results

3D isotropy yields dramatic simplifications, as

$$\hat{R}_{rj}(\boldsymbol{p}) = e(p)P_{rj}(\boldsymbol{p})$$

in Eq. (3.147), and

$$T^e(k) = \tau_{ii}(k)$$

from (3.114). The transfer term $T^{(e)}$ is therefore found as

$$T^{(e)}(k,t) = \iiint 2kp\theta_{kpq}e(q,t)\left[A(k,p,q)e(p,t) - B(k,p,q)e(k,t)\right]d^3\boldsymbol{p} \quad (3.154)$$

with

$$P_{inm}(\boldsymbol{k})P_{sm}(\boldsymbol{q})P_{irs}(\boldsymbol{k})P_{rn}(\boldsymbol{p}) = k^2 A(k,p,q)$$

and

$$2P_{inm}(\boldsymbol{k})P_{sm}(\boldsymbol{q})P_{nrs}(\boldsymbol{p})P_{ri}(\boldsymbol{k}) = 2kpB(k,p,q).$$

Because $kpB(k,p,q) + kqB(k,q,p) = k^2A(k,p,q)$, $A(k,p,q)$ can be replaced with $B(k,p,q)$ in the preceding equation. In addition, it is simpler to express this unique coefficient in terms of the cosines of the internal angles of the triangle of sides k, p, q:

$$x = \cos\alpha = \frac{p^2 + q^2 - k^2}{2pq}, \quad y = \cos\beta = \frac{q^2 + k^2 - p^2}{2qk},$$

$$z = \cos\gamma = \frac{k^2 + p^2 - q^2}{2kp}.$$

Another relevant geometric term is C_{kpq}, which was already found in Eq. (3.110).

Because $C^2B(k,p,q) = kp - q^2z$, $B(k,p,q) = \sin\alpha\sin\beta - z\sin^2\gamma$, $= xy + z - z(1-z^2)$, and finally

$$B(k,p,q) = xy + z^3,$$

the simplified expression follows:

$$T^{(e)} = \iiint 2kp\theta_{kpq}(xy+z^3)e(q,t)\left[e(p,t) - e(k,t)\right]d^3\boldsymbol{p}. \quad (3.155)$$

It is now possible to use the integration variables p, q, and λ as in Eq. (3.111). Because the integrand depends on only k, p, q, and not on λ, performing the integration with respect to λ simplifies as a multiplication by 2π, so that

$$T^{(e)} = \iint_{\Delta_k} 4\pi p^3 q^2 \theta_{kpq}(xy+z^3)e(q,t)\left[e(p,t) - e(k,t)\right]\frac{dpdq}{pq}. \quad (3.156)$$

A last equation is found reintroducing $E(k) = 4\pi k^2 e(k)$ and $T(k) = 4\pi k^2 T^{(e)}(k)$ as

$$T(k,t) = \iint_{\Delta_k} \theta_{kpq}(xy+z^3)E(q,t)\left[E(p,t)pk^2 - E(k,t)p^3\right]\frac{dpdq}{pq}. \quad (3.157)$$

This is the conventional form of the isotropic EDQNM model. Instead of deriving this equation from (3.95), it is also possible to start from (3.98). The "*Byzantine use of projectors*" (Leaf Turner) is the classical way to calculate geometric coefficients, but the same result can be obtained in terms of helical modes and related amplitudes.

Isotropic turbulence allows for dramatic simplifications for all statistical theories or models, and therefore is one of the most interesting canonical flows of reference. For instance, all classical two-point triadic closure theories have the same structure, because they express $T(k)$ as a nonlocal function of $E(k)$.

Different versions of statistical theories differ from only the expression of the damping factor η in (3.149), which adds nonlinear readjustment of the response function.

As shown by Orszag (1970), the use of

$$\eta(k,t) \sim k\sqrt{kE(k,t)}$$

yields a satisfactory behavior of E when solving numerically the Lin equation, with the establishment of a Kolmogorov inertial zone. Another variant (Pouquet et al., 1975) is

$$\eta(k,t) = A\sqrt{\int_0^k p^2 E(p,t)dp}, \qquad (3.158)$$

which amounts to choosing η as the inverse of the Corrsin time scale, the constant A (André and Lesieur, 1977) being fixed by a given value of the Kolmogorov constant.

Results of the EDQNM model in pure decaying (unforced) HIT are subsequently presented.

3.5.7.1 Well-Documented Experimental Data, Moderate Reynolds Number

Comparisons with (Comte-Bellot and Corrsin, 1966) experimental data by Vignon and Cambon (1980) illustrate the relevance of the EDQNM model at moderate Reynolds numbers (see Fig. 3.11). The experimental data are very comprehensive, with access to $E(k,t)$ at different sections downstream of the grid (the downstream distance $x - x_0$ divided by the mean advection velocity U is equivalent to an elapsed time), the energy spectrum is calculated from its 1D counterpart assuming isotropy. In addition, the dissipation spectrum is derived, and finally even the transfer term $T(k,t)$ is captured, comparing measures at two close sections for estimating $\Delta E/\Delta t$.

3.5.7.2 Transfer Term at Increasing Reynolds Number

When the Reynolds number is increased, a large inertial zone is easily constructed for the energy spectrum, but, somewhat surprisingly, the zone of the zero-transfer term is much shorter, as shown in Fig. 3.12. Particularly, the flat zone of zero transfer appears only for huge values of Re_λ (typically $Re_\lambda \geq 10^4$), whereas a significant

3.5 Advanced Analysis of Energy Transfers in Fourier Space

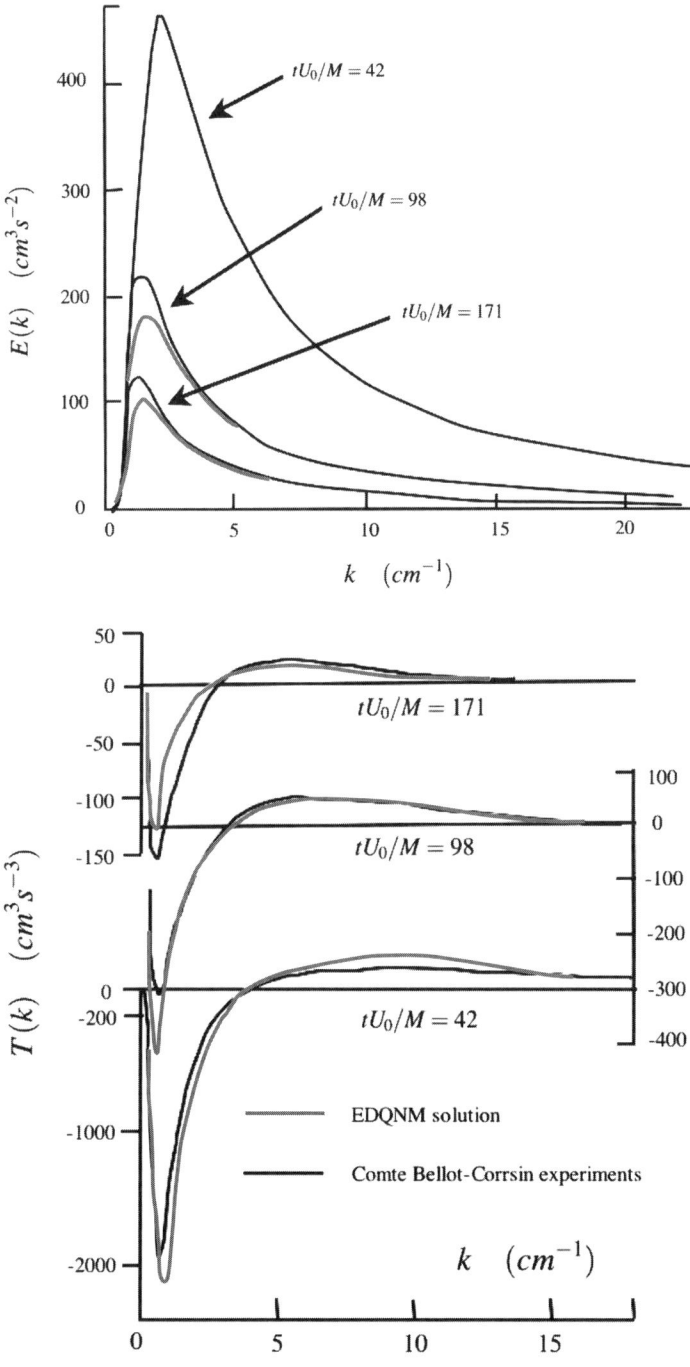

Figure 3.11. Comparisons of EDQNM and experimental data in decaying. Top: turbulence kinetic-energy spectrum $E(k)$ at three different locations/elapsed times. Bottom: spectral-energy transfer function $T(k)$ at some locations. Data taken from Vignon and Cambon (1980) and Cambon et al. (1981).

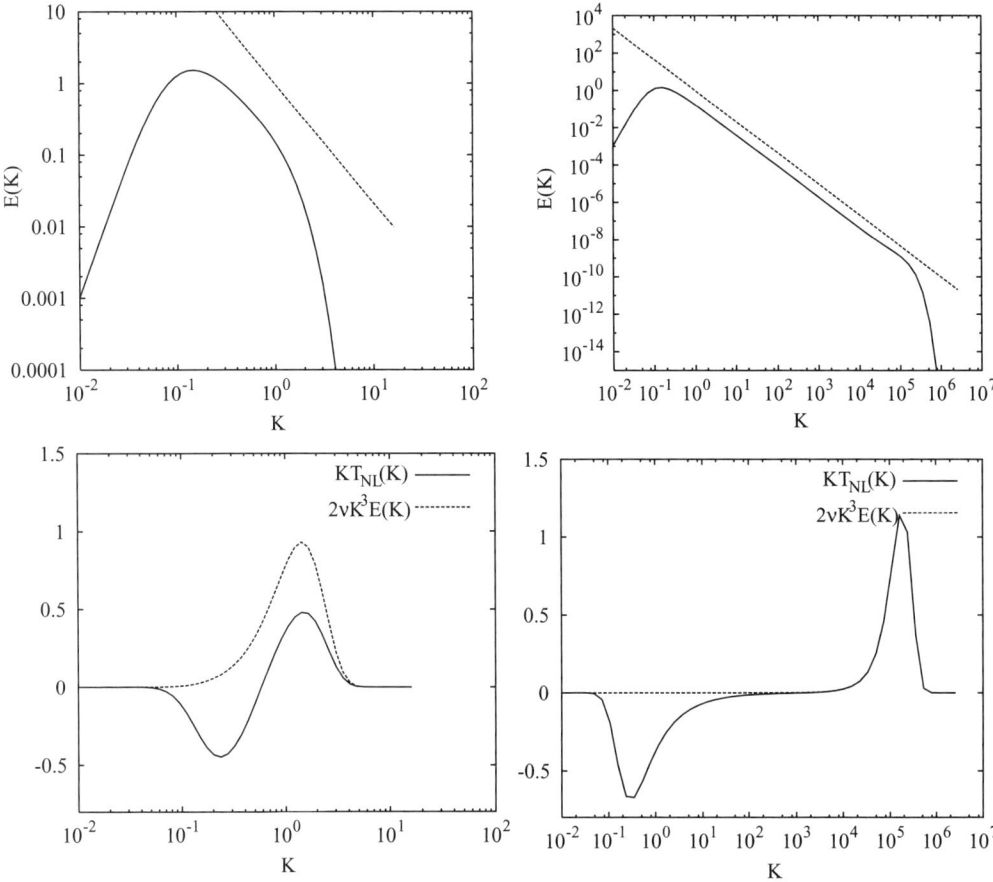

Figure 3.12. Typical spectra (top), nonlinear transfer and viscous dissipation (bottom) in isotropic turbulence at $Re_\lambda = 30$ (left) and $Re_\lambda = 10^5$ (right); x and y scales are chosen arbitrarily. The straight lines are related to the Kolmogorov $-5/3$ slope. Courtesy of W. Bos.

inertial zone appears in the energy spectrum for $10^2 < Re_\lambda < 10^3$. This result is consistent with experimental studies, in which the 4/5 Kolmogorov law for the third-order structure function was recovered only at unexpectedly high Re_λ. Therefore it is seen that the definition of the inertial range deserves further discussion. All wavenumbers located within the inertial range in the energy spectrum do not have a vanishing $T(k)$, and are therefore dynamically sensitive to production and/or dissipation. Modes that are not directly sensitive to production and viscous effects, i.e., modes that are governed by the sole triadic nonlinear transfer terms, are modes with wave numbers such that $T(k) = 0$. This dynamical definition is much more stringent than the one based on the existence of a self-similar zone in the kinetic-energy spectrum.

These observations mean that the EDQNM model can be used to obtain additional results about statistics in physical space, as second- and third-order structure functions, using an isotropic relationship, which is well documented in Mathieu and Scott (2000), because many recent experiments focused on these statistics. However,

3.5 Advanced Analysis of Energy Transfers in Fourier Space

it should be borne in mind that $E(k)$ and $T(k)$ are very informative, as they allow us to compute various statistics, and they are accurately predicted by an EDQNM model at almost any Reynolds number. In Fig. 3.12, the transfer term is multiplied by k in order to preserve the zero value of the integral when k is expressed in logarithmic scale, according to the relation

$$kT(k)d(\ln k) = T(k)dk.$$

For the sake of clarity, the enstrophy (or dissipation up to a factor 2ν) spectrum is also multiplied by k. The positive part of the transfer and the dissipation spectrum are observed to coincide only when the transfer function exhibits a significant plateau.

3.5.7.3 Toward an Infinite Reynolds Number

EDQNM calculations can be started with zero molecular viscosity, initializing the Lin equation with a narrowband energy spectrum. In this case, the inertial zone develops well and extends more and more toward larger and larger k's. It is conjectured that the inertial zone could reach an infinite wavenumber, say $k_{\max} = \infty$, in a *finite* time, yielding a finite dissipation rate at zero viscosity: This is sometimes called the *energetic catastrophe* in the turbulence community. Unfortunately, this cannot be completely proven, because, in practice, the Lin equation closed by the EDQNM model cannot be solved analytically, so that a numerical solution, with discretized k and finite $k_{\max}$ is needed. Nevertheless, very large $k_{\max}$, related to a constant logarithmic step $\Delta k/k = $ constant, can be used, without a possible counterpart in DNS. As very classical behavior, at least in DNS, spectral energy tends to accumulate near the cutoff wavenumber $k_{\max}$, so that a viscous term ought to be introduced in order to avoid an energy peak at the highest wave vector. The only advantage of the EDQNM model with respect to DNS in this case is the huge value of $k_{\max}$ related to a huge (but not infinite) Reynolds number, which can be reached with modest computational ressources.

Very recently, following a calculation of truncated inviscid Euler equations by Brachet et al. (Cichowlas et al., 2005), Bos and Bertoglio (2006a) used the conventional EDQNM model to study the accumulation of spectral energy at a given (very high) $k_{\max}$ with zero viscosity. As a nice result, both a thermalized[¶] tail following a k^2 law and a large inertial range with $k^{-5/3}$ behavior arise, separated by a sink, as shown in Fig. 3.13.

This sink induces a kind of conventional dissipative range – but at zero laminar viscosity – probably mediated by the nonlocal eddy viscosity (Kraichnan, 1971, 1976; Lesieur and Schertzer, 1978), and is even clearer in the EDQNM model than in inviscid truncated DNS. Here, the smallest scales act as the molecular motion in real viscous flows, giving a nice illustration of the turbulent-eddy-viscosity concept.

[¶] *Thermalized* is used here by analogy with the random molecular motion from which the macroscopic quantities such as temperature and pressure originate.

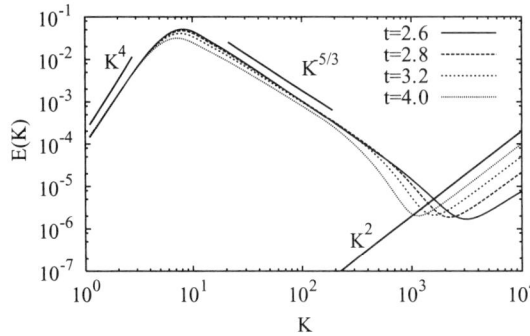

Figure 3.13. Time evolution of the kinetic-energy spectrum in the purely inviscid case using the classical EDQNM model. (Courtesy of W. Bos and J.P. Bertoglio.)

3.5.7.4 Very Recent Improvements

A recent improvement, which renders the EDQNM model closer to a self-consistent theory, consists of evaluating the ED $\mu(k, t)$ using an additional dynamical equation for a velocity-displacement cross correlation (Bos and Bertoglio, 2006b). As shown in Fig. 3.14, a realistic value of the Kolmogorov constant $K_0 \sim 1.73$ is derived, without need to specify it *a priori* in the model for μ, as in Eq. (3.158) by means of A.

It is also interesting to calculate by the EDQNM procedure, not only the contribution of triple correlations to the transfer term (a typical cubic moment at two points), which also generates the third-order structure function, but more complex cubic statistics in three points, which are very difficult to obtain from experiments or even from DNS/LES because they are very noisy quantities. For instance, triple-vorticity (not only velocity) correlations at three points (which are related to their detailed distribution in terms of triads) can be calculated in a systematic way, only from the given energy spectrum. Applications to the statistics of vorticity, with an answer from statistical theory to the problem of cyclonic–anticyclonic asymmetry in rotating turbulence, is presented in Chapter 4.

3.5.7.5 On Instantaneous Energy Transfers

Most of results previously presented dealing with the kinetic-energy spectrum and the energy transfers [e.g., $E(k)$ and $T(k)$ profile] are related to ensemble-averaged data and therefore should be interpreted as time-averaged results (providing that the ensemble average can be seen as a time average thanks to the ergodicity theorem) in the forced HIT case.

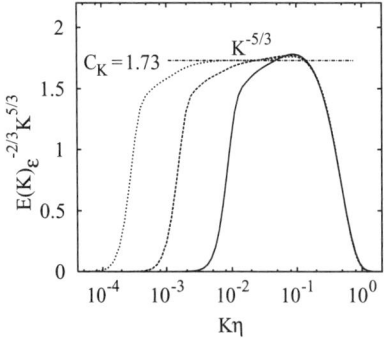

Figure 3.14. Compensated spectrum $E(k)\varepsilon^{-2/3}k^{5/3}$ in isotropic turbulence computed using the EDQNM model with self-consistent ED. The plateau corresponds to the value of the Kolmogorov constant K_0 (denoted as C_K in the figure). (Courtesy of W. Bos and J.P. Bertoglio.)

3.5 Advanced Analysis of Energy Transfers in Fourier Space

DNSs have provided information dealing with the main features of the nonaveraged, instantaneous energy transfers (Kida and Ohkitani, 1992a, 1992b) in forced isotropic turbulence. It is observed that both $E(k,t)$ and $T(k,t)$ fluctuate around their mean values and that the energy transfer function takes both positive and negative values at the same wavenumber, depending on time. As a consequence, the kinetic-energy cascade process is to be understood as an ensemble-averaged concept, which can be difficult to identify in instantaneous fields.

Kida and Okhitani observed that the standard deviation of the energy transfer function, $\sqrt{\overline{T(k,t)^2}}$, scales as k^{-1}. By tracking "blobs" of kinetic energy in the (k,t) plane, they found that the time for energy to be transferred from wavenumber k_0 to wavenumber $k = \alpha k_0$ is equal to

$$T_{k_0 \to k} = \left(\frac{\alpha^{2/3}}{\alpha^{2/3} - 1} \right) \left[(\bar{\varepsilon} k_0^2)^{-1/3} - (\bar{\varepsilon} k^2)^{-1/3} \right], \quad (3.159)$$

where $\bar{\varepsilon}$ is related to the ensemble-averaged value of the dissipation. The value $\alpha \simeq 1.4$ leads to the best fit of the numerical data, indicating that the net energy transfer is mostly local.

It is worth noting that expression (3.159) has been obtained with the Kolmogorov-type expression for the characteristic time τ_k for the energy to be transferred across the wavenumber k:

$$\tau_k = \left(\bar{\varepsilon} k^2 \right)^{-1/3}. \quad (3.160)$$

3.5.7.6 Nonlinear Cascade Time Scale, Equilibrium, and Dissipation Asymptotics

The possible existence of a universal value of the normalized dissipation rate C_ε in high-Reynolds-number turbulent flows has been addressed by several authors, and is sometimes referred to as the *zeroth law of turbulence*. This nondimensional coefficient is defined as

$$C_\varepsilon = \frac{\varepsilon L}{u'^3}, \quad (3.161)$$

where L and $u' = \sqrt{\frac{2}{3}\mathcal{K}}$ are the integral length scales (see Subsection 3.4.1) and a turbulent velocity scale, respectively. It appears in commonly used scaling laws related to Kolmogorov's theory, e.g.,

$$Re_\lambda = \sqrt{\frac{15}{C_\varepsilon} Re_L}. \quad (3.162)$$

Both experimental data and numerical simulations exhibit a significant scatter in the values of C_ε. The sensitivity on the nature of the flow (freely decaying turbulence or forced turbulence) and on the Reynolds number is observed to be large. A rationale for these discrepancies, based on both EDQNM simulations and an analytical analysis based on a simplified model spectrum, has been proposed by Bos, Shao, and Bertoglio (2007).

The first important conclusion is that the asymptotic value of C_ε explicitly depends on the existence of a turbulence-production mechanism at large scales. The key point is that one must distinguish between several characteristic quantities to get an accurate description of kinetic energy dynamics in isotropic turbulence:

- The *production rate*, i.e., the rate at which the turbulent kinetic energy $\mathcal{K}$ is injected at scales of order L. This production rate is characterized by $u'^3(t)/L(t)$. The rate at which kinetic energy leaves the large scales is denoted $\varepsilon_f(t)$, with

$$\varepsilon_f(t) = C_\varepsilon^f \frac{u'^3(t)}{L(t)}, \tag{3.163}$$

where C_ε^f is the proportionality constant.

- The *cascade time*, T_c, which measures the time it takes for an amount of energy initially injected at scale L to reach the dissipative Kolmogorov scale η. Considering a simplified Kolmogorov inertial range, one obtains $T_c = T(1 - \beta^{-2/3})$, where $T = L/u'$ is the integral time scale and $\beta = L/\eta$.

- The *dissipation rate*, $\varepsilon(t)$, which characterizes the transformation of kinetic energy into heat at very small scales.

In forced turbulence with constant injection rate, a statistically stationary state can be reached, in which the production rate is equal to both the cascade transfer rate and the dissipation rate, i.e., $\varepsilon_f(t) = \varepsilon(t)$. The associated value nondimensional dissipation parameter is denoted by $C_\varepsilon = C_\varepsilon^{\text{forced}}$.

In freely decaying turbulence, both u' and L vary in time, yielding a time-dependent production and cascade rate. A packet of kinetic energy injected at time t will be dissipated once it has reached the dissipative scales, i.e., at time $t + T_c$. Therefore, the equilibrium equality between $\varepsilon_f(t)$ and $\varepsilon(t)$ found in the forced turbulence case no longer holds, and one must write $\varepsilon_f(t) = \varepsilon(t + T_c) \neq \varepsilon(t)$, or equivalently

$$\varepsilon(t + T_c) = C_\varepsilon^{\text{forced}} \frac{u'^3(t)}{L(t)}. \tag{3.164}$$

Introducing the time-decay exponent n such that $\mathcal{K}(t) \propto t^{-n}$ and $\varepsilon(t) \propto nt^{-n-1}$, one has $L(t) \propto t^{1-n/2}$ and $T \propto t$, yielding

$$\varepsilon(t + T_c) = C_\varepsilon^{\text{forced}} \frac{u'^3(t + T_c)}{L(t + T_c)} \left(\frac{t}{t + T_c}\right)^{-n-1}$$

$$= C_\varepsilon^{\text{decay}} \frac{u'^3(t + T_c)}{L(t + T_c)} \tag{3.165}$$

and therefore

$$\frac{C_\varepsilon^{\text{decay}}}{C_\varepsilon^{\text{forced}}} = \left(1 + \frac{T_c}{t}\right)^{n+1} = \left[1 + A_c(1 - \beta^{-2/3})\right]^{n+1}, \tag{3.166}$$

showing that the normalized dissipation coefficient cannot cannot be the same in forced and freely decaying turbulence. Another important fact is that the decay exponent n is known to be flow dependent, because it is a function of the kinetic-energy spectrum shape at very large scales [see Eq. (3.41) and Table 3.3]. For large values of β, i.e., for large values of Re_L, a very good agreement with EDQNM results is obtained taking $A_c = 0.2$.

One can find an expression for $C_\varepsilon^{\text{forced}}$ by considering a simplified model spectrum. Using the model

$$E(k) = \begin{cases} Ak^\sigma & kL \leq 1 \\ K_0\varepsilon^{2/3}k^{-5/3} & kL \geq, k\eta \leq 1 \\ 0 & k\eta > 1 \end{cases}, \quad (3.167)$$

where A is an arbitrary positive parameter, one obtains

$$C_\varepsilon^{\text{forced}} = \frac{\pi\left[(3\sigma+5)/5\sigma - \frac{3}{5}\beta^{-5/3}\right]}{2K_0^{3/2}\left[(3\sigma+5)/(3\sigma+3) - \beta^{-2/3}\right]^{5/2}}, \quad (3.168)$$

along with

$$Re_L = \frac{\pi K_0^{3/2}\left[(3\sigma+5)/\sigma - 3\beta^{-5/3}\right]\left[3\beta^{4/3} - (3\sigma+5)/(\sigma+3)\right]}{20\sqrt{(3\sigma+5)/(3\sigma+3) - \beta^{-2/3}}}. \quad (3.169)$$

Relations (3.168) and (3.169) lead to an implicit expression of $C_\varepsilon^{\text{forced}}$ as a function of Re_L, whose asymptotic value is

$$\lim_{Re_L \to +\infty} C_\varepsilon^{\text{forced}} = \frac{\pi(3\sigma+3)^{5/2}}{10K_0^{3/2}\sigma(3\sigma+5)^{3/2}}. \quad (3.170)$$

This asymptotic expression is observed to fit EDQNM results for $Re_L \geq 10^3$. As a general conclusion, let us emphasize that no universal value for C_ε can exist.

Mazelier and Vassilicos (2008) reached similar conclusions without using a cascade time scale or a production rate. The gist of their conclusions can be summarized as follows. A self-similar pattern is one in which the small number of large scales is directly reflected in the large number of small scales. Zero crossings of turbulent velocity correlations form such a pattern. As a consequence, the averaged distance between consecutive zero crossings is strongly influenced by a nondimensional parameter C'_s, which is some sort of number of large-scale eddies within an integral scale. The parameter C_ε is then related to the preceding parameter by $C_\varepsilon = f(\log Re_\lambda)C'^3_s$, in which the dimensionless function tending to 0.26 in the limit of $\log Re_\lambda \gg 1$. In addition to the variability in terms of moderate Re_λ, the topological structure of large eddies governs the parameter C'_s, leading to a lack of universality of the dimensionless dissipation coefficient.

3.6 Topological Analysis, Coherent Events, and Related Dynamics

As already mentioned, it has been known since the direct observations by Siggia (1981) that coherent structures exist in isotropic turbulence.**

** Although the observation of these structures is recent, it is worth noting that the idea that turbulent dissipation can be tied to a random distribution of vortex tubes and vortex sheets goes back to Townsend in 1951.

It must be noticed that the term "coherent structures" can be fuzzy. In the following, we follow only the conventional use, provisionally avoiding the question of temporal coherence – are they known to preserve themselves temporally? – of these structures. Only considering instantaneous spatial coherence, these structures may be grouped into two classes: *vortex tubes* (also referred to as *worms* or *vortex filaments*) and *vortex sheets*. The former are identified as elongated, tubelike vortices mainly subjected to an axial strain, whereas the latter are related to vorticity sheets that experience a plain strain.

The existence of these events raises several important questions for both the analysis of isotropic turbulence study and the general turbulence theory:

1. How do we define these events, or, more precisely, how do we define them unequivocally?
2. What is the dynamics of these events: How are they generated? What is their life cycle? Do they exhibit some universal features?
3. What is their role in isotropic turbulence dynamics? How are they related to well-known features such as the kinetic-energy cascade, the turbulent kinetic-energy dissipation, and the internal intermittency?

Recent results dealing with these issues are subsequently surveyed. But let us emphasize here that, despite the impressive amount of efforts devoted to the analysis of isotropic turbulence, a global complete theory for the coherent events it contains is still lacking.

3.6.1 Topological Analysis of Isotropic Turbulence

The topological analysis of isotropic turbulence first brings in the problem of defining the various coherent events. A huge amount of works have been devoted to this problem. The proposed techniques can be divided roughly into the following two classes.

The first approach consists of projecting the instantaneous turbulent field onto objects (sometimes referred to as the "cartoons of turbulence") whose definitions are given analytically. It involves a local tuning of the control parameters that appear in the analytical model to obtain the best fit with the local turbulent field, leading to the definition of a pattern-tracking algorithm. A complete survey of analytical solutions for an isolated viscous vortex was recently performed by Rossi (2000). Two useful analytical models, namely Burgers' vortex and Burgers' vortex-sheet models, are subsequently given.

Burgers' vortex is a model for an axially stretched viscous vortex. Denoting z as the direction of the vortex axis, Γ as its circulation, α as the time-independent rate of strain, and ν as the viscosity, the cylindrical velocity components are given by

$$u_z = 2\alpha z, \quad u_r = -\alpha r, \quad u_\theta = \frac{\Gamma}{2\pi r}\left(1 - e^{-\zeta}\right), \tag{3.171}$$

3.6 Topological Analysis, Coherent Events, and Related Dynamics

where $\zeta = r^2/4\delta^2$, and

$$\delta^2 = \frac{\nu}{\alpha} + \left(\delta_0^2 - \frac{\nu}{\alpha}\right)e^{-\alpha t}, \tag{3.172}$$

with $\delta(0) = \delta_0$, and t denotes the time. The axial vorticity is found to be equal to (other components are identically zero)

$$\omega_z = \frac{\Gamma}{\pi\delta^2}e^{-\zeta}. \tag{3.173}$$

The induced kinetic-energy dissipation field is

$$\varepsilon = 12\nu\alpha^2 + \frac{\nu\Gamma^2}{16\pi^2\delta^4}\left(e^{-\zeta} - \frac{1-e^{-\zeta}}{\zeta}\right)^2. \tag{3.174}$$

An asymptotic equilibrium solution is found for large times, i.e., for $\delta^2 = \nu/\alpha$. For this solution, diffusion and convection are balanced and the total dissipation is found to be independent of the viscosity ν. It is worth noting that the dissipation is negligible outside a circular area of the order of δ^2, whereas its peak is proportional to $\nu\Gamma^2/\delta^4$. The total rate of vortex-induced dissipation per unit length scales as $\nu\Gamma^2/\delta^2$.

Burgers' vortex sheet is defined as the superposition of a plane potential flow and a plane shear layer. It corresponds to a diffusing vortex sheet with stretched vortex lines. Let us consider the case in which the shear-layer vorticity is along the z axis and varies in the y direction. The Cartesian components of the potential flow field are given by

$$u_p = 0, \quad v_p = -\alpha y, \quad w_p = \alpha z. \tag{3.175}$$

The vorticity field of the Burgers' vortex sheet is given by

$$\omega_z = -\frac{4}{\sqrt{\pi}}\frac{\Delta U}{\delta}e^{-y^2/\delta^2}, \tag{3.176}$$

where ΔU is the velocity jump across the shear layer and δ is defined as

$$\delta^2 = \frac{2\nu}{\alpha}\left(1 - e^{-2\alpha t}\right). \tag{3.177}$$

The equilibrium solution corresponds to $\delta^2 = 2\nu/\alpha$.

Both Burgers' vortex model and Burgers' vortex-sheet model have been observed to compare favorably with local features of a simulated turbulent field and can therefore be used as theoretical models to describe turbulence dynamics.

Before discussing other definitions, let us first recall some results dealing with the topological analysis of instantaneous, incompressible, isotropic turbulent fields. Most analyses rely on the relation that exists between the vorticity vector and the eigenvectors of the strain-rate tensor **S**. Let us denote $\hat{e}_i$ ($i = 1, 2, 3$) as the three eigenvectors of **S** and $\hat{\lambda}_i$ as the corresponding eigenvalues. In the following, the eigenvalues are reordered so that $\hat{\lambda}_1 \geq \hat{\lambda}_2 \geq \hat{\lambda}_3$. The incompressibility constraint yields

$$\hat{\lambda}_1 + \hat{\lambda}_2 + \hat{\lambda}_3 = 0, \tag{3.178}$$

meaning that there is at least one positive ($\hat{\lambda}_1$) and one negative ($\hat{\lambda}_3$) eigenvalue. The intermediate eigenvalue $\hat{\lambda}_2$ can be either negative or positive. Both numerical and experimental data show that the vorticity vector is preferentially aligned with $\hat{e}_2$. Lund and Rogers (1994) defined the following nondimensional parameter:

$$\hat{\lambda}^* = -\frac{3\sqrt{6}\hat{\lambda}_1\hat{\lambda}_2\hat{\lambda}_3}{\left(\hat{\lambda}_1^2\hat{\lambda}_2^2\hat{\lambda}_3^2\right)^{3/2}}, \qquad (3.179)$$

which has the remarkable property that it ranges from -1 to 1 and that its pdf is uniform for a Gaussian random velocity field. This parameter is a measure of the local deformations caused by the strain-rate tensor. Axisymmetric extension and axisymmetric contraction occur when $\hat{\lambda}^* = 1$ and $\hat{\lambda}^* = -1$, respectively, and plane shear corresponds to $\hat{\lambda}^* = 0$. Lund and Rogers observed in DNS data that the most probable case in isotropic turbulence is axisymmetric extension and that this state is well correlated with regions of high dissipation.

The preferential alignment of ω with $\hat{e}_2$ is a pure kinematic effect. Jimenez (1992) showed that in the vicinity of a vortex whose maximum vorticity is large with respect to that in the background flow the vorticity is automatically aligned with the intermediate eigenvector. It can also be shown (Horiuti, 2001; Andreotti, 1997; Nomura and Post, 1998) that this alignment is the result of the crossover of the eigenvalues at a certain distance from the vortex center in Burgers' analytical vortex model.[††]

The second approach for finding reliable definitions of coherent events relies on the local analysis of the velocity-gradient tensor $\nabla u = \mathbf{S} + \mathbf{W}$, intuition telling us that a vortex will be a region where the vortical part dominates over the irrotational part of the strain.

The first general, Galilean-invariant 3D vortex criterion was proposed by Hunt and co-workers (1988). This criterion, referred to as the *Q-criterion*, defines a vortex as a spatial region where the second invariant of the velocity-gradient tensor is positive:

$$Q = \frac{1}{2}\left(|\mathbf{W}|^2 - |\mathbf{S}|^2\right) = -\frac{1}{2}\mathrm{tr}\left(\mathbf{S}^2 + \mathbf{W}^2\right) > 0, \qquad (3.182)$$

where $|\mathbf{W}|$ and $|\mathbf{S}|$ are Euclidian norms. The *Q-criterion* can be related to the pressure field, because Q has the same sign as the Laplacian of the pressure field. Using this criterion is equivalent to saying that vortices are regions where $\nabla_p^2 > 0$. It is worth noting that, in 2D flows, this criterion is equivalent to the Okubo–Weiss

---

[††] This can be directly seen by looking at the analytical expressions of the eigenvalues obtained for the Burgers' vortex:

$$\hat{\lambda}_{\pm} = \frac{\alpha}{2}\left\{-1 \pm Re_\Gamma\left[\frac{4\nu}{\alpha r^2}\left(1 - e^{-\alpha r^2/4\nu}\right) - e^{-\alpha r^2/4\nu}\right]\right\}, \qquad (3.180)$$

$$\hat{\lambda}_z = \alpha, \qquad (3.181)$$

where $Re_\Gamma = \Gamma/4\pi\nu$ is the circulation-based Reynolds number. If Re_Γ is high enough, the crossover between $\hat{\lambda}_+$ and $\hat{\lambda}_z$ occurs, i.e., there exists a region with $\hat{\lambda}_+ \geq \hat{\lambda}_z$.

3.6 Topological Analysis, Coherent Events, and Related Dynamics

criterion derived independently by Okubo in 1970 and Weiss in 1991. Tanaka and Kida (1993) observed that the criterion given by Eq. (3.182) does not allow us to distinguish between vortex-tube cores and curved vortex sheets (subsequently discussed). To isolate vortex-tube cores, they used the threshold $|\mathbf{W}|^2 > 2|\mathbf{S}|^2$.

Another 3D criterion is the Δ-*criterion* (Chong, Perry, and Cantwell, 1990). Here, a vortex is a region where

$$\Delta = \left(\frac{Q}{3}\right)^2 + \left(\frac{\det(\nabla \boldsymbol{u})}{2}\right)^2 > 0 \qquad (3.183)$$

The *swirling-length* criterion defined by Zhou and co-workers (Zhou et al., 1999) is an extension of the Δ-*criterion*. It relies on the observation that in regions where the tensor ∇u has two complex-conjugate eigenvalues $\tilde{\lambda}_{cr} \pm i\tilde{\lambda}_{ci}$ and a real eigenvalue $\tilde{\lambda}_c$, $\tilde{\lambda}_{ci}$ and $\tilde{\lambda}_{cr}$ can be interpreted as having a measure of the local swirling rate inside the vortex (in the plane defined by the eigenvectors associated with the complex eigenvalues) and a local stretching/compression strength along the last eigenvector. A vortex tube is defined as a region satisfying the Δ-*criterion* and in which $\tilde{\lambda}_{ci}$ is above an arbitrary threshold.

This idea of using the local frame associated with the eigenvectors of the velocity-gradient tensor was further developed by Chakraborty and co-workers (2005), who proposed the *enhanced-swirling-strength criterion*. Following this criterion, a vortex is a region where

$$\tilde{\lambda}_{ci} \geq \epsilon \quad \text{and} \quad -\delta' \leq \frac{\tilde{\lambda}_{cr}}{\tilde{\lambda}_{ci}} \leq \delta, \qquad (3.184)$$

where ϵ, δ, and δ' are positive threshold values.

Another popular criterion, referred to as the λ_2-*criterion*, was proposed by Jeong and Hussain (1995). According to this criterion, a vortex is defined as a region where the intermediate eigenvalue (denoted as λ_2 if the eigenvalues are reordered in decreasing order) of the symmetric matrix $\mathbf{S}^2 + \mathbf{W}^2$ is negative:

$$\lambda_2 < 0. \qquad (3.185)$$

A more recent criterion was proposed by Horiuti (2001), which can be seen as an improvement of the λ_2 criterion. The three eigenvalues of the tensor $\mathbf{S}^2 + \mathbf{W}^2$ are renamed as λ_z, λ_+, and λ_-, where λ_z corresponds to the eigenvector that is the most aligned with the vorticity vector, and λ_+ and λ_- are the largest and smallest remaining eigenvalues, respectively. Using these definitions, Horiuti defines a vortex as region where

$$0 > \lambda_+ \geq \lambda_-. \qquad (3.186)$$

This criterion isolates the vorticity-dominated region similar to the core region of a Burgers' vortex tube.

Although these criteria perform similarly well in simple flows, their use in turbulent shear flows and flows submitted to strong rotation is more problematic, as it is not always possible to separate the mean-flow contribution from the turbulent one.

The case of vorticity sheets seems to be more difficult to handle and received less attention than the vortex case. A reason for that is certainly that these structures are more disorganized and less stable than vortex tubes. Therefore their observation is more difficult. Another difficulty is that the category of vortex sheets encompasses different objects. Horiuti (2001) makes the distinction between *flat sheets* similar to Burgers' vortex layer and *curved sheets* that exist in the circumference of the core region of a vortex tube. These two kinds of vorticity sheets may have different dynamical features, as both vorticity and strain are dominant in flat sheets, whereas strain is predominant in curved sheets. The flat sheets are also referred to as *strong vortex layers* by Tanaka and Kida (1993) who defined them as regions where both vorticity and strain rate take large comparable values.[††] The criterion used by Tanaka and Kida is

$$\frac{1}{2} < \frac{|\mathbf{W}|^2}{|\mathbf{S}|^2} < \frac{4}{3}. \tag{3.187}$$

From the same reordering of the eigenvalues of the symmetric tensor $\mathbf{S}^2 + \mathbf{W}^2$ as for the vortex-tube definition given in Eq. (3.186), Horiuti (2001) defines curved sheets as regions where

$$\lambda_+ \geq \lambda_- > 0, \tag{3.188}$$

whereas a flat-sheet definition is

$$\lambda_+ \geq 0 \geq \lambda_-. \tag{3.189}$$

This definition is observed to educe vortex sheets, but also vortex-tube cores in some cases. To get a more accurate definition, Horiuti and Takagi (2005) proposed a new definition based on the eigendecomposition of the symmetric second-order velocity-gradient tensor $\mathbf{SW} + \mathbf{WS}$. Denoting λ_z^s, λ_+^s, and λ_-^s as the eigenvalues associated with the eigenvector that is maximally aligned with the vorticity vector, the largest and the smallest remaining eigenvalue, respectively, it is observed that vortex sheets can be educed using the criterion

$$\lambda_+^s > \epsilon, \tag{3.190}$$

where ϵ is an arbitrary positive threshold. The vortex-sheet normal vector is accurately computed as $\nabla \lambda_+^s$.

3.6.2 Vortex Tube: Statistical Properties and Dynamics

Vortex tubelike structures have been extensively analyzed by both DNSs and laboratory experiments. Probability density functions of vortex-tube main features are displayed in Figs. 3.15–3.18. These data were obtained by Jimenez and Wray (1998) from DNSs of isotropic turbulence for Taylor-scale-based Reynolds numbers Re_λ

[††] These authors also define a *strong vortex tube* as a region with large vorticity and small strain rate.

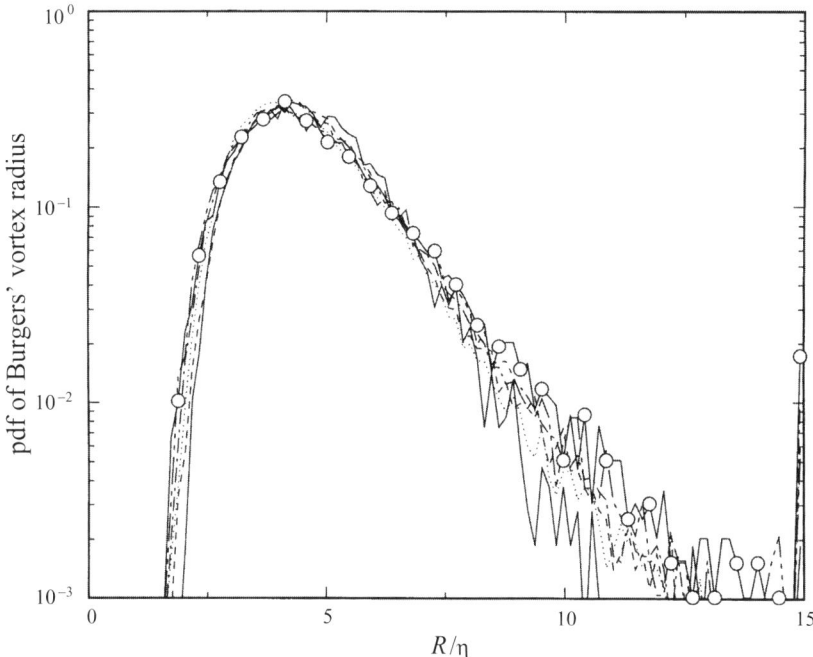

Figure 3.15. The pdf of the radius R of the vortex tube normalized with the Kolmogorov scale η. Different lines and symbols are related to different values of the Reynolds number. From Jimenez and Wray (1998) with permission of CUP.

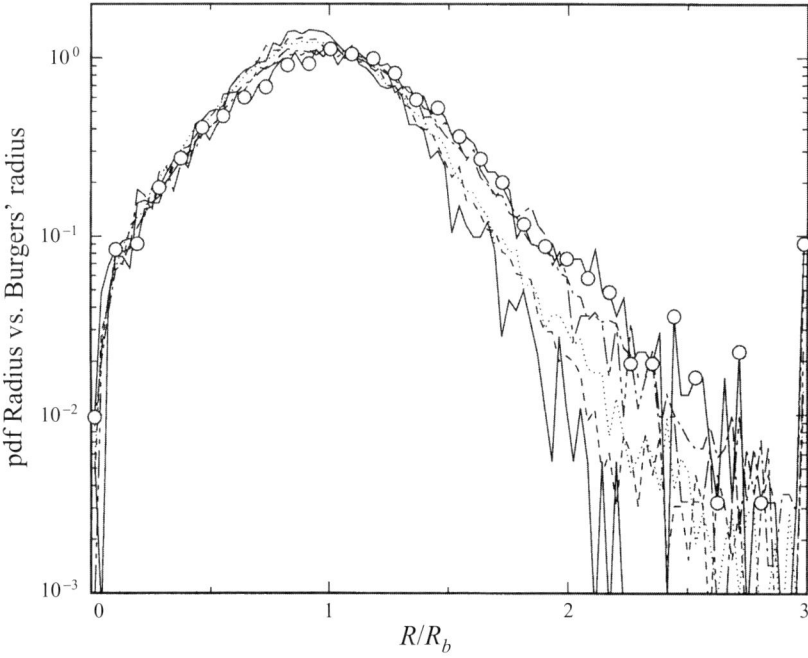

Figure 3.16. The pdf of the radius R of the vortex tube normalized with the local equilibrium Burgers' radius R_b. Burgers' radius is defined as $R_b = 2\sqrt{\nu/\alpha}$, where ν is the viscosity and α is the local axial stretching. Different lines and symbols are related to different values of the Reynolds number. From Jimenez and Wray (1998) with permission of CUP.

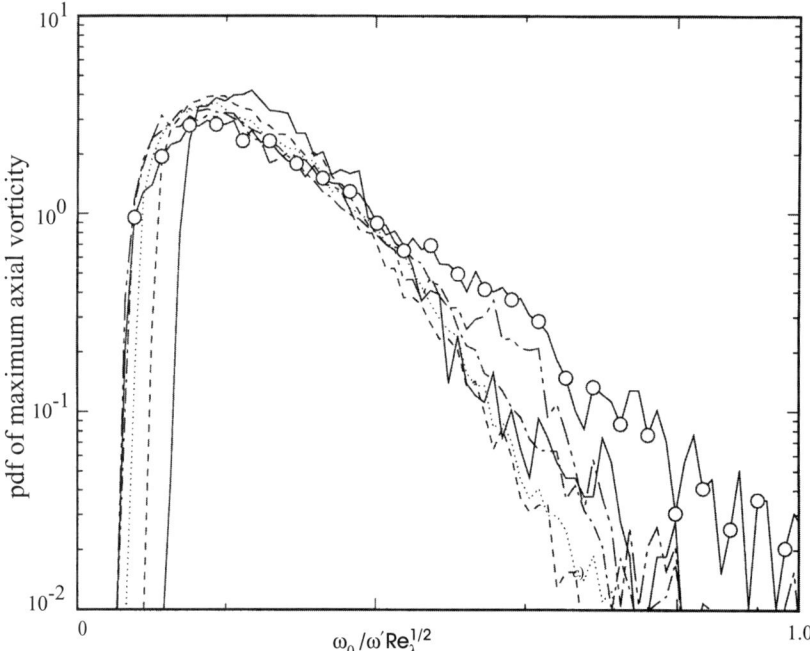

Figure 3.17. The pdf normalized maximum axial vorticity of the vortex tube. Different lines and symbols are related to different values of the Reynolds number. From Jimenez and Wray (1998) with permission of CUP.

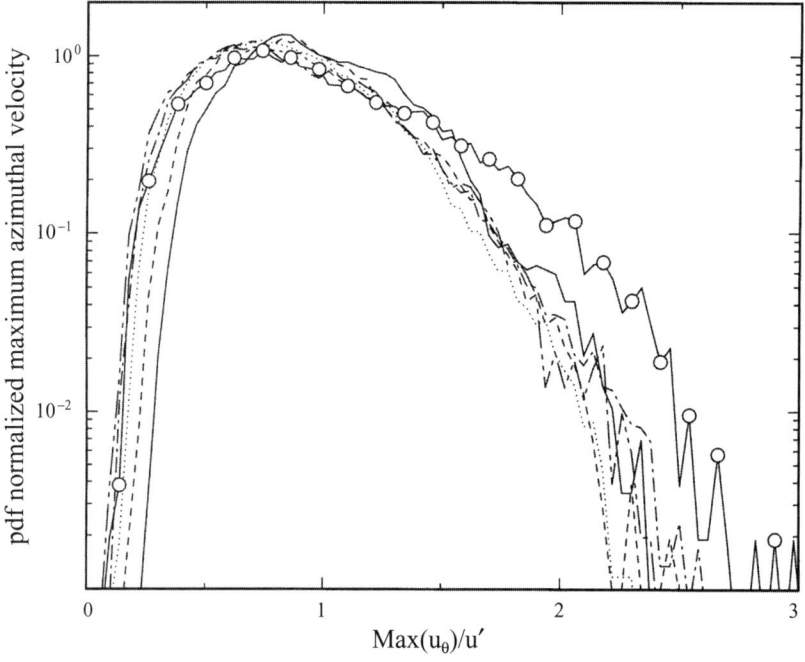

Figure 3.18. The pdf maximum azimuthal velocity u_0 of the vortex tube. Assuming that the vorticity profile is Gaussian, one has $u_0 = 0.319 R \omega_0$. Different lines and symbols are related to different values of the Reynolds number. From Jimenez and Wray (1998) with permission of CUP.

3.6 Topological Analysis, Coherent Events, and Related Dynamics

ranging from 37 to 168; they used a vortex-tracking method that relies on the projection of the instantaneous field onto the Burgers' vortex model. It is worth noting that, although the normalized peak values are Reynolds number independent (showing that the vortex tubes exhibit some universal features), the pdf tails are sometimes observed to be sensitive to the Reynolds number (showing that some extreme events do not have the same dependency with respect to the Reynolds number as the "mean" vortex tubes).

The main conclusions of Jimenez and co-workers are as follows:

- The equilibrium Burgers' vortex model is adequate to describe vortex tubes found in isotropic turbulence, as shown by the peak in the pdf displayed in Fig. 3.16. This point is also supported by results dealing with joint pdf's of stretching and radius and of radius and azimuthal velocity.
- The radius of a vortex tube scales as the Kolmogorov scale η, a typical value being $R \simeq 4 - 4.2\eta$.
- The mean stretching experienced by the vortex tubes scales with ω' independently of Re_λ. The statistics of the stretching along the vortex-tube axis are the same as in the background turbulent flow, showing that the latter is responsible for the main part of vortex stretching.

  The maximum of the axial strain felt by the vortices scales as $O(\omega' Re_\lambda^{1/2})$. Because it is Reynolds number dependent, it is believed to be due to self-stretching.[§§]
- The maximum vorticity in the vortex-tube core scales as $O(\omega' Re_\lambda^{1/2})$. This is in agreement with the idea that vortex tubes are more intense at higher Reynolds numbers.
- The azimuthal velocity, or, equivalently, the azimuthal velocity increment Δu across the vortex-tube diameter, scales with turbulent intensity u'. Because u' is associated with large-scale energy-containing scales, this scaling law is inconsistent with Kolmogorov scaling, which states that the velocity increment across distances of $O(\eta)$ should be $O(u' Re_\lambda^{-1/2})$. The Re_λ-independent upper bound for the azimuthal velocity is approximately $2.5u'$, this limit being reached by vortex tubes with the smallest radii. A rationale for that is subsequently given.
- The circulation-based Reynolds number of the vortices observed in Jimenez and Wray (1998) is about $20Re_\lambda^{1/2}$.
- The vortex-tube length, defined in terms of the autocorrelation of some vortex-tube property ϕ as

$$L_\phi = \int_0^{s_0} \frac{\overline{\phi(s' + s)\phi(s')}}{\overline{\phi^2(s')}} ds, \qquad (3.191)$$

where s_0 denotes the point where the autocorrelation first vanishes, depends on the quantity ϕ. Results show that two groups must be distinguished. The lengths based on vortex radius and axial vorticity are $O(\eta Re_\lambda^{1/2})$, i.e., they scale with the Taylor microscale λ, whereas the one based on the axial stretching varies as the Kolmogorov scale η.

[§§] Another possible physical process for this scaling law, the interactions between vortex tubes, is shown to be much weaker than self-stretching.

The fact that the correlation length of axial stretching is of the order of the vortex-tube diameter (i.e., of the Kolmogorov scale, which is also the correlation length of the velocity gradient in the whole flow) shows that the main stretching experienced by the vortex tubes originates in the background flow.

The existence of the second scale λ can be understood as follows. Let us consider a vortex tube of length $l \ll \lambda$. The line integral of the vorticity stretching is given by

$$\underbrace{\int_0^l \boldsymbol{t} \cdot \mathbf{S} t \, dl}_{O(\omega' L)} = \boldsymbol{u} \cdot \boldsymbol{t} \|_0^l - \underbrace{\int_0^l \frac{\boldsymbol{u} \cdot \boldsymbol{n}}{\mathcal{R}} dl}_{O(Lu'\mathcal{R}^{-1})}, \qquad (3.192)$$

where $\boldsymbol{n}$, $\boldsymbol{t}$, $\boldsymbol{u}$, $\mathbf{S}$, and $\mathcal{R}$ are the unit normal vector and tangent vector, the velocity vector, the stretching tensor, and the local radius of curvature, respectively. To enforce homogeneity between the left- and right-hand sides of Eq. (3.192), one must have $\mathcal{R} = O(u'/\omega') = O(\lambda)$. The physical consequence is that the vortex tube must be geometrically complex over a length larger than the Taylor microscale.

A higher upper bound for the vortex length is found by use of a vortex-tube-tracking algorithm: The length of the most intense tubes is of the order of the velocity integral scale defined as L_ϵ, where ϵ is the dissipation. The difference between the tube length and the axial length of the vortex properties (radius, etc.) can be explained by the existence of axial Kelvin waves driven by the pressure fluctuation along the vortex axis.

- The total volume fraction filled by the vortex tubes decreases as Re_λ^{-2}, corresponding to a total length that increases as Re_λ, leading to a increasing intermittency.

The fact that this upper bound depends on large-scale scale quantities only while the maximum vorticity depends on $Re_\lambda^{1/2}$ is not consistent by the classical dynamical scheme of a stretched vortex with fixed circulation. A possible explanation, based on the stability analysis of a columnar vortex, is that there exists a natural limit beyond which a vortex tube of finite length cannot be stretched without becoming unstable. This instability induces axial currents that counteract the external stretching. This mechanism, studied in the case of Burgers' vortex by Jimenez and co-workers, limits the maximum azimuthal velocity to be of the same order as the straining velocity differences applied along the vortex axis. The straining field being induced by the background turbulent flow, one recovers an $O(u')$ upper limit. As a consequence, the vorticity can be amplified by the stretching while at the same time the maximum azimuthal velocity remains bounded. This implies that the length of the vortex tube with a azimuthal velocity close to u' must be large enough to have an edge-to-edge velocity difference of that order, i.e., it must be of the order of the velocity integral scale L_ϵ, in agreement with the numerical data.

The dynamics of vortex-tube formation is another fundamental issue. A first point is that the vortex tubes are part of the $O(\omega')$ background vorticity and therefore must be seen as particular extreme cases of a more general population of weaker vortical structures. The latter have been observed in numerical simulations

3.6 Topological Analysis, Coherent Events, and Related Dynamics

to originate in the roll-up of vortex sheets because of Kelvin–Helmholtz-type instabilities. In the absence of a mean-flow gradient, vortex tubes are created by straining of the weaker vorticity structures. Dimensional analysis shows that the large-scale strain u'/L_ϵ yields the creation of Burgers' vortices with an equilibrium radius of the order of the Taylor microscale λ, whereas the small-scale strain, which is equal to the inverse of the Kolmogorov time scale and to the rms vorticity $\omega' = \sqrt{\epsilon/\nu}$, generates Burgers' vortices with a radius of the order of the Kolmogorov length η. One can see that the classical dynamical picture, which is in agreement with the Kolmogorov scaling, is associated with the dynamics of the uncoherent part of the turbulent field. The existence of high-intensity vortex tubes that escape the Kolmogorov scaling is discussed in the next section.

The creation of vortex tubes with a length of the order of the integral scale cannot be explained by the usual vortex-stretching mechanism. Jimenez made the hypothesis that they originate in the connection of shorter precursors. It has also been shown (Verzicco, Jimenez, and Orlandi, 1995; Jimenez and Wray, 1998) that infinitely long vortices can be maintained by axially inhomogeneous locally compressive strains. Because similar axial fluctuations of the vorticity have been observed in vortex tubes, this mechanism may explain that these very long vortex tubes are sustained in isotropic turbulence over long times.

3.6.3 Bridging with Turbulence Dynamics and Intermittency

The internal Reynolds number of the vortex tubes being of the order of $O(Re_\lambda^{1/2})$, these vortices can be unstable at high Reynolds numbers. The numerical data suggest that the maximum strain felt by the vortices, which scales like $O(\omega' Re_\lambda^{1/2})$, originates in the first stage of this instability process. The instability process leads to vortex-tube deformation and the creation of small pinched segments whose length is of the order of the diameter of the parent vortex.

This vortex instability led Jimenez and co-workers to suggest the existence of a *coherent Δu cascade*. According to that theory, vortex instability yields the existence of a hierarchy of coherent stretched vortices, the circulation being preserved while the upper bound $\Delta u \sim O(u')$ holds at each level. With Burgers' vortex used as a model, two consecutive levels n and $n-1$ are related by

$$\alpha_n \sim \frac{u'}{R_{n-1}}, \quad l_n \sim R_{n-1}, \quad R_n \sim \sqrt{\frac{\nu}{\alpha_n}} \sim \sqrt{\frac{\nu R_{n-1}}{u'}}, \qquad (3.193)$$

where l_n, R_n, and α_n denote the length, radius, and axial strain of the vortex tubes at the nth level of the coherent cascade, respectively. The limit of the cascade is obtained as the asymptote $n \longrightarrow \infty$:

$$\alpha_\infty \sim \frac{u'^2}{\nu} \sim \omega' Re_\lambda, \quad l_\infty \sim R_\infty \sim \frac{\nu}{u'} \sim \eta Re_\lambda^{-1/2}. \qquad (3.194)$$

The limit value of the circulation-based Reynolds number is 1. An interesting feature of the preceding physical scheme is that it involves scales smaller than the Kolmogorov scale η. Because they originate in vortex-tube instabilities, the

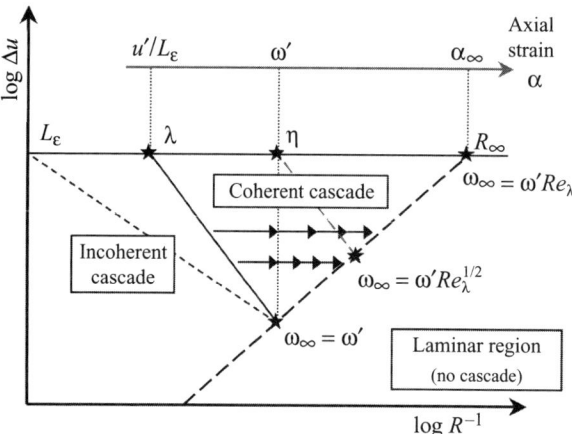

Figure 3.19. Schematic representation of the two turbulent cascade mechanisms. Adapted from Jimenez and Wray (1998).

structures at a given level of the cascade are not space filling but are concentrated in small volumes, leading to a natural interpretation of the internal intermittency of turbulence at small scales.

It is also worth noting that strong vortex tubes with $\Delta u = O(u')$ must be subject to more complex instability mechanisms, which will be compatible with the fact that the circulation Γ is invariant and that the velocity increment $\Delta u \sim \Gamma/R$ is upper bounded. A possible mechanism (compatible with both numerical and experimental observations) is that when a vortex is so strained that its azimuthal velocity would become higher than the driving axial-velocity difference, vorticity is expelled into a cylindrical vorticity sheet. The thickness of this sheet is equal to the Burgers' length of the driving strain. It is unstable, and Kelvin–Helmholtz-type instabilities will lead to its breakup and the formation of longitudinal vortices whose circulation and radius will be such that the global circulation is equal to that of the parent vortex.

The full global dynamical scheme proposed by Jimenez and co-workers consists of two different cascade mechanisms (see Fig. 3.19):

- The *incoherent cascade* associated with space-filling structures such that $\Delta u/R < \omega'$ (i.e., incoherent structures) that fulfill the Kolmogorov scaling $\Delta u = O(R^{1/3})$. The key physical mechanism at play here is the stretching of noncoherent structures by the background vorticity.
- The *coherent Δu cascade* previously described, which is associated with vortex tubes that are not space filling. The governing physical mechanism is the dynamic response of the vortex tubes to the stretching they experience.

The global physical picture is the following. Large-scale uncoherent vortical structures[¶¶] are stretched by the background vorticity, leading to the existence of

---

[¶¶] Uncoherent structures are defined here as structures with a characteristic vorticity weaker than the background vorticity ω'.

smaller structures and the kinetic-energy cascade. Once the cumulated stretching is strong enough, coherent vortex tubes arise, with typical radii ranging from the Taylor microscale to the Kolmogorov scale. Each coherent vortex tube is then subject to the coherent Δu cascade mechanism, leading to the generation of a hierarchy of thinner and thinner tubes. The dynamical scheme previously described does not account for possible interactions between vortex hierarchies generated by the coherent cascades. Some exchanges are *a priori* possible by means of phenomena such as vortex connection or imperfect braiding.

Numerical data reveal that the $O(\omega')$ background vorticity is concentrated in large-scale vortex sheets that separate the energy-containing eddies at the integral scales. This background vorticity is observed to be responsible for almost 80% of the total turbulent dissipation in existing numerical simulations, whereas it fills only 25% of the total volume of the flow.

Vortex tubes are not responsible for the global dynamics of flow and play almost no role in global physical mechanisms like the kinetic-energy cascade in the inertial range or the turbulent dissipation. This point will be further discussed in Section 3.7. Previous scaling laws show that their total energy scales as $O(Re_\lambda^{-2})$, whereas they induce a kinetic-energy dissipation that decreases as $O(Re_\lambda^{-1})$. They are possibly responsible for the intermittency observed in higher-order statistics and for extreme values found in the tails of pdf's of many turbulent quantities. It is to be noted that no satisfactory link between coherent-event dynamics and inertial-range intermittency has been established up to now. Vortex tubes are certainly a source of intermittency, but mostly at small scales. The trend of vortex tubes to form large-scale clusters reported by Moisy and Jimenez (2004) might be a source for large-scale intermittency, but no definitive evidence is available at the present time. Other mechanisms, like the persisting long-range coupling between large and small scales, may also contribute to the inertial-range intermittency.

3.7 Nonlinear Dynamics in the Physical Space

3.7.1 On Vortices, Scales, Wavenumbers, and Wave Vectors – What are the Small Scales?

The analysis of isotropic turbulence dynamics, as done in this chapter, is usually carried out concurrently in both Fourier and physical space, a very difficult issue being to bridge between these two different approaches.

It is important to emphasize here that several shortcomings usually occur that are misleading. Fourier analysis is based on the use of *wave vectors*, which are not equivalent to *scales*, because a wave vector also carries information dealing with orientation. The associated *wavenumber*, defined as a Euclidian norm of the wave vector, has the dimension of the inverse of a length. A large part of the information is now lost, such as the mode polarity in the helical-mode decomposition denoted by the parameter *s* in Eq. (2.86).

Another problem is to switch from the scale concept to classical objects of fluid dynamics like *vortices*. Small scales are very often understood as "small vortices," which is wrong. The three reasons are as follows:

1. Neither Fourier analysis, which introduces wave vectors, nor the scale-dependent analysis in the physical space (based on structure functions, scale-dependent increments, etc.) involves the concept of coherent events such as a vortex. It is worth noting that none of the recent definitions of a vortex or a vortex sheet (see Subsection 3.6.1) is based on the the scale concept.
2. Modes in Fourier space are nonlocal in space, whereas the very concept of a vortex is intrinsically local in physical space because it is associated with a given object.
3. As seen in Subsection 3.6.2, 3D vortices (as defined according to one of the available definitions) cannot be defined by a single length scale. This is obviously the case of vortex tubes, whose axial lengths are much higher than their typical diameters.

Therefore one must be very cautious when "translating" or "extrapolating" results coming from Fourier analysis in physical space (and vice versa).

What definition of *small scales* can be used in physical space? Such a definition should rely on the flow dynamics. It is commonly agreed that most of kinetic-energy dissipation ε occurs at modes with high wavenumbers* because it is equal to

$$\varepsilon = 2\nu \int_0^{+\infty} k^2 E(k) dk, \qquad (3.195)$$

and that scales dominated by viscous effects are the small scales. Because the right-hand side of Eq. (3.195) is proportional to the square of the L_2 norm of the velocity gradient ∇u, one can see that *small scales of turbulence in the physical space should be defined as scales associated with large gradients of the velocity field*. On the opposite side, *large scales in physical space are the ones that carry most of the turbulent kinetic energy*. Because

$$\mathcal{K} = \int_0^{+\infty} E(k) dk, \qquad (3.196)$$

and $E(k) \geq 0, \forall k$, one can see that modes with dominant contributions to $\mathcal{K}$ and ε are not the same, the latter having larger wavenumbers than the former at high Reynolds numbers. In this sense, one can establish a link between wavenumbers and scales in physical space.

Let us conclude this section by emphasizing that velocity gradients, from which one can define the small scales in physical space, include both symmetric and antisymmetric parts, i.e., both turbulent strain **S** and vorticity **ω**.

---

* *High* is to be understood as a relative notion, the reference being the wavenumbers at which turbulent kinetic energy is injected or created by external forcing or hydrodynamic instabilities.

3.7 Nonlinear Dynamics in the Physical Space

It is worthy noting that the true exact local expression for the dissipation in physical space is $\varepsilon = 2\nu S_{ij} S_{ij}$, i.e., dissipation is a function of strain, not vorticity. Introducing the vorticity, one obtains

$$\varepsilon(x,t) \equiv 2\nu S_{ij}S_{ij} = \nu \omega_i \omega_i + \nu \frac{\partial^2}{\partial x_i \partial x_j}(u_i u_j), \qquad (3.197)$$

showing that, in an unbounded or periodic domain, the following usual volume or statistical averaged relations hold:

$$\varepsilon \equiv 2\nu \overline{S_{ij}S_{ij}} = \nu \overline{\omega_i \omega_i}. \qquad (3.198)$$

Therefore mean dissipation can be tied to the mean enstrophy through a purely kinematic relation in isotropic turbulence. But such a relation is meaningless from a local point of view, leading to the conclusion that the strain field is the right quantity to describe the dissipation process.

3.7.2 Is There an Energy Cascade in the Physical Space?

Although the kinetic-energy energy cascade is a well-established result in Fourier space and in an ensemble-averaged sense, its "translation" in physical space is not straightforward. The Navier–Stokes equations just tell us that momentum and kinetic energy are transported in the physical space, global kinetic energy being invariant in a fully periodic domain in the absence of viscous effects and external forcing. Exact equivalences between terms appearing in Fourier and physical space formulations are only global, nonlocal expressions, which do not make it possible to have direct access to single-wave-vector-related information in physical space. Therefore *the energy cascade concept is not relevant in physical space from a rigorous viewpoint*. It is directly related to the projection of the Navier–Stokes equations onto basis functions that intrinsically bear the information related to scale dependency (such as Fourier, but also wavelets, *hp* bases in finite-element methods, etc.). This point was emphasized a long time ago by von Neumann and Onsager in 1949.

A very common picture deals with the kinetic-energy cascade being the results of a hierarchy of vortex-breakdown phenomena, with each vortex generating smaller vortices. This phenomenological picture, very often presented as the Richardson cascade, is wrong: Experimental and numerical results show that vortices observed in isotropic turbulence do not behave this way and that the transfer of kinetic energy does not originate in the instability of the vortices. As emphasized by Tsinober (2001), this flawed physical picture originates in a too-rapid reading of the famous sentence written by Richardson in 1922: "*We thus realize that: big whirls have little whirls that feed on their velocity, and little whirls have lesser whirls and so on on to viscosity – in the molecular sense.*" It is to be noticed that Richardson never made further use of this picture and that the term *cascade* was coined by Onsager two decades later in the 1940s.

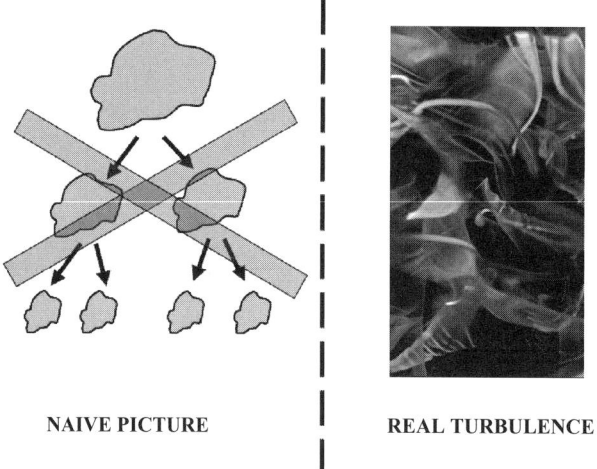

NAIVE PICTURE | **REAL TURBULENCE**

Figure 3.20. Nonphysical simplified view of the turbulent kinetic cascade in physical space (left) and true structure of the instantaneous vorticity field computed by means of high-resolution numerical simulation (right). It is seen that the simplified picture based on hierarchical breakup process has no physical ground. Right picture reproduced with courtesy of Laboratory for Computational Science and Engineering.

Therefore the question arises of the existence of a mechanism in the physical space that can be interpreted as the counterpart of the turbulent kinetic-energy cascade in the Fourier space. In physical space, one observes that the injection of turbulent kinetic energy at a given scale yields the generation of velocity gradients and turbulent kinetic-energy dissipation. Using the definition previously given for the small scales in physical space, one can see that *the turbulent kinetic cascade in Fourier space must be replaced with the generation of velocity gradients (i.e., both vorticity and strain) in physical space.*

It is also important to note that some oversimplified pictures of the cascade that illustrate this process as a hierarchical breakup of structures in smaller ones in physical space is misleading (see Fig. 3.20). One gains a much more realistic picture by looking at the true topology of the turbulent field, revealing that the basic mechanisms are vorticity stretching, vortex-sheet folding/rolling up, vortical-blob reconnection, etc.

3.7.3 Self-Amplification of Velocity Gradients

In agreement with the definition of the small scales in physical space as previously given and the observation that the kinetic-energy cascade picture does not hold in physical space, the dynamics of turbulence should be investigated by looking at the dynamics of velocity gradients. Therefore strain and vorticity fields should be able to be used to describe turbulence dynamics. Another reason is that they are much more sensitive to internal intermittency than velocity and kinetic energy. Historically, Taylor pointed out the importance of vorticity in 1937, and the role of strain was emphasized by Kolmorogov in 1941. These two quantities must be considered

3.7 Nonlinear Dynamics in the Physical Space

in parallel, as they are weakly correlated in isotropic turbulence and they are tied by a strongly nonlocal relation.

Let us recall some fundamental evolution equations for the vorticity ω, the strain **S**, the enstrophy $\omega^2/2$, and the total strain $S^2 = S_{ij}S_{ij}$. In the absence of external forcing, one has

$$\frac{\partial \omega_i}{\partial t} + u_j \frac{\partial \omega_i}{\partial x_j} = \omega_j S_{ij} + \nu \nabla^2 \omega_i, \tag{3.199}$$

$$\frac{1}{2}\frac{\partial \omega^2}{\partial t} + u_j \frac{\partial \omega^2}{\partial x_j} = \omega_i \omega_j S_{ij} + \nu \omega_i \nabla^2 \omega_i, \tag{3.200}$$

$$\frac{\partial S_{ij}}{\partial t} + u_j \frac{\partial S_{ij}}{\partial x_j} = -S_{ik}S_{kj} - \frac{1}{4}\left(\omega_i \omega_j - \omega^2 \delta_{ij}\right) - \frac{\partial p}{\partial x_i x_j} + \nu \nabla^2 S_{ij}, \tag{3.201}$$

$$\frac{1}{2}\frac{\partial S^2}{\partial t} + u_j \frac{\partial S^2}{\partial x_j} = -S_{ik}S_{kj}S_{ij} - \frac{1}{4}\omega_i \omega_j S_{ij} - S_{ij}\frac{\partial^2 p}{\partial x_i x_j} + \nu S_{ij}\nabla^2 S_{ij}. \tag{3.202}$$

Now, with the analysis restricted to isotropic turbulence, the evolution of mean enstrophy and mean total strain are governed by the following equations:

$$\frac{1}{2}\frac{\partial \overline{\omega^2}}{\partial t} = \overline{\omega_i \omega_j S_{ij}} + \nu \overline{\omega_i \nabla^2 \omega_i}, \tag{3.203}$$

$$\frac{1}{2}\frac{\partial \overline{S^2}}{\partial t} = -\overline{S_{ik}S_{kj}S_{ij}} - \frac{1}{4}\overline{\omega_i \omega_j S_{ij}} + \nu \overline{S_{ij}\nabla^2 S_{ij}}. \tag{3.204}$$

Two of the most distinctive features of 3D turbulence are as follows:

1. Enstrophy production by means of vortex stretching is positive in the mean,

$$\overline{\omega_i \omega_j S_{ij}} > 0, \tag{3.205}$$

as hypothesized by Taylor in 1938. Using Lin's equation (3.88) for the evolution of $E(k)$, it is seen that this term is exactly equal to $\int_0^\infty k^2 T(k) dk$.

Numerical simulations show that this term is 2 orders of magnitude larger than other terms that appear on the right-hand side of Eq. (3.203). It is important to note that this term happens to take negative values locally. The positive mean value comes from the fact that its pdf is strongly positively skewed. More details about the enstrophy production will be given later in this section, but let us emphasize here that the positivity on the mean of enstrophy production cannot be explained when vortex lines are considered as material lines. This is a misconception, because material lines and vorticity lines have very different behaviors, which are due to the fact that vorticity is not a passive scalar (it reacts back on the velocity field). These discrepancies are exhaustively discussed in Tsinober (2001).

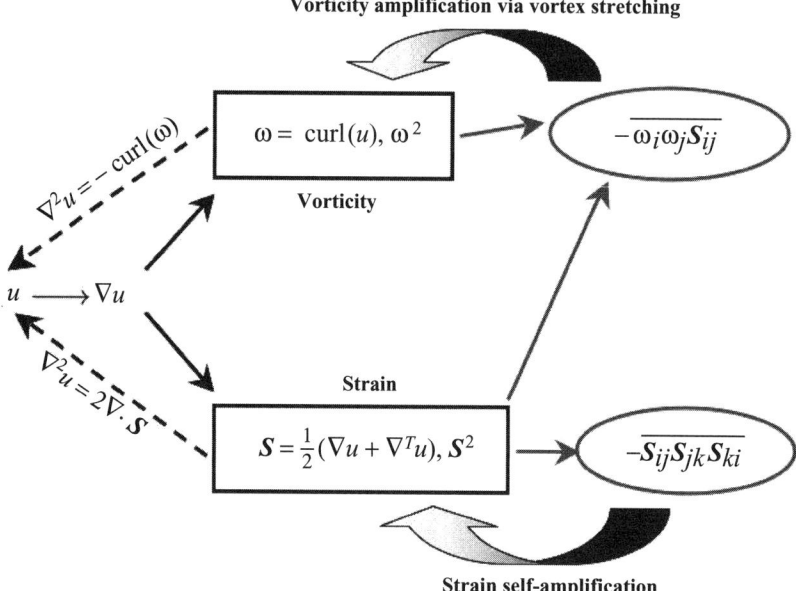

Figure 3.21. Schematic view of the velocity-gradient self-amplification process in isotropic turbulence.

2. Total strain production is positive in the mean. Using the nonlocal kinematic equality†

$$\overline{\omega_i \omega_j S_{ij}} = -\frac{4}{3}\overline{S_{ik} S_{kj} S_{ij}}, \qquad (3.206)$$

one observes that the characteristic feature of turbulence is that

$$-\overline{S_{ik} S_{kj} S_{ij}} > 0. \qquad (3.207)$$

This term is observed to be larger by 2 orders of magnitude than the viscous term in the balance equation for $\overline{S^2}$.

A few important observations can be drawn from Eqs. (3.203) and (3.204):

1. Enstrophy production results from the *interaction of vorticity with the strain field*, whereas the production of total strain mainly comes from *self-amplification of the strain field*. This is illustrated in Fig. 3.21.
2. In regions where $\omega_i \omega_j S_{ij} > 0$, the production of total strain is decreased because the two terms have opposite signs [see Eq. (3.206)], i.e., vortex stretching tends to suppress production of strain, at least in a direct way. On the opposite side, vortex compression (i.e., regions where $\omega_i \omega_j S_{ij} < 0$) aids the production of total strain. Now, identifying the dissipation and its production as the counterpart of the kinetic-energy cascade in physical space, one arrives at the conclusion that *turbulence dynamics in the physical space is associated with*

---

† This equality, found by Betchov (1956), is valid in homogeneous, but not necessarily isotropic, turbulence, as recently noted by J.N. Gence (private communication).

3.7 Nonlinear Dynamics in the Physical Space

Table 3.5. *Individual contributions of eigenmodes of the strain tensor* **S** *to the production of velocity gradient*

Nonlinear term	$i = 1$	$i = 2$	$i = 3$
$\overline{\hat{\lambda}_i^3}$	1.2–1.62	0.05	−2.67 – −2.25
$\overline{\omega^2 \hat{\lambda}_i^2 \cos^2(\omega, \hat{e}_i)}$	0.52–0.53	0.12–0.15	0.32–0.36

Note: Ranges of variations are taken from Tsinober (2001), from $Re_\lambda = 75$ (direct numerical simulation) to $Re_\lambda = 10^4$ (measurements in the atmospheric boundary layer). There is no summation over repeated indices here.

the predominant production of the rate of strain by means of strain self-amplification and vortex compression rather than with vortex stretching. The latter is observed to resist the production of dissipation, and therefore to decrease the intensity of turbulent nonlinear dynamics in some sense.

DNSs provide a deep insight into the dynamics of the generation of total strain and vorticity. Among other results, they make it possible to identify the regions of space and the physical events responsible for the production mechanisms previously presented. As in Subsection 3.6.1, let us denote as $\hat{e}_i (i = 1, 2, 3)$ the three eigenvectors of **S** and as $\hat{\lambda}_i$ the corresponding eigenvalues. Simple algebra yields

$$-S_{ik}S_{kj}S_{ij} = -(\hat{\lambda}_1^3 + \hat{\lambda}_2^3 + \hat{\lambda}_3^3) = -3\hat{\lambda}_1\hat{\lambda}_2\hat{\lambda}_3. \quad (3.208)$$

One knows that $\hat{\lambda}_1 > 0$. Numerical simulation shows that $\hat{\lambda}_2$ is positively skewed, yielding $\overline{\hat{\lambda}_2^3} > 0$, and that $\hat{\lambda}_3$ is negatively skewed, ensuring the positive production in Eq. (3.207). Typical values are displayed in Table 3.5. Therefore the *nonlinear dynamics, understood as the generation of dissipation and small scales, is directly associated with regions in which* $\hat{\lambda}_3 < 0$, *i.e., with regions of vortex compression.*

The vortex-stretching term can be rewritten as follows:

$$\omega_i \omega_j S_{ij} = \omega^2 \hat{\lambda}_i \cos^2(\omega, \hat{e}_i). \quad (3.209)$$

Numerical data reveal that the largest contribution to positive enstrophy production comes from regions where ω tends to align with $\hat{e}_1$ (see Table 3.5, in which typical values of the contributions to $|\omega \mathbf{S}|^2$ are displayed). But, as mentioned in Subsection 3.6.1, it is known that, in vortex tubes, ω is mainly aligned with $\hat{e}_2$. This result indicates that *vortex tubes are not responsible for the main part of enstrophy production, which originates in regions with larger strains than enstrophy, and with large curvature of vorticity lines.* In the latter, enstrophy production is maximal and is much larger than viscous destruction. On the contrary, vortex tubes are axial structures with low curvature and maximal enstrophy. In these tubes, modeled as Burgers-like vortices, one observes an approximate equilibrium between enstrophy production and viscous effects. Because their vorticity field is mostly concentrated on the axial component, they are not able to react back on the strain field that stretches them. In this sense, the nonlinearity is reduced in these objects, yielding a long lifetime.

It is important to note that enstrophy production mainly originates in strain-dominated regions. Two types of such regions are found:

- *Strain-dominated regions with small curvature of vorticity lines.* These regions are mostly located around vorticity-dominated regions (vortex tubes), in which the vorticity lines wrap around the vortices, leading to a preferential alignment of ω with $\hat{e}_2$. These regions are not associated with the maximal enstrophy production.
- *Strain-dominated regions with large curvature of vorticity lines.* In these regions, large enstrophy production is associated with a large magnitude of $\hat{\lambda}_3$ and large negative values of the enstrophy production rate $\hat{\lambda}_i^2 \cos^2(\omega, \hat{e}_i)$. Predominant mechanisms are vortex compression and vortex tilting (change of orientation).

3.7.4 Non-Gaussianity and Depletion of Nonlinearity

The non-Gaussian character of turbulence, pointed out in Subsection 3.1.2, is intrinsically tied to dynamics of turbulence. One can understand this by looking at enstrophy and total strain production processes, which can be seen as the counterpart of the turbulent energy cascade in the physical space. A striking feature is that production terms in Eqs. (3.203) and (3.204) are third-order moments, which should be identically zero if the turbulent field were a Gaussian random field. Production of enstrophy and total strain are non-Gaussian features of turbulence. Therefore non-Gaussianity originates in the very dynamics of turbulence dictated by the Navier–Stokes equations.

The strategy that consists of describing Navier–Stokes turbulence by comparing it with the properties of a Gaussian random velocity field is appealing, as many theoretical results are available for the latter. Kraichnan and Panda (1988) suggested comparing the values of several key nonlinear terms that are involved in the description of nonlinear dynamics in physical space, and introduced the notion of *depletion of nonlinearity*. This term was coined to account for the fact that some even moments related to nonlinear mechanisms are larger in the Gaussian case than in Navier–Stokes turbulence; e.g., the ratio

$$\frac{|u\nabla u + \nabla p|_{\text{Navier–Stokes}}}{|u\nabla u + \nabla p|_{\text{Gaussian}}} \simeq 0.5\text{--}0.6 \quad (3.210)$$

is inferred from available numerical data, where $|\,.\,|$ is related to the rms value, and is different from $\langle.\rangle$, which yields an identically zero value. These results could be interpreted as a sign that the nonlinearities are depleted in Navier–Stokes turbulence. Of course, this idea must be considered with care, because, looking at odd moments, the Navier–Stokes turbulence appears to be infinitely more nonlinear than its Gaussian approximation.

As previously mentioned, it is also observed that both enstrophy and strain production are reduced in regions dominated by enstrophy with respect to strain-dominated regions. Accordingly, vorticity-dominated regions, and more specifically vortex tubes, are regions in which the nonlinear effects are less intense and can be considered as locally depleted.

3.8 What are the Proper Features of Three-Dimensional Navier–Stokes Turbulence?

We now address the following question: Among all the features of turbulence previously presented, which are the ones that are proper characteristics of 3D Navier–Stokes incompressible turbulence in the sense that they are not shared by other systems?

3.8.1 Influence of the Space Dimension: Introduction to d-Dimensional Turbulence

A first question deals with the influence of the space dimension on turbulence dynamics. Although 1D incompressible turbulence does not exist,[†] the dynamics of isotropic turbulence in two (see Lesieur, 1997, for a detailed discussion of 2D turbulence), three, or even four dimensions has been investigated, both theoretically (Fournier and Frisch, 1978) and numerically (Suzuki et al., 2005). The main results are subsequently summarized:

- The turbulent kinetic-energy spectrum exhibits an inertial range at small scales if the Reynolds number is high enough. But it is worth noting that the spectrum shape depends on the space dimension. In 2D turbulence, two inertial ranges are detected. A first inertial range with $E(k) \propto k^{-5/3}$ is followed by a second one at higher wavenumbers, in which $E(k) \propto k^{-3}$. On the other hand, a single inertial range with $E(k) \propto k^{-5/3}$ is observed in three and higher dimensions.
- A kinetic-energy cascade is observed in all cases, even in the 2D case in which the vortex-stretching term in the vorticity equation is identically zero. But, in agreement with Waleffe's instability assumption (see Subsection 3.5.4), because F-type distant interactions are almost absent, the net ensemble-averaged dominant mechanism is a reverse energy cascade from large- to small-wavenumber modes. In the 2D case, this reverse cascade is easily interpreted in terms of vortex dynamics, as vortices are observed to merge, generating larger and larger structures. In both 3D and 4D cases, the forward energy cascade is observed to be dominant at large wavenumbers. Theoretical analyses show that 2D turbulence is a singular point and that the forward cascade is dominant in spaces with dimension $d \geq 3$.
- Self-similar decay regimes exist in all dimensions. An extension of the analysis presented in Subsection 3.1.3 shows that a self-similar decay regime in the d-dimensional case exists if the kinetic-energy spectrum at a small wavenumber behaves like

$$E(k,t) \propto C^{(d)}(t) k^{d+1} \quad \text{(small wavenumbers)}. \tag{3.211}$$

[†] In the 1D case, the divergence-free constraint simplifies into a null-space derivative constraint, leading to uniform solutions in space.

Assuming that $C^{(d)}(t)$ is constant (i.e., assuming that the PLE assumption holds), one obtains the following law for the decay of turbulent kinetic energy:

$$\mathcal{K}(t) \propto t^{-n}, \quad n = \frac{2(d+2)}{(d+4)}, \quad d \geq 2. \tag{3.212}$$

The decay coefficient n is an increasing function of the space dimension d. This fact is interpreted by Suzuki as evidence that energy transfer is more efficient in higher dimensions.

- Comparisons between 3D and 4D isotropic turbulence (Suzuki et al., 2005) show that the total dissipation is less and less intermittent whereas intermittency is stronger on velocity increments as the dimension increases. The reason is the change in balance between pressure and convection terms as the dimension d increases. The role of pressure and incompressibility becomes weaker in higher dimensions, because the system has more degrees of freedom. As a consequence, the velocity field is less constrained and a larger intermittency can exist. The enhanced energy transfer in higher dimensions is also tied to this weakening of the pressure influence: Because more persistent straining of the small scales by the large-scale strain is allowed, the energy transfer toward a small scale is enhanced.
- The role of coherent vortices in a kinetic-energy cascade is less and less important as the dimension d is increased.

3.8.2 Pure 2D Turbulence and Dual Cascade

2D turbulence without forcing can be characterized by the following kinetic-energy spectrum,

$$E(k) \sim \overline{\omega^2} k^{-3},$$

in which $\overline{\omega^2}$ is the total enstrophy and k holds for the wavenumber component in the wave plane normal to the direction of the missing velocity component. 2D turbulence is often considered in the presence of some forcing, for instance with spectral energy injected at a wavenumber k_0, and two situations must be distinguished: The previous law is valid for $k > k_0$, whereas a conventional $k^{-5/3}$ law prevails for $k < k_0$.

The k^{-3} was proposed, probably independently, by Batchelor, Leith, and Kraichnan. Such a law is not really recovered by DNS, but a modified one,

$$E(k) \sim \overline{\omega^2} k^{-3} [\ln(k/k_0)]^{1/3},$$

is consistently derived from theoretical (Kraichnan, 1967) and numerical (Ishihara and Kaneda, 2001) results.

In physical space, a pure K41 law is predicted for nth-order structure functions, without need for curvature of the straight line of exponents as for 3D turbulence. Only the prefactor of the Kolmogorov law is changed, yielding

$$\langle \delta u_\parallel^3 \rangle = \frac{3}{2} \varepsilon r$$

3.8 What are the Proper Features of 3D Navier–Stokes Turbulence?

for $k_0 r \ll 1$. The absence of curvature is interpreted as an absence of intermittency, whereas the change of sign of the prefactor is linked to the inverse energy cascade. It is also conjectured that there is no "dissipation anomaly," so that a vanishing viscosity would mean a vanishing dissipation rate. On the other hand, the constant-enstrophy flux is reflected by a mixed velocity–vorticity third-order structure function, scaling as

$$\langle [\delta u_\|(r)][\delta \omega(r)]^2 \rangle = -\frac{4}{3}\overline{\omega^2} r$$

for $k_0 r \gg 1$.

In view of realistic flow cases, in which the flow is partially 2D but filling in a volume, pure 2D turbulence can be seen as a limiting case of 3D axisymmetric turbulence, so that the previously mentioned (averaged over circles) energy spectrum $E(k), k = k_\perp$ is connected with the fully dimensional one $e(\mathbf{k})$ as

$$e(\mathbf{k}) = \frac{E(k_\perp)}{2\pi k_\perp}\delta(k_\|), \quad (3.213)$$

where the Dirac delta is related to the invariance of the velocity field with respect to the coordinate $x_\|$ in physical space. If two-dimensionality is related to the latter invariance only, the Fourier component of the velocity consists of both components $u^{(1)}$ and $u^{(2)}$ in the Craya–Herring reference frame in Fourier space, but restricted to $k_\| = 0$ (horizontal wave plane). In this case, $u^{(1)}$ corresponds to the horizontal vortical component, and $u^{(2)}$ to the vertical "jetal" velocity component. Classical 2D-2C (two-dimensional two-component) turbulence is characterized only by the $u^{(1)}$-related velocity.[§] The counterpart of 3D isotropic equations for a single triad is (Fjortoft, 1953; Kraichnan, 1967; Waleffe, 1992)

$$\dot{u}_k^{(1)} = (p^2 - q^2)\frac{\imath s}{2} C_{kpq} u_p^{(1)*} u_q^{(1)*}, \quad (3.214)$$

$$\dot{u}_p^{(1)} = (q^2 - k^2)\frac{\imath s}{2} C_{kpq} u_q^{(1)*} u_k^{(1)*}, \quad (3.215)$$

$$\dot{u}_q^{(1)} = (k^2 - p^2)\frac{\imath s}{2} C_{kpq} u_k^{(1)*} u_p^{(1)*}, \quad (3.216)$$

where C_{kpq} is given by Eq. (3.110). The sign s is equal to $+1$ for any even permutation of the vectors $\mathbf{k}, \mathbf{p}, \mathbf{q}$ of the triad and -1 for an odd permutation. It is clear that each interaction independently conserves energy and enstrophy. Without further quantitative statistical analysis, it is immediately shown that only (R) triads are concerned. Compared with the instability principle expressed in terms of helical modes for 3D isotropic turbulence, the analogy with the Euler stability problem of a solid body that rotates around its principal axes of inertia is even more striking, replacing I_1, I_2, I_3 by k^2, p^2, q^2 in Eqs. (3.141)–(3.143). Only positive terms are involved, without the need for looking at signs (i.e., polarities of helical modes) as before.

A last important result is that the triad instability principle is found to be consistent with the concept of a *dual cascade* observed in 2D turbulence, i.e., a dominant

---

[§] Exact relations are $u^{(1)} = -\hat{\omega}_\|/k$ and $u^{(2)} = -\hat{u}_\|$ in the horizontal ($k_\| = 0$) wave plane.

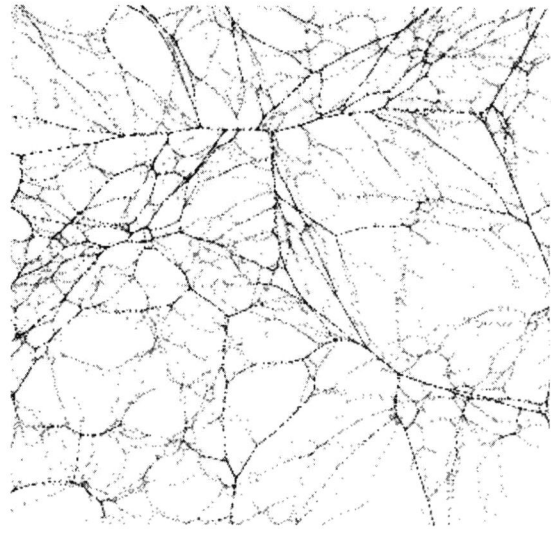

Figure 3.22. Instantaneous "turbulent" solution of the 2D Burgers' equations. Shocks are observed. Courtesy of A. Noullez.

inverse cascade for energy from large to small wavenumbers, and a direct enstrophy cascade from small to large wavenumbers.

3.8.3 Role of Pressure: A View of Burgers' Turbulence

We use here the results dealing with the turbulence-like solutions of Burgers' equations, also referred to as Burgers' turbulence or "Burgulence," to discuss in the role of the pressure. This model,

$$\frac{\partial \boldsymbol{u}}{\partial t} + \boldsymbol{u}\nabla \boldsymbol{u} = \nu \nabla^2 \boldsymbol{u}, \qquad (3.217)$$

can be interpreted as an asymptotic model for hydrodynamics, in which pressure has no feedback on the velocity field. Because it is the pressure gradient that enforces the incompressiblity, Burgers' equations correspond to an infinitely compressible fluid. It is worth noting that the vorticity equation obtained by application of the curl operator to Eq. (3.217) is similar to the usual one derived from Navier–Stokes equations. But vorticity will remain identically zero for irrotational initial conditions and *ad hoc boundary conditions*, because a velocity potential exists.

Extensive analyses of both forced and decaying isotropic Burgers' turbulence have been carried out, with different space dimensions (Girimaji and Zhou, 1995; Noullez and Pinton, 2002; Noullez et al., 2005). The main observations are as follows:

- The Burgers' velocity field is composed of planar viscous shocks (see Fig. 3.22) and does not exhibit vortices as in the Navier–Stokes case. This important fact puts the emphasis on the role of pressure, which is responsible for the existence of coherent vortices (as defined in Subsection 3.6.1). A consequence is that the analysis of the sole vorticity equation is not relevant to characterize Navier–Stokes turbulence. It is also to be noted that this observation is consistent with

3.8 What are the Proper Features of 3D Navier–Stokes Turbulence?

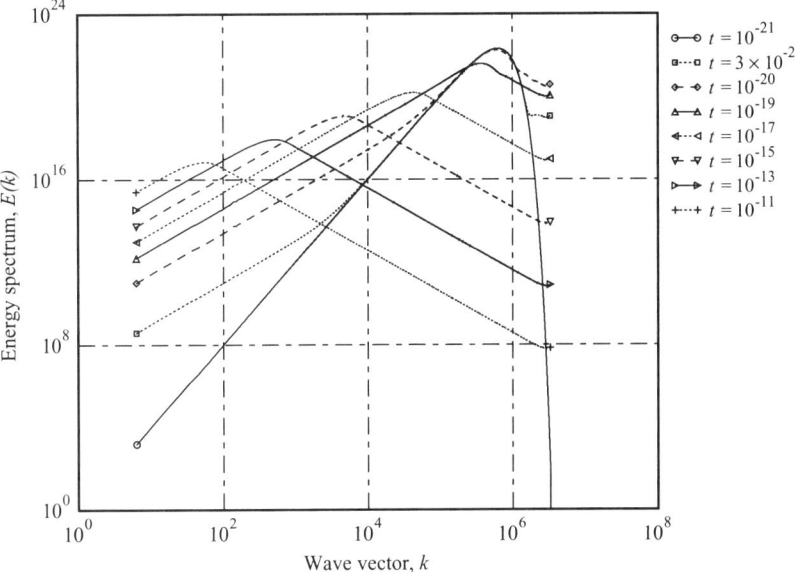

Figure 3.23. Time evolution of the turbulent kinetic-energy spectrum in freely decaying 2D Burgers' turbulence. The occurrence of a self-similar solution with a k^{-2} inertial range is observed. Reproduced from Noullez et al. (2005) with permission of the American Physical Society.

the one dealing with the weakening of both pressure effects and vortices' roles in d-dimensional Navier–Stokes turbulence for increasing d (see the preceding section).

- At high Reynolds numbers, Burgers' turbulence exhibits an inertial range in the kinetic-energy spectrum. Both theoretical and numerical results agree on an $E(k) \propto k^{-2}$ behavior (see Fig. 3.23). The difference from the $E(k) \propto k^{-5/3}$ behavior of the Navier–Stokes turbulence originates in the nature of the small-scale events. Whereas in the Navier–Stokes case the small scales are completely characterized by the molecular viscosity ν and the dissipation rate ε, they are determined by the velocity jump across the shock $[[u]]$ and the characteristic shock separation length L in the Burgers' case. The dissipation rate is therefore estimated as

$$\varepsilon = \frac{[[u]]^3}{24L}. \qquad (3.218)$$

Because $[[u]] \sim \sqrt{12\mathcal{K}}$ and L is approximately equal to the velocity autocorrelation length scale, it is seen that, in Burgers' turbulence, small scales are determined by large-scale parameters.

- As in Navier–Stokes turbulence, the dominant mechanism within the inertial range is a kinetic-energy cascade toward the high wavenumbers, and a reverse cascade drives the small-wavenumber dynamics. Within the inertial range, the energy transfer is local in spectral space. The triadic interactions causing the most energetic transfers are distant ones, whereas most of the net kinetic-energy transfer is induced by local triadic interactions. Therefore the global picture is close to the one found in Navier–Stokes turbulence, despite the very important difference

in the topology of the velocity field, showing that the spectral features of Navier–Stokes dynamics previously mentioned are not intrisically due to pressure effects and the existence of vortices.
- Burgers' turbulence exhibits intermittency, as Navier–Stokes turbulence: Tails of the velocity-fluctuation pdf's have the same non-Gaussian behavior, whereas velocity-increment pdf's exhibit a strong departure from the normal distribution. This shows that intermittency, as a general phenomenon, is not a consequence of the existence of coherent vortices in Navier–Stokes turbulence; neither is it a pressure effect. It is due to the nonlinearity of the governing equations and to the existence of strong nonlocal interactions in Fourier space.

3.8.4 Sensitivity with Respect to Energy-Pumping Process: Turbulence with Hyperviscosity

We now address the influence of the energy-pumping process on the self-similar decay and the inertial-range behavior of isotropic turbulence. This question was investigated by Borue and Orszag (1995a, 1995b), who performed some simulations in the 3D case using the following hyperviscous generalization of the Navier–Stokes equations:

$$\frac{\partial \boldsymbol{u}}{\partial t} + \boldsymbol{u}\nabla \boldsymbol{u} = -\nabla p + \nu_p \nabla^{2p} \boldsymbol{u}, \qquad (3.219)$$

$$\nabla \cdot \boldsymbol{u} = 0, \qquad (3.220)$$

where ν_p is a hyperviscosity. The usual Navier–Stokes equations are recovered by setting $p = 1$. Borue and Orszag used $p = 8$. Their results, in both forced and freely decaying isotropic turbulence, suggest that inertial-range dynamics may be independent of the particular mechanism that governs dissipation at high wavenumbers. The usual inertial-range behavior was recovered, along with the main features of intermittency and non-Gaussianity. But, because the dissipation induced by the hyperviscosity is concentrated at higher wavenumbers than the physical one, the inertial range is observed to be larger in the former case than in the latter.

Bibliography

ANDRÉ, J.-C. AND LESIEUR, M. (1977). Influence of helicity on the evolution of isotropic turbulence at high Reynolds number, *J. Fluid Mech.* **81**, 187–207.

ANDREOTTI, B. (1997). Studying Burgers' models to investigate the physical meaning of the alignments statistically observed in turbulence, *Phys. Fluids* **9**, 735–742.

ANTONIA, R. A. (2007). Talk given at summer school on "Small-scale turbulence: theory, phenomenology and applications," August 2007, Cargèse, France.

AOYAMA, T., ISHIHARA, T., KANEDA, Y., YOKOKAWA, M., ITAKURA, K., AND UNO, A. (2005). Statistics of energy transfer in high-resolution direct numerical simulation of turbulence in a periodic box, *J. Phys. Soc. Jpn.* **74**, 3202–3212.

BENNEY, D. J. AND NEWELL, A. C. (1969). Random wave closure, *Stud. Appl. Math.* **48**, 29–53.

BETCHOV, R. (1956). An inequality concerning the production of vorticity in isotropic turbulence, *J. Fluid Mech.* **1**, 497–504.

BORUE, V. AND ORSZAG, S. A. (1995a). Self similar decay of three-dimensional homogeneous turbulence with hyperviscosity, *Phys. Rev. E* **51**, R856–R859.

BORUE, V. AND ORSZAG, S. A. (1995b). Forced three-dimensional homogeneous turbulence with hyperviscosity, *Europhys. Lett.* **29**, 687–692.

BOS, W. J. T. AND BERTOGLIO, J.-P. (2006a). Dynamics of spectrally truncated inviscid turbulence, *Phys. Fluids* **18**, 071701.

BOS, W. J. T. AND BERTOGLIO, J.-P. (2006b). A single-time two-point closure based on fluid particle displacements, *Phys. Fluids* **18**, 031706.

BOS, W. J. T., SHAO, L., AND BERTOGLIO, J. P. (2007). Spectral imbalance and the normalized dissipation rate of turbulence, *Phys. Fluids* **19**, 045101.

BRASSEUR, J. G. AND WEI, C. H. (1994). Interscale dynamics and local isotropy in high Reynolds number turbulence within triadic interactions, *Phys. Fluids* **6**, 842–870.

CAMBON, C. AND JACQUIN, L. (1989). Spectral approach to non-isotropic turbulence subjected to rotation, *J. Fluid Mech.* **202**, 295–317.

CAMBON, C., JEANDEL, D., AND MATHIEU, J. (1981). Spectral modelling of homogeneous non-isotropic turbulence, *J. Fluid Mech.* **104**, 247–262.

CHAKRABORTY, P., BALACHANDAR S., AND ADRIAN, R. J. (2005). On the relationships between local vortex identification schemes, *J. Fluid Mech.* **535**, 189–214.

CHONG, M. S., PERRY, A. E., AND CANTWELL, B. J. (1990). A general classification of three-dimensional flow fields, *Phys. Fluids* **2**, 765–777.

CICHOWLAS, C., BONATI, P., DEBBASCH, F., AND BRACHET, M. (2005). Effective dissipation and turbulence in spectrally truncated Euler flows, *Phys. Rev. Lett.* **95**, 264502.

CLARK, T. T. AND ZEMACH, C. (1998). Symmetries and the approach to statistical equilibrium in isotropic turbulence, *Phys. Fluids* **31**, 2395–2397.

COLEMAN, G. N. AND MANSOUR, N. N. (1991). Modeling the rapid spherical compression of isotropic turbulence, *Phys. Fluids* **3**, 2255–2259.

COMTE-BELLOT, G. AND CORRSIN, S. (1966). The use of a contraction to improve the isotropy of grid-generated turbulence, *J. Fluid Mech.* **25**, 657–682.

DAVIDSON, P. A. (2004). *Turbulence. An Introduction for Scientists and Engineers*, Oxford University Press.

EYINK, G. L. AND THOMSON, D. J. (2000). Free decay of turbulence and breakdown of self-similarity, *Phys. Fluids* **12**, 477–479.

FALKOVICH, G. AND LEBEDEV, V. (1997). Single-point velocity distribution in turbulence, *Phys. Rev. Lett.* **79**, 4159–4161.

FJORTOF, R. (1953). On the changes in spectral distribution of kinetic energy for two-dimensional, non-divergent flow, *Tellus* **5**, 225–230.

FOURNIER, J. D. AND FRISCH, U. (1978). d-dimensional turbulence, *Phys. Rev. A* **17**, 747–762.

FRISCH, U. (1995). *Turbulence: The Legacy of A.N. Kolmogorov*, Cambridge University Press.

GIRIMAJI, S. S. AND ZHOU, Y. (1995). Spectrum and energy transfer in steady Burgers turbulence, *ICASE Rep. No. 95-13*.

HORIUTI, K. (2001). A classification method for vortex sheet and tube structures in turbulent flows, *Phys. Fluids* **13**, 3756–3774.

HORIUTI, K. AND TAKAGI, Y. (2005). Identification method for vortex sheet structures in turbulent flows, *Phys. Fluids* **17**, 121703.

HUNT, J. C. R., WRAY, A. A., AND MOIN, P. (1988). Eddies, stream, and convergence zones in turbulent flows, *Center for Turbulence Research Report* CTR-S88, 193–208.

ISHIDA, T., DAVIDSON, P. A., AND KANEDA, Y. (2006). On the decay of isotropic turbulence, *J. Fluid Mech.* **564**, 455–475.

ISHIHARA, T. AND KANEDA, Y. (2001). Energy spectrum in the enstrophy transfer range of two-dimensional forced turbulence, *Phys. Fluids* **13**, 544–547.

ISHIHARA, T., KANEDA, Y., YOKOKAWA, M., ITAKURA, K., AND UNO, A. (2005). Energy spectrum in the near dissipation range of high resolution DNS of turbulence, *J. Phys. Soc. Jpn.* **74**, 1464–1471.

JEONG, J. AND HUSSAIN, F. (1995). On the identification of a vortex, *J. Fluid Mech.* **285**, 69–94.

JIMENEZ, J. (1992). Kinematic alignment effects in turbulent flows, *Phys. Fluids* **4**, 652–654.

JIMENEZ, J. (2000). Intermittency and cascades, *J. Fluid Mech.* **409**, 99–120.

JIMENEZ, J. AND WRAY, A. (1998). On the characteristics of vortex filaments in isotropic turbulence, *J. Fluid Mech.* **373**, 255–285.

JIMENEZ, J., WRAY, A., SAFFMAN, P. G. AND ROGALLO, R. (1993). The structure of intense vorticity in isotropic turbulence, *J. Fluid Mech.* **255**, 65–90.

KANEDA, Y. (1993). Lagrangian and Eulerian time correlations in turbulence, *Phys. Fluids A* **5**, 2835–2845.

Bibliography

KANEDA, Y. (2007). Lagrangian renormalized approximation of turbulence, *Fluid Dyn. Res.* **39**, 526–551.

KIDA, S. AND OHKITANI, K. (1992a). Spatiotemporal intermittency and instability of a forced turbulence, *Phys. Fluids* **4**, 1018–1027.

KIDA, S. AND OHKITANI, K. (1992b). Fine structure of energy transfer in turbulence, *Phys. Fluids* **4**, 1602–1604.

KRAICHNAN, R. H. (1967). Inertial ranges in two-dimensional turbulence, *Phys. Fluids* **10**, 1417–1423.

KRAICHNAN, R. H. (1971). Inertial-range transfer in two- and three-dimensional turbulence, *J. Fluid Mech.* **47**, 525–535.

KRAICHNAN, R. H. (1976). Eddy-viscosity in two and three dimensions, *J. Atmos. Sci.* **33**, 1521–1536.

KRAICHNAN, R. H. AND PANDA, R. (1988). Depression of nonlinearity in decaying isotropic turbulence, *Phys. Fluids* **31**, 2395–2397.

LESIEUR, M. (1997). *Turbulence in Fluids*, 3rd revised ed., Kluwer Academic.

LESIEUR, M. AND SCHERTZER, D. (1978). Amortissement auto-similaire d'une turbulence à grand nombre de Reynolds, *J. Méc.* **17**, 609–646.

LUND, T. S. AND ROGERS, M. M. (1994). An improved measure of strain rate probability in turbulent flows, *Phys. Fluids* **6**, 1838–1847.

MANLEY, O. P. (1992). The dissipation range spectrum. *Phys. Fluids* **4**, 1320–1321.

MATHIEU, J. AND SCOTT, J. (2000). *An Introduction to Turbulent Flow*, Cambridge University Press.

MAZELIER, N. AND VASSILICOS, J. C. (2008). The turbulence dissipation constant is not universal because of its universal dependence on large-scale flow topology, *Phys. Fluids* **20**, 015101.

MILLIONSCHIKOV, M. D. (1941). Theory of homogeneous isotropic turbulence, *Dokl. Akad. Nauk SSSR* **32**, 22–24.

MOHAMED, M. S. AND LARUE, J. C. (1990). The decay power law in grid-generated turbulence, *J. Fluid Mech.* **219**, 195–214.

MOISY, F. AND JIMENEZ, J. (2004). Geometry and clustering of intense structures in isotropic turbulence, *J. Fluid Mech.* **513**, 111–133.

MONIN, A. S. AND YAGLOM, A. M. (1975). *Statistical Fluid Mechanics*, MIT Press, Vol. 1.

NOMURA, K. K. AND POST, G. K. (1998). The structure and the dynamics of vorticity and rate of strain in incompressible homogeneous turbulence, *J. Fluid Mech.* **377**, 65–97.

NOULLEZ, A. GURBATOV, S. N., AURELL, E., AND SIMDYANKIN, S. I. (2005). Global picture of self-similar and non-self-similar decay in Burgers turbulence, *Phys. Rev. E* **71**, 056305.

NOULLEZ, A. AND PINTON, J. F. (2002). Global fluctuations in decaying Burgers turbulence, *Eur. Phys. J. B* **28**, 231–241.

NOULLEZ, A. AND VERGASSOLA, M. (1994). A fast Legendre transfrom algorithm and applications to the adhesion model, *J. Sci. Comput.* **9**, 259–281.

NOULLEZ, A., WALLACE, G., LEMPERT, W., MILES, R. B., AND FRISCH, U. (1997). Transverse velocity increments in turbulent flow using the RELIEF technique, *J. Fluid Mech.* **339**, 287–307.

OBERLACK, M. (2002). On the decay exponent of isotropic turbulence, *Proc. Appl. Math. Mech.* **1**, 294–297.

O'Brien, E. F. and Francis, G. C. (1963). A consequence of the zero fourth cumulant approximation, *J. Fluid Mech.* **13**, 369–382.

Ogura, Y. (1963). A consequence of the zero fourth cumulant approximation in the decay of isotropic turbulence, *J. Fluid Mech.* **16**, 33–40.

Orszag, S. A. (1970). Analytical theories of turbulence, *J. Fluid Mech.* **41**, 363–386.

Pao, Y. M. (1965). The structure of turbulent velocity and scalar fields at large wave numbers, *Phys. Fluids* **8**, 1063–1075.

Piquet, J. (2001). *Turbulent Flows. Models and Physics*, 2nd ed., Springer.

Pouquet, A., Lesieur, M., André, J.-C., and Basdevant, C. (1975). Evolution of high Reynolds number two-dimensional turbulence, *J. Fluid Mech.* **75**, 305–319.

Proudman, I. and Reid, W. H. (1954). On the decay of a normally distributed and homogeneous turbulent velocity field, *Philos. Trans. R. Soc. London A* **297**, 163–189.

Rossi, M. (2000). Of vortices and vortical layers: An overview, in *Vortex Structure and Dynamics*, Maurel, A. and Petitjeans, P., eds., Lecture Notes in Physics, Springer, pp. 40–123.

Siggia, E. D. (1981). Numerical study of small scale intermittency in three dimensional turbulence, *J. Fluid Mech.* **107**, 375–406.

Skrbek, L. and Stalp, S. R. (2000). On the decay of homogeneous isotropic turbulence, *Phys. Fluids* **12**, 1997–2019.

Sreenivasan, K. R. (1995). On the universality of the Kolmogorov constant, *Phys. Fluids* **7**, 2778–2784.

Suzuki, E., Nakano, T., Takashi, N., and Gotoh, T. (2005). Energy transfer and intermittency in four-dimensional turbulence, *Phys. Fluids* **17**, 081702.

Tanaka, M. and Kida, S. (1993). Characterization of vortex tubes and sheets, *Phys. Fluids* **5**, 2079–2082.

Tsinober, A. (2001) *An Informal Introduction to Turbulence*, Kluwer Academic.

Verzicco, R., Jimenez, J., and Orlandi, P. (1995). Steady columnar vortices under local compression, *J. Fluid Mech.* **299**, 367–388.

Vignon, J. M. and Cambon, C. (1980). Thermal spectral calculation using EDQNM theory, *Phys. Fluids* **23**, 1935–1937.

Waleffe, F. (1992). The nature of triad interactions in homogeneous turbulence, *Phys. Fluids* **4**, 350–363.

Waleffe, F. (1993). Inertial transfers in the helical decomposition, *Phys. Fluids* **5**, 677–685.

Zakharov, V. E., Lvov, V., and Falkowitch, G. (1992). *Wave Turbulence*, Springer.

Zhou, J., Adrian, R. J., Balachandar, S., and Kendall, T. M. (1999). Mechanisms for generating coherent packets of hairpin vortices, *J. Fluid Mech.* **387**, 353–396.

4 Incompressible Homogeneous Anisotropic Turbulence: Pure Rotation

4.1 Physical and Numerical Experiments

Rotation of the reference frame is an important factor in certain mechanisms of flow instability, and the study of rotating flows is interesting from the point of view of turbulence modeling in fields as diverse as engineering (e.g., turbomachinery and reciprocating engines with swirl and tumble), geophysics, and astrophysics. Effects of mean curvature or of advection by a large eddy can be tackled by use of similar approaches. In this chapter, the emphasis is put on the dynamics of turbulence subjected to solid-body rotation with constant angular velocity. Considering rotation with angular velocity Ω around the axis e_3, the mean-flow-gradient matrix and mean-flow-displacement-gradient matrix are given by the following expressions (see Subsection 2.1.6):

$$\mathbf{A} = \begin{bmatrix} 0 & -\Omega & 0 \\ \Omega & 0 & 0 \\ 0 & 0 & 0 \end{bmatrix}, \quad \mathbf{F} = \begin{bmatrix} \cos \Omega t & -\sin \Omega t & 0 \\ \sin \Omega t & \cos \Omega t & 0 \\ 0 & 0 & 1 \end{bmatrix}. \quad (4.1)$$

This expression for **A** is obtained by setting $S = 0$ in Eq. (2.48).

Some commonly agreed statements have been drawn from several experimental, theoretical, and numerical studies, in which rotation is suddenly applied to homogeneous turbulence. The main results are summarized as follows:

- Rotation inhibits the energy cascade, so that the dissipation rate is reduced (Bardina, Ferziger, and Rogallo, 1985; Jacquin et al., 1990). This is illustrated in Fig. 4.1.
- The initial 3D isotropy is broken, so that a moderate anisotropy, consistent with a transition from a 3D to a 2D state, can develop. Anisotropy is more reflected in integral length scales with various components than in Reynolds stresses (Jacquin et al., 1990; Cambon, Mansour, and Godeferd, 1997). Typical results are shown in Fig. 4.2.
- Elongated vortical structures are generated with asymmetry in terms of cyclonic and anticyclonic axial vorticity (Bartello, Métais, and Lesieur, 1994), structures with cyclonic vorticity being observed to be dominant. *Cyclonic eddies* are characterized by a positive fluctuating vorticity in the axial direction $\omega_\parallel = \omega_i n_i > 0$ $(= \omega_3$ with a particular choice of frame), seen in the rotating frame: They

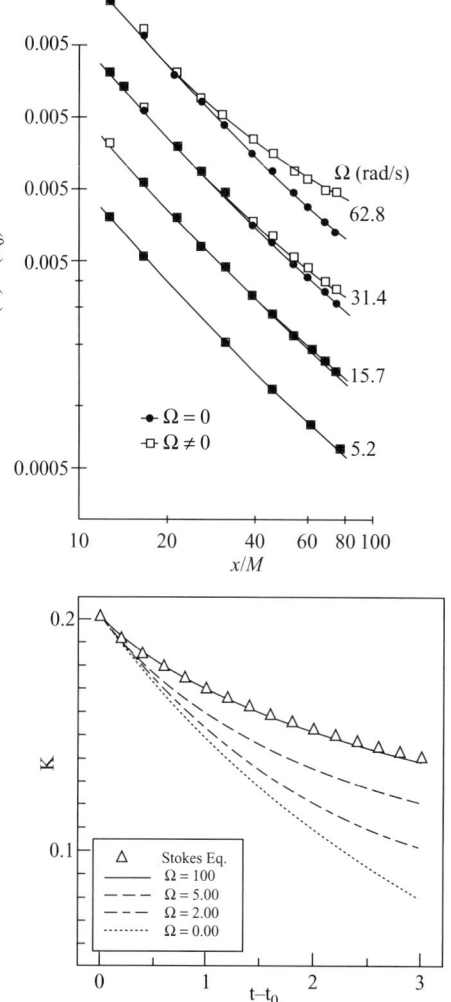

Figure 4.1. Time evolution of turbulent kinetic energy in initially isotropic turbulence submitted to solid-body rotation. Top: experimental data from Jacquin et al. (1990); Bottom: DNS data from Morinishi, Nakabayashi, and Ren (2001). The decay rate is observed to be a decreasing function of the rotation rate Ω. At a very high rotation rate, DNS results perfectly match the decay recovered considering the linear Stokes equations, showing that the nonlinear mechanisms are totally inhibited. The rotation rate Ω is not expressed in the same units in the two plots.

correspond to eddies rotating with the same sense as system vorticity. Negative axial vorticity characterizes *anticyclonic eddies* in the same conditions.

- If the turbulence is initially anisotropic, the "rapid" effects of rotation (i.e., the linear dynamics described by the RDT approach) conserve a part of the anisotropy

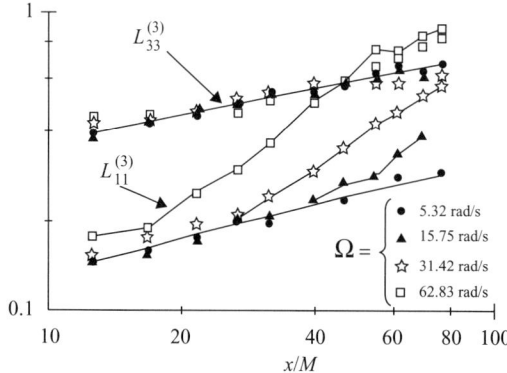

Figure 4.2. Time evolution of the turbulence integral length scales in initially isotropic turbulence submitted to solid-body rotation. Experimental data from Jacquin et al. (1990).

4.1 Physical and Numerical Experiments

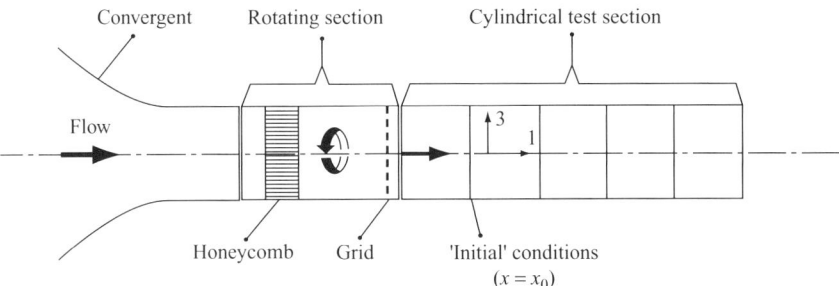

Figure 4.3. ONERA's experimental setup for initialy decaying isotropic turbulence submitted to solid-body rotation. (Courtesy of O. Leuchter).

[called *directional anisotropy* $b_{ij}^{(e)}$] and damp the other part [called *polarization anisotropy* $b_{ij}^{(z)}$], resulting in a spectacular change of the anisotropy b_{ij} of the RST (Cambon, Jacquin, and Lubrano, 1992; Kassinos, Reynolds, and Rogers, 2001).

These effects, which are not at all taken into account by current one-point second-order closure models (from $\mathcal{K} - \varepsilon$ to $\overline{u'_i u'_j} - \varepsilon$ models), have motivated new single-point modeling approaches by Cambon, Jacquin, and Lubrano (1992) and Mansour, Cambon, and Speziale (1991), and to a lesser extent by Kassinos, Reynolds, and Rogers (2001) for linear (or "rapid") effects only. It is worth noticing that the modification of the dynamics by the rotation ultimately comes from the presence of inertial waves (Greenspan, 1968). Inertial waves have an anisotropic dispersion law. They are capable of changing the initial anisotropy of the turbulent flow and also can affect the nonlinear dynamics. This explains the relevance of spectral theory to study HAT with mean-flow rotation.

4.1.1 Brief Review of Experiments, More or Less in the Configuration of Homogeneous Turbulence

A first class of experiments consists of decaying grid turbulence in a wind tunnel, in which a rotation generator creates a constant angular velocity in the streamwise direction. In the experiment by Traugott (1958), the rotation was imposed on the annular region between two coaxial rotating disks. This study is mentioned for our record, but very little information can be obtained because of the nonuniformity of the mean flow and the very short length of the duct. The experiment by Wigeland and Nagib (1978) introduced a much better rotation generator. In this setup, solid-body rotation was enforced thanks to a cylindrical rotating honeycomb, with the grid just behind and attached to the rotating cylinder. This rotating device was successfully used by Jacquin et al. (1990) (see Fig. 4.3), but in a much larger wind tunnel, fulfilling homogeneity with an excellent accuracy and yielding reliable results about kinetic-energy decay, Reynolds stress components, integral length scales, and 1D energy spectra. The results can be analyzed as in a conventional grid-generated turbulence experiment, the streamwise spatial coordinate playing the role of elapsed time. A large range of "initial" (in the first transverse section chosen closely downstream of the grid) Rossby numbers was covered. The only drawback to this experiment was the initially moderate value, and monotonic decrease, of the Reynolds number, as is usual in grid-generated turbulence.

A second class of experiments was carried out in a rotating tank, a diffusive turbulence being created by an oscillating grid (Hopfinger, Browand, and Gagne, 1982; Dickinson and Long, 1984) near the bottom of the tank. The turbulence is essentially inhomogeneous in the vertical axial direction and statistically steady. The steady state is different in different horizontal sections, moving away from the grid. In the experiment by Ibbetson and Tritton (1975), an unsteady turbulence was created by the initial motion of perforated plates in a rotating torical annulus with square sections.

Another interesting experiment was carried out by McEwan (1976). A radially uniform, small-scale mixing pattern without advection in the radial plane was obtained in a rotating transparent cylinder by injecting and pumping fluid from the perforated plane-ended bottom. Polystyrene beads were suspended in the fluid and illuminated by a stroboscope to visualize the flow motion. Without rotation, the jets caused randomly shaped wiggly particle paths, whereas intense vortices, always cyclonic, were observed in the presence of rotation.

The development of particle image velocimetry (PIV) measurements was beneficial to these types of experiments, with renewed interest in the spatial structure of the rotating flow. The recent experiment by Baroud et al. (2003) used a very special forcing by jets in an annulus in the rotating tank. Despite specific nonhomogeneity and anisotropy generated by the forcing process, in a way difficult to control and to compare with the sole effect of rotation, useful conclusions about scaling laws were drawn from the experimental results. Some of these results were consistent with those of the experiment by Simand (2002) dealing with turbulence near the core of a strong vortex; they will be discussed at the end of this chapter (see Subsection 4.7.5). Another experiment by Praud, Sommeria, and Fincham (2006), carried out in the Coriolis platform in Grenoble, deserves attention. It will be discussed in Chapter 8, as it combined solid-body rotation and vertical stable stratification. In addition, the use of a rake instead of a grid yielded preferential forcing of the horizontal motion.

The recent experiment by Morize, Moisy, and Rabaud (2005) has something to do with the experiment by Ibbetson and Tritton (1975), with the grid being moved in only the first phase, before the free decay is studied in a rotating tank. This experiment aimed to reproduce initially homogeneous isotropic turbulence suddenly set into solid-body rotation, but at a higher Reynolds number than in Jacquin's experiment, and using modern PIV anemometry. Nonhomogeneity and anisotropic forcing are prevented, as far as possible, by moving the grid, only in the phase of generation of initial turbulence, in the whole vertical extent of the tank. Another experimental study by Staplehurst, Davidson, and Dalziel (2008) is in progress, and appears to fulfill even better homogeneous conditions than the latter one, especially in removing some mean-flow spurious components.

Detailed results of these experiments are not discussed in this section. Results dealing with quasi-homogeneous turbulence are emphasized and discussed throughout this chapter; inhomogeneous effects are briefly considered in the last section.

4.2 Governing Equations

4.2.1 Generals

The problem of turbulence subjected to solid-body rotation can be directly related to the case of turbulence in the presence of a mean flow with space-uniform gradients, provided a purely antisymmetric mean-velocity-gradient matrix is chosen, i.e., $A_{ij} = \epsilon_{ikj}\Omega_k$, where Ω is the angular velocity, in agreement with Eq. (4.1). But it is simpler to work with a coordinate system and velocity vectors defined in the steadily rotating frame. In this *non-Galilean* frame, rotation of the frame only introduces inertial forces, namely centrifugal and Coriolis forces. Because the former can be incorporated into the pressure term, only the latter has to be explicitly taken into account when the Navier–Stokes equations are written in the rotating frame:

$$\frac{\partial \mathbf{u}}{\partial t} + 2\mathbf{\Omega} \times \mathbf{u} + \nabla p = \nu \nabla^2 \mathbf{u} - \mathbf{u} \cdot \nabla \mathbf{u}. \tag{4.2}$$

As usual in incompressible fluid dynamics, the pressure term is completely determined by the solenoidal condition $\nabla \cdot \mathbf{u} = 0$. Taking the divergence and the curl of these equations yields

$$\nabla^2 p - 2\mathbf{\Omega} \cdot \boldsymbol{\omega} = -\frac{\partial u_j}{\partial x_i}\frac{\partial u_i}{\partial x_j} \tag{4.3}$$

and

$$\frac{\partial \omega_i}{\partial t} - 2\Omega_l \frac{\partial u_i}{\partial x_l} = \omega_l \frac{\partial u_i}{\partial x_l} + \nu \nabla^2 \omega_i. \tag{4.4}$$

In all the equations just given, nonlinear and viscous terms are gathered on the right-hand side.

4.2.2 Important Nondimensional Numbers. Particular Regimes

By use of a reference velocity scale U and a reference length scale L, so that $u_i = \tilde{u}_i U$, $x_i = \tilde{x}_i L$, $\nabla = \tilde{\nabla}\frac{1}{L}$, Eq. (4.2) becomes

$$\frac{\partial \tilde{\mathbf{u}}}{\partial \tilde{t}} + \frac{1}{Ro}\mathbf{n} \times \tilde{\mathbf{u}} - \frac{1}{Re}\tilde{\nabla}^2 \tilde{\mathbf{u}} + \frac{L}{U^2}\nabla p = -\tilde{\mathbf{u}} \cdot \tilde{\nabla}\tilde{\mathbf{u}}, \tag{4.5}$$

in which only nondimensional quantities appear (except in the pressure term, which is subsequently further discussed). The unit vector $\mathbf{n}$ is chosen so that $\mathbf{\Omega} = \mathbf{n}\Omega$. In addition to the Reynolds number Re, the Rossby number,

$$Ro = \frac{U}{2\Omega L}, \tag{4.6}$$

is displayed. In the latter equation, the time scale was taken equal to L/U. Another possibility is to choose $1/2\Omega$ as the time scale, leading to $t = 2\Omega \tilde{t}$,* and

$$\frac{\partial \tilde{\mathbf{u}}}{\partial \tilde{t}} + \mathbf{n} \times \tilde{\mathbf{u}} - \frac{Ro}{Re}\tilde{\nabla}^2 \tilde{\mathbf{u}} + \frac{1}{2\Omega U}\nabla p = -Ro\,\tilde{\mathbf{u}} \cdot \tilde{\nabla}\tilde{\mathbf{u}}, \tag{4.7}$$

---

* Possible slightly different scalings are $t = \tilde{t}\frac{\Omega}{2\pi}$ and $t = \tilde{t}\frac{2\Omega}{2\pi}$.

so that the Rossby number affects only the nonlinear term. The term Ro/Re is the inverse of the Ekman number. The linear inviscid limit is recovered by discarding the right-hand-side, assuming a very low Rossby number and a very high Ekman number. The pressure term is not so easy to treat. Following Eq. (4.3), it must be split into a linear part and a nonlinear part that scale as $\frac{1}{Ro}U^2/L$ and U^2/L, respectively.

Applying the same scaling to Eq. (4.4) yields

$$\frac{\partial \tilde{\omega}_i}{\partial \tilde{t}} - \frac{1}{Ro} n_l \frac{\partial \tilde{u}_i}{\partial \tilde{x}_l} = \tilde{\omega}_l \frac{\partial \tilde{u}_i}{\partial \tilde{x}_l} + \frac{1}{Re} \tilde{\nabla}^2 \tilde{\omega}_i \tag{4.8}$$

in the first case, and

$$\frac{\partial \tilde{\omega}_i}{\partial \tilde{t}} - n_l \frac{\partial \tilde{u}_i}{\partial \tilde{x}_l} = Ro \tilde{\omega}_l \frac{\partial \tilde{u}_i}{\partial \tilde{x}_l} + \frac{Ro}{Re} \tilde{\nabla}^2 \tilde{\omega}_i \tag{4.9}$$

in the second case. The *Proudman theorem* is conventionally derived from Eq. (4.8), in the limit of zero-Rossby number, so that

$$n_j \frac{\partial \tilde{u}_i}{\partial \tilde{x}_j} = 0 \tag{4.10}$$

characterizes a 2D state, in the sense that the dependency of velocity on the axial coordinate $x_\parallel = \boldsymbol{x} \cdot \boldsymbol{n}$ vanishes at a high rotation rate. Accordingly, the velocity equation reduces to the *geostrophic balance* in the same conditions, or

$$\tilde{\nabla} \tilde{p} = \boldsymbol{n} \times \tilde{\boldsymbol{u}}. \tag{4.11}$$

On the other hand, if Ω^{-1} is chosen as the time scale, the zero-Rossby limit leads to only the linear regime, i.e.,

$$\frac{\partial \tilde{\boldsymbol{u}}}{\partial \tilde{t}} + \boldsymbol{n} \times \tilde{\boldsymbol{u}} + \tilde{\nabla} \tilde{p} = 0, \tag{4.12}$$

$$\tilde{\nabla}^2 \tilde{p} - \boldsymbol{n} \cdot \tilde{\boldsymbol{\omega}} = 0, \tag{4.13}$$

$$\frac{\partial \tilde{\omega}_i}{\partial \tilde{t}} - n_l \frac{\partial \tilde{u}_i}{\partial \tilde{x}_l} = 0. \tag{4.14}$$

It is important to note that the conditions for having a complete two-dimensionalization are very stringent, as both linear and steady limits must be reached at the same time. Taylor columns were found in beautiful historical experiments of rotating laminar flows (Taylor, 1921), for instance when Taylor *slowly* pushed a coin in the bottom of his rotating tank. In a rapidly rotating turbulent flow, it is clear that these conditions are not fulfilled at small but nonzero-Rossby numbers because nonlinear effects, even weak at a given time, can accumulate over a long time and induce a modified cascade, which is not necessarily the conventional 2D cascade. Anyway, the transition from a 3D to a 2D structure is essentially an unsteady – transition requires evolution! – and nonlinear process, as will be subsequently seen. The linear regime consists of unsteady wave motion, which becomes steady (zero dispersion frequency) in only the 2D limit. Consequently, the Taylor–Proudman theorem (Proudman, 1916; Taylor, 1921) will be used in a restricted sense here: The steady mode of the motion is the 2D mode in the linear regime of rapidly rotating flow. Nondimensional equations will no longer be used in this chapter, but

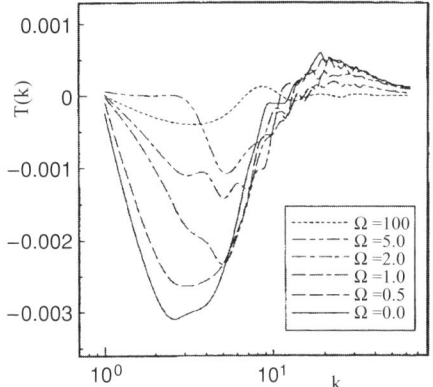

Figure 4.4. Spherically averaged energy transfer function $T(k)$ in initially decaying isotropic turbulence submitted to solid-body rotation for different rotation rates. From Morinishi, Nakabayashi, and Ren (2001), with permission of the American Institute of Physics.

the choice of relevant scales, L, U, and Ω, remains an important issue, allowing us to define different, e.g., macro and micro, Rossby and Reynolds numbers.

4.3 Advanced Analysis of Energy Transfer by DNS

The striking decrease in the dissipation rate of kinetic energy has been intensively investigated by use of DNS. The main findings are subsequently summarized before being analyzed in the rest of this chapter through linear and nonlinear theories.

The main observations are as follows:

- Rotation induces a deep modification of the kinetic-energy transfer function $T(k)$. Spherically averaged profiles of $T(k)$ are displayed in Fig. 4.4. Both its shape and amplitude are drastically modified, resulting in a dramatic reduction of the kinetic-energy cascade.
- Both the forward and the reverse energy cascades are affected by rotation. They both vanish, as illustrated in Fig. 4.5.
- This modification is due to the so-called *phase-scrambling* phenomenon,[†] which originates in the fact that the transfer function $T(k)$ is generated by triadic contributions that are differentially affected by oscillations, depending on the angle θ between the wave vector $\boldsymbol{k}$ and the rotation vector $\boldsymbol{\Omega}$, not to mention similar effects on the other vectors of each triad. This is illustrated in Fig. 4.6, which displays $T(k, \cos\theta)$ for different values of Ω. The usual dynamical picture is recovered in the case $\Omega = 0$, whereas in the cases $\Omega \neq 0$, regions with negative–positive values of $T(k)$ are more and more mixed, leading to a weakening of the kinetic-energy cascade.
- The effect of rotation is visible for a certain range of Rossby numbers only. A very small rotation rate yields a negligible influence of rotation, whereas very high rotation rates lead to an almost complete inhibition of the nonlinear kinetic-energy cascade, resulting in a "frozen" field submitted to linear viscous effects. Partial two-dimensionalization and two-componentalization, resulting from fully nonlinear dynamics, are illustrated in Fig. 4.7, which presents the evolution of

---

[†] This effect could be perhaps better denoted as phase mixing, in connection with the very angle-dependent dispersion relation of inertial waves, as we will discuss on the grounds of basic equations.

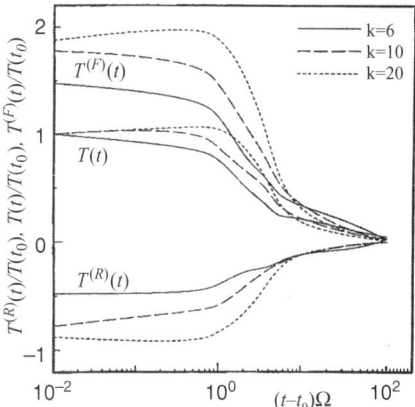

Figure 4.5. Time evolution of the full transfer function $T(k)$, the forward transfer function $T^{(F)}(k)$, and the backward transfer function $T^{(R)}(k)$ in initially decaying isotropic turbulence submitted to solid-body rotation for different rotation rates. One has $T(k) = T^{(F)}(k) + T^{(R)}(k)$. Different values of k are considered. From Morinishi, Nakabayashi, and Ren (2001), with permission of the American Institute of Physics.

directional anisotropy component $b_{33}^{(e)}$ and the polarization anisotropy coefficient $b_{33}^{(z)}$ as functions of the Rossby number. The meaning and the behavior of these descriptors will be rediscussed at length throughout this chapter.

Of course, present DNS results are limited in terms of Reynolds numbers and elapsed time, and a very long evolution time Ωt is required for capturing nonlinear

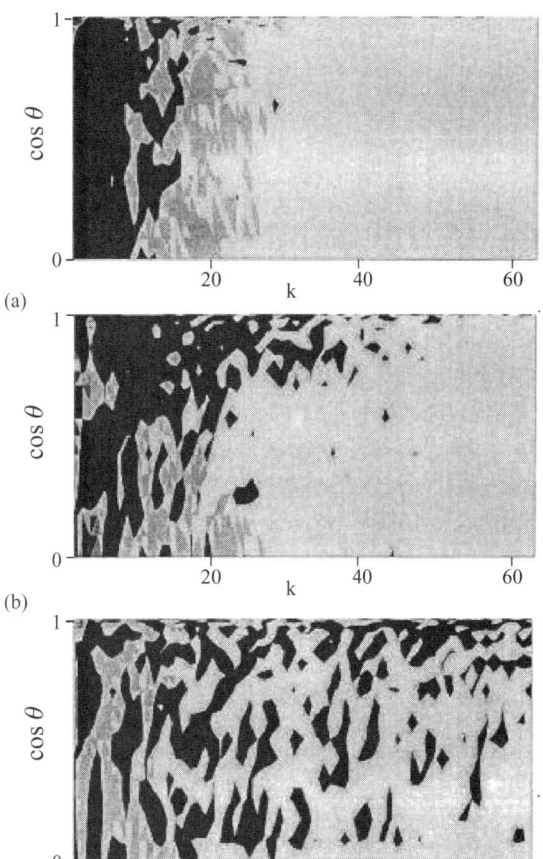

Figure 4.6. Instantaneous plot of the transfer function $T(k)$ as a function of k and θ, which is defined as the angle between $\boldsymbol{k}$ and $\boldsymbol{\Omega}$. Top, no rotation ($\Omega = 0$); middle, medium rotation rate; bottom, strong rotation rate. The dark region corresponds to negative value, and the other regions to positive values. From Morinishi, Nakabayashi, and Ren (2001), with permission of the American Institute of Physics.

4.4 Balance of RST Equations. A Case Without "Production."

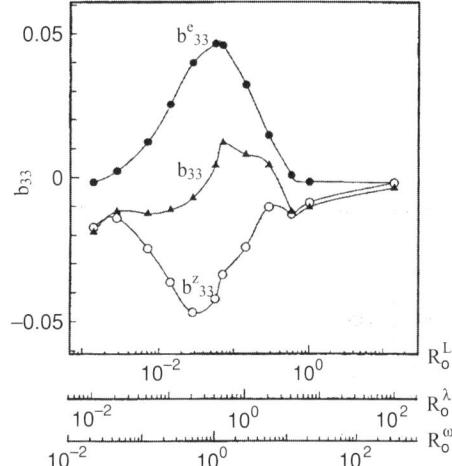

Figure 4.7. Evolution of the directional anisotropy component $b_{33}^{(e)}$) and the polarization anisotropy component $b_{33}^{(z)}$) as functions of the Rossby number. Three Rossby numbers are shown: the macro-Rossby number $Ro^L = u'/2\Omega L$, the Taylor micro-Rossby number $Ro^\lambda = u'/2\Omega\lambda$, and the micro-Rossby number $Ro^\omega = \omega'/2\Omega$, in which $L = \mathcal{K}^{3/2}/\varepsilon$ and ω' denotes the rms vorticity fluctuation. From Morinishi, Nakabayashi, and Ren (2001), with permission of the American Institute of Physics.

effects at very low Rossby numbers. This explains some of the discrepancies observed among DNS, LES, and statistical theory, as discussed later on. Fortunately, a consistent core of agreed statements arises from this threefold approach, not to mention experimental data.

4.4 Balance of RST Equations. A Case Without "Production." New Tensorial Modeling

The Reynolds stress equations for HAT submitted to solid-body rotation defined by Eq. (4.1) are

$$\frac{\partial \overline{u'_i u'_j}}{\partial t} + \overline{u}_k \frac{\partial \overline{u'_i u'_j}}{\partial x_k} = -\Omega \begin{bmatrix} -2\overline{u'_1 u'_2} & \overline{u'_1 u'_1} - \overline{u'_2 u'_2} & -\overline{u'_2 u'_3} \\ \overline{u'_1 u'_1} - \overline{u'_2 u'_2} & 2\overline{u'_1 u'_2} & \overline{u'_1 u'_3} \\ -\overline{u'_2 u'_3} & \overline{u'_1 u'_3} & 0 \end{bmatrix} + \Pi_{ij} - \varepsilon_{ij}. \quad (4.15)$$

A careful examination of Eq. (4.15) reveals that the production term is identically zero if the turbulent field is isotropic at the initial time. Therefore explicit coupling between the mean flow and the turbulent field is not responsible for the triggering of the departure from isotropy; pressure effects are responsible for this.

Considering only the most relevant Reynolds stress components (this is more general than specifying initial isotropy or initial anisotropy), so that $\overline{u'_\alpha u'_3} = \overline{u'_3 u'_\alpha} = 0$, $\alpha = 1, 2$, system (4.15) simplifies as

$$\begin{cases} \dfrac{d\overline{u'_1 u'_1}}{dt} = +2\Omega \overline{u'_1 u'_2} & +\Pi_{11} - \varepsilon_{11} \\ \dfrac{d\overline{u'_2 u'_2}}{dt} = -2\Omega \overline{u'_1 u'_2} & +\Pi_{22} - \varepsilon_{22} \\ \dfrac{d\overline{u'_3 u'_3}}{dt} = & \Pi_{33} - \varepsilon_{33} \\ \dfrac{d\overline{u'_1 u'_2}}{dt} = \Omega(\overline{u'_2 u'_2} - \overline{u'_1 u'_1}) +\Pi_{12} - \varepsilon_{12} \end{cases} \quad (4.16)$$

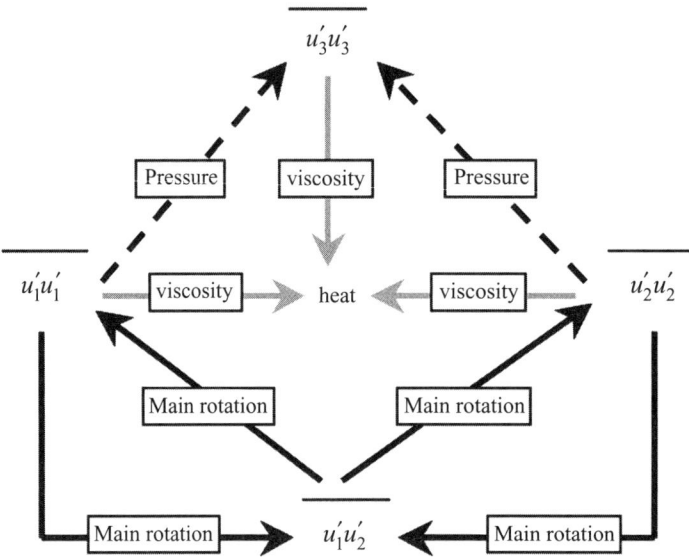

Figure 4.8. Couplings between the different nonvanishing Reynolds stresses in the pure rotation case. Arrows indicate the production process, their color being related to the physical quantity at play (mean strain, pressure, viscosity).

The different couplings are illustrated in Fig. 4.8.

These equations can be rearranged to diagonalize the production term, introducing two deviatoric components (Cambrano, Jacquin, and Lubrano, 1992):

$$A = \overline{u'_3 u'_3} - \frac{1}{2}(\overline{u'_1 u'_1} + \overline{u'_2 u'_2}); \quad B = \frac{1}{2}\overline{(u'_1 + \iota u'_2)^2}.$$

The preceding system of equations results in

$$\begin{cases} \dfrac{d\mathcal{K}}{dt} = -\varepsilon \\ \dfrac{dA}{dt} = \Pi_A^{(r)} + \Pi_A^{(s)} \\ \dfrac{dB}{dt} = 4\iota\Omega B + \Pi_B^{(r)} + \Pi_B^{(s)} \end{cases} \quad (4.17)$$

in which pressure–strain-rate and dissipation-rate components are derived in a trivial way, following the rules for deriving $\mathcal{K}, A, B$ from the Reynolds stress original components. For convenience, pressure–strain-rate contributions are split into a "rapid" linear [superscript (r)] and a "slow" nonlinear [superscript (s)] contribution. Possible deviatoric contributions from the dissipation tensor are included in the "slow" term.

Almost all the principles for single-point modeling are questioned in the case of rotating turbulence. Looking at the turbulent kinetic energy, the exact $\mathcal{K} - \varepsilon$ equations do not include any explicit additional term with respect to the isotropic non-rotating case, because the Coriolis force produces no work [as evidenced from the first equation of (4.17)]. The only way to take into account alteration of the kinetic-energy decay is to modify the $C_{\varepsilon 2}$ constant in the evolution equation for turbulent

4.4 Balance of RST Equations. A Case Without "Production."

dissipation (3.68). Empirical ways to render this constant sensitive to the Rossby number (Bardina, Ferziger, and Rogallo, 1985; Aupoix, 1983) are discussed in Cambon, Jacquin, and Lubrano (1992).

A more rational way consists of modeling the imbalance between the production (nonlinear gradient self-amplification) and the destruction (dissipation) in the ε-equation, as reported in Cambon, Mansour, and Godeferd (1997). From the results of Subsection 3.7.3, the exact equation for the dissipation rate is

$$\frac{d\varepsilon}{dt} = 2\nu \overline{\omega'_i \omega'_j u'_{i,j}} - 2\nu^2 \overline{\omega'_{i,j} \omega'_{i,j}},$$

in rotating and nonrotating homogeneous turbulence, without any *explicit* contribution from the Coriolis force. By use of an adequate scaling, the velocity-derivative skewness S_k is linked to the enstrophy-production term by the following relation:

$$S_k = \frac{6\sqrt{15}}{7} \nu \overline{\omega'_i \omega'_j u'_{i,j}} \frac{\mathcal{K}}{\varepsilon^2} Re^{-1/2},$$

in which Re is a macro-Reynolds number and the numerical prefactor comes from the conventional definition of the skewness used by experimentalists in isotropic turbulence. A second nondimensional parameter was defined by Mansour, Cambon, and Speziale (1991) to account for the departure of the enstrophy-destruction term from its conventional evaluation in the nonrotating case:

$$G = \frac{3\sqrt{15}}{7} \left[2\nu^2 \overline{\omega'_{i,j} \omega'_{i,j}} - C_{\varepsilon 2}(Re) \frac{\varepsilon^2}{\mathcal{K}} \right] \frac{\mathcal{K}}{\varepsilon^2} Re^{-1/2}.$$

A modified ε-equation is then recovered:

$$\frac{d\varepsilon}{dt} = \left[\frac{7}{3\sqrt{15}} (S_k - G) Re^{1/2} - C_{\varepsilon 2}(Re) \right] \frac{\varepsilon^2}{\mathcal{K}}, \quad (4.18)$$

where S_k is the only term that accounts for the triple correlations directly affected by rotation. The nonrotating case is simply recovered taking $S_k = G$, whereas a four-equation model (whose unknowns are $\mathcal{K}$, ε, S_k, and G) was proposed by Mansour, Cambon, and Speziale (1991) in the rotating case. This model was supported by the EDQNM model and full DNS in Cambon, Mansour, and Godeferd (1997), with the following asymptotic model for the velocity-gradient skewness:

$$S_k = \frac{S_k(0)}{\sqrt{1 + 2/Ro_\omega^2}}, \quad Ro_\omega = \frac{\omega'}{2\Omega},$$

where Ro_ω is the micro-Rossby number and $S_k(0) \sim -0.49$ is the asymptotic value in isotropic turbulence without rotation. An almost perfect collapse onto this curve was also recovered in the experimental study by Morize, Moisy, and Rabaud (2005).

The accuracy of description, prediction, or both, provided by conventional single-point models is even worse regarding the anisotropy. In spite of the strong anisotropy evidenced in two-point (or spectral) descriptions, which is mainly reflected by the integral length scales in physical space, the deviatoric part of the RST

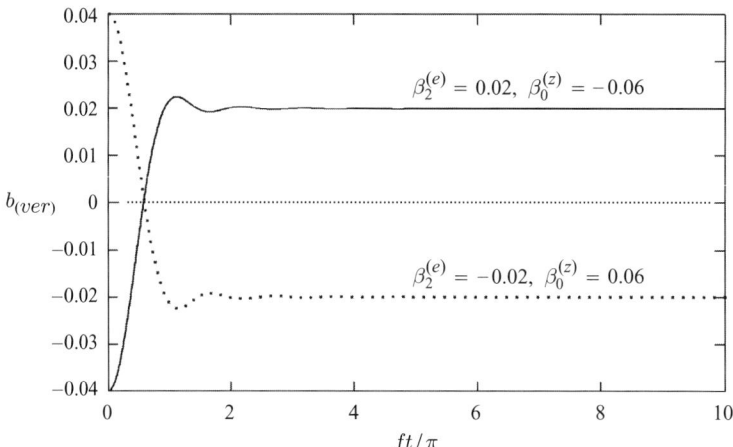

Figure 4.9. Linear evolution of b_{33} (= $b_{(ver)}$) under rapid rotation. $\beta_2^{(e)} = b_{33}^{(e)}(0)$, $\beta_0^{(z)} = b_{33}^{(z)}$. Reproduced from Salhi and Cambon (2007), with permission of AIP.

is a very poor indicator. The condition of statistical axisymmetry implies that

$$b_{ij} = -\frac{3}{2} b_{in} n_i n_j \left(\frac{\delta_{ij}}{3} - n_i n_j \right), \tag{4.19}$$

so that a single-component $b_{ij} n_i n_j$ [b_{33} in the present case, proportional to A in Eq. (4.17)] is enough to describe the full anisotropy tensor (and therefore only the corresponding line in Lumley's map is needed). A similar relationship is valid for any trace-free single-point tensor. Restricting our attention to the linear regime, rapid rotation applied to an initially anisotropic flow yields conservation of directional anisotropy $b_{ij}^{(e)}$ and rapid damping of polarization anisotropy $b_{ij}^{(z)}$. This effect is completely missed in any conventional single-point closure model, in which only b_{ij} is used. The latter effect was called "rotational randomization" by Kassinos, Reynolds, and Rogers (2001), but can be more physically related to anisotropic phase mixing induced by dispersive inertial waves (e.g., Cambon, Jacquin, and Lubrano, 1992; Kaneda and Ishida, 2000). As an illustration, the case of axisymmetric initial anisotropy is shown in Fig. 4.9.

The rapid change of the relevant anisotropy ratio corresponds to the evolution from the initial state in which $b_{33}^{(e)}(0) = -\frac{1}{2} b_{33}(0)$ and $b_{33}^{z}(0) = \frac{3}{2} b_{33}(0)$ (e.g., as in the flow generated by an axisymmetric duct) to a final state in which $b_{33} = b_{33}^{(e)}(0)$ [$= -\frac{1}{2} b_{33}(0)$], because of the conservation of $b_{33}^{(e)}$ and rapid (about a quarter of a revolution) damping of $b_{33}^{(z)}$. In the same "rapid" limit, no Reynolds stress model, even the most sophisticated one, yields an evolution of anisotropy. As a matter of fact, the initial anisotropy is conserved, because there is no production and any conventional closure of the rapid pressure–strain tensor as a function of the sole Reynolds stress tensor yields zero contribution in rotating axisymmetric homogeneous turbulence.

For instance, in the axisymmetric case, only the first two equations of (4.17) are relevant, and all classical closure models yield $\Pi_A^{(r)} = 0$, with $\Pi_A^{(s)} = 0$, in the rapid inviscid limit. Some improvements were independently proposed by Cambon,

4.5 Inertial Waves. Linear Regime

Jacquin, and Lubrano (1992) and Kassinos, Reynolds, and Rogers (2001), using an implicit splitting in terms of directional and polarization anisotropy. Finally, the role conventionally attributed in Reynolds stress models to "rapid" and "slow" pressure–strain tensors is completely wrong in rotating turbulence: In the true rotating homogeneous turbulence case, the rapid (linear) part contributes to a partial return to isotropy [i.e., a damping of $b_{33}^{(z)}$], whereas the slow (nonlinear) part must generate a mild anisotropy associated with the component b_{33} [or equivalently A in Eq. (4.17)] indirectly connected to the transition from a 3D to a 2D structure.

4.5 Inertial Waves. Linear Regime

4.5.1 Analysis of Deterministic Solutions

Linearized inviscid equations, written in a nondimensional form at the end of Section 4.2, are revisited here for velocity, pressure, and vorticity. Equations are rewritten in dimensional form for physical discussions:

$$\frac{\partial \boldsymbol{u}}{\partial t} + 2\boldsymbol{\Omega} \times \boldsymbol{u} + \nabla p = 0, \quad \nabla \cdot \boldsymbol{u} = 0, \tag{4.20}$$

$$\nabla^2 p - 2\Omega \boldsymbol{n} \cdot \boldsymbol{\omega} = 0, \tag{4.21}$$

$$\frac{\partial \omega_i}{\partial t} - 2\Omega \frac{\partial u_i}{\partial x_\parallel} = 0. \tag{4.22}$$

Because the Coriolis force is not divergence free, the pressure term has a nontrivial contribution to enforce the incompressibility constraint. A closed subsystem of equations can be used for $u_\parallel = \boldsymbol{u} \cdot \boldsymbol{n}$, p and $\omega_\parallel = \boldsymbol{\omega} \cdot \boldsymbol{n}$. When the axial components of velocity and vorticity are eliminated in the latter subsystem, the following closed equation is found for p:

$$\frac{\partial^2}{\partial t^2}\left(\nabla^2 p\right) + 4\Omega^2 \nabla_\parallel^2 p = 0, \tag{4.23}$$

with $\nabla_\parallel^2 = \frac{\partial^2}{\partial x_\parallel^2}$. Even if the primitive Poisson equation $\nabla^2 p = f$ is an elliptic one, Eq. (4.23) admits propagating-wave solutions. Surprising properties of these inertial waves are illustrated by the St. Andrew's cross-shaped structures observed in experiments by McEwan (1970) and Mowbray and Rarity (1967) (see Fig. 4.10). If local harmonic forcing with frequency σ_0 takes place in a tank rotating at angular velocity Ω, simplified solutions can be sought by use of a normal-mode decomposition. Considering normal modes of the form $p = e^{i\sigma_0 t}\mathcal{P}$, the spatial part is governed by

$$[\sigma_0^2 \nabla_\perp^2 + (\sigma_0^2 - 4\Omega^2)\nabla_\parallel^2]\mathcal{P} = 0,$$

which shows the possible transition from an elliptic to a hyperbolic problem when σ_0 crosses the threshold 2Ω by decreasing values. This transition explains the sudden appearance of the cross-shaped structures for $\sigma_0 < 2\Omega$. In spite of the rather complex geometry, one can assume, in addition, that the disturbances are plane waves,

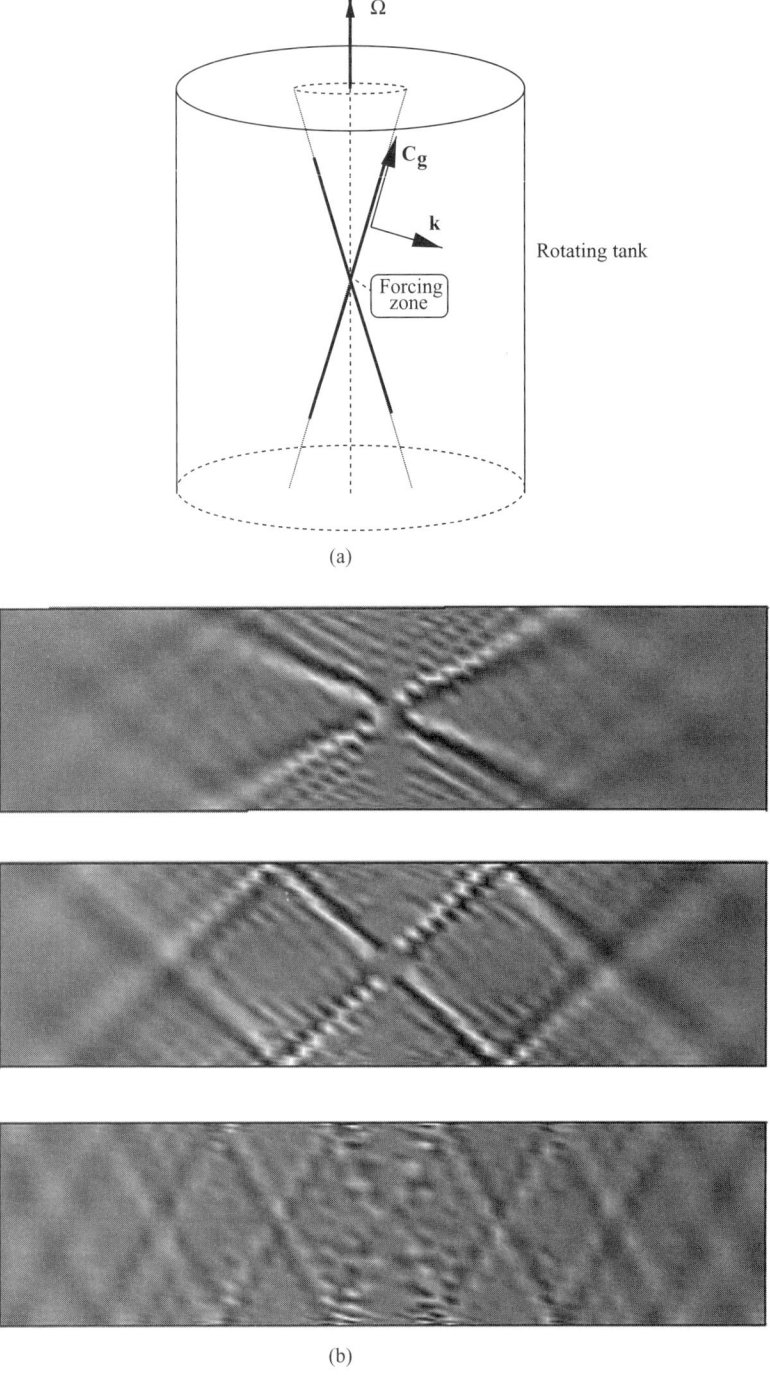

Figure 4.10. Saint Andrew's cross-shaped structures in a rotating flows: (a) sketch of the experiments by McEwan (1970) and Mowbray and Rarity (1967); (b) results from DNS in a plane channel, rotating around the vertical axis. From top to bottom, $2\Omega/\sigma_0 = 1.10, 1.33, 2$. Reproduced from (Godeferd and Lollini, 1999) with permission of CUP.

4.5 Inertial Waves. Linear Regime

i.e., $p \sim e^{i(\mathbf{k} \cdot \mathbf{x} - \sigma_k t)}$. When this solution is injected into Eq. (4.23), the classical dispersion law of inertial waves is recovered:

$$\sigma_k = \pm 2\Omega \frac{k_\parallel}{k} = \pm 2\Omega \cos\theta. \tag{4.24}$$

The phase and group velocity of these inertial waves, denoted as $C_p(\mathbf{k})$ and $C_g(\mathbf{k})$, respectively, are given by

$$C_p(\mathbf{k}) = 2\frac{\Omega \cos\theta}{k^2}\mathbf{k} = \frac{\sigma_k}{k^2}\mathbf{k}, \tag{4.25}$$

$$C_g(\mathbf{k}) = (\nabla \sigma_k) \cdot \mathbf{k} = \frac{2}{k^3}\mathbf{k} \times (\mathbf{\Omega} \times \mathbf{k}) = 2\frac{\Omega \sin\theta}{k}\mathbf{e}^{(2)}, \tag{4.26}$$

where $\mathbf{e}^{(2)}$ is the vector of the local Craya–Herring frame in Fourier space defined in Eq. (2.67).

If one interprets the rays emanating from the small forcing zone in the figure as traces of isophase surfaces, so that the wave vector is normal to them, Eq. (4.24) with $\sigma_k = \sigma_0$ gives the angle θ (defined as the angle between $\mathbf{k}$ and the rotation axis), in excellent agreement with the directions of the rays.

It is important to note that if pressure effects are omitted (leading to the definition of a *pressure-released problem*) only the horizontal part of the flow is affected by the circular periodic (with constant frequency 2Ω) motion, but propagating waves cannot occur. Hence fluctuating pressure (through its linkage with the incompressibility constraint) is responsible both for anisotropic dispersivity and horizontal–vertical coupling.

Going back to velocity, an equation similar to (4.23) can be found for both poloidal and toroidal potentials defined in Eq. (2.65). Without forcing and boundary conditions, the specific initial-value linear problem takes the form

$$\frac{\partial \hat{u}_i}{\partial t} + 2\Omega P_{in}\epsilon_{n3j}\hat{u}_j = 0. \tag{4.27}$$

This equation is simpler than the generic one for the RDT problem addressed in the next chapter, because $\mathbf{x}$ and $\mathbf{u}$ in physical space are projected onto the rotating frame, so that there is no advection by the mean flow, and therefore no time shift in the wave vector.[‡] Given the incompressibility constraint $\hat{\mathbf{u}} \cdot \mathbf{k} = 0$, it is easier to project the equation onto the local frame $(\mathbf{e}^{(1)}, \mathbf{e}^{(2)})$ normal to $\mathbf{k}$ defined by Eq. (2.67). The solution expresses that the initial Fourier component $\hat{\mathbf{u}}(\mathbf{k}, 0)$ is rotated about the axis $\mathbf{k}$ by an angle $2\Omega t k_\parallel / k = \sigma_k t$. The linear solution for the two-component velocity vector $u^{(\alpha)}$, $\alpha = 1, 2$, is

$$\begin{bmatrix} u^{(1)}(\mathbf{k}, t) \\ u^{(2)}(\mathbf{k}, t) \end{bmatrix} = \begin{bmatrix} \cos\sigma_k(t - t') & -\sin\sigma_k(t - t') \\ \sin\sigma_k(t - t') & \cos\sigma_k(t - t') \end{bmatrix} \begin{bmatrix} u^{(1)}(\mathbf{k}, t') \\ u^{(2)}(\mathbf{k}, t') \end{bmatrix}. \tag{4.28}$$

[‡] Of course, a strictly equivalent problem is defined by the equations of Section 2.1 written in a Galilean frame of reference, for a pure antisymmetric gradient matrix $A_{ij} = \Omega \epsilon_{i3j}$, with $\mathbf{k}(\Omega t)$ following the solid-body rotating motion.

The corresponding linear solution in the fixed frame of reference for the initial-value problem is

$$\hat{u}_i(\boldsymbol{k}, t) = G_{ij}(\boldsymbol{k}, t, t')\hat{u}_j(\boldsymbol{k}, t'), \tag{4.29}$$

in which the Green's function is expressed as a function of the two complex eigenvectors $\boldsymbol{N} = \boldsymbol{e}^{(2)} - \imath \boldsymbol{e}^{(1)}$ and $\boldsymbol{N}^* = \boldsymbol{N}(-\boldsymbol{k}) = \boldsymbol{e}^{(2)} + \imath \boldsymbol{e}^{(1)}$ in the plane normal to $\boldsymbol{k}$:

$$G_{ij}(\boldsymbol{k}, t, t') = \sum_{s=\pm 1} N_i(s\boldsymbol{k}) N_j(-s\boldsymbol{k}) e^{\imath s \sigma_k (t-t')}. \tag{4.30}$$

The diagonal decomposition is particularly useful in the context of pure rotation, because $\boldsymbol{N}$ and $\boldsymbol{N}^*$ more generally generate the eigenmodes of the curl operator and directly appear in the (e, Z, h) decomposition [see Eq. (2.93)].

The main features of the inertial waves are subsequently summarized. An inertial wave is

1. *a plane wave* that propagates along $\boldsymbol{k}$,
2. *a transverse wave*, because $\hat{\boldsymbol{u}}(\boldsymbol{k}, t) \perp \boldsymbol{k}$,
3. *a dissipative wave*. The damping factor associated with $\hat{\boldsymbol{u}}(\boldsymbol{k}, t)$ is equal to $e^{-\nu k^2 t}$, as deduced from a trivial extension of the inviscid linear analysis previously discussed.

Complete linear solutions are often referred to as RDT solutions. Even if the previously mentioned Green's function is a particular case of the ones defined in the general RDT theory, the terminology RDT is misleading in the case of rotating turbulence. First, there is no space distortion: Even in the Galilean frame of reference, strictly circular characteristic lines (i.e., mean trajectories) are found in physical space. This is easily seen by writing the equation for these lines: $x_i = Q_{ij}(\Omega t) X_j$. A similar result is obtained in spectral space, because $k_i = Q_{ij}(\Omega t) K_j$, where Q, which is equal to $\mathbf{F}$ in Eq. (4.1), is an orthogonal matrix. Therefore the transformation has isometric properties. Second, the linear solution can be valid for a very long time, because the appearance of a significant nonlinear cascade is delayed with respect to the nonrotating case. The occurrence of phase mixing that is due to interactions between dispersive inertial waves is the best explanation for this depletion of nonlinearity (recall that nonlinear effects do vanish in some DNS results, but keep in mind limitations in terms of Reynolds number and in terms of elapsed time Ωt).

The linear regime of inertial waves has interesting properties, which can be discussed independently of any statistical treatment:

- The dispersion frequency is modulated by the angle-dependent term $\cos\theta$. This modulation reflects the role of fluctuating pressure in connection with $\boldsymbol{k} \cdot \hat{\boldsymbol{u}} = 0$, with a variation of σ_k from 0 (wave plane normal to $\boldsymbol{\Omega}$) to 2Ω (wave vector parallel to $\boldsymbol{\Omega}$). This wide range of dispersion frequencies allows for parametric resonances, either for linear processes (as for elliptical-flow instability with weak

4.5 Inertial Waves. Linear Regime

additional strain discussed in Chapter 8) or for weakly nonlinear interactions (e.g., the wave-turbulence approximation, discussed in Section 4.6).

- The zero frequency is found for the wave plane normal to $\mathbf{\Omega}$. This illustrates the fact that the sole steady mode (i.e., zero-frequency mode) is the 2D mode ($k_\parallel = 0$ corresponds to $\partial/\partial x_\parallel = 0$ in physical space), in agreement with (our restricted use of) the Taylor–Proudman theorem.
- The fact that the dispersion frequency depends on the orientation but not on the modulus of the wave vector is a very particular situation, encountered in other cases of *purely transverse pressure and vorticity waves*, like the internal gravity waves addressed in Chapter 7. As a consequence, phase velocity and group velocity are orthogonal to each other. In the same way, the group velocity is found maximal and in the axial direction, when the phase velocity is near zero, close to the wave plane normal to $\mathbf{\Omega}$.

4.5.2 Analysis of Statistical Moments. Phase Mixing and Low-Dimensional Manifolds

Linear equations can be used to compute various statistical moments of the solution. For instance, the linear solution for the second-order spectral-tensor equation is

$$e(\mathbf{k}, t) = e(\mathbf{k}, t_0), \quad h(\mathbf{k}, t) = h(\mathbf{k}, t_0), \quad Z(\mathbf{k}, t) = e^{4\imath\sigma_k(t-t_0)} Z(\mathbf{k}, 0). \quad (4.31)$$

As a first consequence, an initially anisotropic flow is altered, with $b_{ij}^{(e)}$ and $b_{ij}^{(z)}$ being conserved and damped, respectively, as illustrated in Fig. 4.9. In counterpart, these equations yield no evolution for isotropic initial data, with $Z = h = 0$.

The concept of *phase mixing* can be understood from Eqs. (4.29) and (4.30) in which the initial data term $\hat{u}(\mathbf{k}, t')$ could be replaced with a new slow time-evolving variable, $\mathcal{U}(\mathbf{k}, \epsilon t, t')$. The impact of the basic Green's function, for instance in breaking 3D isotropy, depends on the order and on the degree of complexity of statistical moments (purely initial values or slowly evolving ones) to which it is applied.

Throughout this book, *manifold* means a subspace of the spatial configuration space. The configuration space is defined in 3D Fourier space for mathematical convenience. For instance, the *2D manifold* (also referred to as the *slow manifold* because $\sigma_k = 0$ for modes belonging to this manifold) corresponds to the wave plane $k_\parallel = 0$ embedded in 3D Fourier space (all $\mathbf{k}$). The *manifold of resonant triads* represents the subspace defined by $(\pm\sigma_k \pm \sigma_p \pm \sigma_q = 0, \mathbf{k} + \mathbf{p} + \mathbf{q} = 0)$, which is embedded in the space of all triads ($\mathbf{k} + \mathbf{p} + \mathbf{q} = 0$) in 6D (all $\mathbf{k}, \mathbf{p}$) Fourier space.

4.5.2.1 Single-Time Second-Order Statistics

When single-time second-order statistics are examined, isotropy is essentially conserved in the linear limit, as time dependency can cancel out by multiplying $e^{\imath\sigma_k t}$ by its complex conjugate. This result is often considered as too general, saying that phase information is lost in homogeneous turbulence (Davidson, Stapelhurst, and Dalziel, 2006).

A refined analysis can be derived from Eq. (4.31). Both kinetic energy and directional anisotropy $b_{ij}^{(e)}$ are conserved in the linear inviscid limit, whereas the polarization anisotropy, given by

$$2\mathcal{K}(0)b_{ij}^{(z)}(t) = \iiint \Re\left[Z(k,x,0)e^{4\imath\Omega xt}N_i N_j\right]d^3k \quad \text{with} \quad x = \cos\theta, \quad (4.32)$$

is essentially damped. This damping is a general effect obtained by summing up terms affected by different oscillations, here from 0 ($x = 0$) to 2Ω ($x = 1$). If initial data are axisymmetric with mirror symmetry, for instance, the relevant integral that illustrates the damping of $b_{ij}^{(z)}$ is

$$I(k, \Omega t) = \int_0^1 C(k, x)\cos(4\Omega xt)dx,$$

which always tends to zero as the nondimensional time Ωt becomes large {$C(k, x)$ is taken equal to $(1 - x^2)\Re[Z(k, x, 0)]$ in Eq. (4.32) for $i = j = 3$}. The only exception is found when $C(x)$ has a nonintegrable singularity: A simple instance is given by 2D-2C initial data such that $Z(\mathbf{k}, 0) = -[E(k)/(2\pi k)]\delta(kx)$. It is clear from our very simple example that the phase mixing, induced by the x-weighting term in the integrand by means of the frequency $4\Omega x$, is responsible for damping, whereas $x = 0$ may characterize a low-dimensional manifold that escapes the damping effect if it is singular.

Unexpected behavior of the RST results from the selective damping of the initial polarization anisotropy $b_{ij}^{(z)}$, as shown in Fig. 4.9.

Useful dynamical properties, however, are recovered for two-time (t, t') second-order statistics, with interesting applications to Lagrangian diffusivity as discussed in Kaneda and Ishida (2000) and Cambon et al. (2004), as time dependency cannot cancel out when multiplying $e^{\pm 2\imath\Omega xt}$ by $e^{\pm 2\imath\Omega xt'}$. These applications are beyond the scope of the present book.

4.5.2.2 Single-Time Third-Order Statistics

In this case the linear operator generates a product of three phase terms, $e^{\pm 2\imath(k_\parallel/k)\Omega t}$, $e^{\pm 2\imath(p_\parallel/p)\Omega t}$, and $e^{\pm 2\imath(q_\parallel/q)\Omega t}$, that are related to the triad $(\mathbf{k}, \mathbf{p}, \mathbf{q})$.

Triple-velocity correlations that govern the nonlinear energy and anisotropy transfers in related Lin-type equations are considered in the next subsection: The effect of phase mixing, considered linear if applied to triple correlations, is interpreted as nonlinear by means of the impact of transfer on energy distribution.

Triple correlations undergoing phase mixing can also be studied per se, as the triple-vorticity correlations revisited in Section 4.8. Let us mention that, in any cubic correlation, which is generated from triadic components $\langle \hat{u}_m(\mathbf{q},t)\hat{u}_n(\mathbf{k},t)\hat{u}(\mathbf{p},t)\rangle$ with $\mathbf{k} + \mathbf{p} + \mathbf{q} = 0$, phase mixing is induced by the term

$$\exp\left[\imath 2\Omega t\left(s\frac{k_\parallel}{k} + s'\frac{p_\parallel}{p} + s''\frac{q_\parallel}{q}\right)\right],$$

which results from the previously mentioned product of three phase terms, the zero value of its phase corresponding to the manifold of resonant triads.

4.6 Nonlinear Theory and Modeling: Wave Turbulence and EDQNM

4.6.1 Full Exact Nonlinear Equations. Wave Turbulence

Simplified equations projected on the local basis of eigenmodes N and N^* can be used for discussing both linear and nonlinear operators, as well as for developing closure theories for rotating turbulence. The starting point is the same as in the previous chapter. Using the associated amplitudes ξ_s, $s = \pm 1$, which are defined by

$$\hat{u}(\mathbf{k}, t) = \xi_+(\mathbf{k}, t)N(\mathbf{k}) + \xi_-(\mathbf{k}, t)N(-\mathbf{k}), \qquad (4.33)$$

one obtains the following evolution equation:

$$\left[\frac{\partial}{\partial t} + \nu k^2 - \iota s \underbrace{\left(2\Omega \frac{k_\|}{k}\right)}_{\sigma_k}\right]\xi_s = \sum_{s',s''=\pm 1}\int_{\mathbf{k}+\mathbf{p}+\mathbf{q}=0} M_{ss's''}(\mathbf{k}, \mathbf{p})\xi^*_{s'}(\mathbf{p}, t)\xi^*_{s''}(\mathbf{q}, t)d^3\mathbf{p}, \qquad (4.34)$$

in which a diagonal form of the linear operator appears and the nonlinear term $M_{ss's''}(\mathbf{k}, \mathbf{p})$ is given by Eqs. (3.99) and (3.100). The linear inviscid solution is

$$\xi_s(\mathbf{k}, t) = \exp\left(2\iota s\Omega t \frac{k_\|}{k}\right)\xi_s(\mathbf{k}, 0), \quad s = \pm 1.$$

Replacing[§] the initial condition with a new function a_s such that

$$\xi_s(\mathbf{k}, t) = \exp\left(2\iota s\Omega t \frac{k_\|}{k}\right) a_s(\mathbf{k}, t), \quad s = \pm 1 \qquad (4.35)$$

one obtains an equation for a_s in which linear operators are absorbed in the nonlinear one as integrating factors:

$$\dot{a}_s = \sum_{s',s''=\pm 1}\int_{\mathbf{k}+\mathbf{p}+\mathbf{q}=0} \exp\left[2\iota\Omega\left(s\frac{k_\|}{k} + s'\frac{p_\|}{p} + s''\frac{q_\|}{q}\right)t\right]$$
$$\times M_{ss's''}(\mathbf{k}, \mathbf{p})a^*_{s'}(\mathbf{p}, t)a^*_{s''}(\mathbf{q}, t)d^3\mathbf{p}. \qquad (4.36)$$

Note that the quantities a_s can be interpreted as *amplitudes of slow variables*, because the contribution of rotation has been removed. This problem can be analyzed with a multiple (two) time-scale technique, by setting $a_s = a_s(\mathbf{k}, \epsilon t)$, where ϵ is a (really) small parameter for asymptotic expansion that can be related to a Rossby number. Such a refined analysis is not needed here, and we will just retain from Eq. (4.36) the importance of resonant triads. These triads are defined by the relation

$$s\sigma_k + s'\sigma_p + s''\sigma_q = 0 \qquad (4.37)$$

[§] This change of variables is referred to as the Poincaré transform.

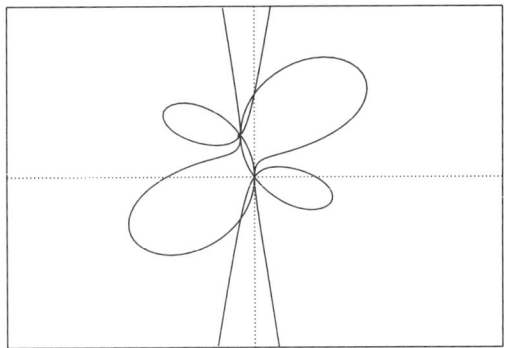

Figure 4.11. Visualization of resonant surfaces of inertial waves, given by Eq. (4.38), for a given orientation of $\mathbf{k}$. The locus of $\mathbf{p}$ is seen in the plane $p_2 = 0$, for $\theta_k = 1.1$. Complex loops appear for $\pi/3 < \theta_k < \pi/2$. Courtesy of F. S. Godeferd.

and correspond to a zero value of the phase term on the right-hand side of Eq. (4.36), leading to

$$s\frac{k_\parallel}{k} + s'\frac{p_\parallel}{p} + s''\frac{q_\parallel}{q} = 0 \quad \text{with} \quad \mathbf{k} + \mathbf{p} + \mathbf{q} = 0. \tag{4.38}$$

These resonant or quasi-resonant triads are found to dominate the nonlinear slow motion, because the effect of the phase term on the left-hand side of Eq. (4.36) is a severe damping of the nonlinearity by *phase mixing*. The complexity of the resonant surfaces is illustrated by Fig. 4.11. The preceding resonance condition can be complemented by projecting the general triad condition on the plane normal to the rotation axis, yielding:

$$k_\parallel + p_\parallel + q_\parallel = k\cos\theta_k + p\cos\theta_p + q\cos\theta_q = 0. \tag{4.39}$$

In that case, why not obtain a simplified model by solving Eq. (4.36) with an integral restricted to the resonant triads? Because the resonant surfaces are sufficiently complex to require very accurate interpolation, rendering the resulting computation efficient for a smooth distribution of the slow amplitudes a_s in Fourier space only. Such a smooth distribution cannot represent turbulence, so that one has to resort to describing naturally smooth quantities like statistical moments instead of the instantaneous solution. It is not forbidden, however, to try to isolate resonant triads in DNS: Some related studies are discussed in Section 4.7.

A qualitative analysis of resonant triads (Waleffe, 1993) deserves attention before quantitative issues dealing with statistical moments are addressed. Going back to the analysis presented in Section 3.5 and introducing the selection rules of resonant triads (4.37) and (4.39), one obtains

$$\frac{\cos\theta_k}{s'q - s''p} = \frac{\cos\theta_p}{s''k - sq} = \frac{\cos\theta_q}{sp - s'k}. \tag{4.40}$$

Applying the instability principle of Waleffe introduced in Subsection 3.5.4, but *restricting the analysis to resonant triads*, equality (4.40) shows that the transfer of energy always goes from a less slanted leg of the triad (with respect to the rotation vector) to a more slanted one. This result comes from the fact that the unstable mode is also the mode whose pulsation σ_k has the larger amplitude and opposite sign with respect to the two other modes of the triad under consideration. Let us consider the triad $(\mathbf{k}, \mathbf{p}, \mathbf{q})$ and assume that $\mathbf{k}$ is the unstable mode. Then one has

4.6 Nonlinear Theory and Modeling

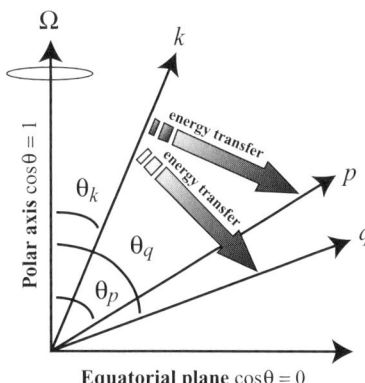

Figure 4.12. Schematic view of kinetic-energy transfer according to Waleffe's instability assumption among resonant modes in rotating turbulence.

$|\sigma_k| > |\sigma_p|, |\sigma_q|$, which leads to $|\cos\theta_k| > |\cos\theta_p|, |\cos\theta_q|$. Accordingly, a drain of energy is predicted toward the wave plane orthogonal to $\mathbf{\Omega}$, as illustrated in Fig. 4.12. Waleffe, however, points out that the rate of energy transfer vanishes exactly when the wave vector reaches the equatorial orientation characterized by $\mathbf{k} \cdot \mathbf{\Omega} = 0$. The latter wave plane is exactly both the slow and the 2D manifold. Its meaning is very different in the discrete case and in the continuous case: For instance, extension of the resonance condition to the exact slow manifold is valid in only the discrete case, and related issues are further discussed in Subsection 4.7.2.

Wave turbulence is an unavoidable aspect in other domains, such as plasma physics, with a very large literature that influenced the "Russian School," although it also addressed turbulence in classical fluids (Zakharov, L'vov, and Falkovich, 1992). Applications can be found even in the physics of solids: Random weakly interacting acoustic waves are considered a gas of phonons in the latter case. In fluid mechanics, the turbulent flow is seen as a sea of random spatiotemporal wave modes, whose nonlinear interactions can be considered as weak in the limit of a small parameter. The deterministic ingredient is the dispersion law, which gives a straightforward link of a "rapid" temporal frequency σ to the spatial wave vector $\mathbf{k}$. All other variables of the flow, such as the amplitudes A of the wave modes and/or some phase terms φ in any generic wave mode of motion $a \exp[\iota(\mathbf{k}\cdot\mathbf{x} - \sigma(\mathbf{k})t + \varphi)]$, are treated as random variables. Even $\mathbf{k}$ could be considered as random, in some models, ranging from linear [kinematic simulation (Cambon et al., 2004)] to weakly nonlinear ones (e.g., Monte Carlo methods for solving statistical closures). The possibility of applying a weakly nonlinear theory relies on a time-scale separation: Amplitudes a are assumed to slowly evolve in time with respect to the rapid temporal oscillations induced by $\sigma(\mathbf{k})t$ (φ phase terms are removed here, or absorbed in a, such as $ae^{\iota\varphi} \to A$). In this sense, wave turbulence is a theory for the evolution of slowly evolving envelopes that modulate high-frequency oscillations.

Rapidly rotating turbulence is an almost perfect case to apply such a theory: The inertial wave modes form a complete basis (they are identical to the helical modes, which are even useful for studying "strong" turbulence); the small parameter that controls the "weak" nonlinearity is not artificial: This is a Rossby number. It is important to notice that to bridge between a small Rossby number and a weak nonlinearity seems to be trivial and tautological at first glance (this argument is used in RDT, for instance, without any refined analysis of the nature of nonlinearity); more important is the fact that the phase mixing induced by rapid oscillations yields a severe damping of nonlinear interactions, so that nonlinearity concentrates on a low-dimension manifold. Among all the triads called into play in the absence of waves, only very few survive in the long-term limit, forming quasi-resonant triads. Finally, when transferring the EDQNM machinery from velocity (spatial) Fourier modes to slow amplitudes, in agreement with an exact Poincaré transform (4.35), the limit of wave turbulence is recovered at vanishing eddy damping, as we will see in the following subsections.

4.6.2 Second-Order Statistics: Identification of Relevant Spectral-Transfer Terms

Second-order correlations are entirely generated by the quantities e, Z, and $\mathcal{H}$, or, equivalently, by $\langle a_s^* a_{s'} \rangle$, $s = \pm 1$, and $s' = \pm 1$. Without any assumption, second-order correlations are governed by the following system of equations:

$$\left(\frac{\partial}{\partial t} + 2\nu k^2 \right) e(k, t) = T^{(e)}(k, t), \tag{4.41}$$

$$\left(\frac{\partial}{\partial t} + + 2\iota \sigma_k + 2\nu k^2 \right) Z(k, t) = T^{(Z)}(k, t), \tag{4.42}$$

$$\left(\frac{\partial}{\partial t} + 2\nu k^2 \right) \mathcal{H}(k, t) = T^{(h)}(k, t), \tag{4.43}$$

in which the nonlinear terms $T^{(e)}(k,t)$, $T^{(Z)}(k,t)$, and $T^{(h)}(k,t)$ are defined starting from the transfer tensor T_{ij} given by Eq. (3.113), as

$$T^{(e)} = \frac{1}{2} T_{ii}, \quad T^{(z)} = \frac{1}{2} T_{ij} N_i^* N_j^*, \quad T^{(h)} = \frac{1}{2} \epsilon_{ijn} \frac{k_i}{k} T_{jn},$$

in full agreement with $e = (1/2)\hat{R}_{ii}$, $Z = (1/2)\hat{R}_{ij} N_i^* N_j^*$, and $h = (1/2)\epsilon_{ijn}(k_i/k)\hat{R}_{ij}$.

It appears that the Coriolis force does not affect the (linear) left-hand sides, except for the polarization parameter $Z(k,t)$. Replacing Z with ζ, where ζ is such that

$$Z(\mathbf{k}, t) = e^{2\iota \sigma_k t} \zeta(\mathbf{k}, t), \tag{4.44}$$

and $T^{(Z)}(\mathbf{k}, t)$ with $e^{2\iota \sigma_k t} T^\zeta(\mathbf{k}, t)$, only the right-hand-side terms, which are linked to triple correlations and mediated by nonlinearity, are possibly rotation dependent. Contributions from triple-velocity correlations are therefore gathered into the generalized spectral-transfer terms $T^{(e,Z,h)}$, which derive from Eqs. (4.41)–(4.43). If the preceding system of equations is initialized with 3D isotropic initial data, i.e., by setting $e(k, 0) = E(k)/(4\pi k^2)$ and $Z = \mathcal{H} = 0$ at the initial time, then the anisotropy that should reflect the transition toward the 2D structure can be created by the nonlinear spectral-transfer terms only. This anisotropy consists of axisymmetry without mirror symmetry, leading to $e = e(k, \cos\theta = \frac{k_\parallel}{k}, t)$ and $Z = Z(k, \cos\theta = \frac{k_\parallel}{k}, t)$, with

4.6 Nonlinear Theory and Modeling

$Z = 0$ if k is parallel to the vertical axis, in agreement with the symmetries of rotating Navier–Stokes equations, which ought to be satisfied by the closure theory.

4.6.3 Toward a Rational Closure with an EDQNM Model

A complete anisotropic EDQN model can be built in terms of the Green's function **G** and the spectral tensor $\hat{\mathbf{R}}$, using products of **G** to solve the linear operators that appear in the equations for third-order correlations (Cambon, 1982; Cambon and Scott, 1999). Related technical details are relegated to Chapter 14. On the grounds of these equations, it is possible to discuss an optimal way to treat the *Markovianization* procedure, i.e., to simplify the time dependency in the integrands that connect the transfer term to second-order correlations. Closed equations display three kinds of time-dependent terms:

1. Viscous, or viscous + damping, terms:

$$\exp\left(\int_{t'}^{t} \mu dt''\right) \to V(t, t').$$

2. Terms involving the RDT Green's function:

$$\mathbf{G}(t, t') \to \exp[\pm \imath \sigma(t - t')].$$

3. Terms from the second-order spectral tensor (through QN assumption):

$$\hat{\mathbf{R}}(t') \to (e, Z, h)(t').$$

According to the Markovianization procedure in classical EDQNM, we can assume that $V(t, t')$ is so rapidly decreasing in terms of time separation $\tau = (t - t')$ that it is concerned only with the time integral in the closure equations, whereas the other terms take their instantaneous value at $t' = t$, so that they are replaced with $\mathbf{G}(t, t)$ and $\hat{\mathbf{R}}(t)$, respectively. In other words, one considers that the only rapid term is $V(t, t')$, the other terms being assumed to be slow ones. This procedure, referred to as EDQNM1, is not relevant for rotating turbulence, as the presence of $\mathbf{G}(t - t')$ in the closure relationship is responsible for the breakdown of the initial isotropy. Using EDQNM1 with isotropic initial data, isotropy is maintained, and no effect of system rotation can appear. A second step, referred to as EDQNM2, consists of simply "freezing" the (e, Z, h) terms by setting $t' = t$ in them, whereas the complete "readjusted" response function, with both $V(t, t')$ and $\mathbf{G}(t, t')$ terms, is conserved in the time integrand with its detailed time dependency. An interesting result is that the time integral of the threefold product of response functions yields a generic closure relationship of the form

$$T^{(e,Z,h)} = \sum_{s=\pm 1, s'=\pm 1, s''=\pm 1} \int \frac{S^{ss's''}(e, Z, h)}{\mu_{kpq} + \imath(s\sigma_k + s'\sigma_p + s''\sigma_q)} d^3\mathbf{p}. \quad (4.45)$$

The nonlinear term $S^{ss's''}(e, Z, h)$ is given in Chapter 14.

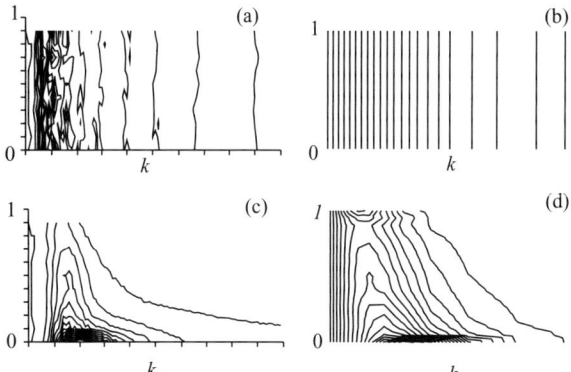

Figure 4.13. Isolines of kinetic energy $e(k, \cos\theta, t)$ for $512 \times 128 \times 128$ LES computations: (a) at $\Omega = 0$ at time $t/\tau = 427$; (b) EDQNM2 with $\Omega = 0$; (c) LES with $\Omega = 1$ at $t/\tau = 575$; and (d) EDQNM2 calculation with $\Omega = 1$ at time $t/\tau = 148$. The vertical axis bears $\cos\theta_k$ (from 0 to 1 upward) and the horizontal one the wavenumber k. Reproduced from Cambon, Mansour, and Godeferd (1997) with permission of CUP.

Results dealing with the rise of directional anisotropy and the description of the transition to 3D isotropy to 2D structure obtained using EDQNM2 are illustrated in Fig. 4.13, in which they are compared with high-resolution $528 \times 128 \times 128$ LES data. It should be borne in mind that the development of angular dependency in $e(k, \cos\theta = \frac{k_\parallel}{k}, t)$, which amounts to a concentration of energy toward the 2D slow manifold (sketched in Fig. 4.16 in the next subsection), results from nonlinear interactions mediated by $T^{(e)}$ in Eqs. (4.41) and (4.45).

The latter procedure can be questioned, in spite of its excellent numerical results, because it is not completely consistent with the basic rapid–slow decomposition suggested by Eq. (4.35). All the terms in the set (e, Z, h) that generate $\hat{R}$ should not be considered as slow terms according to the RDT solution (4.31). Therefore it is necessary to use the decomposition defined by Eq. (4.44), so that only ζ appears as a slow variable, in complete agreement with Eq. (4.35). The resulting optimal procedure, referred to as EDQNM3, yields freezing $e(t') = e(t)$, $h(t') = h(t)$, $\zeta(t') = \zeta(t)$ while keeping the t' dependency under the integral for $Z(t') = \exp(2\iota\sigma t')\zeta(t)$, $V(t, t')$, and $\mathbf{G}(t, t')$, as before. This EDQNM3 version differs only slightly from EDQNM2, but presents decisive advantages. It is completely consistent with building EDQNM in terms of slow amplitudes using relation (4.35). Another advantage is that an asymptotic expansion can be obtained in the limit $\mu_{kpq} \ll 2\Omega$, which exactly coincides with the Eulerian wave-turbulence theory (see Galtier, 2003). It is proved that realizability is enforced in this limit, whereas it is not in the EDQNM2 version.

4.6.4 Recovering the Asymptotic Theory of Inertial Wave Turbulence

Ignoring the $\mathcal{H}$ and ζ contributions for the moment, EDQNM3 [or equivalently EDQNM2 (Cambon and Jacquin, 1989; Cambon, Mansour, and Godeferd, 1997)

4.6 Nonlinear Theory and Modeling

as the two versions differ only in treating Z] yields the following closure for the Lin equation:

$$T^{(e)} = \sum_{s',s''=\pm 1} \int\int\int \frac{A(k, s'p, s''q)}{\mu_{kpq} + \iota(\sigma_k + s'\sigma_p + s''\sigma_q)} e(q)\left[e(p) - e(k)\right] d^3p, \quad (4.46)$$

where the exact forms of A and μ_{kpq} are given in Chapter 14.

The denominator reflects the time integration of a product of three ED Green's functions derived from Eq. (4.30).

In the limit of a very high rotation rate, or at a vanishing Rossby number, the asymptotic version of this equation is obtained with the following Riemann–Lebesgue relationship for distributions (also sometimes referred to as the Plemelj or Sokhotsky formula):

$$\frac{1}{\mu + \iota x} \to \pi\delta(x) + \mathcal{P}\left(\frac{1}{x}\right) \quad \text{when} \quad \mu \to 0,$$

in which $\mathcal{P}$ holds for the principal value in the complete integral expression [such as (4.46)].

The resulting *asymptotic quasi-normal Markovian* (AQNM) closure is expressed as

$$T^{(e)} = \sum_{s',s''=\pm 1} \int\int_{\mathfrak{M}_{s's''}} \pi \frac{A(k, s'p, s''q)}{s'\boldsymbol{C}_g(p) - s''\boldsymbol{C}_g(q)} e(q)\left[e(p) - e(k)\right] d^2p, \quad (4.47)$$

in which $\mathfrak{M}_{s's''}$ is the family of resonant surfaces and $\boldsymbol{C}_g(k)$ is the group velocity of inertial waves. The damping factor μ no longer appears in the final equation, and the denominator accounts for the fact that the reduction from a volume to a surface integral brings in the gradient of resonant surfaces, whence the occurrence of the group velocity. The reader is referred to Cambon, Rubinstein, and Godeferd (2004) for a presentation of the full EDQNM3 equations (without $\mathcal{H}$) and to Bellet et al. (2006) for AQNM equations for $e, \zeta, \mathcal{H}$.

Starting from isotropic initial data, with a narrowband energy spectrum, an inertial zone is constructed that solves the AQNM equation for $e(k, \theta)$ at a vanishing Rossby number and infinite Reynolds number, until the inertial range reaches the maximum wavenumber. At this stage of the computation, the laminar viscosity is reintroduced,¶ and a self-similar spectrum is obtained. The spherically averaged energy spectrum $E(k)$ is constructed with a k^{-3} slope, as shown in Fig. 4.14, but the prefactor is $E(k) \sim \frac{\Omega}{t} k^{-3}$. Axisymmetric shape, with strong directional anisotropy, is found for the angle-dependent spectrum $4\pi k^2 e(k, \cos\theta, t)$, as shown in Fig. 4.15. This directional anisotropy, mediated only by nonlinear transfer, is consistent with

---

¶ Unfortunately, it is not possible to continue the computation at infinite Reynolds number at this stage, as was done in the isotropic case without rotation in Chapter 3, because the ED is no longer present in the AQNM equation, and especially because accumulation of spectral energy at $k_{\max}$ is no longer possible, because of typical oscillations emanating from the largest wave vectors as numerical instabilities.

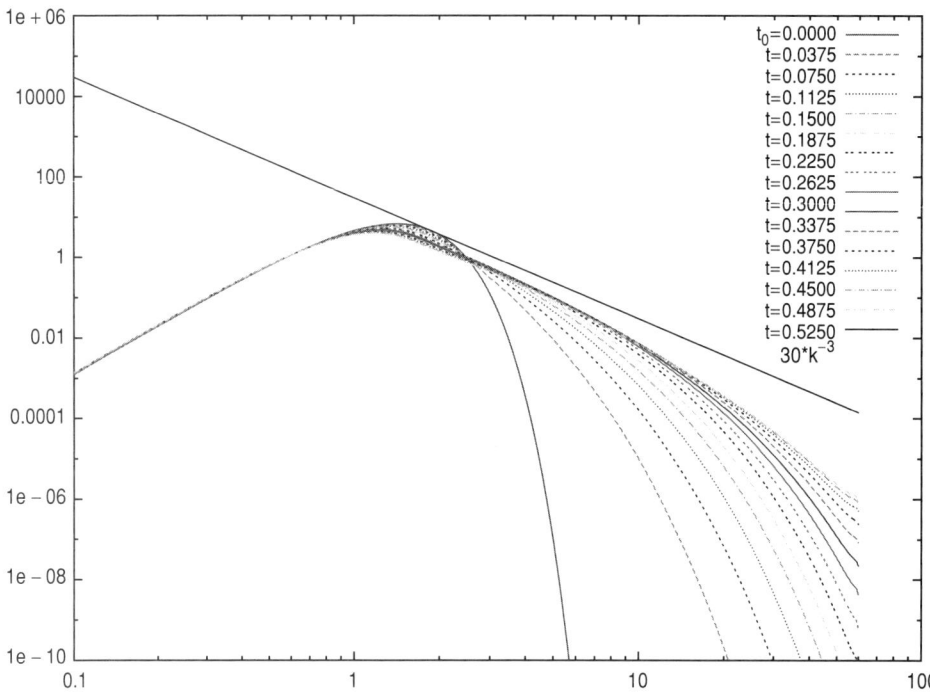

Figure 4.14. Construction of the spherically averaged spectrum in AQNM. The straight line gives the k^{-3} slope. Reproduced from Bellet et al. (2006) with permission of CUP.

the sketch displayed in Fig. 4.16, and with all previous theoretical and numerical studies by Cambon and Jacquin (1989), Waleffe (1993), and Cambon, Mansour, and Godeferd (1997). That illustrates a transition from a 3D (e equidistributed on spherical shells) to a 2D structure (e concentrated on the horizontal wave plane). Nevertheless, the two-dimensionalization is limited to large k, and is never fully achieved. The k^{-3} slope for E results from the averaging of various slopes for $4\pi k^2 e$, ranging from k^{-2} (for quasi-horizontal wave vectors) to k^{-5} (for quasi-vertical wave vectors). The relevance of this asymptotic result is perhaps marginal, because the time τ needed to obtain the inertial zone built by means of weak wave-turbulence

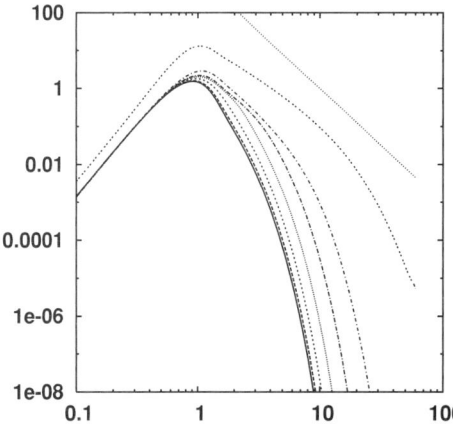

Figure 4.15. Asymptotic angular-dependent spectra from AQNM. Spectral energy density for different angles, from bottom to top, $\theta/(\pi/2) = 1/300$ (what we call the vertical mode), $\theta/(\pi/2) = 51/300$, $\theta/(\pi/2) = 101/300$, $\theta/(\pi/2) = 151/300$, $\theta/(\pi/2) = 201/300$, $\theta/(\pi/2) = 251/300$, and $\theta/(\pi/2) = 299/300$ (the "horizontal" mode). The straight line gives the k^{-2} slope. Reproduced from Bellet et al. (2006) with permission of CUP.

4.7 Fundamental Issues: Solved and Open Questions

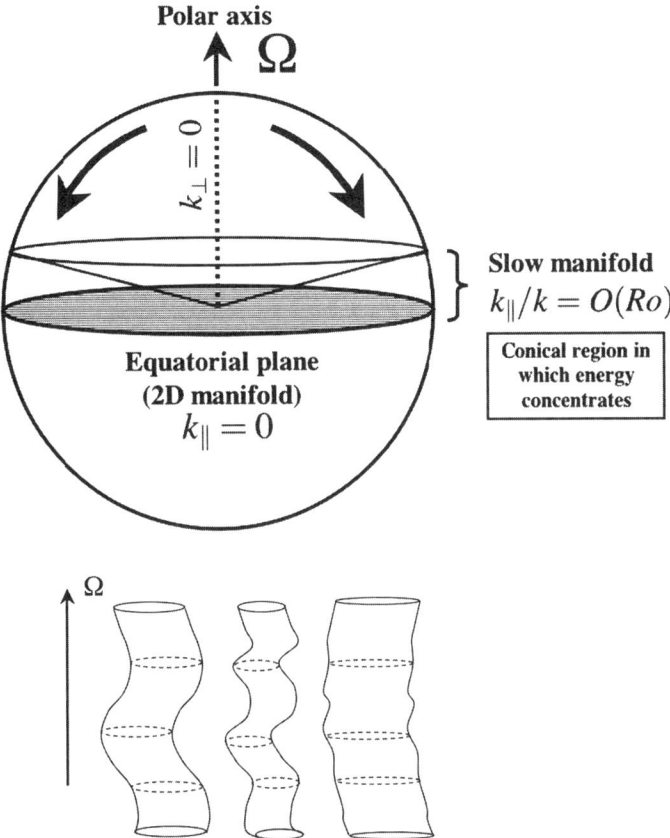

Figure 4.16. Top: Schematic view of the net energy spectral transfers in the pure rotation case. Bottom: Sketch of emerging large-scale coherent structures. Notice that the conical region in which the energy concentrates remains finite, even if the size of the slow manifold tends to zero at vanishing Rossby numbers.

dynamics is very high, as $\Omega\tau \sim O(Ro^{-2})$ at a very small Rossby number Ro. In this context, it is interesting to note that a similar result was obtained by a high-resolution (512^3) DNS, therefore at moderate Ro, Re, and elapsed time, as shown in Fig. 4.17.

4.7 Fundamental Issues: Solved and Open Questions

4.7.1 Eventual Two-Dimensionalization or Not

It is clear that the trends toward a 2D structure saturate at very long time, at least in the continuous case. The anisotropic state, which eventually becomes self-similar, is consistent with power-law decay for single-point statistics. The turbulent kinetic energy is observed to decay as $\mathcal{K}(t) \sim t^{-0.86}$ in AQNM. Full two-dimensionalization requires very strong conditions of axisymmetric angular distribution for e:

$$e(k_\perp, k_\parallel) = \frac{E(k_\perp)}{2\pi k_\perp} \delta(k_\parallel), \qquad (4.48)$$

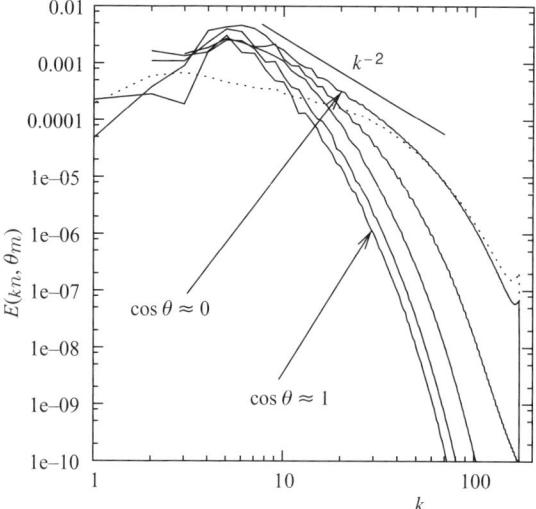

Figure 4.17. Angular-dependent spectra of purely rotating turbulence. A comparable isotropic spectrum of the same quantity is shown as a black dotted curve. Data taken from Liechtenstein, Godeferd, and Cambon (2005).

or equivalent conditions expressed with $k = \sqrt{k_\perp^2 + k_\parallel^2}$ and $\cos\theta = k_\parallel/k$, where δ denotes the Dirac delta function. An additional condition possibly brings in the polarization anisotropy Z, leading to

$$Z(k_\perp, k_\parallel) = -\frac{E(k_\perp)}{2\pi k_\perp}\delta(k_\parallel) \qquad (4.49)$$

in order to ensure that the contribution to vertical velocity is identically zero. The first equation characterizes a 2D state only, whereas both characterize a 2D-2C state (Cambon, Mansour, and Godeferd, 1997). In contrast, the asymptotic state of weak inertial turbulence predicted by the AQNM model is consistent with an *integrable singularity* at $k_\parallel = 0$ for $e(k_\perp, k_\parallel)$, and with a zero Z. Near the 2D manifold, the distribution is consistent with

$$e(k_\perp, k_\parallel) \sim k_\perp^{-7/2} k_\parallel^{-1/2}, \quad \text{or} \quad e(k, \cos\theta) \sim (\cos\theta)^{-1/2} k^{-4}, \qquad (4.50)$$

as analytically obtained by Galtier (2003). The k^{-4} law at smallest $\cos\theta$'s is consistent with the AQNM result (Fig. 4.15) or the k^{-2} slope after multiplication by k^2.

Things can be different in the discrete case, for instance when the velocity field is chosen to be periodic with a finite wavelength in one, two, or three directions. On the one hand, some mathematical theorems can predict decoupled dynamics and eventual dominance of the slow manifold. Such a "nonlinear Proudman theorem" relies on smoothness assumptions about the initial velocity field and emphasizes the role of purely 2D particular resonant triads that are sometimes referred to as *catalytic triads*. On the other hand, some underresolved DNSs or LESs (for instance DNS with hyperviscosity) discussed in Subsection 4.7.3 seem to predict two-dimensionalization, in agreement with essentially decoupled dynamics of the slow manifold, in which the energy is eventually concentrated.

4.7 Fundamental Issues: Solved and Open Questions

4.7.2 Meaning of the Slow Manifold

Both the definition and the relative weight of the slow manifold depend on the discretization in conventional pseudo-spectral DNS and LES. In any case, the underlying assumptions of weak turbulence are no longer valid in the domain $k_\parallel/k = O(Ro)$, because the time-scale separation between "slow" amplitudes $a_{\pm 1}$ and "rapid" phases $\exp(\pm 2\Omega t k_\parallel/k)$ no longer holds. In DNS and LES, $k_\parallel/k$ cannot be smaller than a typical mesh ratio Δ_k/k, so that the apparent thickness of the slow manifold is fixed independently of the Rossby number, which questions any calculation at too small a Rossby number.

Even if EDQNM3 and AQNM equations deal with the continuous case, their numerical resolution needs discretization in Fourier space, but the angular step can be much smaller than in DNS/LES. The exact limit $k_\parallel = 0$ cannot be afforded by AQNM equations in any case, and the AQNM numerical code is used only until the smallest nonzero value of this parameter is reached.

The consequence is twofold:

1. The contribution of the neighborhood of $k_\parallel = 0$ is singular. But, because this singularity is integrable, any quantity that involves an integral over the whole angle-dependent wave space can be accurately computed. Examples of such quantities are the spherically averaged energy spectrum $E(k)$ and the Reynolds stress components.
2. The system of AQNM equations has to be complemented in order to take into account the slow manifold per se. This is needed to evaluate statistical quantities that involve only the $k_\parallel = 0$ wave plane, as the 2D energy components:

$$\overline{u_3^2} L_{33}^{(3)} = 2\pi^2 \int_0^\infty [e(k_\perp, k_\parallel = 0) + Z(k_\perp, k_\parallel = 0)] k_\perp dk_\perp \quad (4.51)$$

and

$$\overline{u_1^2} L_{11}^{(3)} = \overline{u_2^2} L_{22}^{(3)} = \pi^2 \int_0^\infty [e(k_\perp, 0) - Z(k_\perp, 0)] k_\perp dk_\perp. \quad (4.52)$$

These quantities are very important. They were measured in Jacquin et al. (1990) and accurately predicted by DNS and EDQNM2 in Cambon, Mansour, and Godeferd (1997). The strong difference in the evolution of these quantities suggests that the polarization anisotropy $Z = \Re Z$ is important in the exact slow manifold $k_\parallel = k_3 = 0$. Generally, $e + \Re Z$ and $e - \Re Z$ give the spectral energy of the poloidal and toroidal modes, respectively. In the equatorial wave plane ($k_\parallel = 0$), they contribute to both horizontal and vertical energy. A refined statistical model ought to match AQNM outside the vicinity of the slow manifold, with $Z = 0$, and full EDQNM3 in the vicinity of the 2D manifold (Cambon, Rubinstein, and Godeferd, 2004). An interesting related problem is that the ED cannot be ignored in the vicinity of the slow manifold, as it is in classical wave-turbulence theory, so that a fully nonlinear statistical theory is needed.

4.7.3 Are Present DNS and LES Useful for Theoretical Prediction?

DNS and LES results have also shown the tendency of rotating turbulence to become anisotropic by spectral transfer toward the horizontal wave plane (Cambon, Mansour, and Godeferd, 1997; Morinishi, Nakabayashi, and Ren, 2001), not to mention qualitative results dealing with the development of vortices elongated in the vertical direction (Bartello, Métais, and Lesieur, 1994). Nevertheless, it is difficult to decide, based on these results, whether the flow becomes 2D in the long time limit, for several reasons. Spatial periodicity of the flow, which is assumed in numerical models, implies that the size of the periodic box must be sufficiently large to avoid spurious confinement effects. In particular, the turbulent correlation length and $C_g t$ must remain small compared with the box size, where C_g is the inertial-wave group velocity given by Eq. (4.26). The latter condition is very stringent for long time simulations, as the evolution time scales as $Ro^{-2}\Omega^{-1}$ at small Rossby numbers.

Regarding RST anisotropy with directional/polarization splitting, the exact equation

$$2\mathcal{K}(t)b_{33}(t) = \underbrace{\iiint \left(e - \frac{E}{4\pi k^2}\right) \sin^2\theta d^3\mathbf{k}}_{b_{33}^{(e)}(t)} + \underbrace{\iiint \Re\left(\zeta e^{-2i\cos\theta t}\right) \sin^2\theta d^3\mathbf{k}}_{b_{33}^{(z)}(t)}, \quad (4.53)$$

along with Eq. (4.41), allows us to discuss some results. Recent DNSs and LESs (Cambon, Mansour, and Godeferd, 1997; Morinishi, Nakabayashi, and Ren, 2001; Yang and Domaradzki, 2004) yield results similar to those of AQNM (Bellet et al., 2006) dealing with the time history of $b_{33}^{(e)}$. A monotonic increase from about 0 (initial isotropy) to a maximum value is observed. This maximum value is never larger than 0.08, and therefore remains far below the theoretical 2D limit that is equal to 1/6 [obtained in injecting Eq. (4.48) into Eq. (4.53) (Cambon, Mansour, and Godeferd, 1997)]. In AQNM, the $b_{33}^{(z)}$ term remains equal to zero, so that $b_{33} = b_{33}^{(e)}$. The vanishing of $b_{33}^{(z)}$ seems to be a very general result, also valid in the nonlinear case if ζ evolves slowly and has integrable singularity at $k_\parallel = 0$. In almost all underresolved DNSs (or LESs), a rapid evolution of $b_{33}^{(z)}$ with negative value can yield a strong departure of b_{33} from $b_{33}^{(e)}$, resulting eventually in a negative value of b_{33} of about -0.2. The latter effect (e.g., Bartello, Métais, and Lesieur, 1994), which means that two-componentalization is much more important than two-dimensionalization, is probably due to the numerical confinement. This discrepancy yields distinguishing the continuous case from the discrete one. In the continuous unbounded case, it is clear that Z can have a physically relevant negative value in the slow manifold, allowing a large increase of the ratio $\overline{u_1^2}L_{11}^{(3)}/\overline{u_3^2}L_{33}^{(3)}$, according to Eqs. (4.51) and (4.52) and Cambon and Jacquin (1989); Jacquin et al. (1990), and Cambon, Mansour, and Godeferd (1997), but its integral contribution to $b_{33}^{(z)}$ must vanish by phase mixing under the conditions previously mentioned on ζ in Eq. (4.53). Finally, Eq. (4.53) illustrates the fact that directional and polarization anisotropies can have

4.7 Fundamental Issues: Solved and Open Questions

opposite effects on the RST anisotropy. Using the terminology introduced by Kassinos, Reynolds, and Rogers (2001), one should say that *dimensionality* [anisotropy of the dimensionality tensor corresponds to $-2b_{ij}^{(e)}$] *and polarization have opposite effects on componentality* (conventionally measured by b_{ij}).

4.7.4 Is the Pure Linear Theory Relevant?

When the analysis is restricted to single-time second-order statistics in homogeneous turbulence, it is clear that anisotropic structuration is possible only through nonlinear mechanisms. As pointed out by Davidson, Stapelhurst, and Dalziel (2006), this does not exclude the fact that formation of organized structures can be mediated by linear mechanisms.

On the one hand, single-time second-order statistics are particular cases, as phase information is essentially lost when combining $e^{\iota\sigma_k t}$ and its conjugate in the definition of the second-order spectral tensor $\hat{R}$, at least looking at its trace. Phase information is recovered considering more complex correlations, even in the homogeneous case. Not to mention two-time second-order statistics with relevant "linear" (so-called RDT) applications (Kaneda and Ishida, 2000), third-order correlations are affected by these phase effects. It is because the linear operator has a deep influence on third-order correlations that the transfer terms $T^{(e,z,h)}$ become rotation dependent and anisotropic in Eq. (4.41), breaking the initial isotropy.

On the other hand, it is suggested that inertial waves propagating from a blob of vorticity can generate elongated structures in the pure linear – but nonhomogeneous – case. This illustrates the fact that the lowest frequencies are linked to the fastest group velocity, which is close to the axial direction. In addition, a very interesting analysis of the angular momentum, with different time scalings depending on the angle of ray propagation, is performed in Davidson, Stapelhurst, and Dalziel (2006). It is also suggested that the strong but transient anisotropy of integral length scales observed in the intermediate range of Rossby numbers by Jacquin et al. (1990) reflects this linear mechanism. Incidentally, this intermediate range was very clearly delineated in Jacquin's experimental study, as subsequently discussed. First, several parameters were defined to describe the flow dynamics: a macro-Rossby number $Ro^L = \frac{u'}{2\Omega L}$ with $L = L_{33}^{(3)}$ and $u' = \sqrt{u_3^2}$ (from the observation that the axial components were the less altered by rotation) and a micro-Rossby number Ro^λ choosing the axial Taylor length scale for λ with $Ro^\lambda \sim \omega'/(2\Omega)$. Second, in the free-decay experiments, both Rossby numbers were initialized with $Ro^\lambda > Ro^L > 1$, and then decreased together, so that two transitions were successively observed. The first one is for $Ro^\lambda > Ro^L = 1$ and the second one for $Ro^\lambda = 1 > Ro^L$, respectively. These two transitions delineate the intermediate range of Rossby numbers, in which anisotropy was observed to develop. A very clear collapse of quantities combining Reynolds stresses and integral length scales from Figs. 4.1 and 4.2 was found in terms of the macro-Rossby number, showing that anisotropy of these quantities is triggered at a macro-Rossby number close to one. Nevertheless, these features were well reproduced by pure homogeneous EDQNM-type models and DNS/LES, in which RDTs for single-time second-order statistics give no anisotropy at all, if started from isotropic initial data.

From very large and old experiences with rotating flows, we consider that the different viewpoints can be reconciled. Instead of opposing linear to nonlinear dynamics, or homogeneous to inhomogeneous flows, we prefer to say that linear and nonlinear processes interact in a subtle way. As also discussed in the last section, it is more important to specify the order and the nature of the correlations to which the linear operator is applied. The fact that these correlations do or do not exhibit quasi-Gaussian properties is perhaps more important than their degree of statistical inhomogeneity.

4.7.5 Provisional Conclusions About Scaling Laws and Quantified Values of Key Descriptors

The kinetic energy decays more slowly in homogeneous rotating turbulence, with an exponent (e.g., -0.86 in AQNM) about one-half of the one observed in the nonrotating case (Squires et al., 2000; Bellet et al., 2006). This situation seems to correspond to high-Reynolds-number and very low-Rossby-number limits, so that different decay laws can be found in DNSs and experiments, such as a purely viscous decay law linked to negligible nonlinearity. On the other hand, a faster decay can be explained by inhomogeneous effects, dissipation of energy carried by inertial waves near the Ekman boundary layers (Ibbetson and Tritton, 1975), and more nonlocal effects of Ekman pumping on organized eddies (Morize, Moisy, and Rabaud, 2005). Directional anisotropy reaches a value of about $b_{33}^{(e)} \sim 0.07$–0.08 (Cambon, Mansour, and Godeferd, 1997; Morinishi, Nakabayashi, and Ren, 2001; Yang and Domaradzki, 2004; Bellet et al., 2006), whereas polarization anisotropy $b_{33}^{(z)}$ remains zero (Bellet et al., 2006), weak (Cambon, Mansour, and Godeferd, 1997; Morinishi, Nakabayashi, and Ren, 2001), or reaches a large negative value, so that $b_{33} \sim -0.2$ (DNS and LES with long elapsed time and low resolution).

The spherically averaged energy spectrum is assumed to scale as $E(k) \sim \sqrt{\Omega \varepsilon} k^{-2}$ (Zhou, 1995), in agreement with an energy transfer scaling as Ω^{-1} but completely ignoring the anisotropy. In the same way, an isotropic scaling of the second-order structure function in terms of r (and not $r^{2/3}$ as in the usual Kolmogorov theory) is invoked by Baroud et al. (2003) and directly connected to a k^{-2} energy spectrum. We consider this proposal as not fully consistent, also because the authors claimed that they have quasi-2D dynamics with an inverse energy cascade. More generally, however, the "anomalous" scaling of the nth-order structure function as $r^{n/2}$ seems to be supported by both Baroud et al. (2003) and Simand (2002). A very clear trend is shown in Fig. 4.18 from the latter author.

A complete scaling of $e(k_\perp, k_\parallel)$ for any wave-vector modulus and direction is possible from the AQNM numerical database (Bellet et al., 2006), or even from DNS/LES with convenient postprocessing (Cambon, Mansour, and Godeferd, 1997; Liechtenstein, Godeferd, and Cambon, 2005), but it is not yet available. It may generalize the scaling law (4.50) given in Galtier (2003).

4.8 Coherent Structures, Description, and Dynamics

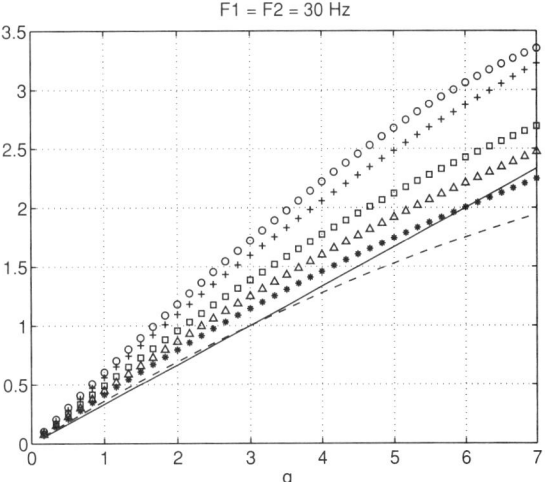

Figure 4.18. $\zeta_q(q)$ exponents for structure functions at positions d closer and closer to the core of an intense vortex ("French washing machine" with corotative disks at 30 Hz): $d = 4.5$ cm, ★; $d = 3.5$ cm, △; $d = 2.5$ cm, □; $d = 1.5$, +; $d = 0.5$, ○; K41 model, solid line; K62 model, dashed curve. (Simand, 2002.)

4.8 Coherent Structures, Description, and Dynamics

Emergence of "cigar-shaped" vortex structures has been observed in several DNS and LES studies. A recent visualization is shown in Fig. 4.19. At least in DNS initialized with conventional "almost Gaussian" realizations with random phase terms, they do not emerge if the nonlinear terms are canceled. This is consistent with the hypothesis of "nonlinear formation of structures," supported by the anisotropic statistical approach of Cambon's team, but relevant criticisms by Peter Davidson must be accounted for. The appearance of these structures depends on the range of Rossby and Reynolds numbers, and also on the resolution and effective confinement of the numerical simulation.

Figure 4.19. Isovorticity surfaces from recent high-resolution DNS. Dominant cyclonic structures are gray. Courtesy of L. Liechtenstein.

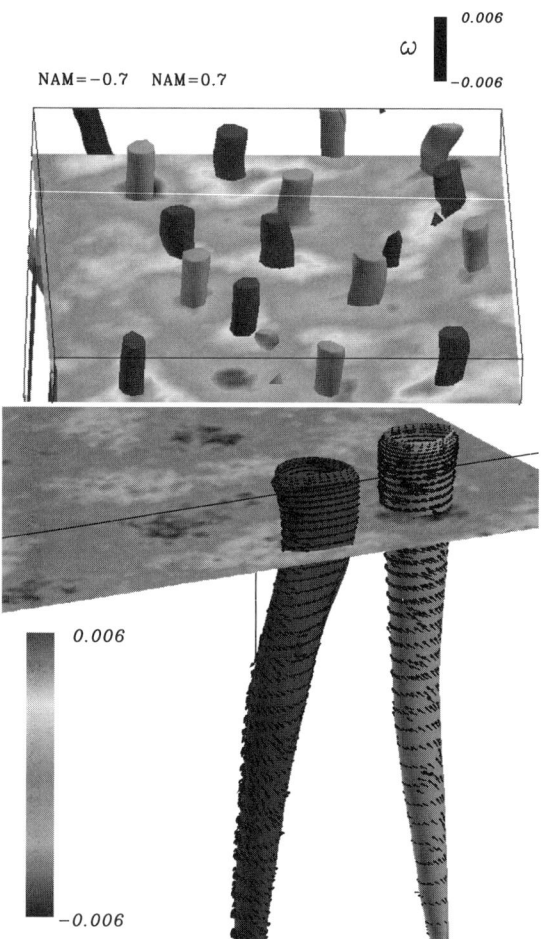

Figure 4.20. Top: Vortex structures identified by normalized angular-momentum (NAM) isovalues (tubes) and horizontal cross section of vorticity isovalues (noisy spots in the bottom plane). NAM value at point M is obtained by averaging $|\mathbf{MP} \times \mathbf{u}(P)|/(|\mathbf{MP}||\mathbf{u}(P)|)$ over point P in a small domain surrounding M. Bottom: Selected pair of cyclonic–anticyclonic eddy structures, identified by NAM $= 0.7$ isosurfaces. Helical lines along them correspond to instantaneous streamlines in the close vicinity of isosurfaces. Reproduced from Godeferd and Lollini (1999) with permission of CUP.

A more realistic confinement is present in the DNS by Godeferd and Lollini (1999) on a plane channel rotating about the vertical direction. In addition to a realistic numerical approach to vertical confinement (pseudo-spectral Fourier–Fourier–Chebyshev code with no-slip boundary conditions), another motivation was to reproduce the main results of the experiment by Hopfinger, Browand, and Gagne (1982), briefly introduced in Subsection 4.1.1. The identification of vortices is illustrated in Fig. 4.20 (top) using both horizontal sections of isosurfaces (noisy spots in the bottom plane of the figure) and isovalue surfaces of a normalized angular momentum, which is defined in the caption. The latter criterion (NAM) was suggested by experimentalist Marc Michard (Lyon) in PIV for obtaining smooth isovalues. Asymmetry in terms of cyclones–anticyclones is mainly induced by the Ekman pumping near the solid boundaries, yielding helical trajectories. This is illustrated in Fig. 4.20 (bottom), in which a cyclone–anticyclone pair is isolated. Even if the Ekman pumping generates a three-component motion, the presence of the horizontal walls and the presence of the forcing in the horizontal plane between them

4.8 Coherent Structures, Description, and Dynamics

are essential for enforcing coherent vortices. Nevertheless, and in contrast with the experimental results, no significant asymmetry between cyclonic and anticyclonic structures was observed in terms of number and intensity. In the same way, the typical distance between adjacent vortices is of the same order of magnitude as that of their diameter, and the Rossby number in their core is close to one. It was expected that, for a given symmetric distribution of more intense and concentrated vortices, the centrifugal and the elliptic instabilities could act in preferentially destabilizing the anticyclones, so that the cyclone could emerge. It seems that the insufficiently high Reynolds number is responsible for the lack of intensity and concentration of cigar vortices.

There is now a general agreement about the fact that cyclonic vertical vorticity is dominant, i.e., $\omega_3 > 0$ on the average, at sufficiently high Reynolds numbers and in an *intermediate range* of Rossby numbers. The intermediate range is not the same, according to the definitions by Jacquin et al. (1990) or by Bourouiba and Bartello (2007). Anyway, the latter definition deals with only micro-Rossby numbers significantly smaller than 1, as in Bartello, Métais, and Lesieur (1994).

It is perhaps puzzling that the approach by Bartello, Métais, and Lesieur (1994) and more recently by Chen et al.(2005) is essentially supported by low-resolution LES (not DNS because of hyperviscosity), whereas the dynamics of vorticity is emphazized. It is well known that LESs cannot accurately capture the small scales that contribute to the enstrophy, except if a sophisticated subgrid-scale model is used to explicitely represent the continuation of scales. The fact that the skewness of axial vorticity in Bartello, Métais, and Lesieur (1994) seems to grow with a positive value until a reasonable level is reached is corroborated by recent experiments (Morize, Moisy, and Rabaud, 2005) (see Fig. 4.21). This suggests that, even if the dimensional value of $\langle \omega_3^3 \rangle$ is likely strongly underestimated in a low-resolution LES, the nondimensional ratio

$$S_\omega = \frac{\langle \omega_3^3 \rangle}{\langle \omega^2 \rangle^{3/2}} \tag{4.54}$$

is probably captured with an acceptable order of magnitude.

In addition, the recent study by Morize, Moisy, and Rabaud (2005) of decaying rotating turbulence shows the relevance of the linear time scale to compare different cases with the same scaling: The vorticity skewness grows as t^α with $\alpha \in [0.7; 0.75]$ in Fig. 4.21. In this figure, the final rapid collapse is attributed to the rise of nonhomogeneous mechanisms, such as Ekman pumping. DNSs by van Bokhoven et al. (2007) yield similar results [Fig. 4.21]. The late-time collapse in Fig. 4.21 can be interpreted as a final stage of linear "triadic" phase mixing because of the absence of strong enough nonlinearity in decaying turbulence at a moderate initial Reynolds number.

Because a significant part of the vorticity statistics, informative for the previously mentioned issue of asymmetry in terms of cyclonic and anticyclonic axial

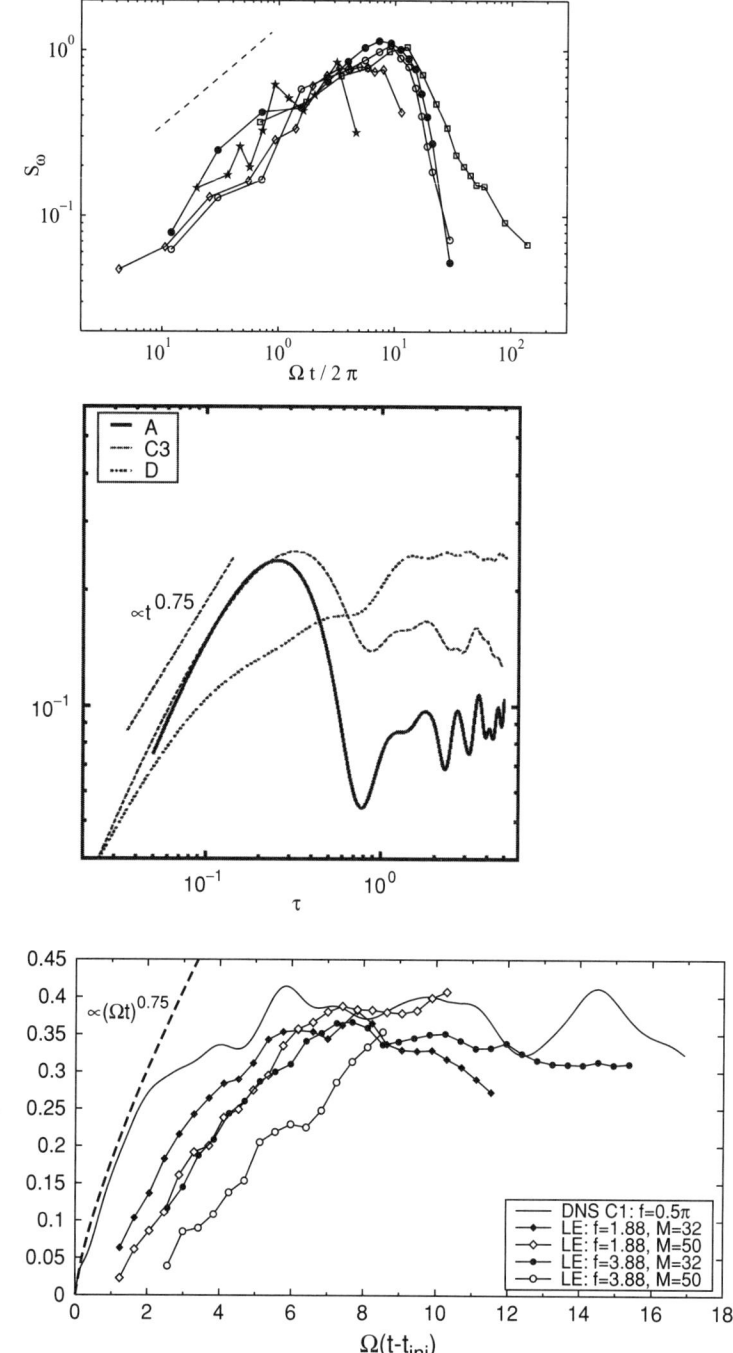

Figure 4.21. Skewness of the vertical vorticity distribution, experiment (top) (Morize, Moisy, and Rabaud, 2005), DNS (middle) (van Bokhoven et al., 2007), DNS run at the largest Rossby number plotted with experimental results (bottom) (Staplehurst, Davidson, and Dalziel, 2007).

4.8 Coherent Structures, Description, and Dynamics

vorticity, deals with triple correlations, statistical theory can be revisited and confronted by arguments from stability analysis, as subsequently discussed.

Triple-vorticity correlations are found as

$$\langle \omega_\parallel^3 \rangle(t) = \sum_{s,s',s''=\pm 1} \int_{R^6} \exp\left[i 2\Omega t \left(s \frac{k_\parallel}{k} + s' \frac{p_\parallel}{p} + s'' \frac{q_\parallel}{q} \right) \right] T_{ss's''}(\mathbf{k}, \mathbf{p}, \epsilon t) d^3\mathbf{k} d^3\mathbf{p}.$$
(4.55)

This equation is exact in the linear (RDT) limit, with $\epsilon = 0$ in the contribution from slowly evolving amplitudes denoted $T_{ss's''}$, even if applying RDT to cubic statistical moments is not usual. It exactly reflects the consequence of Poincaré transform (4.35) at the level of cubic moments if ϵ is not zero.

To compute this integral, it is necessary to know the contribution from initial, or slowly evolving, triple correlations for all triads.** But the problem is much better documented than in physical space because robust spectral theories such as EDQNM provide a systematic way to express initial, isotropic $T_{ss's''}$ in terms of the initial scalar-energy spectrum $E(k)$. More generally, more advanced EDQNM versions can be used to solve the full nonlinear problem, not only for generating isotropic initial data in Eq. (4.55) with $\epsilon = 0$. Common to the linear and nonlinear formulations, the phase term controlling phase mixing appears in Eq. (4.55) and is zero when triads are in exact resonance.

As another result of statistical theory, it can be shown that the triple-vorticity correlation $\langle \omega_3^3 \rangle$ is necessarily produced with a positive value (corresponding to net production of cyclonic vertical vorticity) when 3D isotropic turbulence is suddenly set into solid-body rotation (Gence and Frick, 2001). This result comes from the Euler equations written in the rotating frame,

$$\frac{d}{dt}\langle \omega_3^3 \rangle = 3\langle \omega_3^2 \omega_j S_{3j} \rangle + 6\Omega \langle \omega_3^2 S_{33} \rangle \quad \text{with} \quad S_{ij} = \frac{1}{2}\left(\frac{\partial u_i}{\partial x_j} + \frac{\partial u_j}{\partial x_i} \right),$$

as the specific rotation-induced "production" term $6\Omega\langle \omega_3^2 S_{33} \rangle$ is essentially positive, as is the classical "nonlinear vortex-stretching term" $\langle \omega_i \omega_j S_{ij} \rangle$. The reader is referred to van Bokhoven et al. (2007) for the statistical analysis and its deep discussion, supported by both DNS and experimental results.

** Or equivalently, in physical space, for any triple correlation at three points, information that cannot be provided by the third-order structure function alone.

Bibliography

Aupoix, B., Cousteix, J., and Liandrat, J. (1983). Effects of rotation on isotropic turbulence (1983), in *Turbulent Shear Flows*, Bradbury, L. J. S., Durst, F., Launder, B. E., Schmidt, F. W., and Whitelaw, J. H., eds., Springer-Verlag.

Bardina, J., Ferziger, J. M., and Rogallo, R. S. (1985). Effect of rotation on isotropic turbulence: Computation and modelling, *J. Fluid Mech.* **154**, 321–326.

Baroud, C. N., Plapp, B. B., and Swinney, H. L. (2003). Scaling in three-dimensional and quasi-two-dimensional rotating turbulent flows, *Phys. Fluids* **15**, 2091–2104.

Bartello, P., Métais, O., and Lesieur, M. (1994). Coherent structures in rotating three-dimensional turbulence, *J. Fluid Mech.* **273**, 1–29.

Bellet, F., Godeferd, F. S., Scott, J. F., and Cambon, C. (2006). Wave-turbulence in rapidly rotating flows, *J. Fluid Mech.* **562**, 83–121.

van Bokhoven, L. J. A., Cambon, C., Liechtenstein, L., Godeferd, F. S., and Clercx, H. J. H. (2008). Refined vorticity statistics of decaying rotating three-dimensional turbulence, *J. Turbulence* **9**, 1–24.

Bos, W. J. T. and Bertoglio, J.-P. (2006). A single-time two-point closure based on fluid particle displacements, *Phys. Fluids* **18**, 031706.

Benney, D. J. and Newell, A. C. (1969). Random wave closure, *Stud. Appl. Math.* **48**, 29–53.

Bourouiba, L. and Bartello, P. (2007). The intermediate Rossby number range and 2D-3D transfers in rotating decaying homogeneous turbulence, *J. Fluid Mech.* **587**, 131–161.

Cambon, C. (1982). Etude spectrale d'un champ turbulent incompressible soumis à des effets couplés de déformation et rotation imposés extérieurement, Thèse de Doctorat d'Etat, Université Lyon I, France.

Cambon, C. (2001). Turbulence and vortex structures in rotating and stratified flows, *Eur. J. Mech. B (Fluids)* **20**, 489–510.

Cambon, C., Godeferd, F. S., Nicolleau, F., and Vassilicos, J. C. (2004). Turbulent diffusion in rapidly rotating flows with and without stable stratification, *J. Fluid Mech.* **499**, 231–255.

Cambon, C. and Jacquin, L. (1989). Spectral approach to non-isotropic turbulence subjected to rotation, *J. Fluid Mech.* **202**, 295–317.

Cambon, C., Jacquin, L., and Lubrano, J-L. (1992). Towards a new Reynolds stress model for rotating turbulent flows, *Phys. Fluids A* **4**, 812–824.

CAMBON, C., MANSOUR, N. N., AND GODEFERD, F. S. (1997). Energy transfer in rotating turbulence, *J. Fluid Mech.* **337**, 303–332.

CAMBON, C., RUBINSTEIN, R., AND GODEFERD, F. S. (2004). Advances in wave-turbulence: Rapidly rotating flows, *New J. Phys.* **6**, 73, 1–29.

CAMBON, C. AND SCOTT, J. F. (1999). Linear and nonlinear models of anisotropic turbulence, *Annu. Rev. Fluid Mech.* **31**, 1–53.

CHEN, Q., CHEN, S., EYINK, G. S., AND HOLM, D. D. (2005). Resonant interactions in rotating homogeneous three-dimensional turbulence, *J. Fluid Mech.* **542**, 139–164.

DAVIDSON, P. A., STAPELHURST, P. J., AND DALZIEL, S. B. (2006). On the evolution of eddies in a rapidly rotating system, *J. Fluid Mech.* **557**, 135–144.

DICKINSON, S. C. AND LONG, R. R. (1984). Oscillating grid-turbulence including effects of rotation, *J. Fluid Mech.* **126**, 315–333.

GALTIER, S. (2003). A weak inertial wave-turbulence theory, *Phys. Rev. E* **68**, 1–4.

GALTIER, S., NAZARENKO, S., NEWELL, A. C., AND POUQUET, A. (2000). A weak turbulence theory for incompressible MHD, *J. Plasma Phys.* **63**, 447–488.

GENCE, J. N. AND FRICK, C. (2001). Naissance des corrélations triples de vorticité dans une turbulence statistiquement homogène soumise à une rotation, *C. R. Acad. Sci. Paris Série II b* **329**(5), 351–356.

GODEFERD, F. S. AND LOLLINI, L. (1999). Direct numerical simulations of turbulence with confinement and rotation, *J. Fluid Mech.* **393**, 257–308.

GREENSPAN, H. P. (1968). *The Theory of Rotating Fluids*, Cambridge University Press.

HOPFINGER, E. J., BROWAND, F. K., AND GAGNE, Y. (1982). Turbulence and waves in a rotating tank, *J. Fluid Mech.* **125**, 505–534.

IBBETSON, A. AND TRITTON, D. (1975). Experiments of rotation in a rotating fluid, *J. Fluid Mech.* **68**, 639–672.

JACQUIN L., LEUCHTER O., CAMBON C., AND MATHIEU J. (1990). Homogeneous turbulence in the presence of rotation, *J. Fluid Mech.* **220**, 1–52.

KANEDA, Y. AND ISHIDA T. (2006). Suppression of vertical diffusion in strongly stratified turbulence, *J. Fluid Mech.* **402**, 311–327.

KASSINOS, S. C., REYNOLDS, W. C., AND ROGERS, M. M. (2001). One-point turbulence structure tensors, *J. Fluid Mech.* **428**, 213–248.

LIECHTENSTEIN, L., GODEFERD, F. S., AND CAMBON, C. (2005). Nonlinear formation of structures in rotating stratified turbulence, *J. Turbulence* **6**, 1–18.

MANSOUR, N. N., CAMBON, C., AND SPEZIALE, C. G. (1991). Single point modelling of initially isotropic turbulence under uniform rotation. *Annual Research Briefs*, Center for Turbulence Research, Stanford University.

MCEWAN, A. D. (1970). Inertial oscillations in a rotating fluid cylinder, *J. Fluid Mech.* **40**, 603–639.

MCEWAN, A. D. (1976). Angular momentum diffusion and the initialization of cyclones, *Nature London* **260**, 126.

MORINISHI, Y, NAKABAYASHI, K., AND REN, S. K. (2001). Dynamics of anisotropy on decaying homogeneous turbulence subjected to system rotation, *Phys. Fluids* **13**, 2912–2922.

MORIZE, C., MOISY, F., AND RABAUD, M. (2005). Decaying grid-generated turbulence in a rotating tank, *Phys. Fluids* **17**, 095105.

MOWBRAY, D. E. AND RARITY, B. S. H. (1967). A theoretical and experimental investigation of the phase configuration of internal waves of small amplitude in a density stratified liquid, *J. Fluid Mech.* **28**, 1–16.

ORSZAG, S. A. (1970). Analytical theories of turbulence, *J. Fluid Mech.* **41**, 363–386.
PEDLOWSKY, J. (1987). *Geophysical Fluid Dynamics*, Springer-Verlag.
PRAUD, O., SOMMERIA, J., AND FINCHAM, A. (2006). Decaying grid turbulence in a rotating stratified fluid, *J. Fluid Mech.* **547**, 389–412.
PROUDMAN, I. (1916). On the motion of solids in a liquid possessing vorticity, *Proc. R. Soc. London Ser. A* **92**, 408.
SALHI, A. AND CAMBON, C. (2007). Anisotropic phase-mixing in homogeneous turbulence in a rapidly rotating or in a strongly stratified fluid: An analytical study, *Phys. Fluid*, **19**, 055102.
SIMAND, C. (2002). Etude de la turbulence au voisinage d' un vortex, *Ph.D. Thesis*, Ecole Normale Supérieure de Lyon, defended on 28/11/2002.
SMITH, L. M. AND WALEFFE, F. (1999). Transfer of energy to two-dimensional large scales in forced, rotating three-dimensional turbulence, *Phys. Fluids* **11**, 1608–1622.
SQUIRES, K. D., CHASNOV, J. R., MANSOUR, N. N., AND CAMBON, C. (2000). The asymptotic state of rotating homogeneous turbulence at high Reynolds number, presented at the *74th Fluid Dynamics Symposium on Application of Direct and Large Eddy Simulation to Transition and Turbulence*, Chania, Greece, AGARD CP 551, 4-1–4-9.
STAPLEHURST, P. J., DAVIDSON, P. A., AND DALZIEL, S. B. (2008). Structure formation in homogeneous, freely-decaying, rotating turbulence, *J. Fluid Mech.* **598**, 81–103.
TAYLOR, G. I. (1921). Experiments on the motion of solid bodies in rotating fluids, *Proc. R. Soc. London Ser. A* **104**, 213.
TRAUGOTT, S. S. (1958). Influence of solid body rotation on screen produced turbulence, *NACA Tech. Rep.* 4135.
WALEFFE, F. (1993). Inertial transfers in the helical decomposition, *Phys. Fluids A* **5**, 677–685.
WIGELAND, R. A. AND NAGIB, H. M. (1978). Grid generated turbulence with and without rotation about the streamwise direction, *IIT Fluids and Heat Transfer Rep.* R 78-1.
YANG, X. AND DOMARADZKI, J. A. (2004). LES of decaying rotating turbulence, *Phys. Fluids* **16**, 4088–4104.
ZAKHAROV, V. E., L'VOV, V. S., AND FALKOWICH G. (1992). *Kolmogoroff Spectra of Turbulence I. Wave Turbulence.* Springer Series in Nonlinear Dynamics. Springer-Verlag.
ZHOU, Y. (1995). A phenomenological treatment of rotating turbulence, *Phys. Fluids* **7**, 2092–2094.

5 Incompressible Homogeneous Anisotropic Turbulence: Strain

5.1 Main Observations

This chapter is devoted to the dynamics of homogeneous turbulent flows submitted to a pure strain. The pure strain case is defined as the case in which the mean-velocity-gradient matrix **A** is symmetric. As discussed in the rest of this chapter, several experimental setups have been designed during the past few decades that lead to different forms for **A**. Kinematic aspects, from the design of ducts in experiments to a first insight into RDT (more details are given in Chapter 13), are also introduced in the general case in which **A** combines a symmetric and an antisymmetric part (mean vorticity) in order to characterize in the simplest way what the specificity is of an irrotational straining process.

Both experiments and numerical simulations lead to the following observations dealing with the dynamics of homogeneous turbulence subjected to pure strain:

- The initially isotropic turbulence becomes anisotropic in the presence of a mean strain, and the principal axes of the RST become identical to those of the **A**, the axis of contraction for **A** corresponding to the direction of maximum amplification for the RST. If the strain is applied for a long enough time, anisotropy reaches an asymptotic state. Typical results are displayed in Fig. 5.1.
- Turbulent kinetic energy $\mathcal{K}(t)$ is a growing function of time at large nondimensional time St (see Fig. 5.2), where S is related to a norm of **A**. This *production of turbulent kinetic energy* is related to the growth of anisotropy. For initially isotropic turbulence, an initial period of decay is observed, which corresponds to the transient phase during which the anisotropy starts raising from zero.
- *Negative production*, i.e., destruction of $\mathcal{K}(t)$ by the mean flow, can be observed over the transient phase for some initially anisotropic turbulent flows (see Fig. 5.3). It is important to note that this phenomenon is not related to a dissipative mechanism involving molecular viscosity or nonlinear cascade. It is due to the same physical mechanisms that are responsible for turbulence production in other cases. The negative–positive character of turbulence production by pure strain is determined by the angle between the principal axes of the RST and those of **A**. In addition, the temporal memory of the straining process, identified by the Cauchy matrix **F** related to **A**, plays an essential role for explaining the response of turbulence to time-dependent processes **A**(t).

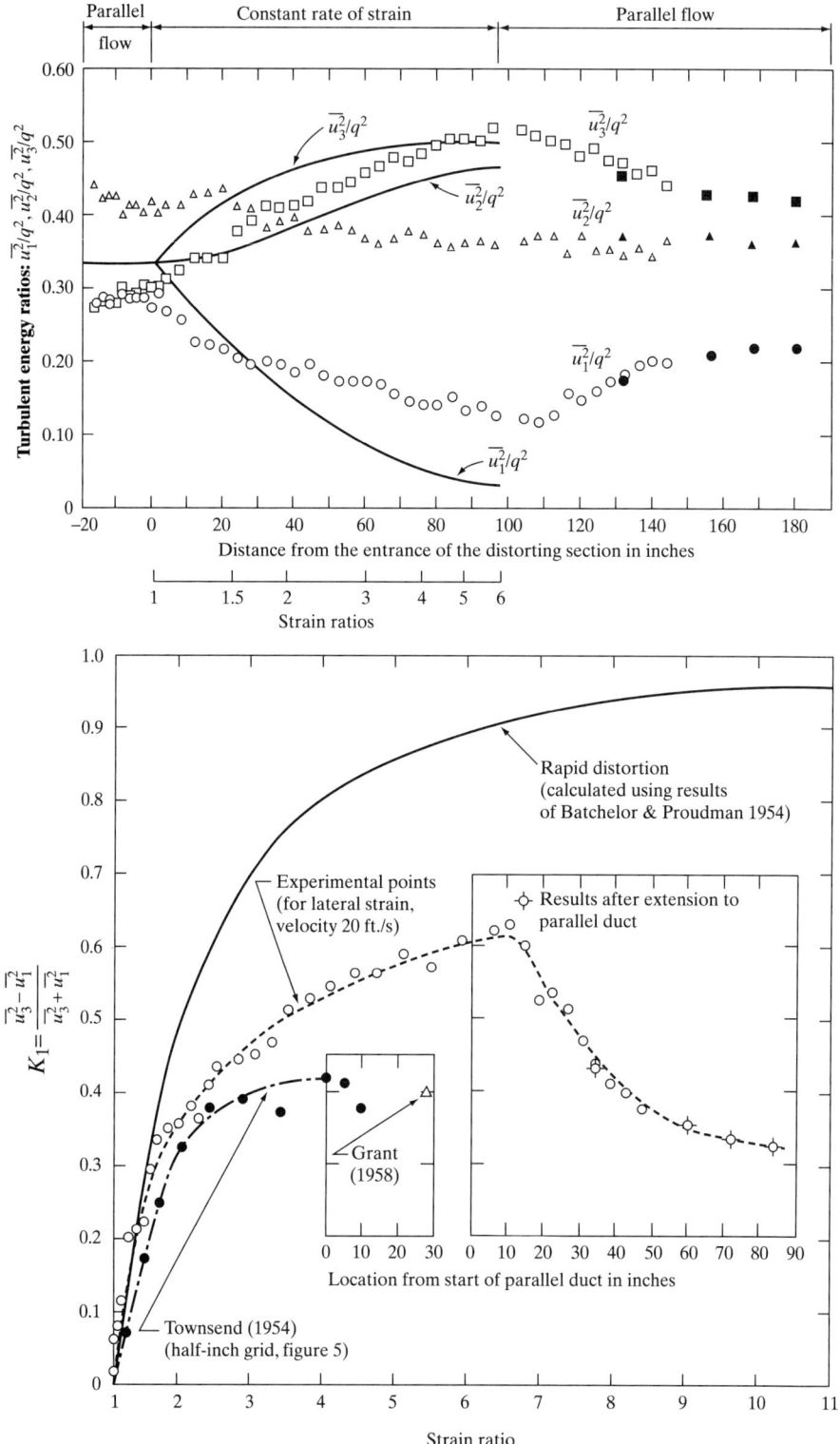

Figure 5.1. Time evolution of anisotropy in the pure strain case with isotropic initial field. Top: evolution of the three components of the total kinetic energy (solid curves: linear RDT prediction; symbols: experiments). Bottom: evolution of the structural anisotropy indicator in different experiments. Reproduced from Tucker and Reynolds (1968) with permission of CUP.

5.2 Turbulence in the Presence of Mean Strain

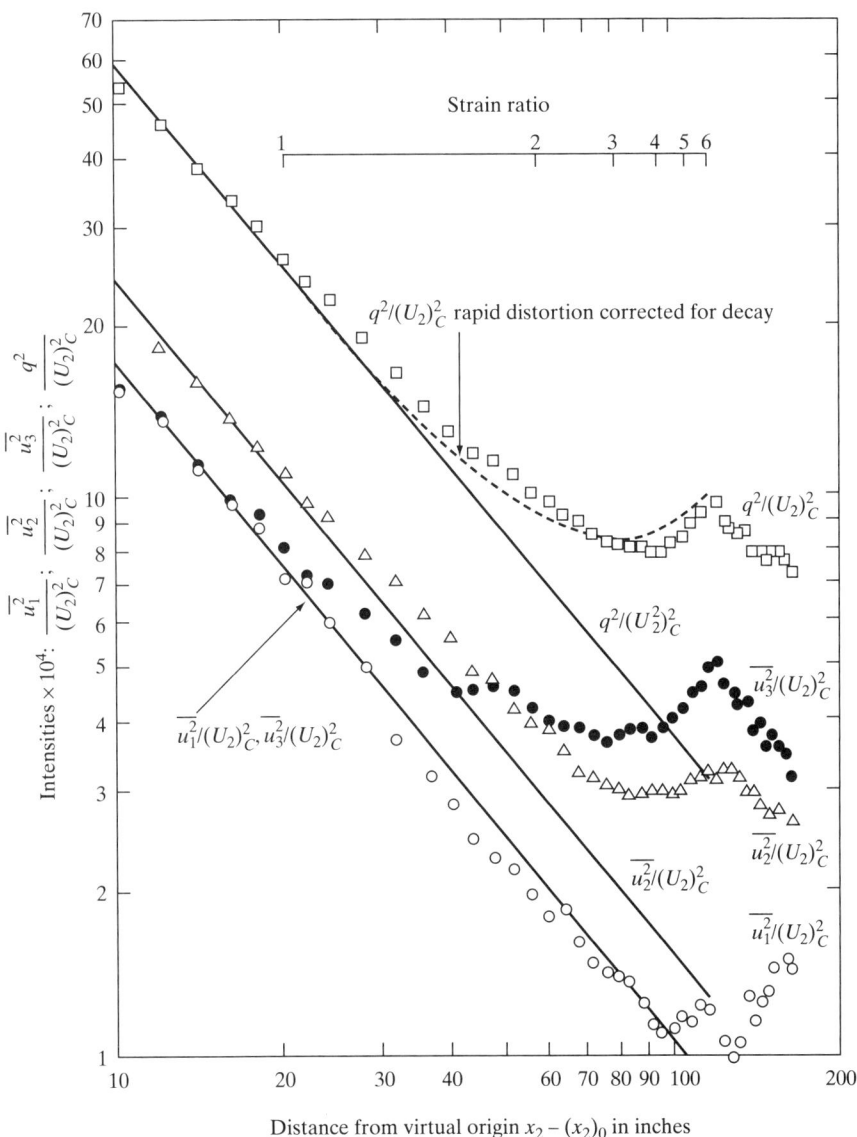

Figure 5.2. Time evolution of turbulent kinetic energy in the pure strain case with isotropic initial field, illustrating the production phenomenon. The turbulent kinetic energy $\mathcal{K}$ is denoted by q^2 here. Reproduced from Tucker and Reynolds (1968) with permission of CUP.

5.2 Experiments for Turbulence in the Presence of Mean Strain. Kinematics of the Mean Flow

In most wind-tunnel experiments, turbulence was generated by a grid and transported along the tunnel by the mean flow. Distorted ducts located downstream of the grid were used to impose the desired strain on the initially isotropic turbulence. The principle is the following: The distorted duct is designed so that its internal

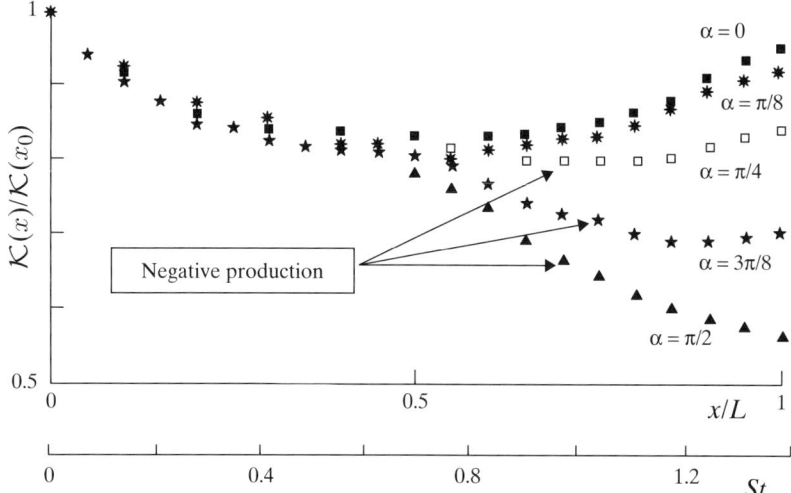

Figure 5.3. Time evolution of turbulent kinetic energy in the pure strain case with anisotropic initial field, illustrating the negative production phenomenon. α is the angle of rotation between two successive coplanar strains, $\alpha = 0$ corresponds to a constant-strain rate without rotation of the strain direction. Reproduced from Gence and Mathieu (1979) with permission of CUP.

surface is coincidental with a stream-tube surface of the desired mean flow, leading to the imposition of the target mean-velocity-gradient field. The Lagrangian formalism introduced in Subsection 2.1.6 is particularly useful. It provides a simple and elegant framework to describe the geometry of stream tubes and to recover a kinematic interpretation of the mathematical operators that appear in the RDT. Accordingly, the kinematic description of the mean "distorting flow" makes use of the Lagrangian and mixed Lagrangian–Eulerian formalism introduced in Chapter 2, but *the quantities related to Lagrangian features of the solution, such as the Lagrangian coordinates X_i, the trajectories, and mixed Eulerian–Lagrangian quantities such as the Cauchy matrix F_{ij}, are restricted to the mean flow only.*

5.2.1 Pure Irrotational Strain, Planar Distortion

The decay of HIT is well reproduced in wind-tunnel experiments, in which turbulence created by a grid is advected downstream without significant distortion (see Subsection 3.1.1). In this case, an equivalent elapsed time is estimated thanks to the Taylor frozen-turbulence hypothesis (see Subsection 3.1.1) as $t = \frac{x_3 - x_3^0}{U_0}$, where x_3 denotes the streamwise coordinates. Here, x_3^0 is a typical "initial" distance from the grid, needed to homogenize the wakes of the rods (about 40 mesh sizes in practice), and U_0 is the mean velocity, which is considered as uniform in the duct, outside the boundary layers.

The additional straining process is obtained with a distorting duct, whose transverse sections have a constant area, in order to conserve U_0, but are more and more elongated as the distance from the grid increases. Rectangular transverse sections

5.2 Turbulence in the Presence of Mean Strain

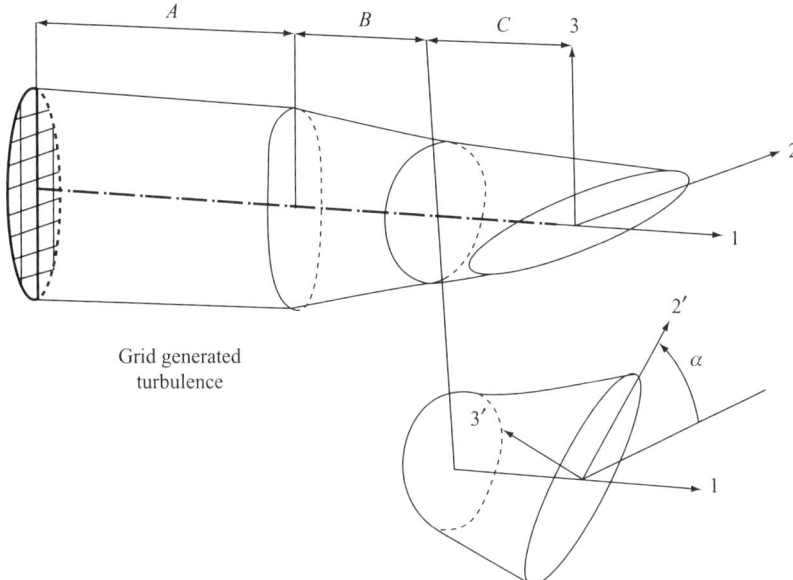

Figure 5.4. Sketch of the duct used by Gence and Mathieu to obtain plane straining of turbulence. Reproduced from Gence and Mathieu (1979) with permission of CUP.

were used by Maréchal (1970) and Tucker and Reynolds (1968), whereas elliptical sections were used by Gence and Mathieu (1979). In all cases, the contour lines have a hyperbolic design in order to reproduce a mean strain with constant rate.

As a typical feature of the experiment by Gence and Mathieu, the initial section of the distorting duct is elliptical with its large axis in the vertical direction, so that the aspect ratio of the ellipse first decreases to reproduce a compression in the vertical direction (say x_2) and a dilatation in the spanwise (x_1) direction. Continuation of the process yields recovering an increasing aspect ratio, with elongation of the elliptical section in the spanwise direction. The change from decreasing to increasing the aspect ratio implies that a *circular* section is obtained at a particular downstream position. Accordingly, the distorting duct is split into two parts, before and after the circular section, so that a sudden change of the principal axes of the straining process can be reproduced only by rotating the second part of the duct with respect to the first part from an angle α (see Fig. 5.4). For instance, in the case of continuous strain, a rotation of $\pi/2$ allows reverting the strain, so that the initial section is exactly recovered at the end of the duct.

Moreover, a duct with a constant section can be added at the end of the distorting one, in order to study the expected return-to-isotropy of turbulence, at least when the distortion results in a large anisotropy of the turbulent flow.

All these experiments illustrate the generation of the mean-velocity gradients **A** in a cross section normal to the uniform streamwise velocity, denoted U_0 and chosen along the direction 3, so that the mean trajectories can be defined by

$$x_i = F_{ij}(t, t_0)X_j + U_0 t \delta_{i3}. \tag{5.1}$$

The initial time t_0 is omitted or denoted by 0 in the following discussion.

The corresponding mean-gradient matrix **A** and mean-displacement matrix **F** are (see Subsection 2.1.6)

$$\mathbf{A} = \mathbf{A}_0 = \begin{bmatrix} -S & 0 & 0 \\ 0 & S & 0 \\ 0 & 0 & 0 \end{bmatrix}, \quad \text{and} \quad \mathbf{F} = \mathbf{F}_0(t) = \begin{bmatrix} e^{-St} & 0 & 0 \\ 0 & e^{St} & 0 \\ 0 & 0 & 1 \end{bmatrix} \quad (5.2)$$

for a constant-straining process, or the first part of the duct in the Gence's experiment, and

$$\mathbf{A} = \tilde{\mathbf{Q}}\mathbf{A}_0\mathbf{Q}, \mathbf{F} = \mathbf{F}_0(t_1) + \tilde{\mathbf{Q}}\mathbf{F}_0(t - t_1)\mathbf{Q} \quad \text{for} \quad t_1 < t < 2t_1 \quad (5.3)$$

with

$$\mathbf{Q} = \begin{bmatrix} \cos\alpha & -\sin\alpha & 0 \\ \sin\alpha & \cos\alpha & 0 \\ 0 & 0 & 1 \end{bmatrix}, \quad (5.4)$$

for the second part of the duct with $t_1 = L/U_0$ corresponding to the location of the circular section at half the length $2L$ of the full Gence's distorting duct.

The distorting duct is built by materializing a stream tube, corresponding to an initial cross section, chosen to be rectangular (Maréchal, 1970; Tucker and Reynolds, 1968) or elliptical (Gence and Mathieu, 1979). True streamlines are expected to be homothetic and to follow the geometry imposed by the duct, in agreement with previous equations, an expectation that appeared to be reasonable in a large part of the duct not too close to solid boundaries.

5.2.2 Axisymmetric (Irrotational) Strain

This case is of particular interest because axial symmetry is the simplest anisotropy. The corresponding mean flow can be reproduced by means of an axisymmetric distorting duct, convergent or divergent. As also discussed in the next subsection, the mean velocity is not constant in the streamwise – and axial – direction, and it is not constant in a given cross section with increasing (divergent duct) or decreasing (convergent duct) surface. Flow separation and specific instabilities can appear, especially in the divergent case, but also in the convergent case (Leclaire, 2006). In spite of the complexity of these experimental issues, this flow is considered as a reference case, at least from a theoretical and numerical viewpoint.

The configuration of axisymmetric strain is approached in convergent and divergent ducts, and only the vicinity of the centerline (axis) is considered for the sake of simplicity. The mean-velocity-gradient matrix is

$$\mathbf{A} = \begin{bmatrix} -S/2 & 0 & 0 \\ 0 & -S/2 & 0 \\ 0 & 0 & S \end{bmatrix}, \quad (5.5)$$

5.2 Turbulence in the Presence of Mean Strain

where S is possibly time dependent and $S > 0$ corresponds to the case of an axisymmetric convergent duct. As previously, when the axial direction is chosen as $n_i = \delta_{i3}$ without lack of generality, the nontrivial components in the Cauchy matrix are

$$F_{33}(t) = \exp\left[\int_0^t S(t')dt'\right] = C(t), \quad F_{11}(t) = F_{22}(t) = \frac{1}{\sqrt{C(t)}}. \tag{5.6}$$

5.2.3 The Most General Case for 3D Irrotational Case

On the other hand, 3D irrotational strain characterized by **A** having three eigenvalues, S_1, S_2, S_3, with zero sum (because of incompressibility constraint), and possibly being time dependent, holds little interest from the viewpoint of homogeneous turbulence and related experimental approaches. Some specific RDT solutions were initially proposed by Courseau and Loiseau (1978), but can be easily generalized and simplified, using the formalism introduced by Cambon (1982) and Cambon, Teissèdre, and Jeandel (1985), using both a reduced Green's function and the Cauchy matrix. *For any analytical result subsequently given, we will specify whether it applies to the 3D general irrotational case.*

5.2.4 More General Distortions. Kinematics of Rotational Mean Flows

A different kind of experimental procedure was initially proposed to obtain a pure mean-shear flow. For instance, in Rose (1966) and Champagne, Harris, and Corrsin (1970), the shear gradient is created in the vertical direction (here x_1) by a pileup of plates, without distortion of the duct. Even if a constant-mean-velocity gradient $\partial \bar{u}_2/\partial x_1$ is obtained throughout the duct, the consistency with statistical homogeneity of turbulence is much more problematic than in the experiments for irrotational strain presented in the preceding section. The streamwise velocity is no longer uniform in a cross section, so that the mean advection time $(x_3 - X_3)/U_3(x_1)$ varies with x_1; it is shorter near the top (largest U_3) than near the bottom (smallest U_3) of the duct. As a consequence, $A_{21} = S$ can be considered as constant in a current cross section, but not $F_{21} = St$.

To obtain a pure shear flow in a more satisfactory way (regarding homogeneity of turbulence), and especially to generalize it to an arbitrary combination of vorticity and strain, a general procedure was defined at ONERA, in close collaboration with the LMFA team (Leuchter, Benoit, and Cambon, 1992). The principle is to generate solid-body rotation in a cylindrical duct, and then to superimpose a convenient irrotational process by a subsequent distorting duct. Jacquin's experiment for pure rotation, presented in the previous chapter, was used for this purpose, replacing the cylindrical duct after the rotation generator (a rotating honeycomb) with a duct designed in exact accordance with Eq. (5.1). Current cross sections of the distorting duct do correspond to a single advection time $(x_3 - X_3)/U_0$. They are ellipses of constant area with both their aspect ratio and the orientation of their axes varying continuously with the streamwise position. In Eq. (5.1), **F** (with corresponding **A**) is easily calculated as follows:

$$\mathbf{A} = \begin{bmatrix} 0 & S-\Omega & 0 \\ S+\Omega & 0 & 0 \\ 0 & 0 & 0 \end{bmatrix}, \quad \mathbf{F} = \begin{bmatrix} \cosh \sigma_0 t & (S-\Omega)\frac{\sinh \sigma_0 t}{\sigma_0} & 0 \\ (S+\Omega)\frac{\sinh \sigma_0 t}{\sigma_0} & \cosh \sigma_0 t & 0 \\ 0 & 0 & 1 \end{bmatrix}, \quad (5.7)$$

with

$$\sigma_0^2 = S^2 - \Omega^2. \qquad (5.8)$$

The hyperbolic case $S > \Omega$ is given here as an example, but the elliptical case $S < \Omega$ is straightforwardly derived by changing σ_0 into $\iota\sigma_0$, yielding a periodic **F**. The analytical solution $F_{\alpha\beta} = \delta_{\alpha\beta} \cosh \sigma_0 t + A_{\alpha\beta} \frac{\sinh \sigma_0 t}{\sigma_0}$ results from $A_{\alpha\gamma} A_{\gamma\beta} = \sigma_0^2 \delta_{\alpha\beta}$, so that $\ddot{F}_{\alpha\beta} = \sigma_0^2 F_{\alpha\beta}$, excluding the value 3 for Greek indices. The particular case of pure shear $S = \Omega$, whose associated shear rate is equal to $\Omega + S = 2S$, is consistently recovered in the limit $\sigma_0 \to 0$.

Elliptical cross sections are analytically derived from the initial (circular) section of the distorting duct, i.e., $X_\alpha X_\alpha = R^2$, so that

$$F^{-1}_{\alpha\beta}(t) F^{-1}_{\alpha\gamma}(t) x_\beta x_\gamma = R^2, \quad t = \frac{x_3 - x_3^0}{U_0}. \qquad (5.9)$$

Typical streamlines, such as those sketched in Fig. 2.1, are recovered in the plane (1, 2) of the mean distortion as the envelope of the moving ellipses. They are hyperboles for $\sigma_0^2 > 0$, straight lines for $\sigma_0 = 0$, and ellipses for $\sigma_0^2 < 0$. Only in the last case is the duct periodic; a typical sketch is shown in Fig. 5.5.

5.3 First Approach in Physical Space to Irrotational Mean Flows

5.3.1 Governing Equations, RST Balance, and Single-Point Modeling

5.3.1.1 Planar Strain

The evolution equations for Reynolds stresses associated with the gradient matrix **A** defined in Eq. (5.2) are

$$\frac{\partial \overline{u'_i u'_j}}{\partial t} + \overline{u}_k \frac{\partial \overline{u'_i u'_j}}{\partial x_k} = S \begin{bmatrix} 2\overline{u'_1 u'_1} & 0 & \overline{u'_1 u'_3} \\ 0 & -2\overline{u'_2 u'_2} & -\overline{u'_2 u'_3} \\ \overline{u'_1 u'_3} & -\overline{u'_2 u'_3} & 0 \end{bmatrix} + \Pi_{ij} - \varepsilon_{ij}. \qquad (5.10)$$

The corresponding equation for the turbulent kinetic energy is

$$\frac{\partial \mathcal{K}}{\partial t} = S \left(\overline{u'_1 u'_1} - \overline{u'_2 u'_2} \right) - \varepsilon. \qquad (5.11)$$

It is seen that the production of kinetic energy by explicit linear effects is due to the anisotropy and more precisely to the difference between the two diagonal

5.3 First Approach in Physical Space to Irrotational Mean Flows

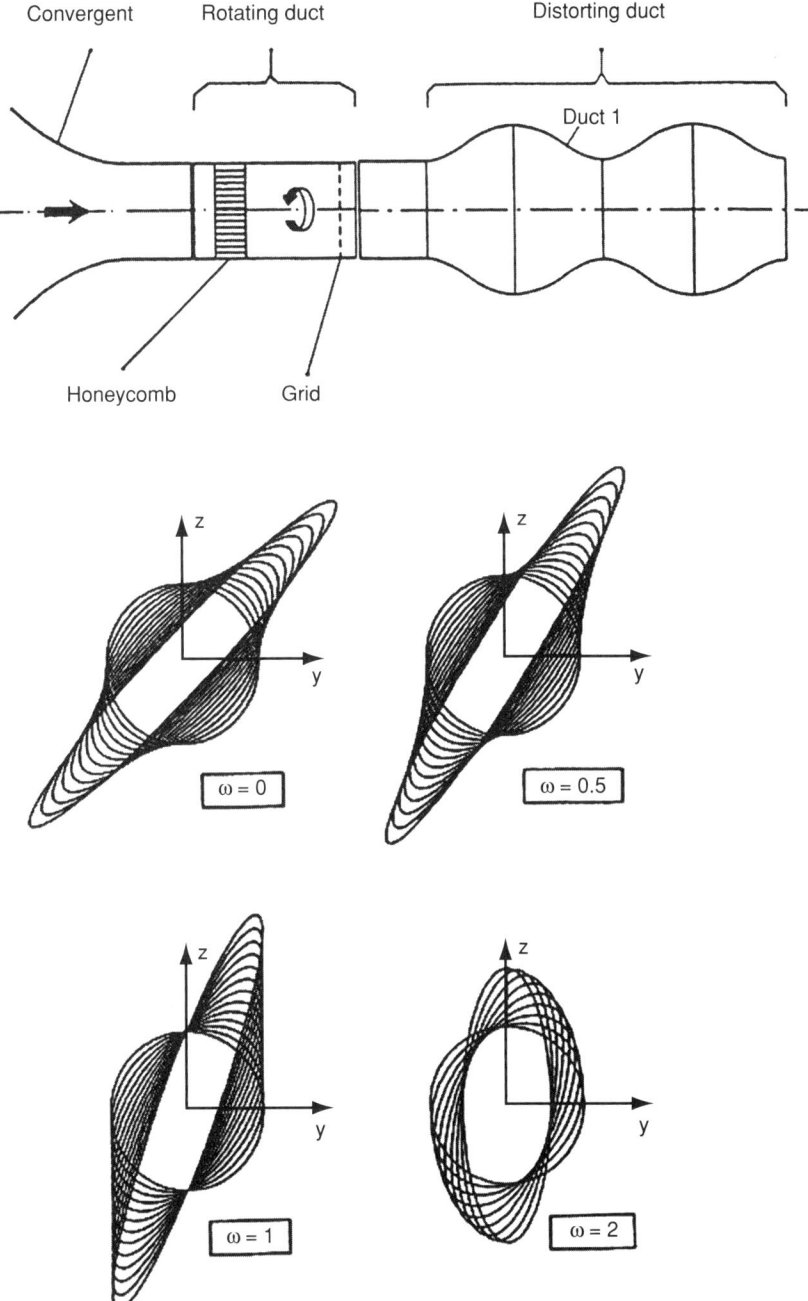

Figure 5.5. Sketch of the distorting ducts in the experiment by Leuchter, Benoit, and Cambon (1992). Top: side view of the experimental facilities, "periodic" case with $\omega = \Omega/S = 2$. Bottom: front view of the duct for different values of $\omega = \Omega/S$.

Reynolds stresses $\overline{u'_1 u'_1}$ and $\overline{u'_2 u'_2}$. This production can be either positive or negative, depending on the signs of S and $\left(\overline{u'_1 u'_1} - \overline{u'_2 u'_2}\right)$. The possible occurrence of a negative production term corresponds to the existence of flows in which the mean irrotational strain will destroy the turbulent kinetic energy. A direct consequence is that the production mechanism escapes the isotropic two-equation turbulence models, in which the differences between the diagonal Reynolds stresses are neglected. The errors committed on turbulent kinetic-energy production in the vicinity of stagnation points in nonhomogeneous flows are directly related to this problem.

Of course, negative production occurs in Gence's experiment, and this is further discussed at the end of this section.

5.3.1.2 Axisymmetric Irrotational Strain

For axisymmetric strain with **A** given by Eq. (5.5), one has

$$\frac{\partial \overline{u'_i u'_j}}{\partial t} + \overline{u}_k \frac{\partial \overline{u'_i u'_j}}{\partial x_k} = S \begin{bmatrix} \overline{u'_1 u'_1} & \overline{u'_1 u'_2} & -\frac{1}{2}\overline{u'_1 u'_3} \\ \overline{u'_1 u'_2} & \overline{u'_2 u'_2} & -\frac{1}{2}\overline{u'_2 u'_3} \\ -\frac{1}{2}\overline{u'_1 u'_3} & -\frac{1}{2}\overline{u'_2 u'_3} & -2\overline{u'_3 u'_3} \end{bmatrix} + \Pi_{ij} - \varepsilon_{ij}. \quad (5.12)$$

The corresponding equation for the turbulent kinetic energy is

$$\frac{\partial \mathcal{K}}{\partial t} = \frac{S}{2}\left(\overline{u'_1 u'_1} + \overline{u'_2 u'_2} - 2\overline{u'_3 u'_3}\right) - \varepsilon. \quad (5.13)$$

It is seen that the production mechanisms are still governed by anisotropy in this configuration, but this time it involves the three diagonal Reynolds stresses.

5.3.1.3 More General Rotational Strains

In the general case corresponding to Eq. (5.7), one has

$$\frac{\partial \overline{u'_i u'_j}}{\partial t} + \overline{u}_k \frac{\partial \overline{u'_i u'_j}}{\partial x_k}$$

$$= -\begin{bmatrix} 2(S-\Omega)\overline{u'_1 u'_2} & (S-\Omega)\overline{u'_2 u'_2} + (S+\Omega)\overline{u'_1 u'_1} & (S-\Omega)\overline{u'_2 u'_3} \\ (S-\Omega)\overline{u'_2 u'_2} + (S+\Omega)\overline{u'_1 u'_1} & 2(S+\Omega)\overline{u'_1 u'_2} & (S+\Omega)\overline{u'_1 u'_3} \\ (S-\Omega)\overline{u'_2 u'_3} & (S+\Omega)\overline{u'_1 u'_3} & 0 \end{bmatrix}$$

$$+ \Pi_{ij} - \varepsilon_{ij} \quad (5.14)$$

and

$$\frac{\partial \mathcal{K}}{\partial t} = 2S\overline{u'_1 u'_2} - \varepsilon. \quad (5.15)$$

5.3 First Approach in Physical Space to Irrotational Mean Flows

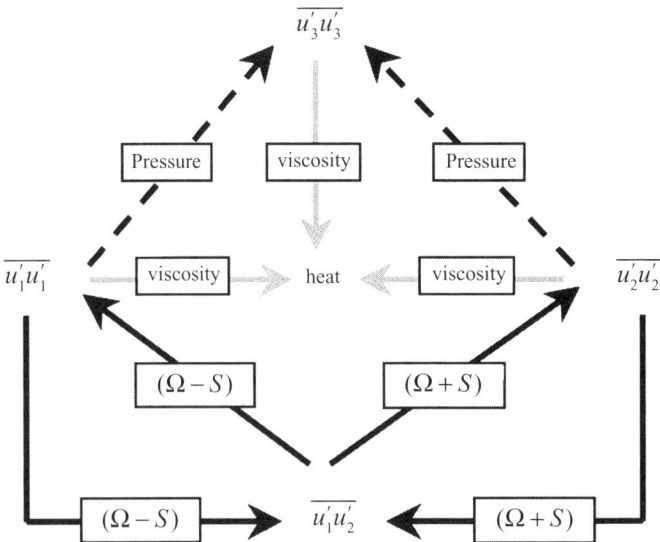

Figure 5.6. Couplings between the different nonvanishing Reynolds stresses in the general strain case. Arrows indicate the production process, their color being related to the physical quantity at play (mean strain, pressure, viscosity).

Considering the case of an initially isotropic field, one has $\overline{u'_\alpha u'_3} = \overline{u'_3 u'_\alpha} = 0$, $\alpha = 1, 2$ so that Eq. (5.14) simplifies to

$$\begin{cases} \dfrac{d\overline{u'_1 u'_1}}{dt} = 2(\Omega - S)\overline{u'_1 u'_2} & +\Pi_{11} -\varepsilon_{11} \\ \dfrac{d\overline{u'_2 u'_2}}{dt} = -2(\Omega + S)\overline{u'_1 u'_2} & +\Pi_{22} -\varepsilon_{22} \\ \dfrac{d\overline{u'_3 u'_3}}{dt} = & \Pi_{33} -\varepsilon_{33} \\ \dfrac{d\overline{u'_1 u'_2}}{dt} = \Omega(\overline{u'_2 u'_2} - \overline{u'_1 u'_1}) + S(\overline{u'_2 u'_2} + \overline{u'_1 u'_1}) & +\Pi_{12} -\varepsilon_{12} \end{cases} \quad (5.16)$$

The different couplings are illustrated in Fig. 5.6. This case is rediscussed from the viewpoint of RDT, first in Subsection 5.4.1 and then in Chapter 8.

5.3.2 General Assessment of RST Single-Point Closures

The most general irrotational flow case, with time-dependent and even 3D (i.e., with three different nonzero eigenvalues) **A** is now considered.

Full Reynolds stress models work satisfactorily to predict the effect of irrotational strain. For instance, the Reynolds stress component in the direction of mean compression is shown to increase and the one in the direction of mean dilatation is shown to decrease, so that increasing anisotropy is created by a monotonic strain.

In addition, directional and polarization anisotropies seem to be closely related, at the level of single-point statistics, by means of

$$b_{ij}^{(e)} = -\frac{1}{2}b_{ij}, \quad b_{ij}^{(z)} = \frac{3}{2}b_{ij}. \tag{5.17}$$

This relationship was quoted in Kassinos, Reynolds, and Rogers (2000) as "*dimensionality* [measured by $-2b_{ij}^{(e)}$] *equals to componentality*" (measured by b_{ij}). Such a relationship will be shown to be consistent with RDT, but only for a very short time and starting from isotropic initial data in Subsection 5.4.2.

On the other hand, $\mathcal{K} - \varepsilon$ modeling is questioned if the straining process is time varying. This can be explained by the fact that even a crude "pressure-released" equation like Eq. (5.20) given in the next subsection is much better for predicting RST anisotropy than the so-called Boussinesq approximation $b_{ij}(t) \propto A_{ij}(t)$. Because the instantaneous Boussinesq relationship used in $\mathcal{K} - \varepsilon$ models, and even in its so-called nonlinear variants, cannot take into account the time history of **A**, they completely miss the quasi-reversible behavior observed in Gence's experiment. Similar conclusions can be drawn for time-periodic strains (relevant for reciprocating engines, for instance) for both full Reynolds stress models and $\mathcal{K} - \varepsilon$ variants (Hadzic, Hanjalic, and Laurence, 2001).

5.3.3 Linear Response of Turbulence to Irrotational Mean Strain

The role of mean vorticity is easily understood by linearizing the vorticity equation:

$$\dot{\omega}'_i = \frac{\partial \bar{u}_i}{\partial x_j}\omega'_j + \frac{\partial u'_i}{\partial x_j}\bar{\omega}_j.$$

Only in the absence of mean vorticity is this equation similar to Eq. (2.21) and admits the solution

$$\omega'_i(\mathbf{x}, t) = F_{ij}(\mathbf{X}, t, t_0)\omega'_j(\mathbf{X}, t_0), \tag{5.18}$$

which is now a true solution, as **F** and **X** are related to only the (irrotational) mean flow, and therefore are *externally given data*. Similarly, a linearized Weber equation can be written, leading to

$$u'_i(\mathbf{x}, t) = F_{ji}^{-1}(\mathbf{X}, t)u'(\mathbf{X}, t_0) + \frac{\partial \phi'}{\partial x_i}. \tag{5.19}$$

The two last equations, which are directly useful in RDT, are no longer valid if the mean flow is rotational.

Generalization to rotational mean flows is possible, using Clebsh potentials, but the method is much less tractable; it is touched on by Hunt (1973) and Goldstein (1978) and used by Nazarenko and Zakharov (1994) in the case of pure plane shear flow.

Interpretation of results from experiments of distortion is easy in the irrotational case. From the vorticity equation, with **F** given by Eq. (5.2), it is seen than the vorticity component is decreased in the direction of compression (direction x_1

5.3 First Approach in Physical Space to Irrotational Mean Flows

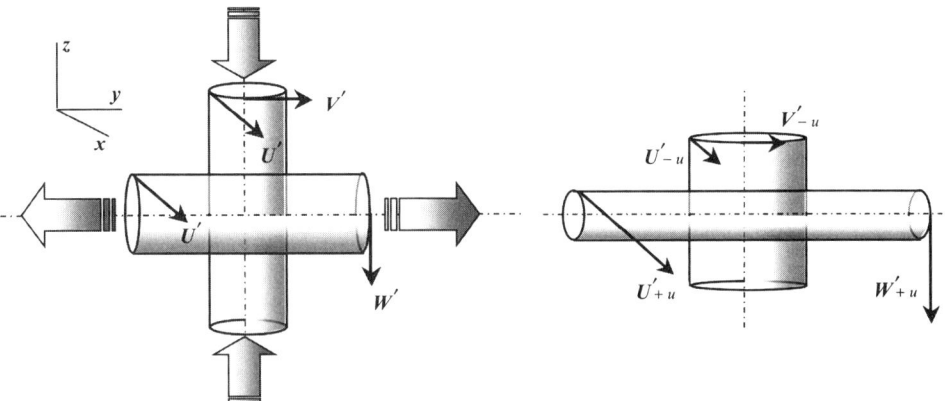

Figure 5.7. Cartoon using two individual vortices aligned with compression–dilatation axes to predict anisotropic trends induced by a mean strain. Left: Two straight vortices with equal diameters and circulation. Right: vortices after compression–dilatation assuming preservation of (i) volume of each vortex and (ii) angular momentum of each vortex. It is seen that the velocity component in the plane normal to the strain is left unchanged, whereas velocity is increased (resp. decreased) in the direction of the compression (resp. dilatation).

here) and increased in the direction of elongation of the mean constant strain. More generally, the RDT solution for vorticity correlations is

$$\overline{\omega'_i \omega'_j}(t) = F_{im}(t) F_{jn}(t) \overline{\omega'_m \omega'_n}(0).$$

Such a solution is not very realistic, as enstrophy involves the smallest scales, and it is much more constraining (than for the RST) to exclude nonlinearity in the response of turbulence to the mean strain. The basic process suggested by Eq. (5.18), however, yielded a qualitative argument to interpret the effect of the strain on velocity fluctuations (Gence and Mathieu, 1979), as given schematically in Fig. 5.7: Vorticity is amplified if a vortex tube is elongated, so that it rotates faster and the velocity components are amplified in the two directions *normal to* the axis of the elongated vortex tube. Conversely, a compressed vortex tube rotates slower and the velocity components are decreased in the two directions normal to the axis of the compressed vortex tube. This effect is sketched starting from two similar vortex tubes (a situation that mimics initial isotropy) with axes in directions x_1 and x_2, the first being compressed and the second being elongated. Accordingly, it is suggested that the velocity fluctuation is amplified in the direction of compression, with $\langle u'^2_1 \rangle$ increasing, and is decreased in the direction of elongation, with $\langle u'^2_2 \rangle$ decreasing. Of course, this simple reasoning accounts for the fact that **F** is present in Eq. (5.18) whereas its inverse **F**$^{-1}$ is present in Eq. (5.19), but does not account for the fact that linear velocity dynamics is also dependent on a "rapid" pressure effect, mediated by the scalar potential term in Eq. (5.19). However, the qualitative prediction remains correct, and the "pressure-released" equation,

$$\overline{u'_i u'_j}(t) = F^{-1}_{mi}(t) F^{-1}_{nj}(t) \overline{u'_m u'_n}(0), \qquad (5.20)$$

for the RST is consistent with the qualitative development of anisotropy.

If the mean strain is irrotational, but with the principal axes of **A** possibly moving with time, as in Gence's experiments, it is possible to use the previous equations with **F** given by Eq. (5.2) for $0 < t < t_1$ and by Eq. (5.3) for $t_1 < 0 < 2t_1$. The quasi-reversibility of the flow anisotropy when $\alpha = \pi/2$, which is associated to $F_{ij}(2t_1) = F_{ij}(0)$, reflects the dominant role of **F**, which returns to its initial value, as the final elliptical section returns to its initial position. Note that **F** is no longer symmetric in the second phase [see Eq. (5.3)], even if **A** is, because of the history of **A**. As a consequence, it is convenient to distinguish between F_{ij} and F_{ji} in the general equations, even for irrotational mean flows. The matrix **F** is continuous in time, generating continuous mean trajectories, even if **A** is discontinuous because of a sudden change of principal axes. The same reversible behavior might be obtained at the end of the more complex distorting duct in the elliptic-flow case, "forgetting" to rotate the honeycomb. This experiment is just mentioned here as a "*gedanken* experiment" as it has never been performed. The design of the distorting duct accounts for a given Ω rate and makes it possible – in principle – to reproduce the elliptical-flow instability at given S, with $S < \Omega$ (see Chapter 8); this possibility appeared more exciting than producing a new reversible irrotational strain with continuous motion of its principal axes.

5.4 The Fundamentals of Homogeneous RDT

An exhaustive presentation of the RDT is given in Chapter 13. The present section aims to provide the key elements of this method that are necessary for understanding the physical analysis subsequently given in this chapter.

The simplest multipoint closure consists of dropping all nonlinear terms in Eq. (2.29) before applying the statistical average operator. Also dropping the viscous term, one obtains the RDT, introduced by Batchelor and Proudman (1954). The RDT model is assumed to be a relevant model for the large scales of turbulence, which are not directly governed by viscous effects at very high Reynolds numbers. This approach was further developed by Townsend (1976) and Hunt (1973). Useful reviews were given by Hunt and Carruthers (1990) and more recently by Cambon and Scott (1999). An effort was made in the latter review to bridge the RDT basic concepts and equations and some studies carried out in the fields of applied mathematics and hydrodynamic stability by Craik and Criminale (1986), Bayly (1986), and other authors.

In neglecting nonlinearity entirely, the effects of the interaction of turbulence with itself are supposed to be small compared with those resulting from mean-flow distortion of turbulence. One often has in mind flows such as weak turbulence encountering a sudden contraction in a channel or flows around an airfoil. The underlying implicit assumption is that the time required for a significant distortion by the mean flow to develop is short compared with that for the turbulent evolution in the absence of distortion effect. Linear theory can also be relevant, at least over short enough times, if physical influences leading to linear terms in the fluctuation equations dominate turbulent flows, such as in a strongly stratified or rotating fluid or a

5.4 The Fundamentals of Homogeneous RDT

conducting fluid in a strong magnetic field. For such cases, the term "rapid distortion theory" is probably a little bit misleading.

Thanks to the linearity assumption, the time evolution of u'_i may be formally written as

$$u'_i(\mathbf{x}, t) = \int \mathcal{G}_{ij}(\mathbf{x}, \mathbf{x}', t, t') u'_j(\mathbf{x}', t') d^3\mathbf{x}', \qquad (5.21)$$

where $\mathcal{G}_{ij}(\mathbf{x}, \mathbf{x}', t, t')$ is a Green's function matrix expressing the evolution from time t' to time t. Whereas u'_i is a random quantity, which varies from one realization to another realization of the flow, $\mathcal{G}_{ij}$ is deterministic and can in principle be calculated for a given $\bar{u}_i(\mathbf{x}, t)$. From $\mathcal{G}_{ij}$ and the initial turbulence, Eq. (5.21) may be used to predict the time evolution.

Another simplifying assumption that is often made is that the size of turbulent eddies, L, is small compared with the overall length scales of the flow, ℓ, which might be the size of a body encountering fine-scale free-stream turbulence (see e.g., Hunt and Carruthers, 1990). In that case, one uses a local frame of reference convected with the mean velocity and approximates the mean-velocity gradients as uniform, but time varying. Thus the mean velocity is approximated by Eq. (2.46) in the moving frame of reference. In the example of fine-scale turbulence encountering a body, one may imagine following a particle convected by the mean velocity, which sees a varying mean-velocity gradient, $A_{ij}(t)$, even when the mean flow is steady.

For the sake of simplicity, the following equations are derived in the case of an extensional mean flow, with mean-velocity gradients uniform in the whole space.* This is referred to as "homogeneous RDT" because statistical homogeneity (invariance by translation of all fluctuating flow statistics) holds for the fluctuating flow, provided some additional conditions of admissibility are imposed to **A**, as introduced in Chapter 2. Recall that the mean flow is a particular solution of Euler equations and is not itself invariant by translation (only its gradient is). In the linear limit, the fluctuating fields (u'_i, p') satisfy modified equation (2.29) with the advection–distortion parts written in terms of $A_{ij}(t)$:

$$\underbrace{\frac{\partial u'_i}{\partial t} + A_{jk} x_k \frac{\partial u'_i}{\partial x_j}}_{\text{advection}} + A_{ij} u'_j + \frac{\partial p'}{x_i} = 0. \qquad (5.22)$$

Its solution is most easily obtained by means of Fourier analysis, with elementary components of the form

$$u'_i(\mathbf{x}, t) = a_i(t) \exp[\imath \mathbf{k}(t) \cdot \mathbf{x}], \qquad (5.23)$$

$$p'(\mathbf{x}, t) = b(t) \exp[\imath \mathbf{k}(t) \cdot \mathbf{x}]. \qquad (5.24)$$

Evolution equations for both the wave vector and the amplitudes are easily obtained from Eq. (5.22) (details are given in Chapter 13). They can be written

---

* One has to keep in mind that essentially the same equations can be used in more realistic flow cases, following Hunt and co-workers and Lifschitz and Hameiri (1991).

as follows, under simple ordinary differential equations (ODEs):

$$\frac{dk_i}{dt} + A_{ji}k_j = 0 \tag{5.25}$$

and

$$\frac{da_i}{dt} = -\underbrace{\left(\delta_{in} - 2\frac{k_i k_n}{k^2}\right) A_{nj}}_{M_{ij}} a_j. \tag{5.26}$$

General solutions that are valid for arbitrary initial data are expressed as follows in terms of linear transfer matrices,

$$k_i(t) = F_{ji}^{-1}(t, t_0)k_j(t_0), \tag{5.27}$$

$$a_i(t) = G_{ij}(t, t_0)a_j(t_0), \tag{5.28}$$

in which the Cauchy matrix appears under a transposed and inverse form [this is a general solution for any Eikonal-type equation; see Eq. (5.25)] and corresponds to the solution for the gradient of a passive scalar (see Chapter 13). The use of a Green's function **G** allows us to get rid of particular initial data for the fluctuating field. In the preceding equations, it is perhaps clearer to specify the wave-vector dependency in *a* and **G**, especially if we combine elementary solutions of the form given by Eq. (5.23) by means of Fourier synthesis. As a consequence, the RDT solution can be expressed as follows:

$$\widehat{u}_i[\mathbf{k}(t), t] = G_{ij}(\mathbf{k}, t, t_0)\widehat{u}_j[\mathbf{k}(t_0), t_0], \tag{5.29}$$

in which the Green's function is eventually determined by the (universal, not flow-dependent) initial conditions:

$$G_{ij}(\mathbf{k}, t_0, t_0) = \delta_{ij} - \frac{K_i K_j}{K^2}, \quad K_i = k_i(t_0). \tag{5.30}$$

At this stage, it may be noticed that homogeneous RDT gathers enough features for solving two problems:

- A deterministic problem, which consists of solving the initial-value linear system of equations for a_i, in the most general way. This is done by determining the spectral Green's function, which is also the key quantity requested in linear-stability analysis.
- A statistical problem that is useful for the prognosis of statistical moments of *u'* and *p'*. Interpreting the initial amplitude $\widehat{u}[\mathbf{k}(t_0), t_0]$ as a random variable with a given dense $\mathbf{k}(t_0)$ spectrum, relation (5.29) yields the prediction of statistical moments though products of the basic Green's function (details are displayed in Chapter 13).

A useful reduced (using the minimum number of components) Green's function can be used in the Craya–Herring frame of reference, as

$$u^{(\alpha)}[\mathbf{k}(t), t] = g_{\alpha\beta}(\mathbf{k}, t, t')u^{(\beta)}[\mathbf{k}(t'), t'], \tag{5.31}$$

5.4 The Fundamentals of Homogeneous RDT

A Green's function is then expressed in terms of only four components, solving the two-component linear system in (2.86) (details are provided in Chapter 13). Using helical modes is less useful, except near the limit of pure rotation or purely antisymmetric **A**.

5.4.1 Qualitative Trends Induced by the Green's Function

Considering a mean flow given by Eq. (5.7), qualitative RDT results are subsequently presented, before being discussed with more details in Chapters 6 and 8.

Irrotational mean flows with $A_{ij} = A_{ji}$ yield simple RDT solutions in which both **F** and **G** display dominant exponential growth (if **A** is not time dependent), reflecting pure stretching of vorticity disturbances, in accordance with the existence of the hyperbolic instability.

Rotational mean flows yield more complicated linear solutions, and only the steady case has received much attention [although Craik and co-workers and Bayly and co-workers made recent developments in unsteady cases; see, e.g., Bayly, Holm, and Lifschitz (1996)].

The steady, rotational case, when axes are chosen appropriately, corresponds to constant $S, \Omega \geq 0$, generating steady plane flows, which combine vorticity 2Ω and irrotational straining S. The related stream function is sketched in Fig. 2.1. The problem with arbitrary S and Ω was analyzed in Cambon (1982) and Cambon, Teissèdre, and Jeandel (1985) with the purpose of extending classical RDT results, which were restricted to pure strain and pure shear. For $S > \Omega$, the mean-flow streamlines are open and hyperbolic, and RDT results are qualitatively close to those of the pure strain case $\Omega = 0$. For $S < \Omega$, the mean-flow streamlines are closed and elliptic about the stagnation point at the origin. This case is the most surprising one, in which **F** is periodic in time whereas **G** is capable of generating exponential growth of fluctuations for **k**-directions concentrated about special angles ($k_3/k \sim \pm 1/2$ if $S \ll \Omega$). The RDT can therefore be relevant for explaining the mechanism of *elliptical-flow instability* (Bayly, 1986) (details are discussed in Chapter 8). The limiting case $S = \Omega$ corresponds to pure shear of straight mean streamlines. The RDT solutions of Townsend (1976) reflect *algebraic growth* in the parlance of stability analysis (see Chapter 6).

5.4.2 Results at Very Short Times. Relevance at Large Elapsed Times

As recalled in Cambon and Rubinstein (2006), the first significant terms of the RDT solution for single-time second-order statistics at a short time, starting from 3D isotropic initial data, involve only the symmetric part **S** of **A**, yielding

$$e(\mathbf{k}, t) = \frac{1}{2}\left[k\frac{\partial e}{\partial k}\bigg|_{t=0} + e(\mathbf{k}, 0)\right] t S_{ij} \frac{k_i k_j}{k^2}$$

and

$$Z(\mathbf{k}, t) = \frac{1}{2} e(\mathbf{k}, 0) t S_{ij} N_i^*(\mathbf{k}) N_j^*(\mathbf{k}),$$

with $e(k, 0) = E(k)/(4\pi k^2)$.

Spherical integration gives

$$H_{ij}^{(e)}(k,t) = -\frac{1}{15}\left(-1 + \frac{k}{E}\frac{dE}{dk}\right)S_{ij}t \quad \text{and} \quad H_{ij}^{(z)}(k,t) = -\frac{2}{5}S_{ij}t$$

for the spherically averaged spectra of $b_{ij}^{(e)}$ and $b_{ij}^{(z)}$, given eventually by

$$b_{ij}^{(e)} = \frac{2}{15}S_{ij}t, \quad b_{ij}^{(z)} = -\frac{2}{5}S_{ij}t, \quad b_{ij} = -\frac{4}{15}S_{ij}t.$$

One recovers the fact that "componentality" is equal to "dimensionality" [see Eq. (5.17) and related discussion], but this result does not persist for a long time. Looking at a nonlinear theory, assuming that weak anisotropy results from a linear response to turbulence perturbated from a fully nonlinear quasi-isotropic state, Ishihara, Yoshida, and Kaneda (2002) and Yoshida et al. (2003) found similar results:

$$H_{ij}^{(e)} = \frac{1}{15}(B-A)k^{-2/3}\varepsilon^{-1/3}S_{ij} \quad \text{and} \quad H_{ij}^{(z)} = \frac{2}{5}Ak^{-2/3}\varepsilon^{-1/3}S_{ij}.$$

One can see that the time scale in short-time RDT is simply the elapsed time t, whereas it is a turbulent time scale in the linear response theory (in fact a fully nonlinear theory, touched on in Chapter 14). Given the values of "constants" A and B, which are evaluated from LRA theory and from DNS data, no simple relationship between $b_{ij}^{(e)}$ and $b_{ij}^{(z)}$ similar to Eq. (5.17) is shown.

5.5 Final RDT Results for Mean Irrotational Strain

5.5.1 General RDT Solution

A complete analytical solution for the velocity in Fourier space can be obtained from its counterpart in terms of vorticity, using Eq. (5.18). An easier way to derive this is to use linearized Weber equation (5.19):

$$G_{ij}(\mathbf{k}, t, t') = P_{in}(\mathbf{k})F_{jn}^{-1}(t, t'). \tag{5.32}$$

This solution is valid for any irrotational straining process, even 3D and time dependent. The Cauchy matrix appears as the only explicit time-dependent tensor in the solution; an implicit time dependency is mediated by $\mathbf{k}$, but it is again governed by the Cauchy matrix, according to Eikonal solution (5.27). Accordingly, this solution is completely time reversible if the history of $\mathbf{F}$ is. This shows that the complete RDT solution shares qualitative properties with the pressure-released oversimplified one introduced in Subsection 5.3.3. Only the additional viscous factor, which is very easy to add, not to mention nonlinear effects, can break the reversibility of such a solution.

In this case, the use of the Craya–Herring frame does not simplify the solution significantly, except if the straining process is axisymmetric, as considered in the next subsection.

5.5 Final RDT Results for Mean Irrotational Strain

5.5.2 Linear Response of Turbulence to Axisymmetric Strain

The mean-velocity-gradient matrix is given by (5.5), general equations (5.32) – any irrotational strain – and (13.13) – any RDT case – are valid, and characteristic lines in Fourier space are

$$k_\alpha = K_\alpha C^{1/2}, \quad k_3 = K_3 C^{-1}. \tag{5.33}$$

The parameter C, which is given by Eq. (5.6), is directly related to the contraction ratio of the stream tube or to the ratio $A(t)/A(0)$ of the area of a current circular section to the initial one in the corresponding axisymmetric duct, because $C^{-2}(t) = A(t)/A(0)$. The linear inviscid solution in the Craya–Herring frame is

$$u^{(\alpha)}(\boldsymbol{k}, t) = \underbrace{e_i^{(\alpha)}(\boldsymbol{k}) F_{ij}^{-1}(t) e_j^{(\beta)}(\boldsymbol{K})}_{g_{\alpha\beta}} u^{(\beta)}(\boldsymbol{K}, 0),$$

and finally a very simple form of $g_{\alpha\beta}$ is obtained:

$$g_{\alpha\beta} = \begin{bmatrix} C^{1/2} & 0 \\ 0 & C^{-1/2}\frac{K}{k} \end{bmatrix}. \tag{5.34}$$

For instance, solutions for the spectral tensor of double correlations are equal to

$$\Phi^{11}(\boldsymbol{k}, t) = C(t) \frac{E(K)}{4\pi K^2}, \quad \Phi^{22}(\boldsymbol{k}, t) = C^{-1}(t) \frac{E(K)}{4\pi k^2}, \tag{5.35}$$

starting from isotropic initial data. Only two nontrivial components are needed, Φ^{11} related to toroidal energy and Φ^{22} related to poloidal energy, taking into account simplifications from axisymmetry (with mirror symmetry), yielding

$$e = \frac{1}{2}(\Phi^{11} + \Phi^{22}), \quad Z = \frac{1}{2}(\Phi^{22} - \Phi^{11})$$

in the expression of the general spectral tensor $\hat{\boldsymbol{R}}$. As a consequence, toroidal and poloidal contributions to the kinetic energy are found equal to

$$\frac{\mathcal{K}^{(tor)}(t)}{\mathcal{K}(0)} = \frac{3}{2} C(t), \quad \frac{\mathcal{K}^{(pol)}(t)}{\mathcal{K}(0)} = C^{-1}(t) \int_0^1 \frac{K_\perp K^2}{k^4} dx.$$

These results use a minimal number of components for generating RDT solutions, and they are consistent with those of Sreenivasan and Narasimha (1978), Lee (1989), and Ribner (1953). In addition, Lee (1989) provided complete useful analytical solutions for the RST (with only two nontrivial axial and transverse components here):

$$\overline{u_3'^2} = \mathcal{K}(0) \frac{C^{4s/3}}{2(1 - C^{2s})} \left[-1 + (2 - C^{2s})\sigma' \right], \tag{5.36}$$

$$\overline{u_1'^2} = \overline{u_2'^2} = \mathcal{K}(0) \frac{C^{4s/3}}{4(1 - C^{2s})} \left[\frac{2 - C^{2s}}{C^{2s}} - C^{2s}\sigma' \right], \tag{5.37}$$

with $s = \pm 1$ being the sign of S, and

$$\sigma' = \frac{1}{2\sqrt{1-C^{-2}}} \text{Ln} \frac{1+\sqrt{1-C^{-2}}}{1-\sqrt{1-C^{-2}}} \quad \text{if} \quad S > 0$$

and

$$\sigma' = \frac{\arctan(\sqrt{C^2-1})}{\sqrt{C^2-1}} \quad \text{if} \quad S < 0,$$

and for various terms in the RST budget.

5.6 First Step Toward a Nonlinear Approach

A "first loop" of nonlinear iteration was offered by Kevlahan and Hunt (1997) for a pure irrotational constant strain.

Issues linked to tradic closures are subsequently introduced (see details in Chapter 14).

Before examining the simplest and most interesting applications of linear and nonlinear theory, it is worthwhile to anticipate the difficulties for passing from (linear) RDT to generalized quasi-normal (nonlinear) closure for HAT in the presence of mean flows given by (5.7).

In the "hyperbolic" and "elliptic" cases, with $0 \neq S \neq \Omega$ in (5.7), the RDT Green's function can display exponential growth, at least for particular angles of $\mathbf{k}$ ($k_3/k \sim 1/2$ in the case $S \ll \Omega$). If the bare zeroth-order response function is only modified by ED, convergence is not ensured for the time integral of the threefold product $\mathbf{GGG}$ in the generic closure relationship.

A less critical situation occurs when $S = \Omega$ (pure plane shear), because the RDT Green's function yields only algebraic growth, so that the viscous term ensures convergence of the time integral involved in the closure. Nevertheless, it is very cumbersome to develop, and especially to solve numerically with enough accuracy, a complete anisotropic EDQNM model in this case. Recall that even calculation of single-point correlations resulting from viscous RDT at high St is not easy (Beronov and Kaneda, private communication), the asymptotic analysis being even complex for inviscid RDT (Rogers, 1991). DNSs suggest that fully nonlinear effects yield exponential growth for the turbulent kinetic energy, but computations are very sensitive to cumulated errors (remeshing, low angular resolution at small k). Such a transition from algebraic growth (linear dynamics at small time) to exponential growth (nonlinear dynamics) is not completely described and explained, but very simple single-point closure models can mimic it. In addition, interesting scaling laws for possible exponential growth follow from an approach by Julian Scott (private communication), which is itself a refinement of Oberlack's approach dealing with symmetries of the Navier–Stokes equations. All these issucs are addressed in the relevant chapter (Chapter 6) and rediscussed in Chapter 14.

5.7 Nonhomogeneous Flow Cases

The linear response theory (in fact a fully nonlinear theory) by Kaneda and co-workers is relevant but only in the limit of small scales and vanishing shear (or strain) rate. General solutions depend on only the symmetric part of the mean strain, and therefore are the same for the three different cases (elliptical, hyperbolical, rectilinear), a result that is at odds with our main approach here.

Only for pure rotation, or $S = 0, \Omega \neq 0$, the most general EDQNM versions were carried out toward complete achievement. In this case, the zeroth-order state consists of superimposed oscillating modes of motion, with no amplification and no interaction: They correspond to neutral dispersive inertial waves. A time integral of a threefold product of the Green's functions converges, provided an infinitesimal viscous (or ED) term is added. In the limit of small interactions, two-point closures and theories of wave turbulence were reconciled (Chapter 4).

This preliminary discussion justifies, to a certain extent, discriminate flows dominated by production from flows dominated by waves, a distinction that is revisited in Chapters 7, 8, and 15. The first class is illustrated by classical shear flows, in which a nonzero "production term" is displayed in the equations governing the RST. This production is often related to the growth of instabilities, when stability analysis is addressed. The second class is illustrated in Chapters 4, 7, and 8, as the more relevant area to apply spectral closures. It is worth noting that the dynamics can be dominated by dispersive waves, which are neutral but for a small part of the configuration space, in which exponential amplification occurs. In the latter case, e.g., for flows with weak ellipticity ($S \ll \Omega$), production of energy is nonzero, but classic single-point closure models are of poor relevance, because only particular orientations in wave space are subjected to parametric instability.

5.7 Nonhomogeneous Flow Cases. Coherent Structures in Strained Homogeneous Turbulence

The RDT for irrotational strain can be extended to analyze the vicinity of stagnation points in connection with the hyperbolic instability. Application to modeling of turbulence impinging on bluff bodies was given by Hunt and co-workers. The mean flow is a potential 2D inviscid flow, and hence strictly irrotational, and equations very similar to the ones found in the homogeneous case are recovered following mean-flow trajectories for short-wave disturbances.

A more advanced theoretical study was performed by Leblanc and Godeferd (1999), on the grounds of the zonal short-wave analysis introduced by Lifschitz and Hameiri (1991). Further theoretical insight into the hyperbolic instability is found in a "nonhomogeneous" case, in which the "mean" flow is a cell of 2D Taylor–Green vortices (the Taylor's four-roller mill). Stretching of the vorticity perturbation along the principal axis of strain leads to the formation of spanwise counterrotating vortices (also sometimes referred to as ribs or braids) in the irrotational stagnation-point region. Beautiful rib (or braid) structures obtained in a DNS with 128^3 grid points are shown in Fig. 5.8.

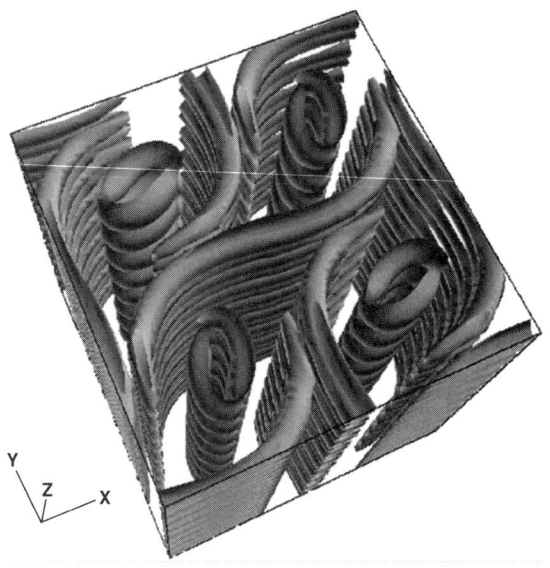

Figure 5.8. Isosurface of the vorticity magnitude of the perturbed flow in the nonlinear regime. Reproduced from Leblanc and Godeferd (1999) with permission of AIP.

Only a very few studies have been devoted to the analysis of coherent structures in homogeneous strained turbulence, because the topology of the flow is simpler than the one observed in other cases (see chapters dealing with rotation, pure shear, etc.) and because flow dynamics is relatively well understood.

Using low-Reynolds-number DNS, Rogers and Moin (1987) also observed that vorticity tends to be statistically aligned with the direction of positive strain. Vorticity occurs in coherent elongated vortex tubes that are stretched and strengthened by the mean strain. Vortex tubes submitted to a compressive effect buckle rather than decrease in strength. The absence of mean-flow rotation is observed to prevent the occurrence of hairpin vortices (which are observed in the pure shear case; see Section 6.5).

A look at instantaneous fields yields the following observations:

- Axisymmetric-contraction flows develop elongated vortex tubes in the stretching direction.
- Axisymmetric-expansion flows develop no unique structures, and a number of ringlike structures are observed.
- Plane-strain flows exhibit a combination of the structures observed in the two previous cases.

Bibliography

BATCHELOR, G. K. (1953). *The Theory of Homogeneous Turbulence*, Cambridge University Press.

BATCHELOR, G. K. AND PROUDMAN I. (1954). The effect of rapid distortion in a fluid in turbulent motion. *Q. J. Mech. Appl. Math.* **7**, 83–103.

BAYLY, B. J. (1986). Three-dimensional instability of elliptical flow, *Phys. Rev. Lett.* **57**, 2160–2163.

BAYLY, B. J., HOLM, D. D., AND LIFSCHITZ, A. (1996). Three-dimensional stability of elliptical vortex columns in external strain flows, *Philos. Trans. R. Soc. London A* **354**, 895–926.

CAMBON, C. (1982). Etude spectrale d'un champ turbulent incompressible soumis à des effets couplés de déformation et rotation imposés extérieurement, Thèse de Doctorat d'Etat, Université Lyon I, France.

CAMBON, C. AND RUBINSTEIN, R. (2006). Anisotropic developments for homogeneous shear flows, *Phys. Fluids* **18**, 085106.

CAMBON, C. AND SCOTT, J. F. (1999). Linear and nonlinear models of anisotropic turbulence, *Annu. Rev. Fluid Mech.* **31**, 1–53.

CAMBON, C., TEISSÈDRE, C., AND JEANDEL, D. (1985). Etude d' effets couplés de rotation et de déformation sur une turbulence homogène, *J. Méc. Théor. Appl.* **5**, 629–657.

CHAMPAGNE, F. H., HARRIS, V. G., AND CORRSIN, S. J. (1970). Experiments on nearly homogeneous turbulent shear flows, *J. Fluid Mech.* **41**, 81–139.

COURSEAU, P. AND LOISEAU, M. (1978). Contribution à l'analyse de la turbulence homogène anisotrope, *J. Méc.* **17**, 245–297.

CRAIK, A. D. D. AND CRIMINALE, W. O. (1986). Evolution of wavelike disturbances in shear flows: A class of exact solutions of Navier–Stokes equations, *Proc. R. Soc. London Ser. A*, **406**, 13–26.

GENCE, J.-N. AND MATHIEU, J. (1979). On the application of successive plane strains to grid-generated turbulence, *J. Fluid Mech.* **93**, 501–513.

GODEFERD, F. S., CAMBON, C., AND LEBLANC, S. (2001). Zonal approach to centrifugal, elliptic and hyperbolic instabilities in Suart vortices with external rotation, *J. Fluid Mech.* **449**, 1–37.

GOLDSTEIN, M. E. (1978). Unsteady vortical and entropic distortions of potential flows round arbitrary obstacles, *J. Fluid Mech.* **89**, 433–468.

HADZIC, I., HANJALIC, K., AND LAURENCE, D. (2001). Modeling the response of turbulence subjected to cyclic irrotational strain, *Phys. Fluid* **13**, 1739–1747.

HERRING, J. R. (1974). Approach of axisymmetric turbulence to isotropy, *Phys. Fluids* **17**, 859–872.

HUNT, J. C. R. (1973). A theory of turbulent flow around two-dimensional bluff bodies, *J. Fluid Mech.* **61**, 625–706.

HUNT, J. C. R. AND CARRUTHERS, D. J. (1990). Rapid distortion theory and the 'problems' of turbulence, *J. Fluid Mech.* **212**, 497–532.

ISHIHARA, T., YOSHIDA, K., AND KANEDA, Y. (2002). Anisotropic velocity correlation spectrum at small scale in a homogeneous turbulent shear flow, *Phys. Rev. Lett.* **88**, 154501.

JACQUIN, L., LEUCHTER, O., CAMBON, C., AND MATHIEU, J. (1990). Homogeneous turbulence in the presence of rotation, *J. Fluid Mech.* **220**, 1–52.

KASSINOS, S. C., REYNOLDS, W. C., AND ROGERS, M. M. (2001). One-point turbulence structure tensors, *J. Fluid Mech.* **428**, 213–248.

KEVLAHAN, N. K. R. AND HUNT, J. C. R. (1997). Nonlinear interactions in turbulence with strong irrotational straining, *J. Fluid Mech.* **337**, 333–364.

LEBLANC, S. AND GODEFERD, F. S. (1999). An illustration of the link between ribs and hyperbolic instability, *Phys. Fluids* **11**, 497–499.

LECLAIRE, B. (2006). Etude théorique et expérimentale d' un écoulement tournant dans une conduite, *Ph.D. thesis,* Thèse de l' Ecole Polytechnique, December 21, 2006.

LEE, M. J. (1989). Distortion of homogeneous turbulence by axisymmetric strain and dilatation, *Phys. Fluids A* **1**, 1541–1557.

LEUCHTER, O., BENOIT, J.-P., AND CAMBON, C. (1992). Homogeneous turbulence subjected to rotation-dominated plane distortion, presented at the symposium on *Turbulent Shear Flow 4*, Delft, The Netherlands.

LEUCHTER, O. AND DUPEUBLE, A. (1992). Rotating homogeneous turbulence subjected to an axisymmetric contraction, in *Proceedings of the International Symposium on Turbulent Shear Flow*, **137**, 1–6.

LIFSCHITZ, A. AND HAMEIRI, E. (1991). Local stability conditions in fluid dynamics, *Phys. Fluids A* **3**, 2644–2641.

MARÉCHAL, J. (1970). Contribution à l' étude de la déformation plane de la turbulence, *Doctorat es Sciences*, Université de Grenoble, France.

MATHIEU, J. AND SCOTT, J. F. (2000). *Turbulent Flows: An Introduction*, Cambridge University Press.

NAZARENKO, S. V. AND ZAKHAROV, V. E. (1994). The role of the convective modes and sheared variables in the Hamiltonian-dynamics of uniform-shear-flow perturbations, *Phys. Lett. A* **191**, 403–408.

NAZARENKO, S., KEVLAHAN, N. N., AND DUBRULLE, B. (1999). A WKB theory for rapid distortion of inhomogeneous turbulence, *J. Fluid Mech.* **390**, 325–348.

ORSZAG, S. A. (1970). Analytical theories of turbulence, *J. Fluid Mech.* **41**, 363–386.

PIQUET, J. (2001) *Turbulent Flows. Models and Physics*, 2nd ed., Springer.

RIBNER, H. S. (1953). Convection of a pattern of vorticity through a shock wave, *Tech. Rep. 1164*, NACA.

ROGERS, M. (1991). The structure of a passive scalar field with a uniform mean gradient in rapidly sheared homogeneous turbulent flow, *Phys. Fluids A* **3**, 144–154.

ROGERS, M. M. AND MOIN, P. (1987). The structure of the vorticity field in homogeneous turbulent flows, *J. Fluid Mech.* **176**, 33–66.

ROSE, H. A. (1966). Results of an attempt to generate a homogeneous turbulent shear flow, *J. Fluid Mech.* **25**, 97–120.

SREENIVASAN, K. R. AND NARASIMHA, R. (1978). Rapid distortion theory of axisymmetric turbulence, *J. Fluid Mech.* **84**, 497–516.

TOWNSEND, A. A. (1976). *The Structure of Turbulent Shear Flow*, 2nd ed., Cambridge University Press.

TUCKER, H. J. AND REYNOLDS, A. J. (1968). The distortion of turbulence by irrotational plane strain, *J. Fluid Mech.* **32**, 657–673.

YOSHIDA, K., ISHIHARA, T., AND KANEDA Y. (2003). Anisotropic spectrum of homogeneous turbulent shear flow in a Lagrangian renormalized approximation, *Phys. Fluids* **15**, 2385–2397.

6 Incompressible Homogeneous Anisotropic Turbulence: Pure Shear

6.1 Physical and Numerical Experiments: Kinetic Energy, RST, Length Scales, Anisotropy

Mean-shear flows are ubiquitous in turbulence. In a real flow, the shear is always created by the no-slip condition on solid walls, except when there is no tangential velocity or when the wall is a belt moving with the same velocity as the flow (shear-free boundary layer). Shear flows are therefore intimately connected with near-wall turbulence dynamics. Nevertheless, many features can be understood in the idealized case of a uniform mean shear in the absence of boundaries, in the context of HAT. The relevance of this idealized model flow was discussed by W. C. Reynolds, among many others. The effect of the wall is to create a mean shear and to block the vertical motion. The arbitrary imposed uniform shear in the HAT framework is also responsible for a reduction of vertical velocity fluctuations (as we shall see with all details in this chapter). Therefore the presence of a wall is not mandatory.

The emphasis in this chapter is put on the departure from isotropy that is due to the application of a constant shear. The main reasons are that it contains all the physical mechanisms present in homogeneous shear flows and that it is the most extensively analyzed flow in this family. The mean flow $\bar{u} = (Sy, 0, 0)$ is characterized by the following space-uniform mean-velocity-gradient matrix $\mathbf{A}$ and Cauchy (or displacement gradient) matrix $\mathbf{F}$,

$$\mathbf{A} = \begin{bmatrix} 0 & S & 0 \\ 0 & 0 & 0 \\ 0 & 0 & 0 \end{bmatrix}, \quad \mathbf{F}(t) = \begin{bmatrix} 1 & St & 0 \\ 0 & 1 & 0 \\ 0 & 0 & 1 \end{bmatrix}, \quad (6.1)$$

and components $i = 1, 2, 3$ classically are referred to as streamwise, cross-gradient (or vertical), and spanwise directions, respectively. In other chapters, permutations of the indices 2 and 3 are used, but the intrinsic "streamwise/cross-gradient/spanwise" nomenclature is kept unchanged.

It is important to note that the shear rate S in Eq. (6.1) is equal to twice the rotation rate Ω and the strain rate S defined in Eq. (5.7), as the pure shear rate is defined as $\Omega = S$. Therefore, in this chapter, the rotation rate and the strain rate are equal to $S/2$.

6.1 Physical and Numerical Experiments

6.1.1 Experimental and Numerical Realizations

Many experiments were designed to reproduce the mean-flow gradient given in Eq. (6.1) while preserving a quasi-homogeneous turbulent state. The experimental setups combining a piling-up of plates in the vertical direction and grid-generated turbulence are the most well known and documented (Rohr et al., 1988; De Souza, Nguyen, and Tavoularis, 1995). As discussed in Chapter 5, in which older experimental studies were also quoted, they suffer a significant drawback in that they do not ensure uniformity of the Cauchy matrix in the whole cross section, or in a simpler way they do not ensure the uniformity of the mean streamwise velocity. As a consequence, the equivalent elapsed time St is not uniform in a cross section, yielding spurious diffusion effects. The specific experimental facility developed at ONERA by Leuchter and co-workers (see Chapter 5), which combines a mean-rotation generator with an angular velocity $S/2$ in the direction normal to the plane of the shear and an additional distorting duct to create the additional straining process in the cross sections is much more satisfactory, but the length of the distorting duct severely limits the maximum elapsed time $St \sim 1.5$. Finally, some relevant experiments aim at reproducing a planar Couette flow, using a moving belt, but "initial" turbulence cannot be created independently, so that such experiments are more devoted to studying hydrodynamic stability than developed turbulence dynamics.

Some relevant DNSs have been performed. Most of them were based on the Rogallo (1981) method, which uses a pseudo-spectral scheme to evaluate nonlinear terms and the mean-Lagrangian system of coordinates to capture the linear effects in an optimal way. Among these numerical studies, many useful results and analyses can be found in the rather old study by Lee, Kim, and Moin (1990), whereas the recent numerical study by Brethouwer (2005) provides one of the most reliable and accurate databases for both homogeneous rotating and nonrotating shear flows.

6.1.2 Main Observations

Looking at the time development of Reynolds stresses, physical and numerical experiments provide a consistent picture that consists of three phases (see Figs. 6.1 and 6.2):

1. Initial data, or upstream data, in a grid-generated turbulence, being quasi-isotropic, the RST is nearly spherical with no significant cross correlation $\overline{u'_1 u'_2}$, so that the first phase of the evolution is close to the decay of unsheared turbulence.
2. In the second phase, anisotropy develops, so that the production of turbulent kinetic energy (which is proportional to $\overline{u'_1 u'_2}$) becomes larger than its dissipation rate, and the turbulent kinetic energy begins to increase. It is worth noting that, if the initial turbulent kinetic energy is too low, the viscous damping may

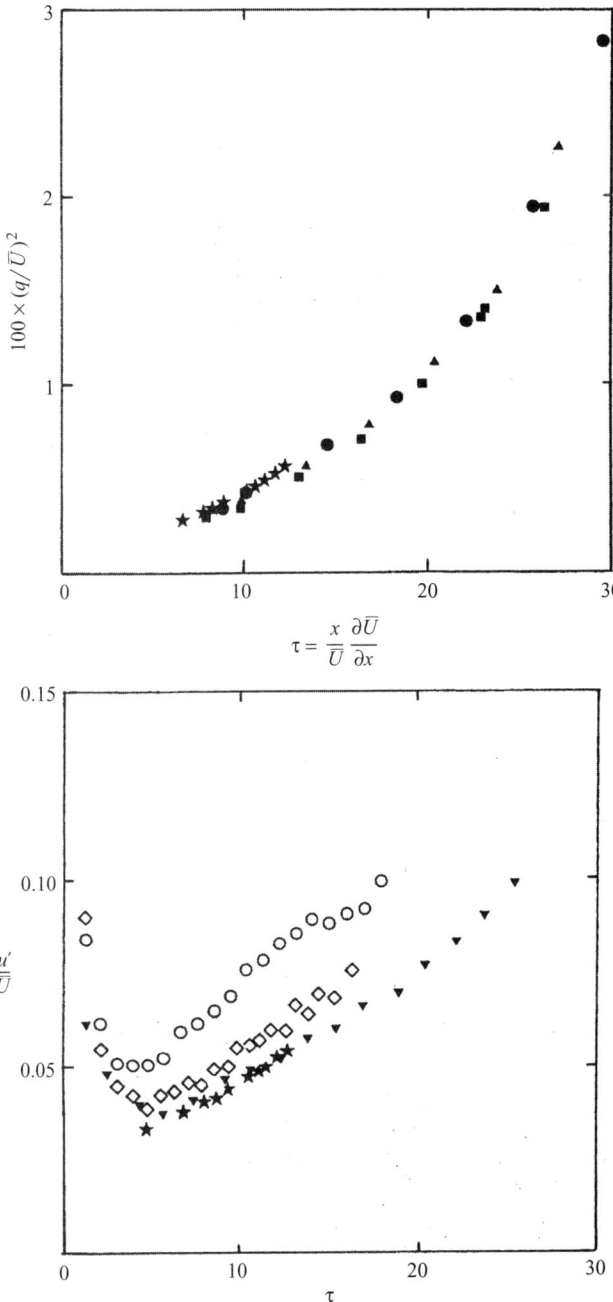

Figure 6.1. Evolution of turbulent kinetic energy, $\mathcal{K}$ (top) and streamwise turbulence intensity (bottom), measured in different laboratory experiments. In all cases, the production phenomenon is clearly observed after the initial decay phase. Reproduced from Rohr et al. (1988) with permission of CUP.

6.1 Physical and Numerical Experiments

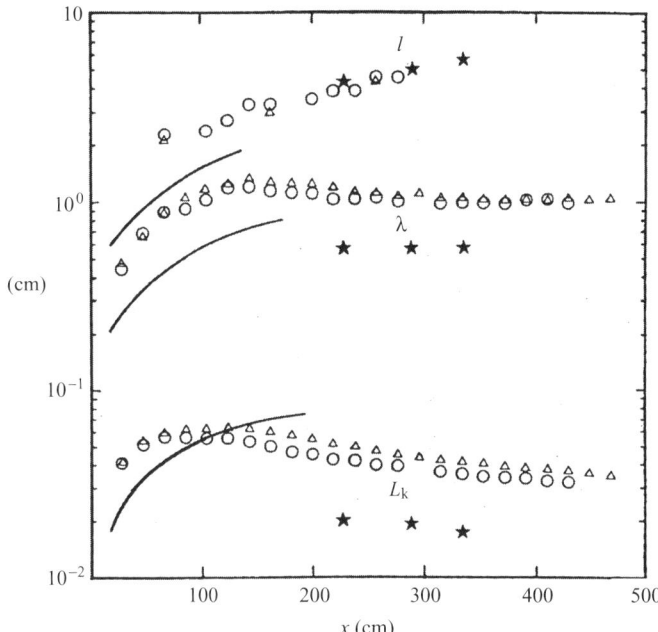

Figure 6.2. Evolution of turbulence characteristic scales in different experiments. Top: integral scale; middle: Taylor scale; bottom: Kolmogorov scale. The ratio L/η is observed to increase, showing that the spectrum is filled by the production mechanism. Solid curves are related to isotropic decay. Reproduced from Rohr et al. (1988) with permission of CUP.

be so high that the flow will be dominated by linear viscous effects, resulting in a monotone decay until all fluctuations have been suppressed.

3. In the final stage, an asymptotic regime is reached in which the turbulent kinetic-energy growth rate can become exponential, but this point is more subtle than is generally admitted. The exponential growth is associated with constant values of the components of the anisotropy tensor, b_{ij}. Both numerical and experimental data indicate that the following nondimensional quantities exhibit constant (but flow-dependent!) values in the asymptotic regime:

$$\frac{S\mathcal{K}}{\varepsilon}, \quad \frac{S\overline{u_1'u_2'}}{\varepsilon},$$

which are the *shear rapidity* (which compares the nonlinear time scale $\tau_{NL} = \mathcal{K}/\varepsilon$ with the shear time scale S^{-1}) and the ratio of the production of turbulent kinetic energy to the dissipation rate. Combining these two quantities, one finds that $\overline{u_1'u_2'}/\mathcal{K}$ is also constant. It is worth noting that the exponential growth regime cannot be sustained for arbitrary long times, as turbulent kinetic energy must remain finite in physical systems. The kinetic-energy growth can be estimated as (Rohr et al., 1988)

$$\mathcal{K}(t) = \mathcal{K}(0)e^{\sigma St}, \quad \sigma = \left(-\frac{\overline{u_1'u_2'}}{\mathcal{K}}\right)\left(1 - \frac{S\overline{u_1'u_2'}}{\varepsilon}\right), \tag{6.2}$$

Table 6.1. *Asymptotic behavior of homogeneous shear turbulence for large St*

Quantity	DNS and experiments	RDT	Pressure-released RDT
$\mathcal{K}(St \gg 1)/\mathcal{K}(0)$	$\propto e^{\sigma St}$	$\propto St$	$\propto (St)^2$
$b_{11}(St \gg 1)$	0.203	2/3	2/3
$b_{22}(St \gg 1)$	−0.143	−1/3	−1/3
$b_{33}(St \gg 1)$	−0.06	−1/3	−1/3
$b_{12}(St \gg 1)$	0 (−0.15)	0	0

Note: The asymptotic value given between parentheses for b_{12} in the first column is the one given in Piquet (2001), which is a plateau value observed for a finite range of St, whereas zero is presumably the true asymptotic value for $St \gg 1$.

where the damping factor σ is constant, flow dependent, and positive. This expression is very useful: It shows that St is not the only parameter that describes the flow, that σ must be taken into account, and that, for low values of σSt, a first-order Taylor series expansion makes it possible to recover the turbulent kinetic-energy linear growth rate reported by some authors.

We now discuss the asymptotic regime in more detail. Anisotropy of the RST develops too, with the streamwise component becoming largely dominant with respect to the two other diagonal ones, and the vertical one being the smallest. Exact values of the components of the anisotropy tensor b_{ij} are difficult to infer from available data, as a nonnegligible dispersion among data is observed. Plausible target values given by Piquet (2001) are displayed in the first column of Table 6.1. It is worth noting that the asymptotic value of b_{12} is observed to be flow dependent, but that the true asymptotic value may be equal to 0. Other values may be in fact intermediate plateau values found in experiments and numerical simulation at moderate St.

It is worth noting that b_{ij} is only one descriptor of anisotropy among many others. Interesting anisotropy indicators also are provided by the integral length scales $L_{ij}^{(n)}$, which are defined as follows:

$$\overline{u_i' u_j'} L_{ij}^{(n)} = \int \overline{u_i'(\mathbf{x}) u_j'(\mathbf{x} + r\mathbf{a}^{(n)})} dr \quad \text{with} \quad a_m^{(n)} = \delta_{mn}, \tag{6.3}$$

with no summation over repeated i, j subscripts, $i, j, n = 1, 3$.*

The longitudinal integral scale $L_{11}^{(1)}$ becomes very large and dominates all other components at large St. Among various components, the (large) ratio $L_{11}^{(1)}/L_{11}^{(3)}$ is particularly informative as it is directly linked to the aspect-ratio (streamwise length to spanwise spacing) streaklike structures (this point will be further developed in Subsection 6.3.2). Incidentally, one can notice that the appearance of streaks is

---

* In isotropic turbulence, all these quantities reduce to a single one, say L_f, with $L_{nn}^{(n)} = L_f$ (any n, no summation on it), $L_{ii}^{(n)} = L_f/2$ if $n \neq i$ (no summation on i), and $L_{ij}^{(n)} = 0$ if $i \neq j$. Accordingly, departure from this simple relationship reflects the anisotropic structure too.

6.2 Reynolds Stress Tensor and Analysis of Related Equations

found in the homogeneous shear case, even if their dynamics and topology are significantly different from the "true" near-wall streaks. Analysis of structures is addressed in much more detail in Subsection 6.3.2 and Section 6.6.

6.2 Reynolds Stress Tensor and Analysis of Related Equations

The equation governing the RST is

$$\frac{d\overline{u'_i u'_j}}{dt} = -S \begin{bmatrix} 2\overline{u'_1 u'_2} & \overline{u'^2_2} & \overline{u'_2 u'_3} \\ \overline{u'^2_2} & 0 & 0 \\ \overline{u'_2 u'_3} & 0 & 0 \end{bmatrix} + \Pi_{ij} - \varepsilon_{ij}, \qquad (6.4)$$

in which the structure of the production term (first term on the right-hand side) has been detailed. Reynolds stress components involving the vertical, or cross-gradient, u'_2 component are present in this term.

Now, considering the case of an initially isotropic field, this system simplifies as

$$\begin{cases} \dfrac{d\overline{u'_1 u'_1}}{dt} = -2S\overline{u'_1 u'_2} + \Pi_{11} - \varepsilon_{11} \\ \dfrac{d\overline{u'_2 u'_2}}{dt} = \phantom{-2S\overline{u'_1 u'_2} +} \Pi_{22} - \varepsilon_{22} \\ \dfrac{d\overline{u'_3 u'_3}}{dt} = \phantom{-2S\overline{u'_1 u'_2} +} \Pi_{33} - \varepsilon_{33} \\ \dfrac{d\overline{u'_1 u'_2}}{dt} = -S\overline{u'_2 u'_2} + \Pi_{12} - \varepsilon_{12}. \end{cases} \qquad (6.5)$$

The different couplings are illustrated in Fig. 6.3. The associated evolution equation for the turbulent kinetic energy is

$$\frac{d\mathcal{K}}{dt} = -S\overline{u'_1 u'_2} - \varepsilon, \qquad (6.6)$$

which shows the importance of the cross correlation $\overline{u'_1 u'_2}$ for the kinetic-energy growth rate.

Reynolds stress models with conventional closure techniques perform satisfactorily in the shear-flow case. One reason is historical and not really rational: Adjustable constants in the closure models were fitted with maximum care in this case only! Another reason is that the dynamics is dominated by a simple production to dissipation balance (or partial imbalance), and it is not very sensitive to the modeling of the pressure–strain-rate tensor, especially to its rapid part, whose single-point modeling is the most difficult task.

For instance, the exponential growth of turbulent kinetic energy obtained after a sufficiently large elapsed time and at large Reynolds number can be predicted, even if not really explained. Equation (6.6) can be rewritten as

$$\frac{1}{S\mathcal{K}} \frac{d\mathcal{K}}{dt} = -\frac{\overline{u'_1 u'_2}}{\mathcal{K}} - \frac{\varepsilon}{S\mathcal{K}}, \qquad (6.7)$$

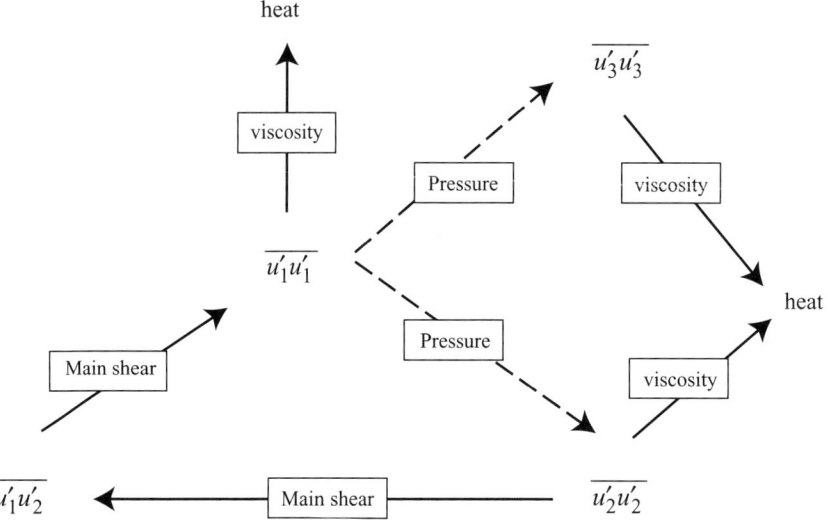

Figure 6.3. Couplings between the different nonvanishing Reynolds stresses in the pure shear case. Arrows indicate the production process, their color being related to the physical quantity at play (mean strain, pressure, viscosity).

leading to $\frac{1}{S\mathcal{K}}\frac{d\mathcal{K}}{dt} =$ constant in the asymptotic regime. A reasonable asymptotic value of the shear-rapidity term is obtained from both the previous equation and the modeled corresponding ε equation,

$$\frac{1}{S\varepsilon}\frac{d\varepsilon}{dt} = C_{\varepsilon 1}(-2b_{12}) - C_{\varepsilon 2}\frac{\varepsilon}{S\mathcal{K}}, \qquad (6.8)$$

provided a correct asymptotic value is assumed for b_{12}. The dependence of the prediction on the two empirical constants is not discussed here. The reasonable asymptotic value for the nondimensional term $-2b_{12}$ previously given is obtained considering the pressure–strain-rate modeling and distinguishing between the rapid (or linear) and the slow (or nonlinear) pressure terms, respectively denoted $\Pi_{ij}^{(r)}$ and $\Pi_{ij}^{(s)}$. In the pure shear case, the rapid and slow time scales are S^{-1} and $\mathcal{K}/\varepsilon$, respectively. Assuming that the dissipation is nearly isotropic for the sake of simplicity, the equation for the cross stress can be rewritten as

$$\frac{d\overline{u_1'u_2'}}{dt} = -S\overline{u_2'^2} + \Pi_{12}^{(r)} + \Pi_{12}^{(s)}, \qquad (6.9)$$

where $\overline{u_2'^2}$ is governed by

$$\frac{d\overline{u_2'^2}}{dt} = \Pi_{22}^{(r)} + \Pi_{22}^{(s)} - \frac{2}{3}\varepsilon. \qquad (6.10)$$

It immediately appears that the growth rate of $\overline{u_1'u_2'}$ is first driven by the vertical correlation $\overline{u_2'^2}$, this effect being modulated by both rapid and slow pressure–strain-rate correlations terms. The effect of the linear term $\Pi_{12}^{(r)}$ is modeled to reduce the

6.3 Rapid Distortion Theory

production, and is perhaps not so important, at least qualitatively. In contrast, the conventional return-to-isotropy effect of the modeled nonlinear term,

$$\Pi_{22}^{(s)} = -C_{(rti)}\varepsilon b_{22}, \tag{6.11}$$

is essential for allowing an exponential growth rate in a fully nonlinear regime. A simple explanation can be offered as follows. In the absence of nonlinear terms (and without significant dissipation), Reynolds stress equations, even if they cannot reproduce the RDT behavior (see Section 6.3), at least are consistent with an algebraic growth of the turbulent kinetic energy: $\mathcal{K}(St) \propto (St)^n$, $1 \leq n \leq 2$. In this regime, $\overline{u_2'^2}$ remains very small. The presence of the nonlinear pressure–strain rate, modeled in agreement with the return-to-isotropy principle, will redistribute the energy between the diagonal components of the RST, therefore feeding the smallest component $\overline{u_2'^2}$. This effect will reinforce the production term $-S\overline{u_2'^2}$ through a strong positive $\Pi_{22}^{(s)}$ term in Eq. (6.10). Even if the term $\Pi_{12}^{(s)}$, being positive, will contribute to damping the growth of $\overline{u_1'u_2'}$, the effect of $\Pi_{22}^{(s)}$ will be the most efficient "nonlinear" one to enhance $\overline{u_2'^2}$ and therefore to allow a dramatic increase of production, consistent with an eventual exponential growth.

6.3 Rapid Distortion Theory: Equations, Solutions, Algebraic Growth

Linearized inviscid equations in physical space are

$$\frac{\partial u_i'}{\partial t} + Sx_2 \frac{\partial u_i'}{\partial x_1} + S\delta_{i1}u_2' = -\frac{\partial p'}{\partial x_i}. \tag{6.12}$$

The pressure (here divided by the mean reference density) term is identified by taking the divergence of the previous equation as

$$\nabla^2 p' = -2S \frac{\partial u_2'}{\partial x_1} \tag{6.13}$$

so that the vertical (cross-gradient) component of the velocity is evidenced as the key component. Combining linearized Navier–Stokes and Poisson equations, it can be easily found that the Laplacian of the vertical velocity fluctuation is simply advected by the mean flow:

$$\frac{D}{Dt}(\nabla^2 u_2') = \left(\frac{\partial}{\partial t} + Sx_2 \frac{\partial}{\partial x_1}\right) \nabla^2 u_2' = 0. \tag{6.14}$$

The fact that $\nabla^2 u_2'$ obeys a decoupled equation and may be chosen as one of the basic variables to study linear solutions in the presence of mean shear is not surprising because in the analyses of Orr–Sommerfeld and Squire, $\nabla^2 u_2'$ and ω_2' (vertical vorticity fluctuation) are the two basic variables. Accordingly, it is not too difficult to find complete solutions in physical space for these variables. Nevertheless, these solutions display nonlocal operators and involve an integrodifferential Green's function in physical space, so that the RDT problem is much more easily solved in Fourier space.

Using the general formalism introduced in Chapter 5 (detailed in Chapter 13), RDT equations can be recast as

$$\dot{\hat{u}}_i + S\left(\delta_{i1} - 2\frac{k_1 k_i}{k^2}\right)\hat{u}_2 = 0 \tag{6.15}$$

and

$$\dot{k}_i + Sk_1\delta_{i2} = 0. \tag{6.16}$$

The latter equation generates the following characteristic lines in Fourier space,

$$k_1 = K_1, \quad k_2 = K_2 - StK_1, \quad k_3 = K_3, \tag{6.17}$$

which are related to the mean trajectories in physical space:

$$x_1 = X_1 + X_2 St, \quad x_2 = X_2, \quad x_3 = X_3. \tag{6.18}$$

It is worth noting that Eqs. (6.17) and (6.18) are a special case of $k_i = F_{ji}^{-1} K_j$ and $x_i = F_{ij} X_j$ using Eq. (6.1). Taking advantage of the decoupling of the equation for $\hat{u}_2$,

$$\dot{\hat{u}}_2 - 2S\frac{k_1 k_2}{k^2}\hat{u}_2 = 0, \tag{6.19}$$

and using $\dot{k}_i k_i = \dot{k}k = -Sk_1 k_2$ from Eq. (6.16), one obtains

$$\frac{D}{Dt}(k^2 \hat{u}_2) = 0, \tag{6.20}$$

which is the exact counterpart of Eq. (6.14). The solution is

$$\hat{u}_2(\boldsymbol{k}, t) = \frac{K^2}{k^2}\hat{u}_2(\boldsymbol{K}, 0). \tag{6.21}$$

Finally, the full solution is expressed as (e.g., Townsend, 1976; Piquet, 2001)

$$\begin{pmatrix}\hat{u}_1(\boldsymbol{k},t)\\ \hat{u}_2(\boldsymbol{k},t)\\ \hat{u}_3(\boldsymbol{k},t)\end{pmatrix} = \begin{bmatrix}1 & G_{12} & 0\\ 0 & \frac{K^2}{k^2} & 0\\ 0 & G_{32} & 1\end{bmatrix}\begin{pmatrix}\hat{u}_1(\boldsymbol{K},0)\\ \hat{u}_2(\boldsymbol{K},0)\\ \hat{u}_3(\boldsymbol{K},0)\end{pmatrix} \tag{6.22}$$

where the two extradiagonal terms are given by

$$G_{12} = -S\int\left(1 - 2\frac{K_1^2}{k^2}\right)\frac{K^2}{k^2}dt, \quad G_{32} = 2S\frac{K_1 K_3}{K^2}\int\frac{K^4}{k^4}dt, \tag{6.23}$$

in which the time dependency is induced by $k^2(t)$ following Eqs. (6.17). Analytical integration is not difficult but rather tedious (a whole page would be needed to write the analytical solutions with only various algebraic and $\tan^{-1}$ terms); see Townsend (1976) and Piquet (2001).

For $K_1 = 0$, the solution drastically simplifies, leading to $K/k = 1$, $G_{12} = -St$, and $G_{32} = 0$.

This solution can be found with the minimum number of components in the Craya–Herring frame of reference. Choosing $\boldsymbol{n}$ (the polar axis of the decomposition)

6.3 Rapid Distortion Theory

in the vertical direction, the two modes $u^{(1)}$ and $u^{(2)}$ are related to vertical vorticity and the Laplacian of vertical velocity, respectively. Therefore they appear to be the spectral normalized counterparts of Orr–Sommerfeld–Squire variables that are commonly used within the linear-instability theory framework. The resulting system of the two equations is

$$\dot{u}^{(\alpha)} + S e_1^{(\alpha)} e_2^{(\beta)} u^{(\beta)} = 0, \quad \alpha, \beta = 1, 2, \tag{6.24}$$

as the terms $\dot{e}_i^{(\alpha)} e_i^{(\beta)}$ identically vanish, although $\mathbf{k}$ itself is time dependent.

As for the solution in the fixed frame of reference, the equation for the poloidal component $u^{(2)}$ is decoupled from the one for the toroidal component because

$$\dot{u}^{(2)} - S \frac{k_1 k_2}{k^2} u^{(2)} = 0.$$

The evolution equation for the toroidal component $u^{(1)}$ reduces to

$$\dot{u}^{(1)} + S \frac{K_3}{k} u^{(2)} = 0$$

so that the complete solution is

$$\begin{pmatrix} u^{(1)}(\mathbf{k}, t) \\ u^{(2)}(\mathbf{k}, t) \end{pmatrix} = \begin{bmatrix} 1 & g_{12} \\ 0 & \frac{K}{k} \end{bmatrix} \begin{pmatrix} u^{(1)}(\mathbf{K}, 0) \\ u^{(2)}(\mathbf{K}, 0) \end{pmatrix}, \tag{6.25}$$

in which the unique extradiagonal term is

$$g_{12} = -S \frac{K_3}{K} \int \frac{K^2}{k^2} dt = \frac{K K_3}{K_1 K_\perp} \left(\tan^{-1} \frac{k_2}{K_\perp} - \tan^{-1} \frac{K_2}{K_\perp} \right), \tag{6.26}$$

with

$$K_\perp = \sqrt{K_1^2 + K_3^2}, \tag{6.27}$$

so that a complete solution is generated that is much simpler than Townsend's in the fixed frame of reference. As before, the particular case $K_1 = 0$ yields $K/k = 1$ and $g_{12} = -St \frac{k_3}{k}$.

6.3.1 Some Properties of RDT Solutions

The role of fluctuating pressure is clearly to reduce the vertical velocity component, and therefore to diminish the production of turbulent kinetic energy. This point is illustrated by the growth rates reported in the first line of Table 6.1. Ignoring the pressure term in the linearized equation, the vertical velocity component is just advected. More generally the *pressure-released RDT* solution is

$$u'_2(\mathbf{x}, t) = u'_2(\mathbf{X}, 0), \quad u'_3(\mathbf{x}, t) = u'_3(\mathbf{X}, 0), \quad u'_1(\mathbf{x}, t) = u'_1(\mathbf{X}, 0) - Stu'_2(\mathbf{X}, 0). \tag{6.28}$$

Of course, this oversimplified solution is not divergence free. The pressure-released RDT solution for the departure from the isotropy problem is

$$\overline{u'_1 u'_1}(t) = \frac{2}{3}\mathcal{K}(0)\left[1 + (St)^2\right], \tag{6.29}$$

$$\overline{u'_2 u'_2}(t) = \overline{u'_3 u'_3}(t) = \frac{2}{3}\mathcal{K}(0), \tag{6.30}$$

$$\overline{u'_1 u'_2}(t) = -\frac{2}{3}St\mathcal{K}(0), \tag{6.31}$$

yielding a quadratic $(St)^2$ growth rate for the kinetic energy. Corresponding asymptotic values of the anisotropy tensor components are presented in Table 6.1.

But it must be borne in mind that this is the Laplacian of the vertical velocity component that is advected in the full RDT solution, so that

$$\frac{\partial^2 u'_2(\boldsymbol{x}, t)}{\partial x_i \partial x_i} = \frac{\partial^2 u'_2(\boldsymbol{X}, 0)}{\partial X_i \partial X_i},$$

leading to a decrease of the vertical component. This effect is quantified in Fourier space by the factor $K^2/k^2(t)$, which tends to zero at large St if K_1 is nonzero. For instance,

$$\overline{u'^2_2} = \iint \frac{K^4}{k^4} \hat{R}_{22}(\boldsymbol{K}, 0) d^3 k,$$

which can be evaluated from isotropic initial data,

$$\hat{R}_{ij}(\boldsymbol{K}, 0) = \frac{E(K)}{4\pi K^2}\left(\delta_{ij} - \frac{K_i K_j}{K^2}\right),$$

with $d^3 k = d^3 K$ coming from the incompressibility constraint,[†] so that

$$\overline{u'^2_2} = \frac{2\mathcal{K}(0)}{3}\frac{1}{4\pi}\iint_{|K|=1}\frac{K^2 K^2_{\perp}}{k^4} d^2\boldsymbol{K}, \tag{6.32}$$

where the surface integral on the initial wavenumber $\boldsymbol{K}$ has to be performed on a spherical shell of radius unity. A system of polar-spherical coordinates can be used for further calculations. The decay with time of the integral results from the growth of $k^4(t)$ for almost all $\boldsymbol{K}$-directions, except $K_1 = 0$.

All the Reynolds stresses can be obtained in a similar way. Let us just mention the general solution for the kinetic energy, as

$$\overline{u'_i u'_i} = \iiint \hat{R}_{ii}(\boldsymbol{k}, t) d^3 k$$

and

$$\hat{R}_{ii}(\boldsymbol{k}, t) = \frac{E(K)}{4\pi K^2} g_{\alpha\beta}(\boldsymbol{k}, t) g_{\alpha\beta}(\boldsymbol{k}, t)$$

[†] It is recalled that, in this case, one has Det F = 1.

6.3 Rapid Distortion Theory

if the initial field is isotropic. Finally, the RDT amplification rate of kinetic energy is found as

$$\frac{\mathcal{K}(t)}{\mathcal{K}(0)} = \frac{1}{2}\frac{1}{4\pi}\iint_{|K|=1} |\mathbf{g}|^2 \, d^2\mathbf{K}, \quad (6.33)$$

where the integral of the square of the norm of $\mathbf{g}$ has to be calculated on a spherical shell of radius unity for the initial wave vector.

Despite the simplicity of the latter integral and the fact that $g_{\alpha\beta}$ is analytically expressed from Eqs. (6.25) and (6.26), the final derivation of the kinetic-energy history is not an easy task. The problem comes from the existence of two different solutions, one for $K_1 = 0$ and one for $K_1 \neq 0$, even if continuity holds. An expansion for high values of St yields a result that is not uniformly valid over the angular domain in k: A substantial contribution to the integral comes from a narrow region of thickness $O\big((St)^{-1}\big)$ near $K_1 = 0$ as St increases. This difficulty persuaded Rogers (1991) to use matched asymptotic expansions to recover the large St behavior of the turbulent kinetic energy. Only the final result is discussed here for the sake of brevity: The growth rate is linear, $\mathcal{K}(St)/\mathcal{K}(0) \sim St$. Such a linear growth rate is satisfactorily recovered in the DNS of Brethouwer (2005), discarding nonlinear terms. More generally, large time contributions were derived for all Reynolds stress components as

$$\frac{\overline{u'_1 u'_2}}{2\mathcal{K}(0)} \to -\ln 2, \quad (6.34)$$

$$\frac{\overline{u'^2_1}}{2\mathcal{K}(0)} \to (2\ln 2)St, \quad (6.35)$$

$$\frac{\overline{u'^2_2}}{2\mathcal{K}(0)} \to 4(St)^{-1}\ln(4St), \quad (6.36)$$

$$\frac{\overline{u'^2_3}}{2\mathcal{K}(0)} \to \frac{\pi^2}{8}\ln(St) - 1.419. \quad (6.37)$$

The corresponding asymptotic values of the components of the anisotropy tensor are given in Table 6.1.

As a last result, it is interesting to calculate some statistical quantities with very simple RDT solutions.

Let us first consider statistical quantities that we define by looking at the sole plane $K_1 = 0$. This plane corresponds to 2D structures averaged in the streamwise direction, so that $\overline{u'_i u'_j} L^{(1)}_{ij}$ (without summation over repeated indices) are immediately found from a RDT integral restricted to $K_1 = 0$, e.g.,

$$\overline{u'^2_1} L^{(1)}_{11} = \frac{2\mathcal{K}(0)}{3} L_f \left[1 + \frac{(St)^2}{3}\right],$$

where L_f is the reference integral scale in isotropic turbulence.

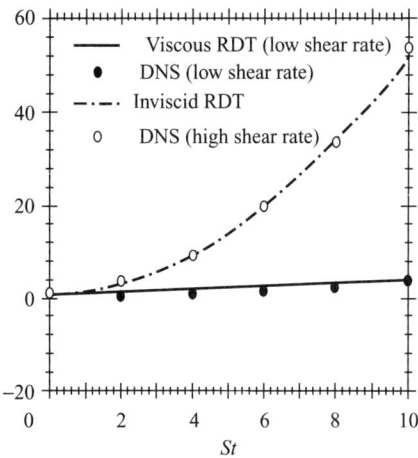

Figure 6.4. Inviscid and viscous RDTs compared with high-shear-rate and low-shear-rate DNSs. DNS came from a joint exploration by C. Cambon and M. J. Lee of the CTR database in 1990 (the same database was used by Lee, Kim, and Moin, 1990). Time histories of the streamwise 2D energy components $\overline{u_1'^2} L_{11}^{(1)}$ are plotted. Reproduced from Salhi and Cambon (1997) with permission of CUP.

The plane $K_3 = 0$ corresponds to 2D structures averaged in the spanwise direction, and similarly simple RDT solutions can be derived for $\overline{u_i' u_j'} L_{ij}^{(3)}$:

$$\overline{u_1'^2} L_{11}^{(3)} = \frac{2\mathcal{K}(0)}{3} \frac{L_f}{2} = \text{constant}.$$

The ratio of the two latter quantities illustrates the fact that a simple RDT analysis can predict an increasing streaky aspect ratio $L_{11}^{(1)}/L_{33}^{(3)}$. The idea of evaluating the integral length scales, or more precisely their product by related Reynolds stresses (called later "2D energy components" by Cambon and co-workers), was introduced by Townsend but applied to RDT only for an irrotational mean strain. Applications to pure shear and to rotating shear cases are reported in Salhi and Cambon (1997). Figure 6.4 shows the excellent agreement between RDT predictions and DNS results if the shear rate is high enough.

6.3.2 Relevance of Homogeneous RDT

RDT can predict qualitative trends, and even quantitative ones for single- and two-point statistical quantities, which are often dimensionless and characterize anisotropy. Most usual quantities are Reynolds stress components $\overline{u_i' u_j'}$, with nondimensional deviatoric tensor b_{ij}, and integral length scales $L_{ij}^{(n)}$ for different velocity components (subscripts i and j) and different directions of two-point separation (superscript n) for them. The anisotropy reflected in the latter length scales can be very different from the Reynolds stress anisotropy and therefore cannot be derived from the knowledge of b_{ij}. The qualitative relevance of RDT solutions can appear even for particular realizations (snapshots) of the fluctuating-velocity field when compared with full DNS. This is illustrated in Fig. 6.5, which compares instantaneous velocity fields obtained in the case of pure plane shear and plane channel flow near the wall. It is concluded that the tendency to create elongated streaky structures by a strong mean shear is inherent in this "homogeneous RDT" operator, independent of the presence of a wall and nonlinear effects.

6.3 Rapid Distortion Theory

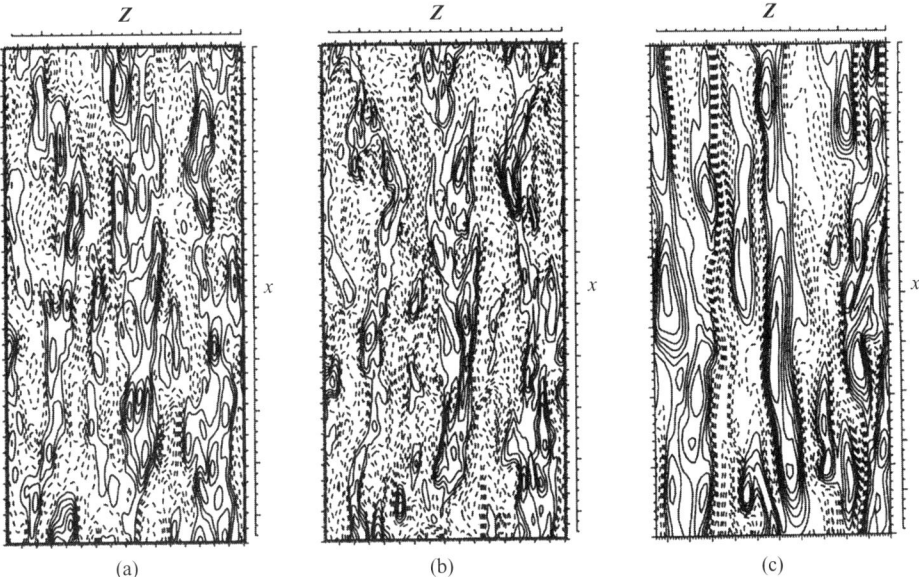

Figure 6.5. Contours of streamwise fluctuating velocity from (a) DNS, (b) RDT calculations for uniformly sheared homogeneous turbulence, and (c) DNS of plane channel flow near a wall horizontal plane $y^+ \sim 10$. The streamwise elongation of turbulent structures resulting from shear appears clearly, as does the strong similarity between RDT and DNS results. From Lee, Kim, and Moin (1990) with permission of CUP.

A detailed analysis of the vortical structures dynamics is given in Section 6.5. We just summarize here the results of Iida, Iwatsuki, and Nagano (2000), who performed a detailed analysis of subtle discrepancies that exist between vortex tubes predicted by RDT and those observed in DNS for a medium shear $0 \le St \le 6$.

DNSs reveal that these longitudinal vortices are inclined in the (x, y) plane and tilted in the (x, z) plane. Vortices with positive (resp. negative) longitudinal vorticity tend to tilt at a positive (resp. negative) angle, while they are all inclined at a positive angle. An important result is that RDT is able to predict the inclination of longitudinal vortices, but not their tilting. Therefore the tilting appears to be a nonlinear phenomenon. This subtle kinematical difference on the vortices' topology has a very large impact on nonlinear dynamics. To measure this effect, Iida and co-workers computed nonlinear terms by using both DNS and RDT velocity fields as inputs. Their main observations are as follows:

- In DNS, the kinematics of longitudinal vortices is deeply affected by the instantaneous strain-rate tensor. They are stretched in the streamwise and spanwise direction and compressed in the vertical direction. These local strains yield the existence of spiral streamlines in the streamwise direction and the production of nonzero instantaneous local Reynolds stress $u'_2 u'_3$. The streamwise fluctuations generated at the sides of the longitudinal vortex are wrapped around it, leading to the existence of negative values of the local fluctuating strain $(\partial u'_1/\partial x_2 + \partial u'_2/\partial x_1)$

inside the vortex. This phenomenon, referred to as *vortex wrapping*, is absent in RDT fields.
- The vortex-wrapping phenomenon and its effect on the kinematics of longitudinal vortices have a strong impact on nonlinear energy transfers. RDT fields lead to a vanishing transfer function in Fourier space for the vertical Reynolds stress $\overline{u'_2 u'_2}$, whereas it contributes to an inverse energy cascade in the DNS field. RDT fields also yield to an underestimation of the forward energy cascade associated with the nonlinear transfers of $\overline{u'_1 u'_1}$ and $\overline{u'_3 u'_3}$. This underestimation is tied to the misprediction of the instantaneous values of $u'_2 u'_3$.

6.4 Evidence and Uncertainties for Nonlinear Evolution: Kinetic-Energy Exponential Growth Using Spectral Theory

Some attempts exist to reproduce both linear and fully nonlinear regimes by a unified spectral theory. EDQNM approaches by Cambon et al. (1981) were limited to moderate anisotropy and were unable to cover a very large St domain. Theoretical derivations from LRA by Ishihara et al. (2002) are even more limited to weak anisotropy and small structures. The general formalisms, called EDQNM1 and EDQNM3, are valid in principle, but no complete solution, with an arbitrary degree of anisotropy, was numerically computed. General issues are discussed in Chapter 14. Instead of a general EDQNM (or LRA) approach, we discuss here how to introduce a self-similar argument in the spectral theory, following a very relevant approach proposed by Julian Scott (private communication).

A different approach was proposed by Nazarenko and Zakharov (1994). A kind of "first loop" is used for evaluating the impact of nonlinearity. On the one hand, this approach includes an interesting formalism, using Clebsch potentials, allowing us to derive a Hamiltonian operator (Hamiltonian formalism is also very important in the "Russian" school of wave turbulence). On the other hand, the basic RDT solution is completely missed, as the authors consider that the asymptotic value of kinetic energy is a (nonzero) constant in inviscid RDT, ignoring all the subtle effects correctly accounted for by Rogers (1991). This last point unfortunately invalidates their main result, which is that turbulent kinetic energy grows as $(St)^2$ in their particular nonlinear regime.

A relevant analysis is proposed by Julian Scott as follows. Large-scale self-similarity can be expressed very similarly as in Eq. (3.38) for the shearless-flow case, as

$$\hat{R}_{ij}(\mathbf{k}, t) = u^2 L^3 \Psi_{ij}(\mathbf{k}L), \tag{6.38}$$

where $u(t)$ and $L(t)$ are velocity and length scales characterizing the self-similar evolution of turbulence, respectively, and Ψ_{ij} is a dimensionless tensor. The RST is therefore given by

$$\overline{u'_i u'_j} = u^2 \iiint \Psi_{ij}(\mathbf{q}) d^3 \mathbf{q}, \tag{6.39}$$

showing that its different components are proportional to the same function, u^2, of time. Thus the ratio of different components is constant, as observed asymptotically. When Eq. (6.38) is used, Craya's equation (2.81) becomes

$$\alpha_2 \left(q_m \frac{\partial \Psi_{ij}}{\partial q_m} + 3\Psi_{ij} \right) - \alpha_1 \Psi_{ij} + \alpha_3 \left(M_{im}\Psi_{mj} + M_{jm}\Psi_{im} - A_{lm}q_l \frac{\partial \Psi_{ij}}{\partial q_m} \right) = \Xi_{ij}, \tag{6.40}$$

where $q = kL$ is the similarity variable, $T_{ij} = u^2 L^3 \Xi(q)$, and the quantities α_1, α_2, and α_3 are given by

$$\alpha_1(t) = \frac{L}{u^3}\frac{du^2}{dt}, \quad \alpha_2(t) = \frac{1}{u}\frac{dL}{dt}, \quad \alpha_3 = \frac{L}{u}. \tag{6.41}$$

Given the fact that we are concerned with large scales, the viscous term in Eq. (2.81) is dropped. From Eqs. (6.41), it follows that the α's are related by

$$\frac{d\alpha_3}{dt} = \alpha_2 + \frac{1}{2}\alpha_1. \tag{6.42}$$

Presuming that the given mean flow does not permit self-similar solutions of the RDT (which is the case for the pure plane shear, but also for all but the pure rotation case), the only possible large-scale self-similarity, allowing for nonlinearity, has constant α's. From Eqs. (6.41) this implies the following exponential behavior,

$$u(t) \sim \exp\left(\frac{\alpha_2}{\alpha_3}t\right), \quad L(t) \sim \exp\left(\frac{\alpha_2}{\alpha_3}t\right), \tag{6.43}$$

by use of $\alpha_1 = -2\alpha_2$. A positive value of α_2/α_3 is consistent with experimental and numerical DNS results.

Of course, we have not shown that large-scale self-similarity occurs, merely that, if it does, it must respect (6.43). The previous analysis has something to do with Oberlack's approach, in the sense that no closure theory is needed, but the discussion of the admissible values for the constants α's relies on a very subtle analysis of asymptotic RDT (not reported here for the sake of brevity).

We also take from Julian Scott the following remark. An exponential growth of L means that, in practice, the large scales in turbulence increase rapidly in size until they encounter inhomogeneities or boundaries of the flow, at which point the preceding model, assuming homogeneous turbulence in an infinite domain, no longer holds.

6.5 Vortical–Structure Dynamics in Homogeneous Shear Turbulence

The statistical behavior of homogeneous shear turbulence previously described can be explained as being the consequence of the growth and collapse of vortical structures. It is worth noting here that these structures govern the dynamics of the flow, although it was seen in Section 3.6 that their influence on the dynamics of isotropic turbulence is weak.

Using numerical simulations, Kida and Tanaka (1994) identify the following scenario for the departure from isotropy:

1. Uncoherent vorticity blobs initially present in isotropic turbulence are transformed into coherent elongated vortex tubes by the imposed mean shear by means of the vortex-stretching process. This linear mechanism, well recovered by the RDT, yields the formation of *longitudinal vortex tubes*. These structures are aligned with the directions of maximal extension of the mean shear flow (i.e., they are inclined at 45° and 225° to the downstream direction). The distance between the longitudinal vortex tubes is determined by the initial field and has not been observed to depend on the mean-shear rate.
2. Longitudinal vortex tubes experience the mean shear and are inclined more and more toward the streamwise direction with a further increase in their vorticity. These two trends are easily understood considering a rectilinear vortex filament that makes an angle θ with the streamwise axis and with axial vorticity Ω. Neglecting viscous effects and assuming that the vortex filament remains rectilinear, one obtains (Brasseur and Wang, 1992)

$$\frac{d\Omega}{dt} = \frac{1}{2} S\Omega \sin(2\theta), \qquad (6.44)$$

$$\frac{d\theta}{dt} = -S \sin^2(\theta). \qquad (6.45)$$

Because of nonlinear effects, vorticity vectors inside the longitudinal vortex tubes are less inclined (by about 10°) than the vortex tubes themselves, leading to the *vortex-wrapping phenomenon*.

3. Longitudinal vortex tubes induce a swirling motion that leads to the formation of *vortex sheets* with a spanwise component. These sheets are generated in planes nearly parallel both to the longitudinal vortex-tube axes and to the spanwise axis.
4. The vortex sheets are linearly unstable and roll up through the Kelvin–Helmholtz instability, leading to the growth of vortex tubes in the spanwise direction. These new vortex tubes are usually referred to as *lateral vortex tubes*.
5. Lateral vortex tubes are subject to the mean-shear effect in the presence of fluctuations, yielding the generation of *hairpinlike vortices*, whose legs correspond to streamwise vortices.
6. All vortical structures present in the flow interact and are subjected to the mean shear, and break down into a disordered field with weak enstrophy.
7. The continuous action of the mean shear leads to the occurrence of a large *oblique stripe structure*, which inclines at 10°–15° with the downstream direction. The growth of this structure leads to a very large decrease of vertical velocity and vorticity fluctuations.

6.6 Self-Sustaining Turbulent Cycle

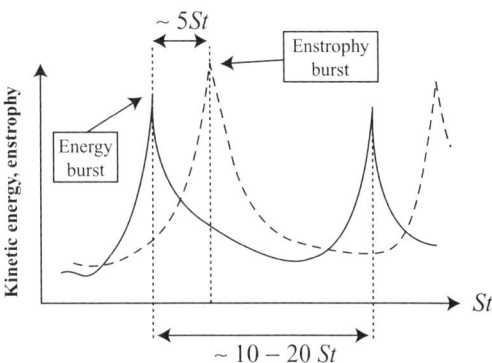

Figure 6.6. Schematic view of the time histories of global turbulent kinetic energy and enstrophy in DNS of homogeneous shear turbulence. One cycle of the SSP is shown.

6.6 Self-Sustaining Turbulent Cycle in Homogeneous Sheared Turbulence

The asymptotic long-time behavior of homogeneous shear turbulence previously discussed in this chapter cannot be sustained for an arbitrarily long time, because turbulent kinetic energy and characteristic length scales must remain bounded. The very large St behavior of homogeneous shear turbulence is usually not observed in wind-tunnel experiments because of the experimental setup characteristics. A cyclic behavior of global turbulent kinetic energy and enstrophy associated with a kind of unsteady equilibrium solution has been observed in numerical simulations at very large St (typically for $St \geq 30$) (Pumir, 1996; Gualtieri et al., 2002). This turbulent cycle involves the existence of a *self-sustaining turbulent mechanism* [also referred to as *self-regenerating* or *autonomous* cycle, or *self-sustaining process* (SSP)]. The typical evolution of global turbulent kinetic energy and enstrophy is displayed in Fig. 6.6 A typical cycle is composed of a spike of energy followed by a spike of enstrophy. These global quantities are observed to exhibit very large relative fluctuations within 40%–50%. The period of the cycle is observed to be of the order of 10–20 S^{-1}.

As stated by Pumir (1996), let us first note that an additional arbitrary length scale must be prescribed in the simulation to allow for the existence of an equilibrium solution.[‡] From Eq. (6.6), it is seen that a steady statistical equilibrium is reached if

$$-S\overline{u'_1 u'_2} = \varepsilon. \qquad (6.46)$$

Now using the usual scaling laws $\varepsilon \propto \mathcal{K}^{3/2}/L$ and $\overline{u'_1 u'_2} \propto \sqrt{\mathcal{K}}$, where L is a characteristic length scale, one sees that equilibrium is possible if and only if $\mathcal{K} \propto SL$, leading to the constraint that L must be finite. In very large St simulations, the length scale L is imposed by the size of the computational domain, which is always finite and represents an upper bound for the large-scale size. But is is worth noting that

---

[‡] It is recalled here that in the homogeneous shear problem a length scale is missing, because the mean shear applies in an infinite domain. In this sense, violation of statistical homogeneity, which is a drawback from a too-rigorous viewpoint, is of interest here.

in such simulations large scales interact with themselves by means of the periodic boundary conditions, resulting in a breakdown of ergodicity.

Numerical simulations have shown that kinetic-energy production is mainly governed by the interaction of the mean shear with the spanwise mode $k_S = (0, 0, \pm 1)$. The energy extracted by this mode from the mean flow is then transferred to other modes by the nonlinear kinetic-energy cascade process, the frequency of the cycle being determined by the dynamical balance between these two processes. Because of the incompressiblity constraint, the associated velocity field is $\hat{\boldsymbol{u}}(k_S) = (\hat{u}, \hat{v}, 0)$. The linear interaction mechanism is described by the following system:

$$\frac{d\hat{u}}{dt} = -S\hat{v} - \nu k_S^2 \hat{u}, \tag{6.47}$$

$$\frac{d\hat{v}}{dt} = -\nu k_S^2 \hat{v}. \tag{6.48}$$

The vertical component $\hat{v}$ is monotonically damped by the viscous effects, whereas the streamwise component $\hat{u}$ is amplified if

$$S\left[\text{Re}(\hat{u})\text{Re}(\hat{v}) + \text{Im}(\hat{u})\text{Im}(\hat{v})\right] < 0. \tag{6.49}$$

The corresponding physical scheme is as follows: The growth of energy results from one of the streamwise velocity fluctuations, which is due to the advection of fluid blobs of high streamwise velocity toward regions of lower velocity by the normal velocity v. This scenario is reminiscent of the *ejection/sweep* mechanism observed in turbulent boundary layers.

The different phases of the self-sustaining cycle are fully compatible with the vortical-structure dynamics observed by Kida and Tanaka (1994), described in the previous section. The energy burst is observed to occur when the legs of the hairpin vortices (i.e., longitudinal vortex tubes) interact with large negative Reynolds stresses, whereas the minimum of kinetic energy is observed at the beginning of the cycle when randomly distributed vorticity blobs experience the mean-shear action. A sketch of the full cycle is displayed in Fig. 6.7. Typical vortical structures at different stages of the cycle are displayed in Fig. 6.8.

6.7 Self-Sustaining Processes in Nonhomogeneous Sheared Turbulence: Exact Coherent States and Traveling-Wave Solutions

The *self-sustaining process* (SSP) in homogeneous shear flows described in the preceding section has been identified in numerical simulations only. Its existence seems to rely on a numerical trick, namely the possibility of enforcing an upper bound for the turbulent integral scale by means of the use of periodic boundary conditions in the simulation. In these simulations, the length scale that is missing because of the homogeneity assumption is recovered by defining the computational box. Nevertheless, previous results show that some SSPs may exist in turbulent shear flows. This phenomenon is found in the inner region of turbulent boundary layers, and modern

6.7 Self-Sustaining Processes in Nonhomogeneous Sheared Turbulence

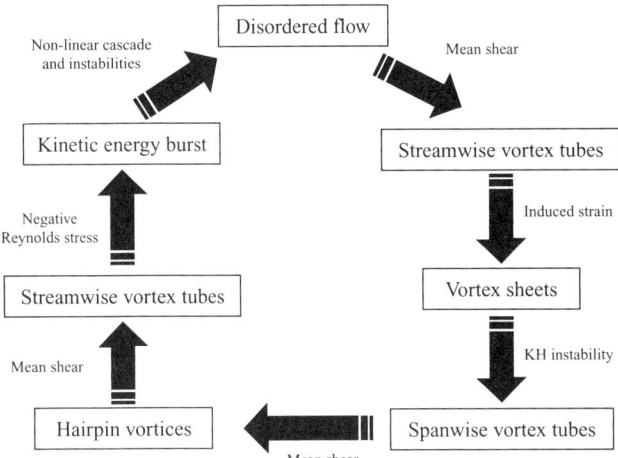

Figure 6.7. Sketch of the self-sustaining turbulent cycle in homogeneous shear flows.

analyses dealing with turbulence control and turbulent drag reduction in these flows rely on the SSP concept.

Because of the huge importance of shear flows in all fields of application, recent theoretical results dealing with SSPs in wall-bounded shear flows are briefly surveyed in this section. The main goal here is to characterize near-wall turbulence in terms of nonlinear exact solutions to incompressible Navier–Stokes equations for Couette, Poiseuille, and Couette–Poiseuille flows. All these solutions look qualitatively similar: a wavy low-velocity streak flanked by staggered streamwise vortices of alternating signs. According to Jimenez and co-workers (Jimenez et al., 2005), these solutions, which correspond to permanent stationary or traveling waves and to limit cycles in autonomous flows, can be classified into upper- and lower-branch families. The upper-branch familiy consists of weak streaks with strong streamwise vortices, whereas the lower-branch solutions have much stronger streaks and weaker vortices.

Emphasis is put on the theory proposed by Waleffe and co-workers (see Hamilton, Kim, and Waleffe, 1995; Waleffe, 1996, 2003, and references given therein), as it is fully consistent and closed from the theoretical point of view and its results correlate satisfactorily with wind-tunnel experiments and numerical simulation. The main discrepancy with the self-sustaining turbulent cycle in homogeneous shear flows discussed in the previous section is that the SSPs in nonhomogeneous shear flows involve a local change in the mean flow because of feedback of the fluctuating field. It is worth recalling here that such a feedback is by definition precluded in homogeneous turbulent flows.

The original purpose of Waleffe and co-workers was to explain the regeneration of turbulent structures observed in the near-wall region in turbulent wall-bounded flow. An important finding is that the results subsequently discussed have been proved to hold for a large variety of shear flows: Couette flow, Poiseuille flow, and the continuum of Couette–Poiseuille solutions. Because the theoretical analysis

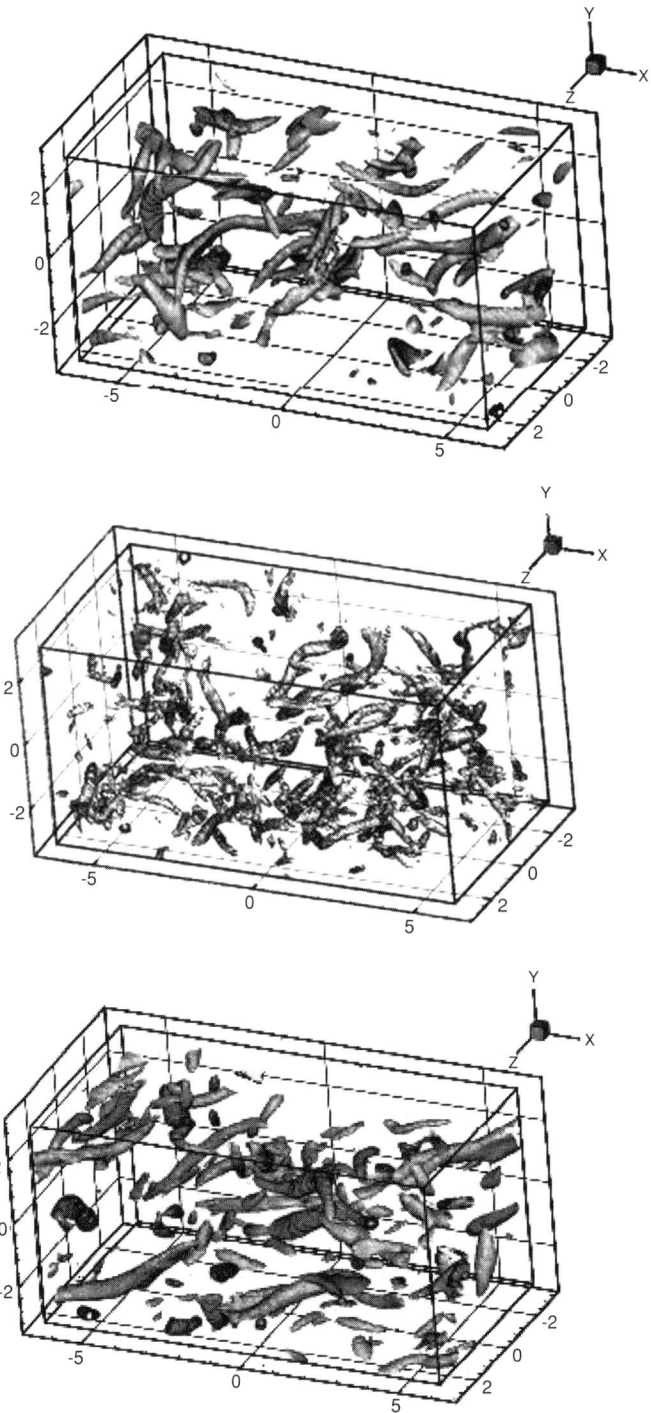

Figure 6.8. Vortical structures at different stages of a self-sustaining turbulent cycle in homogeneous shear flows. Top, during the kinetic-energy growth phase; middle, after the energy burst; bottom, when kinetic energy reaches a minimum before a new burst. Reproduced from Gualtieri et al. (2002) with permission of AIP.

6.7 Self-Sustaining Processes in Nonhomogeneous Sheared Turbulence

reveals that the SSP is not sensitive to the boundary conditions imposed on the fluctuations (either free-slip or no-slip conditions can be used), it can be conjectured that the main role of the solid boundary is to sustain a mean shear, and that a similar SSP might develop in free-shear flows.

Waleffe's SSP theory is essentially a weakly nonlinear theory of a 3D process about a base shear flow that has an $O(1)$ spanwise modulation $U(y, z)$, but it is not weakly nonlinear about a 1D laminar base flow $U(y)$. It relies on *exact traveling-wave solutions* of the incompressible Navier–Stokes equations of the form $u(x, t) = u(x - ct e_x, 0)$, where c and e_x are the constant wave velocity and the unit vector in the streamwise direction, respectively. The full velocity field, including the base shear flow, is decomposed by means of a Fourier transform in the streamwise direction, leading to

$$u(x, t) = u_0(y, z) + \left[e^{i\alpha\zeta} u_1(y, z) + \text{c.c.} \right] + \cdots, \tag{6.50}$$

where $u_0(y, z) = [u_0(y, z), v_0(y, z), w_0(y, z)]$ is the base shear flow and $\zeta = x - ct$. The base shear flow can be decomposed as the sum of a 1D mean-shear flow $\bar{u}(y)$ (defined as the streamwise velocity averaged over the periodic x and z directions) and streaky structures responsible for the modulation $u_0(y, z) - \bar{u}(y)$. These streaky structures are assumed to represent elongated streamwise blobs of rapid and slow fluids observed in the near-wall region of turbulent wall-bounded flows. They are modeled by means of streamwise rolls, which are longitudinal vortices with alternating streamwise vorticity sign and low $O(Re^{-1})$ amplitude. The SSP theory consists of three main steps (see Fig. 6.9):

1. **Formation of the streaky flow**. The existence of weak streamwise rolls $[0, v_0(y, z), w_0(y, z)]$ redistributes the streamwise momentum, leading to large spanwise fluctuations in the streamwise velocity, $[u_0(y, z), 0, 0]$. If the rolls are strong enough, an inflexional streamwise velocity profile can be generated.
2. **Instability of the streaky flow**. The existence of a locally inflexional streamwise velocity profile leads to wakelike instability in which a 3D disturbance develops.
3. **Nonlinear feedback on the rolls**. The *streak eigenmode* $[e^{i\alpha\zeta} u_1(y, z) + \text{c.c.}]$ is the first harmonic in the streamwise direction of the disturbance. Its quadratic self-interaction $[e^{2i\alpha\zeta} u_1(y, z) u_1(y, z) + u_1^*(y, z) u_1(y, z) + \text{c.c.}]$ generates a second harmonic $[e^{2i\alpha\zeta} u_2(y, z) + \text{c.c.}]$. But, more important, the nonlinear self-interaction term $u_1^*(y, z) u_1(y, z)$ is observed to extract energy from the streak and to reenergize the original streamwise rolls, leading to the definition of a closed nonlinear feedback loop.

Therefore the three necessary ingredients of the SSP are streamwise rolls, streaks, and streak eigenmode. An additional element is the mean shear, which provides the overall energy. It is worth noting that the streak instability extracts energy from the streaks and then cannot directly sustain them. It sustains the rolls, which sustain the streaks. The destruction of one of these key elements would lead to a breakdown of the SSP, and therefore to a possible deep modification of turbulence, as done in several turbulent drag-reduction strategies.

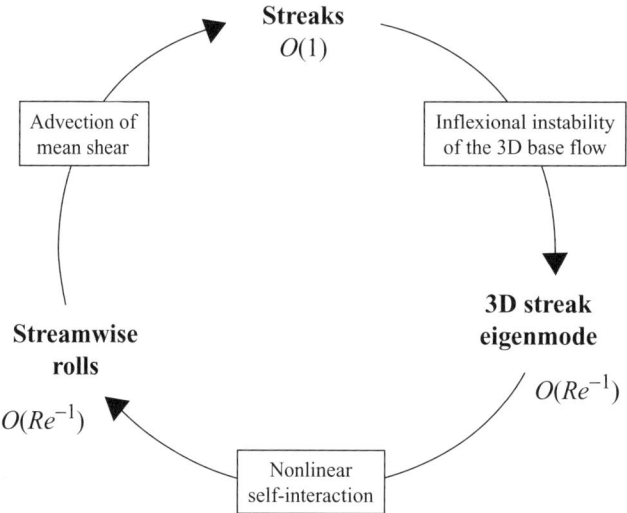

Figure 6.9. Schematic representation of Waleffe's SSP in nonhomogeneous shear flows.

6.8 Local Isotropy in Homogeneous Shear Flows

The question of the validity of Kolmogorov's local isotropy hypothesis in homogeneous shear flows has been addressed by several authors, using both experimental and simulation data. A first observation is that, as expected, the mean shear induces a breakdown of global isotropy. Looking at pdfs of velocity increments (see Fig. 6.10), one can see that the effect of the shear is scale dependent. Its influence on small-scale anisotropy is strong, leading to a noticeable departure from the isotropic turbulence case, whereas the effects at larger scales are weaker.

Therefore the question arises of the existence of a range of scales for which the local isotropy may hold. Experimental data show that, under certain circumstances, the turbulent energy spectrum $E(k)$ exhibits two different inertial ranges. More precisely, there exists a wavenumber k_S such that, within the inertial range,

$$E(k) \propto \begin{cases} k^{-1} & k < k_S \\ k^{-5/3} & k > k_S \end{cases}. \tag{6.51}$$

This bifurcation can be recovered by means of a simple dimensional analysis (Gualtieri et al., 2002). Introducing the *shear scale* $L_S \simeq 1/k_S$,

$$L_S = \sqrt{\frac{\varepsilon}{S^3}}, \tag{6.52}$$

one can see that at scales larger than L_S the dynamics is expected to be dominated by linear shear effects, whereas scales much smaller than L_S should be governed by nonlinear effects. Local isotropy may hold for the latter range of scales, but it is not observed in all cases, as subsequently discussed. Experimental data show that the bifurcation between the two inertial ranges occurs for $kL_S \sim 0.5$–1.

6.8 Local Isotropy in Homogeneous Shear Flows

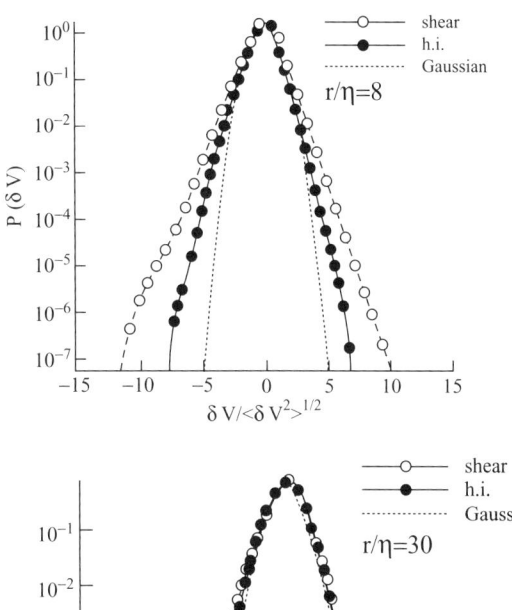

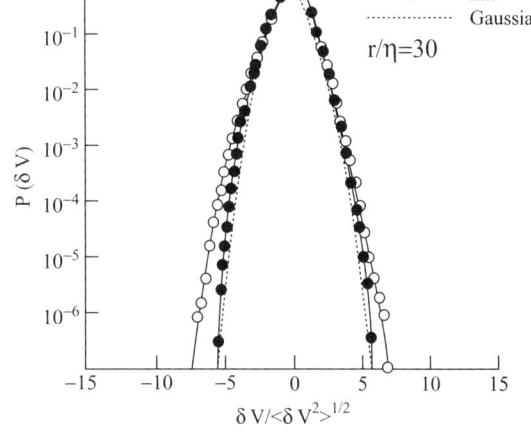

Figure 6.10. Probability density function of the velocity increment in isotropic turbulence (solid circles) and homogeneous shear turbulence (open circles) for $r/\eta = 8$ (top) and $r/\eta = 30$ (top), where η is the Kolmogorov scale. The dashed curve is related to the Gaussian distribution. Reproduced from Gualtieri et al. (2002) with permission of AIP.

The preceding criterion is not sufficient to account for the full complexity of the problem, as viscous effects are neglected and these effects can preclude the occurrence of the quasi-isotropic inertial range. One needs to define two nondimensional parameters, S_i and S_d, to describe the full problem. The first one measures the relative importance of nonlinear inertial mechanisms and linear shear effects:

$$S_i = \frac{2S\mathcal{K}}{\varepsilon} = \left(\frac{l}{L_S}\right)^{2/3}, \quad l = \frac{(2\mathcal{K})^{3/2}}{\varepsilon}. \qquad (6.53)$$

The second one is defined as the ratio of the dissipative and shear effects:

$$S_d = S\sqrt{\frac{\nu}{\varepsilon}} = \left(\frac{\eta}{L_S}\right)^{2/3}. \qquad (6.54)$$

One can expect to observe a pseudo-isotropic inertial range for small values of S_d only. For large values of S_i, most of the scales in the inertial range are dominated by the mean-shear effects, precluding the existence of scales compatible with local isotropy. A consequence is that true local isotropy, if it exists, can be recovered at a very high Reynolds number only. Such high values have not been reached up to now, and available data make it possible to identify only trends. Available results suggest that, with an increase in the Reynolds number, small scales in homogeneous turbulence come closer to isotropy, but that some anisotropy persists, even at $Re_\lambda = 660$ (Ferchichi and Tavoularis, 2000). An open issue is the existence, even at very high Reynolds number, of a pseudo-isotropic state of the small scales, in which some anisotropy would remain.

Bibliography

BRASSEUR, J. G. AND WANG, Q. (1992). Structural evolution of intermittency and anisotropy at different scales analyzed using three-dimensional wavelet transforms, *Phys. Fluids* **4**, 2538–2554.

BRETHOUWER, G. (2005). The effect of rotation on rapidly sheared homogeneous turbulence and passive scalar transport. Linear theory and direct numerical simulation, *J. Fluid Mech.* **542**, 305–342.

CAMBON, C. (1990). Contribution to single and double point modelling of homogeneous turbulence, *Annual Research Briefs*, Center for Turbulence Research, Stanford University and NASA Ames.

CAMBON, C., JEANDEL, D., MATHIEU, J. (1981). Spectral modelling of homogeneous non-isotropic turbulence, *J. Fluid Mech.* **104**, 247–262.

DE SOUZA, F. A., NGUYEN, V. D., AND TAVOULARIS, S. (1995). The structure of highly sheared turbulence, *J. Fluid Mech.* **303**, 155–167.

FERCHICHI, M. AND TAVOULARIS, S. (2000). Reynolds number effects on the fine structure of uniformly sheared turbulence, *Phys. Fluids* **12**, 2942–2953.

GARG, S. AND WARHAFT, Z. (1998). On the small scale structure of simple shear flow, *Phys. Fluids* **10**, 662–673.

GUALTIERI, P., CASCIOLA, C. M., BENZI, R., AMATI, G., AND PIVA, R. (2002). Scaling laws and intermittency in homogeneous shear flows, *Phys. Fluids* **14**, 583–596.

HAMILTON, J. M., KIM, J., AND WALEFFE, F. (1995). Regeneration mechanisms of near-wall turbulence structures, *J. Fluid Mech.* **287**, 317–348.

IIDA, O., IWATSUKI, M., AND NAGANO, Y. (2000). Vortical turbulence structure and transport mechanism in a homogeneous shear flow, *Phys. Fluids* **12**, 2895–2905.

ISHIHARA, T., YOSHIDA, K., AND KANEDA, Y. (2002). Anisotropic velocity correlation spectrum at small scale in a homogeneous turbulent shear flow, *Phys. Rev. Lett.* **88**, 154501.

JIMENEZ, J., KAWAHARA, G., SIMENS, M. P., NAGATA, M. AND SHIBA, M. (2005). Characterization of near-wall turbulence in terms of equilibrium and bursting solutions, *Phys. Fluids* **17**, 015105.

KIDA, S. AND TANAKA, M. (1994). Dynamics of vortical structures in a homogeneous shear flow, *J. Fluid Mech.* **274**, 43–68.

LEE, M. J., KIM, J., AND MOIN, P. (1990). Structure of turbulence at high shear rate, *J. Fluid Mech.* **216**, 561–583.

NAZARENKO, S. V. AND ZAKHAROV, V. E. (1994). The role of the convective modes and sheared variables in the Hamiltonian-dynamics of uniform-shear-flow perturbations, *Phys. Lett. A* **191**, 403–408.

PIQUET, J. (2001) *Turbulent Flows. Models and Physics*, 2nd ed. Springer.

PUMIR, A. (1996). Turbulence in homogeneous shear flows, *Phys. Fluids* **8**, 3112–3127.

PUMIR, A. AND SHRAIMAN, B. I. (1995). Persistent small scale anisotropy in homogeneous shear flows, *Phys. Rev. Lett.* **75**, 3114–3117.

ROGERS, M. (1991). The structure of a passive scalar field with a uniform mean gradient in rapidly sheared homogeneous turbulent flow, *Phys. Fluids A* **3**, 144–154.

ROHR, J. J., ITSWEIRE, E. C., HELLAND, K. N., AND VAN ATTA, C. W. (1988). An investigation of the growth of turbulence in a uniform-mean-shear flow, *J. Fluid Mech.* **187**, 1–33.

SALHI, A. AND CAMBON, C. (1997). An analysis of rotating shear flow using linear theory and DNS and LES results, *J. Fluid Mech.* **347**, 171–195.

TOWNSEND, A. A. (1976). *The Structure of Turbulent Shear Flow*, 2nd ed. Cambridge University Press.

WALEFFE, F. (1996). On a self-sustaining process in shear flows, *Phys. Fluids* **9**, 883–900.

WALEFFE, F. (2003). Homotopy of exact coherent structures in plane shear flows, *Phys. Fluids* **15**, 1517–1534.

7 Incompressible Homogeneous Anisotropic Turbulence: Buoyancy and Stable Stratification

7.1 Observations, Propagating and Nonpropagating Motion. Collapse of Vertical Motion and Layering

Turbulent flows can transport passive scalars, such as temperature or concentration. In important applications, such scalar (e.g., temperature, salinity) fluctuations generate a buoyancy force in the presence of gravity, which directly affects the velocity field. In addition, the transport of such "active" scalars by turbulence is altered by a mean-density gradient – intimately related to a mean-scalar gradient – in many applications, especially in atmospheric and oceanic research.

A first sketch of what stable and unstable stratifications are can be understood from a simple displaced-particle argument, as follows. Considering a vertical negative mean-density gradient (the heaviest flow is at the bottom), as in the scheme in Fig. 7.1, if a fluid particle is displaced upward, keeping its density and initially in hydrostatic equilibrium, it must experience a lighter fluid environment: The imbalance between (smaller) buoyancy and (same) weight will result in a downward force. The opposite situation occurs if the particle is moved downward, the imbalance buoyancy–weight will result in a upward force. Accordingly, the buoyancy force acts as a restoring force in this situation of negative mean-density gradient. Vertical oscillations with a typical frequency N are expected (as subsequently rediscussed).

The same reasoning holds for explaining unstable stratification. The mean-velocity gradient is now positive: A particle that is displaced upward will experience a heavier fluid environment, so that the buoyancy will result in a upward force, forcing the particle to continue to move up. This case of instable stratification, which includes important instances of thermal convection, is not addressed in this book.

Only stable stratification is considered in this chapter. This situation is current in the ocean, except in a (neutral) mixing layer near the surface. All situations, stable, unstable, and neutral, are encountered in the atmosphere, with a persistent case of the inversion of the temperature gradient, which yields a stable case in the tropopause and low stratosphere.

The previously mentioned vertical oscillations are the simplest mechanism for generation of internal gravity waves. Gravity waves have a strong analogy to the inertial waves introduced in Chapter 4. On the one hand, because the velocity field remains divergence free, pressure fluctuations are responsible for both anisotropic dispersivity and vertical–horizontal interchange of motion by the gravity waves, in a

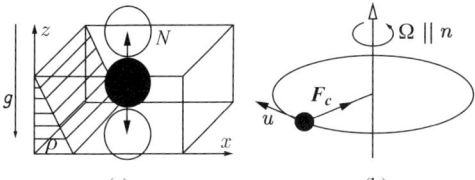

Figure 7.1. Sketch of basic oscillations, vertical displacement in stable stratification (left), horizontal displacement in rotating-flow case (right).

way very similar to what occurs for inertial waves. Similar St. Andrew's cross-shaped structures are found in both (rotating and stratified) cases. On the other hand, it is possible to define a potential energy for gravity waves only, as we will see further: The wave kinetic energy is essentially the poloidal kinetic energy, whereas the potential energy is proportional to the variance of density (or buoyancy) fluctuation. In this sense, buoyant flows with stable stratification illustrate another case of *flow dominated by wavy effects* with *zero-energy production*, as introduced in Chapter 4. For pure rotating turbulence, there is zero production of kinetic energy and for stratified (and stratified rotating) turbulence there is zero production of total energy, the latter being defined as the sum of kinetic energy and potential energy.

The essential difference between the rapidly rotating case and the strongly stratified one is the existence of an important nonpropagating mode of motion in the latter case. The toroidal part of the motion is unaffected by the gravity waves in the linear limit of a small Froude number. It is perhaps useful to recall the definition of Froude numbers, which are very similar to macro-Rossby numbers in Chapter 4, only replacing 2Ω with N and using alternatively large horizontal $L_\perp$ or vertical $L_\parallel$ length scales:

$$Fr_\perp = \frac{u'}{NL_\perp}; \quad Fr_\parallel = \frac{u'}{NL_\parallel}. \tag{7.1}$$

More generally, the Ertel theorem yields a nonlinear definition of the nonpropagating mode (rediscussed in Chapter 8, in agreement with general definition of potential vorticity). In the geophysical community, this mode is referred to as the *quasi-geostrophic (QG) mode* that is generally defined in the presence of additional Coriolis effects (see Chapter 8). Many wave-vortex decompositions, which are often neither intrinsic nor general, exist in the literature, so that particular care is given in this book to the very definition of the modes of motion. For instance, the wave-vortex decomposition by Riley et al. (1981) is meaningful only in the absence of additional rotation. The same decomposition was recently coined "vortical-divergent" by Brethouwer et al. (2007): This terminology is more confusing than the former, because the whole velocity field is divergence free (it would be relevant, however, in connection with the true Helmholtz decomposition for compressible flows, for instance in Chapter 9). The "vortex" (Riley, Metcalfe, and Weissman, 1981) or "vortical" (Brethouwer et al., 2007) mode is better qualified as the toroidal one anyway.

The main effect of stable stratification is to inhibit the vertical motion. Nevertheless, this stabilizing effect is not necessarily true for the horizontal motion, as recently shown in several studies devoted to the *zig-zag instability*, following Billant and Chomaz (2000). Considering vertical columnar vortices, strong stratification

7.1 Observations, Propagating and Nonpropagating Motion

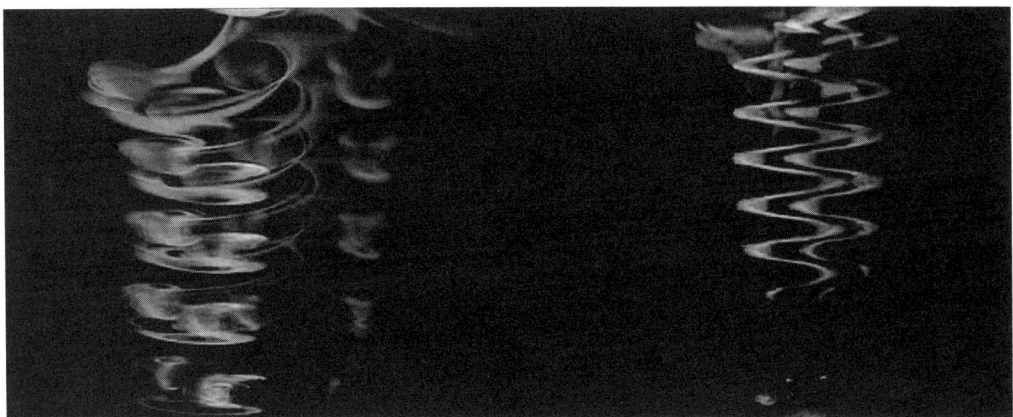

Figure 7.2. Experimental illustrations of the zig-zag instability. Courtesy of J.-M. Chomaz and P. Billant.

partly inhibits elliptical and/or centrifugal instabilities (touched on in Chapter 8) but breaks the vertical coherence of the columns by creating alternate, tangling, horizontal motion by means of zig-zag instability (Fig. 7.2).

This instability is even invoked to explain the horizontal layering of strongly stratified flows: We think that this is only part of the full answer. In the same way, the scaling of the vertical length scale U/N, where U [or u' in Eq. (7.1)] is a typical horizontal velocity scale and N is the gravity-wave typical frequency, is suggested by the zig-zag instability (Billant and Chomaz, 2001) but also by many other arguments disconnected from it.

Finally, the morphology of a strongly stratified flow is essentially a piling-up of velocity and density horizontal "pancakes" that can be observed in Fig. 7.6 in Section 7.5. A similar topology was reported from the observation of instantaneous isovorticity surfaces by Kimura and Herring (1996).

The flow is really quasi-horizontal but far from being 2D. Even with a random forcing of 2D modes, a DNS by Herring and Métais (1989) exhibits a clear tendency of forming horizontal layers, with a limited vertical thickness.

Gravity waves are present in the layered flow, but perhaps only in the limit of low frequencies (low dispersion frequency at a given high frequency N is obtained in the limit of quasi-vertical wave vectors, forming a *vertically sheared horizontal flow* – VSHF – mode, as subsequently shown). A clear analysis of such flows cannot be made without rigorous terminology, avoiding the confusion among horizontal, 2D, and toroidal motions. In addition to an accurate description of the morphology, dynamical arguments must be discussed: What is the mechanism that controls the thickness of the pancakes?

In addition to the mathematical–numerical decomposition by Riley, Metcalfe, and Weissman (1981), another related aspect is the scaling arguments for small (horizontal) Froude numbers, which supports the idea of almost vertically decorrelated thin horizontal pancake layers. The scaling by Riley, Metcalfe, and Weissman (1981) seems to hold for laboratory experiments, for internal waves (because the vertical scale of U/N is generally not of importance), and possibly for larger-scale flows (if their dynamics is not controlled by the vertical scale U/N). Note that strongly stratified flows (low Froude numbers in either sense) can exhibit waves over a broad range of frequencies. They are not limited by scaling or dynamical arguments to vertical scales of the order of U/N. There can be significant vertical motion associated with the waves.

Applications are very important for flows in the atmosphere and the ocean, in which stable stratification limits the vertical motion and makes the flow mainly horizontal. The problem of the sense of the kinetic-energy cascade (forward or backward) in such flows is still controversial, even if a global consensus is now emerging against the idea of a classical 2D inverse cascade. On the one hand, the analogy between QG and 2D dynamics, with conservation of potential vorticity, was investigated by Charney (1971). This analogy was revisited by Bartello (1995) with a refined analysis of pure QG interactions, also in the line of the Waleffe's instability principle. Regarding applications, Lilly (1983) proposed that the kinetic-energy spectra observed in the atmosphere at mesoscales (i.e., very low wavenumbers) are a manifestation of this 2D mechanism. In spite of this questionable speculation about upscale energy transfer, Lilly (1983) was probably the first to sugggest that vertical layering and instabilities would result from strong stratification (at low Froude numbers). At the bottom of page 755 and the top of page 756 of his paper he states:

> The second and more difficult problem is concerned with the continuing validity of the RMW [Riley, Metcalfe, and Weissman, 1981] scale analysis over a long period of time. The predicted decoupling of the dynamics at adjacent vertical levels can be maintained only as long as the local Richardson number is large. Both the inherent instability of turbulent flows and the existence of any mean vertical shear will decorrelate the vertical flow structure and produce locally small Richardson numbers. The subsequent regeneration of small-scale three-dimensional turbulence then modifies the stratified turbulence evolution to a yet uncertain degree.

Recently, Lindborg and Cho (Lindborg and Cho, 2001; Cho and Lindborg, 2001) deduced from analysis of third-order statistical moments that the energy cascade is in the direct sense, i.e., from small to large wavenumbers. This observational evidence was further supported by a dimensional analysis related to the zig-zag instability (Billant and Chomaz, 2001), showing that the vertical scale is necessarily limited by a local buoyancy length scale $L_B = U/N$, where U is the horizontal velocity scale and N is the Brunt–Väisälä frequency. Several DNSs (or rather LESs because of the use of a hyperviscosity instead of the classical molecular viscosity) were carried out by Lindborg and co-workers to investigate such a forward cascade. In these computations, the 2D-2C modes, and only these modes, are randomly forced, and the horizontal length scales are *a priori* chosen much larger than the vertical ones, using flattened boxes. Even if these studies present interest for atmospheric flows, their contribution to a better conceptual understanding of turbulence is limited by both geometric constraints and artificial forcing: No refined analysis of the anisotropy of the flow is performed, as was done in the case of rotating turbulence.

7.2 Simplified Equations, Using Navier–Stokes and Boussinesq Approximations, With Uniform Density Gradient

Navier–Stokes equations, with buoyancy force $b = b\mathbf{n}$ within the Bousinesq assumption, are subsequently given in the presence of a uniform mean-density gradient. For the sake of simplicity, the mean flow is restricted to a "stabilizing" uniform vertical gradient of density, whose strength is given by N, the Brunt–Väisälä frequency:

$$(\partial_t + \mathbf{u} \cdot \nabla)\mathbf{u} + \nabla p - \nu \nabla^2 \mathbf{u} = b\mathbf{n}, \quad (7.2)$$

$$(\partial_t + \mathbf{u} \cdot \nabla)b - \kappa \nabla^2 b = -N^2 \mathbf{n} \cdot \mathbf{u}, \quad (7.3)$$

$$\nabla \cdot \mathbf{u} = 0. \quad (7.4)$$

Dependent variables are the fluctuating velocity $\mathbf{u}$, the pressure p divided by a mean reference density, and the buoyancy force b. The vector $\mathbf{n}$ denotes the vertical unit upward direction aligned with the gravitational acceleration $\mathbf{g} = -g\mathbf{n}$. The Boussinesq approximation (Boussinesq, 1876) (the reader is referred to the large literature on geophysics) preserves the solenoidal property for the velocity, but allows the density to fluctuate. In the basic continuity equation, $\dot{\rho} + \rho u_{i,i} = 0$, it amounts to considering separately $\dot{\rho} = 0$ and $u_{i,i} = 0$. The first condition generates b-equation (7.3), whereas only the right-hand-side of momentum equation (7.2) calls b into play.

Because a large amount of literature is devoted to the turbulent transport of the passive scalar, a short discussion of this case cannot be avoided. Let us consider first that the fluctuating concentration of a passive scalar, say c, is addressed, instead of b, and that a vertical mean gradient $(\partial C/\partial x_3)\mathbf{n}$ exists for scalar concentration. The classical advection–diffusion equation, with additional "mean production" is

$$(\partial_t + \mathbf{u} \cdot \nabla)c - \kappa \nabla^2 c + \frac{\partial C}{\partial x_3} \mathbf{u} \cdot \mathbf{n} = 0,$$

which is essentially the same as Eq. (7.3). The only difference with the passive scalar equation is the presence of the right-hand side in (7.2) that reflects an "active" feedback from scalar concentration to velocity field.

The use of the buoyancy variable b allows us to have the same equations, with the unique frequency N, for a liquid or for a gas. For a liquid, $b = g\rho'/\rho_0$, where ρ' denotes a small fluctuation and ρ_0 is the mean reference density. The definition $N = \sqrt{g\Gamma/\rho_0}$, where Γ is the mean vertical gradient of density, presents a strong analogy with the frequency of a pendulum, with $(\Gamma/\rho_0)^{-1}$ playing the role of the length of the pendulum. For a gas, b is proportional to the fluctuating potential temperature τ, as $b = \beta g \tau$, where β is the thermometric expansivity. Accordingly, one has $N = \sqrt{g\beta\gamma}$, where γ is the mean vertical gradient of temperature.

Finally, the different flow cases are distinguished only by the diffusivity coefficient for b. Because there is no meaning for a diffusive density, κ must be considered as the diffusivity of the stratifying agent, for instance the temperature for a gas (κ/ν is a Prandtl number) or the salinity for a liquid (κ/ν is a Schmidt number). Without loss of generality, the fixed frame of reference is subsequently chosen such that $n_i = \delta_{i3}$. Therefore u_3 is the vertical velocity component.

7.2.1 Reynolds Stress Equations With Additional Scalar Variance and Flux

As for the case of a passive scalar, single-point second-order correlations include not only the RST $\overline{u_i u_j}$ (there is no mean velocity here, so that $u = u'$), but also the scalar variance $\overline{b^2}$ and the scalar flux $\overline{bu_i}$. Transport equations for the latter correlations are standard (passive scalar in the presence of a mean gradient of passive scalar of magnitude N^2). The Reynolds stress equations must be affected by the *active* buoyancy term, yielding

$$\frac{d\overline{u_i u_j}}{dt} = \Pi_{ij}^{(s)} + \overline{bu_j}\delta_{i3} + \overline{bu_i}\delta_{j3} - \varepsilon_{ij}.$$

All Reynolds stresses with a vertical component are therefore altered by buoyancy fluxes. Buoyancy fluxes $\overline{bu_i}$ are themselves governed by

$$\frac{d\overline{u_i b}}{dt} = N^2 \overline{u_i u_3} + \Pi_{ib}^{(r)} + \Pi_{ib}^{(s)} - \epsilon_{ib},$$

whereas the buoyancy variance $\overline{b^2}$ satisfies

$$\frac{d\overline{b^2}}{dt} = 2N^2 \overline{bu_3} - \epsilon_b.$$

These equations immediately result from combining basic equations (7.2) and (7.3). They can be found in Craft and Launder (2002) up to slightly different notation, with additional nonhomogeneous diffusive terms, and replacing d/dt with D/Dt (i.e., considering fully convective terms). Even in strictly homogeneous turbulence, some additional contributions from triple correlations, neglected in conventional modeling, may appear. They result from the fact that conservation laws (with zero contribution of related nonlinear transfer terms) are valid for separately considered toroidal and wave (poloidal + potential) energies, but not for horizontal, vertical, and potential energies [those that are tractable in only the Reynolds stress model (RSM) framework, with potential energy proportional to $\overline{b^2}$]. This will be evident when we look at generalized Lin equations in Section 7.5.

Craft and Launder developed one of the most sophisticated full RSM models (details are not given here for the sake of brevity; see Craft and Launder, 2002), with application in the two-component limit (TCL), *not to be confused with the 2D limit*, relevant here, but also (for different reasons) for near-wall turbulence. In particular cases, such as the pure decay of homogeneous stably stratified turbulence, the model mimics the damping of oscillations induced by gravity waves, without significant

7.2 Simplified Equations

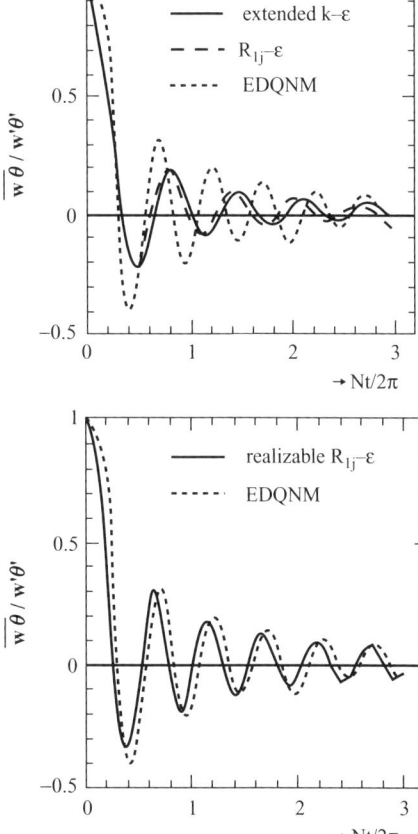

Figure 7.3. Normalized vertical buoyancy flux $\overline{u_3 b}$. Realizable $R_{ij} - \epsilon$ means the sophisticated TCL model. The EDQNM is essentially RDT here; the damping of oscillations reflects the phase mixing that is due to anisotropic dispersivity of gravity waves. Courtesy of L. Van Haren.

dissipation (see Fig. 7.3). The damping effect results from phase mixing of dispersive waves, and this physical effect cannot be directly incorporated into a single-point closure model: The correct behavior of the model is even more surprising and is probably due to the high complexity of the linear pressure–strain-rate tensors, such as $\Pi_{3b}^{(r)}$, along with strong constraints imposed, such as realizability. Recall, however, that in similar conditions, all single-point closure models miss the effects of inertial waves.

7.2.2 First Look at Gravity Waves

Analysis of the linear limit, mathematical treatment of equations in terms of eigenmodes, and closure methods for statistics in HAT, can be developed as for the particular case of pure rotation in Chapter 4.

Without pressure fluctuation, the additional buoyancy and stratification terms yield oscillations for vertical velocity and buoyancy terms, with frequency N. This simple motion reflects that the buoyancy force acts as a restoring force in the case of stable stratification. As for the case of pure rotation, the pressure term in (7.2) is needed to satisfy (7.4), and its role in the complete linear solution consists of coupling vertical and horizontal motion and of generating dispersive inertia–gravity waves.

The counterpart of Eq. (4.23) for inertial waves is

$$\frac{\partial^2}{\partial t^2}\left(\nabla^2 p\right) + N^2 \nabla_\perp^2 p = 0. \tag{7.5}$$

It is needed only to replace 2Ω with N and to replace the vertical component $\nabla_\parallel$ of the Laplacian operator with its horizontal $\nabla_\perp$ counterpart. The same treatment (i.e., normal-mode decomposition and derivation of the dispersion relation) shows that the threshold value to trigger the St. Andrew's cross-shaped structures with a local harmonic forcing σ_0 (Mowbray and Rarity, 1967) is $\sigma_0 = N$. The dispersion frequency in an unbounded domain is defined as

$$\sigma_k = N \frac{k_\perp}{k} = N \sin\theta. \tag{7.6}$$

As for the case of rapid rotation, a zero-frequency mode exists for gravity waves, but it corresponds to the vertical wave-vector direction, forming the *1D* VSHF mode, instead of the 2D (Taylor–Proudman) mode (linked to $k_\parallel = 0$) for inertial waves. As stated before, another even more important difference with the rotating-flow case is that a part of the horizontal motion remains steady in the linear limit and therefore decoupled from 3D wave motion.

7.3 Eigenmode Decomposition. Physical Interpretation

In the unbounded case, or for periodic boundary conditions, the different modes, wavy and steady, are easily found in Fourier space, and a tractable RDT solution is found in terms of them. Pressure fluctuation is removed from consideration in the Fourier-transformed equations by use of the local Craya–Herring frame of reference in the plane normal to the wave vector, taking advantage of Eqs. (7.4) and (2.67), so that the problem with five components (u_1, u_2, u_3, p, b) in physical space is reduced to a problem with three components in Fourier space, namely two solenoidal velocity components $[u^{(1)}, u^{(2)}]$ and a component for $\hat{b}$. The three-component set $[u^{(1)}, u^{(2)}, \hat{b}]$ is not a true vector, and this can complicate further mathematical developments in terms of its eigenmodes and statistical correlations. So it is more convenient to gather these three components into a new vector $\hat{v}$, whose inverse 3D Fourier transform, v, is real. $\hat{v}$ can be written as*

$$\hat{v} = \hat{u} + \imath \frac{1}{N}\hat{b}\frac{k}{k}, \tag{7.7}$$

---

* The term $\imath k \hat{b}$ corresponds to the *gradient* of the fluctuating buoyancy term in physical space. As in other studies dealing with the passive scalar, it can be better to use the scalar gradient than the scalar itself.

7.3 Eigenmode Decomposition

so that its three components are $u^{(1)}$, $u^{(2)}$, and $u^{(3)} = \imath \frac{1}{N}\hat{b}$ in the Craya–Herring frame, using also its third direction $e^{(3)} = \boldsymbol{k}/k$, even if it is more usually related to a divergent part of the velocity flow. The scaling of the contribution of the buoyancy force allows one to define twice the total spectral-energy density as

$$\widehat{v}_i^* \widehat{v}_i = \widehat{u}_i^* \widehat{u}_i + N^{-2}\widehat{b}^*\widehat{b}. \tag{7.8}$$

Linear inviscid equations are easily found as

$$\begin{pmatrix} \dot{u}^{(1)} \\ \dot{u}^{(2)} \\ \dot{u}^{(3)} \end{pmatrix} = \begin{bmatrix} 0 & 0 & 0 \\ 0 & 0 & -N\frac{k_\perp}{k} \\ 0 & N\frac{k_\perp}{k} & 0 \end{bmatrix} \begin{pmatrix} u^{(1)} \\ u^{(2)} \\ u^{(3)} \end{pmatrix}. \tag{7.9}$$

Linear – improperly called RDT – solutions are easily found, with constant $u^{(1)}$ and oscillating $u^{(2)}$–$u^{(3)}$. In any orthonormal frame of reference, linear solutions can be found in terms of the three eigenmodes,

$$\hat{v} = \xi^{(0)} N^{(0)} + \xi^{(1)} N^{(1)} + \xi^{(-1)} N^{(-1)}, \tag{7.10}$$

or

$$\hat{v} = \sum_{s=0,\pm 1} a_s(\boldsymbol{k}, t) \exp(\imath s \sigma_k t) N^{(s)}, \tag{7.11}$$

in which the eigenmodes $N^{(s)}, s = 0, \pm 1$ are simple linear combinations of the vectors in Eq. (2.67). $N^{(0)}$ reduces to the toroidal mode $e^{(1)}$ here. The reader is referred to Chapter 8 and to Cambon (2001) for a more general QG-AG (ageostrophic) decomposition, also valid in the presence of additional rotation. Essentially the same decomposition was introduced by Bartello (1995) in the geophysical context. Of course, a_s are constants given by initial data in the strict linear limit. A Green's function similar to (4.30) is derived as

$$G_{ij}(\boldsymbol{k}, t, t_0) = \sum_{s=0,\pm 1} N_i^s(\boldsymbol{k}) N_j^{-s}(\boldsymbol{k}) \exp[\imath s \sigma_k (t - t_0)], \tag{7.12}$$

and the nonlinear equations can be expressed in terms of the eigenmodes, as in Chapter 4, if time dependency is reintroduced in $a_s, s = 0, \pm 1$, as in the more general version of (7.11). The new element, with respect to (4.36), is the presence of a "strong" nonlinearity, related to a term $M_{(000)}$, which does not reduce to "weak" wave turbulence, as do the other coupling terms such as $M_{(00\pm 1)}$, $M_{(0\pm 1\pm 1)}$, and $M_{(\pm 1\pm 1\pm 1)}$ (only the latter being present in rotating turbulence). Because of the form of the eigenvectors and of the dispersion law, the structure of $\mathbf{G}$ in (7.12) is consistent with axisymmetry around the axis of reference (chosen to be vertical here), with mirror symmetry, and where $k_\parallel$ and $k_\perp$ hold for axial (along the axis) and transverse (normal to the axis) components of $\boldsymbol{k}$, respectively.

Anisotropy can be significantly broken through the axisymmetrical response function for triple correlations only, or possibly for two-time second-order statistics (whose analysis is beyond the scope of this book), but the linear limit exhibits no interesting creation of structural anisotropy in classic RDT for predicting second-order single-point statistics. However, in practice, there is a (partial) two-dimensionalization in rotating turbulence and a horizontal layering tendency in the stably stratified case. In other words, RDT alters only phase dynamics and conserves exactly the spectral density of typical modes (full kinetic energy for the rotating case, total energy and toroidal energy for the stably stratified case), so that two-dimensionalization or "two-componentalization" (horizontal layering), which affect the distribution of this energy, are typically nonlinear effects.

Nevertheless, the eigenmodes of the linear regime form a useful basis for expanding the fluctuating-velocity–buoyancy field, even when nonlinearity is present, and nonlinear interactions can be evaluated and discussed in terms of triadic interactions between these eigenmodes. Accordingly, the complete anisotropic description of two-point second-order correlations can be related to spectra and cospectra of these eigenmodes.

Finally, it is important to recall that the spectral mode related to the first vector $e^{(1)}$ of the Craya–Herring frame of reference is linked to the toroidal mode in physical space only if the wave-vector direction differs significantly from n. This mode matches the VSHF mode if k is vertical. The same property holds for the second vector $e^{(2)}$, which corresponds essentially to poloidal motion but also matches the VSHF mode for vertical k. The VSHF mode, or $u_\perp(k_\parallel, t)$ in physical space, is not really a wavy mode, even if it corresponds to zero frequency of gravity waves. In addition, the coupling with buoyancy, which is the main characteristic of linear gravity waves (poloidal velocity coupled with buoyancy) vanishes for this mode, so that b is again a passive scalar in the VSHF limit, and strong departure from equipartition in terms of kinetic and potential wave energy is possible. Regarding vorticity, the VSHF mode has no contribution to vertical vorticity and contributes to horizontal vorticity, whereas the toroidal mode (sometime called "vortex" or "vortical" mode) generates the vertical vorticity component. The Craya–Herring decomposition allows us to incorporate in a very tractable geometrical way the toroidal–poloidal decomposition, with very different limits given by 2D Taylor–Proudman modes (horizontal wave vectors) and 1D VSHF modes (vertical wave vectors). Near the 2D limit, the toroidal mode corresponds to horizontal velocity and vertical vorticity, and vice versa for the poloidal mode. More generally, toroidal velocity corresponds to poloidal vorticity and vice versa [the Craya–Herring frame is also a useful cyclic basis; see Eq. (7.15)]. Only near the VSHF limit are vorticity and velocity both quasi-horizontal.

The main nonlinear mechanism in quasi-homogeneous unsteady stratified flows consists of concentrating energy toward more and more vertical wave vectors, as shown in the cartoon from Godeferd and Cambon (1994) in Fig. 7.5 in Section 7.5. This anti-2D (compared with the case of rotation; see Fig. 4.16 in Chapter 4) trend can be explained by the toroidal cascade, independently of wave-turbulence "weak" nonlinearity, and without invoking specific instabilities to prexisting and/or forced coherent vertical vortices.

7.4 The Toroidal Cascade as a Strong Nonlinear Mechanism Explaining the Layering

Looking at the velocity equation, under a slightly different form (the inviscid case is considered for the sake of simplicity), one finds

$$\frac{\partial \boldsymbol{u}}{\partial t} + \boldsymbol{\omega} \times \boldsymbol{u} + \nabla\left(p + \frac{1}{2}u^2\right) = b\boldsymbol{n}, \tag{7.13}$$

in which projection onto the $\boldsymbol{e}^{(1)}$ mode removes both the "divergent" term (total pressure here) because of solenoidal property and the b term because it is vertical. The toroidal (or toroidal + VSHF) equation is therefore

$$\frac{\partial u^{(1)}}{\partial t} + \boldsymbol{e}^{(1)} \cdot \widehat{\boldsymbol{\omega} \times \boldsymbol{u}} = 0,$$

and it is possible to extract the pure toroidal contribution in the nonlinear term as

$$\frac{\partial u^{(1)}(\boldsymbol{k}, t)}{\partial t} + \iota \frac{k}{2} \sum_\Delta \boldsymbol{e}^{(1)}(\boldsymbol{k}) \cdot \left[\boldsymbol{e}^{(2)}(\boldsymbol{p}) \times \boldsymbol{e}^{(1)}(\boldsymbol{q}) + \boldsymbol{e}^{(2)}(\boldsymbol{q}) \times \boldsymbol{e}^{(1)}(\boldsymbol{p})\right]$$

$$\times u^{(1)*}(\boldsymbol{p}, t) u^{(1)*}(\boldsymbol{q}, t) + CCC, \tag{7.14}$$

using

$$\hat{\boldsymbol{u}} = u^{(1)}\boldsymbol{e}^{(1)} + u^{(2)}\boldsymbol{e}^{(2)}, \quad \hat{\boldsymbol{\omega}} = \iota k \left[u^{(1)}\boldsymbol{e}^{(2)} - u^{(2)}\boldsymbol{e}^{(1)}\right]. \tag{7.15}$$

The CCC term denotes the contribution of other quadratic terms, those that correspond to $u^{(1)*}u^{(2)*}$ and to $u^{(2)*}u^{(2)*}$. Some are identically zero because, for instance, the triple scalar product in terms of $\boldsymbol{e}^{(1)}$ for $\boldsymbol{k}, \boldsymbol{p}, \boldsymbol{q}$ is zero [$\boldsymbol{e}^{(1)}$ being always horizontal]. More generally, the decomposition in terms of eigenmodes shows that any $u^{(2)}$ contribution in CCC involves a "rapid" phase factor $e^{\iota\sigma t}$, as seen from Eq. (7.11). These rapid factors result in an efficient damping of the nonlinearity by anisotropic phase mixing, unless their time dependency is canceled out by means of resonance conditions. Accordingly, in the limit of strong stratification, the CCC terms survive only through "weak" resonant-wave interactions (exactly as for rotating turbulence in Chapter 4). The new fact is that all triads involving $u^{(1)}$ only have a nonvanishing contribution without any wave-resonance constraint. On the other hand, resonant-wave interactions represent a low-dimensional manifold, so that magnitude of the CCC contributions can be considered as being of the order of the Froude number. Discarding CCC terms is therefore relevant if the Froude number is very small (but with very high Reynolds number in order to allow significant nonlinearity, purely toroidal here) and the elapsed time is not too high, with in addition a special care to investigate the VSHF limit.

The relevance of a pure toroidal cascade, or of a pure QG cascade, revisited in Chapter 8, deserves further discussion in the geophysical context. But let us note that this study can be almost disconnected from the context of geophysical applications, as it deals with the basic nonlinearity of Euler equations, seen by means of triad interactions in 3D Fourier space. Given the success of Waleffe's triad instability principle for predicting the energy cascades, a similar approach is now applied to Eq. (7.14) neglecting the CCC term.

Restricting this equation to a single triad, one obtains

$$\dot{u}_k^{(1)} = (p_\perp^2 - q_\perp^2) G u_p^{(1)*} u_q^{(1)*}, \tag{7.16}$$

$$\dot{u}_p^{(1)} = (q_\perp^2 - k_\perp^2) G u_q^{(1)*} u_k^{(1)*}, \tag{7.17}$$

$$\dot{u}_q^{(1)} = (k_\perp^2 - p_\perp^2) G u_k^{(1)*} u_p^{(1)*}, \tag{7.18}$$

where

$$G = \frac{\imath}{2} C_{kpq} \frac{kpq}{k_\perp p_\perp q_\perp} \frac{\bm{k} \times \bm{p}}{|\bm{k} \times \bm{p}|} \cdot \bm{n} \tag{7.19}$$

and

$$C_{kpq} = \frac{|\sin \widehat{(\bm{p}, \bm{q})}|}{k} = \ldots \text{sym}(k, p, q). \tag{7.20}$$

New detailed conservation laws can be identified in the present case. The factor G is invariant with respect to any even permutation of the vectors $\bm{k}, \bm{p}, \bm{q}$ of the triad and changes its sign for an odd permutation. Therefore it is clear that triadic interactions within a single triad conserve toroidal energy, because $\dot{u}_k^{(1)} u_k^{(1)*} + \dot{u}_p^{(1)} u_p^{(1)*} + \dot{u}_q^{(1)} u_q^{(1)*} = 0$ and vertical contributions to toroidal enstrophy, as

$$k_\perp^2 \dot{u}_k^{(1)} u_k^{(1)*} + p_\perp^2 \dot{u}_p^{(1)} u_p^{(1)*} + q_\perp^2 \dot{u}_q^{(1)} u_q^{(1)*} = 0. \tag{7.21}$$

The analogy with the 2D case is very strong; see also Waleffe (1992), and especially his Appendix A, and pioneering papers by Fjortoft (1953) and Kraichnan, but it must be noticed that the 2D-2C limit requires the additional condition that $k_\| = p_\| = 0$ in Eqs. (7.18). Without further quantitative statistical analysis (next section), it is imediately shown that only (R) triads are concerned, but in terms of $k_\perp$ only. Compared with the instability principle expressed in terms of helical modes for 3D isotropic or rapidly rotating turbulence, the analogy of (7.18) with the stability of a solid rotating around its principal axes of inertia (Euler problem) is even more striking. In contrast with the helical case, in which terms $sk, s'p, s''q$ play the role of (positive) principal inertia moments I_1, I_2, I_3, with the additional difficulty linked to various signs (polarities of helical modes $s = \pm 1, s' = \pm 1, s'' = \pm 1$), now really positive terms ($k_\perp^2, p_\perp^2, q_\perp^2$) play these roles.

In short, the presence of only reverse interactions could suggest an inverse cascade, at least in terms of $k_\perp$ wave-vector components and therefore in terms of cylinders. Strong anisotropy allows for very rich modalities of cascade, with various senses depending on shell-to-shell (direct), cylinder-to-cylinder, or angle-to-angle

7.5 The Viewpoint of Modeling and Theory: RDT, Wave Turbulence, EDQNM

The case of stably stratified turbulence is different from the one of pure rotation, even if the gravity waves present strong analogies with inertial waves. An additional element is the presence of the toroidal mode, which is steady and decoupled from gravity-wave modes, at least in the linear limit.

As for the case of rotating turbulence, exact generalized Lin equations are easily found:

$$\left(\frac{\partial}{\partial t} + 2\nu k^2\right) U^{(\text{tor})} = T^{(\text{tor})}, \qquad (7.22)$$

$$\left(\frac{\partial}{\partial t} + 2\nu k^2\right) U^{(w)} = T^{(w)}, \qquad (7.23)$$

$$\left(\frac{\partial}{\partial t} + 2\nu k^2 + 2\imath N \frac{k_\perp}{k}\right) Z' = T^{(z')}, \qquad (7.24)$$

in which

$$U^{(\text{tor})} = \frac{1}{2} u^{(1)} u^{(1)*}, \quad U^{(\text{pol})} = \frac{1}{2} u^{(2)} u^{(2)*}, \quad U^{(\text{pot})} = \frac{1}{2} u^{(3)} u^{(3)*}. \qquad (7.25)$$

The total energy of gravity waves (in the linear limit of eigenmode decomposition) is given by

$$U^{(w)} = U^{(\text{pol})} + U^{(\text{pot})}. \qquad (7.26)$$

The nonlinear term Z' quantifies the imbalance between poloidal kinetic and potential (buoyancy) parts of the total wave energy: Its real part is $(1/2)[U^{(\text{pol})} - U^{(\text{pot})}]$ and its imaginary part is related to the poloidal buoyancy flux (details are given in Godeferd and Cambon, 1994, and Godeferd and Staquet, 2003). Closures for the transfer terms [right-hand side of Eqs. (7.22)–(7.24)] are found in terms of the basic set of the previously mentioned spectra, depending on both $k_\perp$ and $k_\parallel$ (or on k and $\cos\theta = k_\parallel/k$) in the simplest statistical way consistent with the symmetries (axisymmetry with mirror symmetry) of the dynamical basic equations. As an example, the contribution to $T^{(\text{tor})}$, which is related to purely toroidal triple correlations $\langle u^{(1)}(\mathbf{k},t) u^{(1)}(\mathbf{p},t) u^{(1)}(\mathbf{q},t) \rangle$ under an integral, involve a term $\theta_{kpq} U^{(\text{tor})}(\mathbf{q},t)[a(\mathbf{k},\mathbf{p}) U^{(\text{tor})}(\mathbf{p},t) - b(\mathbf{k},\mathbf{p}) U^{(\text{tor})}(\mathbf{k},t)]$ once closed by the anisotropic EDQNM procedure.

About global and detailed conservation laws, it is important to stress that Eqs. (7.22) and (7.23) are always exact, with their right-hand side having zero integral. On the other hand, *detailed conservation laws per triad, for both toroidal kinetic energy and for vertical enstrophy, are valid only if the poloidal contribution*

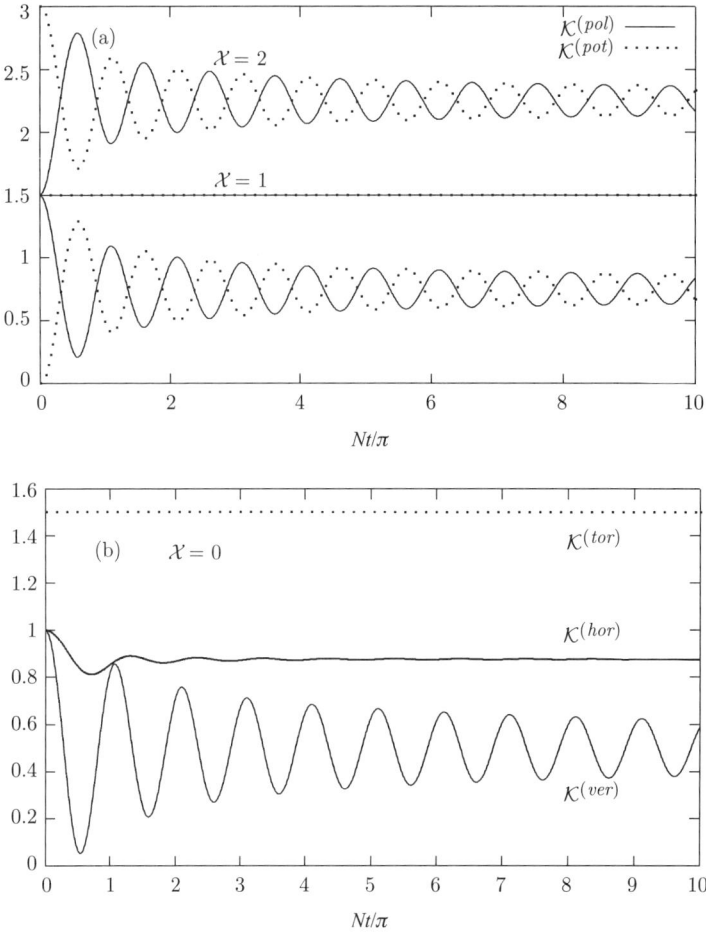

Figure 7.4. Stable stratification, purely linear inviscid calculation, isotropic initial data with initially unbalanced poloidal/potential energy of ratio χ; the toroidal component remains constant and equal to the initial poloidal component. From Salhi and Cambon (2007) with permission of the American Physical Society.

to the velocity field is discarded. Removal of poloidal components in the detailed triadic budget for the toroidal mode amounts to neglecting weak gravity-wave turbulence with respect to strong turbulence, a conjecture that is reasonable at a very small Froude number and a moderate elapsed time.

Discarding the right-hand side in the preceding system of equations (and viscosity, even if it is easily acounted for) yields the so-called RDT limit for second-order single-time statistics. When the spectra and cospectra are integrated over Fourier space, toroidal energy is strictly conserved, as is $U^{(\text{tor})}$ at any k, whereas poloidal- and potential-energy components asymptotically equilibrate after a transient phase made of damped oscillations with opposing phases (see Salhi and Cambon, 2007, for details, and Fig. 7.4). The damping originates in phase mixing, because the integral

7.5 The Viewpoint of Modeling and Theory

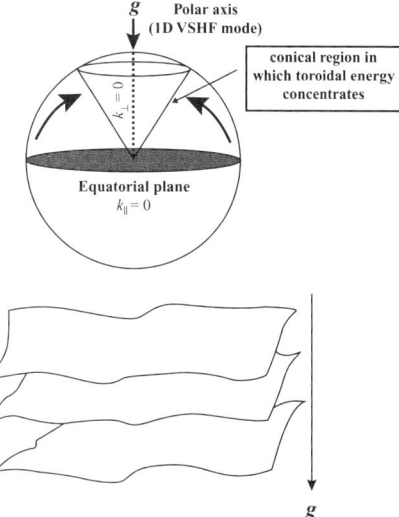

Figure 7.5. Stable stratification, sketch of the nonlinear cascade (top, angular drain in Fourier space) and corresponding layering effect in physical space (bottom).

of Z' over the polar angle $\sin\theta = k_\perp/k$ tends toward zero because of the weighting factor $e^{2iNt\sin\theta}$ coming from RDT.

Detailed equations and DNS/EDQNM comparisons, including the angle-dependent spectra, are given in Godeferd and Staquet (2003). The EDQNM2 model in Godeferd and Cambon (1994) yielded the angular drain of energy that condenses the energy toward vertical wave vectors, in agreement with the collapse of vertical motion and layering (see the sketch in Fig. 7.5). Recall that, because of the incompressibility constraint ($\boldsymbol{k}\cdot\hat{\boldsymbol{u}} = 0$), both contributions to velocity and vorticity become almost horizontal if the spectral density of energy is concentrated near the vertical wave vectors. In terms of directional and polarization anisotropy, polarization becomes marginal and all anisotropic features depend on the sole directional anisotropy, including the collapse of vertical motion.

The latter effect is reflected in physical space by a pancake structure, sketched in Fig. 7.5(bottom) and illustrated in Fig. 7.6, in which isovalues of velocity gradients are obtained from a snapshot of instantaneous DNS data. This layering can be statistically quantified by the development of two different integral length scales, as shown in Fig. 7.7 (from EDQNM2) and Fig. 7.8 (from DNS), with excellent agreement. The integral length scale related to horizontal velocity components and horizontal separation $L_{11}^{(1)}$ is shown to develop similarly to isotropic unstratified turbulence, whereas the one related to vertical separation $L_{11}^{(3)}$ is blocked. In the same conditions, with initial equipartition of potential and wave energy, linear calculation (RDT) exhibits no anisotropy, i.e., $L_{11}^{(1)} = 2L_{11}^{(3)}$.

The EDQNM2 procedure was made as simple as possible in Cambon et al. (2007) in order to focus on pure toroidal interactions and to reach very high Reynolds numbers Re at very low Froude numbers Fr and long elapsed times, a range of parameters not presently accessible to DNS. A typical shape of strongly anisotropic transfer related to $T^{(\mathrm{tor})}(k,\cos\theta)$ in Eq. (7.24) is shown in Fig. 7.9.

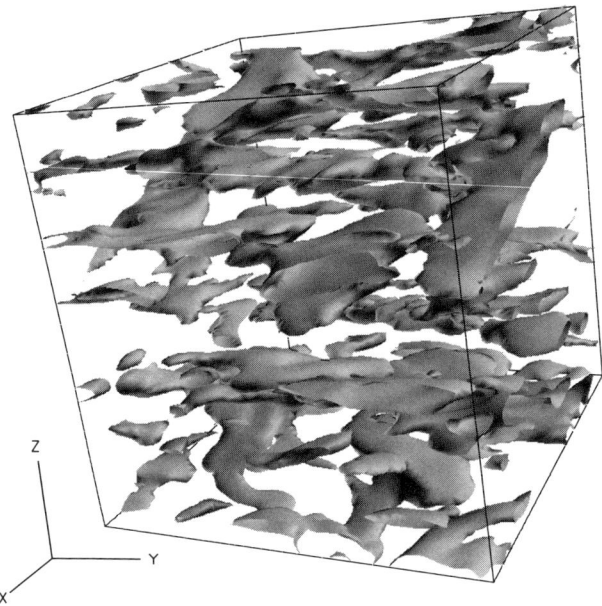

Figure 7.6. Isovalues of the vertical gradient of horizontal velocity fluctuation. Pure stratification. DNS with 256^3 grid points and isotropic initial data. Reproduced from Godeferd and Staquet (2003) with permission of CUP.

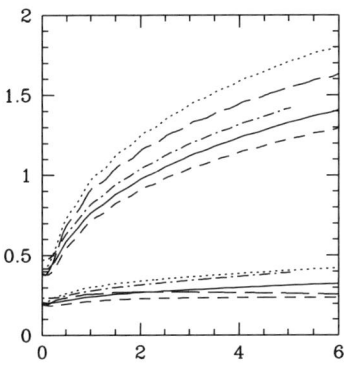

Figure 7.7. Development of typical integral length scales from EDQNM2. $L_{11}^{(1)}$ (top) and $L_{11}^{(3)}$ (bottom), where indices 1 and 3 denote horizontal and vertical directions, respectively. (Initial) isotropy implies $L_{11}^{(1)} = 2L_{11}^{(3)}$. Reproduced from Godeferd and Staquet (2003) with permission of CUP.

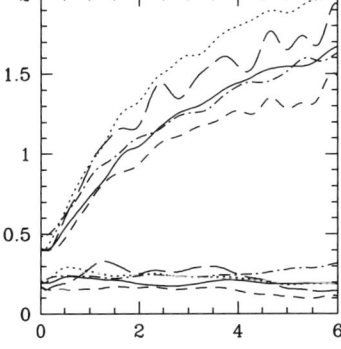

Figure 7.8. Same as Fig. 7.7, from DNS with 256^3 grid points. Reproduced from Godeferd and Staquet (2003) with permission of CUP.

7.6 Coherent Structures

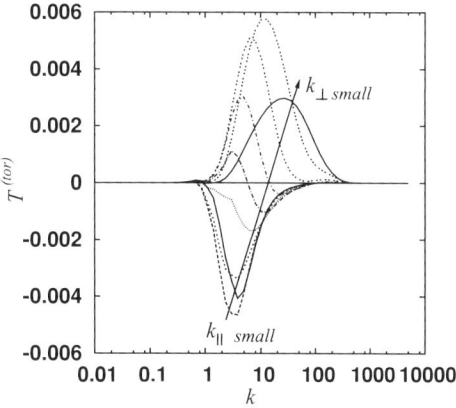

Figure 7.9. Angle-dependent toroidal transfer term. Each figure displays its k-dependence at a given polar angle $\theta = \widehat{(\mathbf{k}, \mathbf{n})}$. EDQNM2 results in stably stratified freely decaying turbulence at initial $Re_\lambda - 145$. Data taken from Cambon et al. (2007).

Cartoons for different types of interactions, and related cascades, are displayed for sphere-to-sphere and cylinder-to-cylinder energy transfers in Fig. 7.10.

7.6 Coherent Structures: Dynamics and Scaling of the Layered Flow, "Pancake" Dynamics, Instabilities

7.6.1 Simplified Scaling Laws

When the detailed anisotropy of the flow is ignored, the simplified scaling laws seem to be valid from examination of results of various numerical and physical experiments. For instance, Lindborg (2006) reported a conventional scaling in DNS/LES

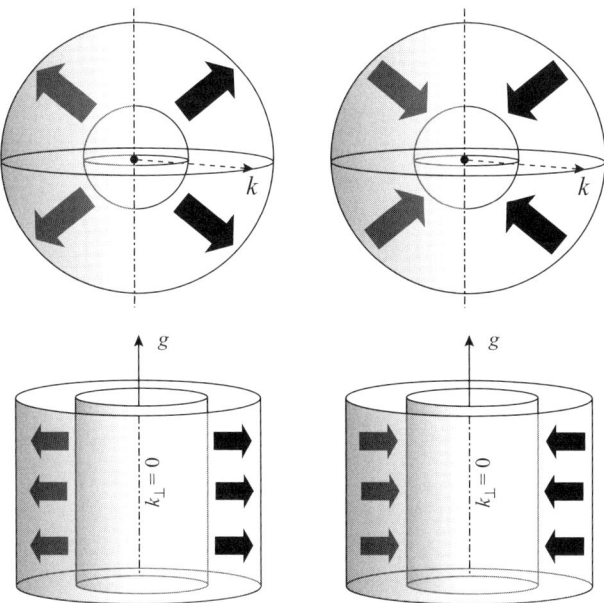

Figure 7.10. Top: isotropic energy drain in spectral space. Direct (left) and inverse (right) cascade. Bottom: cylinder-to-cylinder cascade.

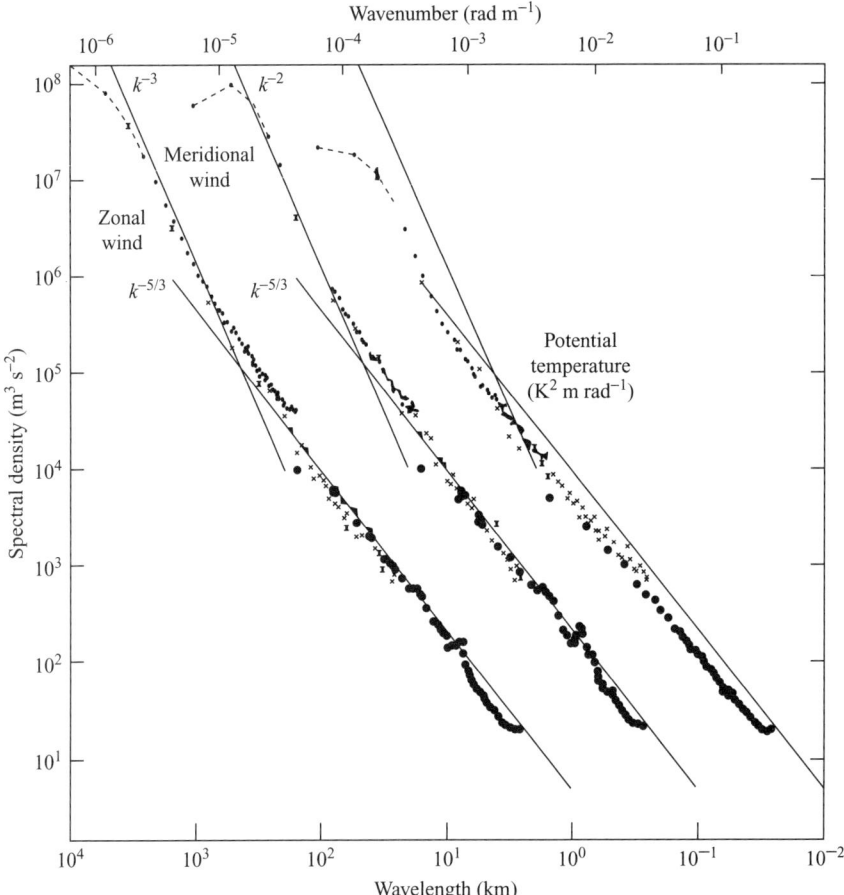

Figure 7.11. From left to right: variance power spectra of zonal wind, meridional wind, and potential temperature near the tropopause from Global Atmospheric Sampling Program aircraft data. The spectra for meridional wind and temperature are shifted one and two decades to the right, respectively. Reproduced from Nastrom and Gage (1985) and Lindborg (2006) with permission of CUP.

in flattened boxes with the strongest stratification for both horizontal kinetic- and potential-energy spectra:

$$E_\perp(k_\perp) = C_1 \varepsilon^{2/3} k_\perp^{-5/3}, \quad E^{(pot)}(k_\perp) = C_2 \varepsilon_p \varepsilon^{-1/3} k_\perp^{-5/3}, \quad (7.27)$$

where ε and ε_p are the dissipation rates of kinetic and potential energy, respectively. These scaling laws are consistent with a vertical Froude number close to the unity, equipartition in terms of potential and kinetic energy,[†] and classical estimates by Taylor for length scales, as in isotropic flows without stratification. The spectral scalings are consistent with the ones by Nastrom and Gage (1985), as shown in Fig. 7.11.

The numerical method can be questioned because of underresolution in the vertical direction (use of a hyperviscosity) and artificial forcing of purely 2D horizontal

---

[†] The equipartition of kinetic and potential energy is not imposed by dynamical equations, except for linear internal waves.

7.6 Coherent Structures

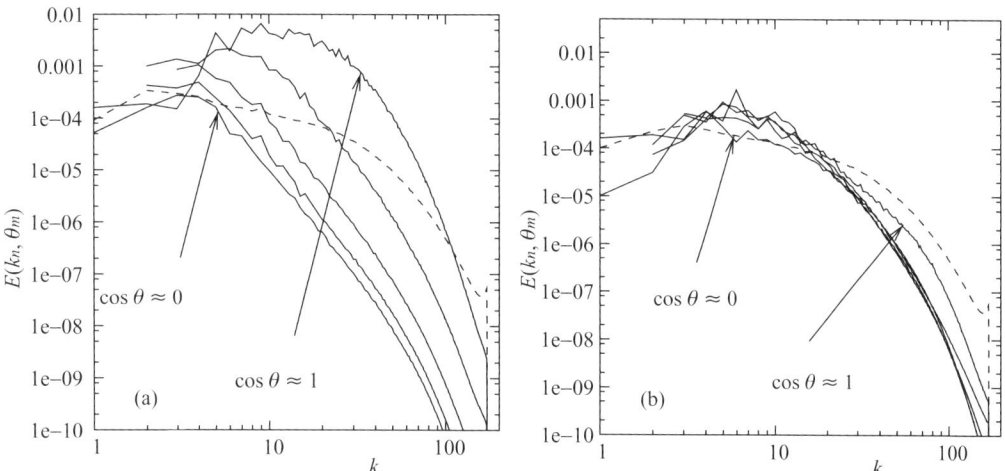

Figure 7.12. Angle-dependent spectra, toroidal (left) and poloidal (right) energy. High-resolution DNS of strongly stratified flow. Courtesy of L. Liechtenstein.

modes, but the very different simulation by Riley and deBruynKops (2003), rediscussed in the next subsection, gives a similar scaling. Such a scaling was expected but not observed in a recent experiment by Praud, Fincham, and Sommeria (2005) – Adam Fincham attributes this to the Reynolds number being too low – carried out in the large Coriolis tank filled with saltwater (without rotation here; the cases with additional rotation are addressed in the next chapter), turbulence being generated by a moving rake.

7.6.2 Pancake Structures, Zig-Zag, and Kelvin–Helmholtz Instabilities

The fact that conventional scaling laws are recovered in a very similar way to what is found in isotropic turbulence without stratification seems to contradict the highly anisotropic organization of the strongly stratified flow. This anisotropic organization can be quantified by various statistical indicators, from single-point correlations to two-component spectra such as the ones presented in Godeferd and Staquet (2003) and Liechtenstein, Godeferd, and Cambon (2005). It is linked to anisotropic structures that are identified in DNS snapshots by means of isovelocity-gradient surfaces. Pancake structures were identified a long time ago in pseudo-spectral DNS (Kimura and Herring, 1996). More recently, it was shown that the horizontal layering with pancake structures essentially modifies the toroidal part of the flow, whereas the poloidal part remains apparently almost isotropic. Angle-dependent spectra $U^{(\text{tor})}(k, \cos\theta)$ and $U^{(\text{pol})}(k, \cos\theta)$ calculated from DNS give consistent, more quantitative, information. The whole result, in both spectral space (see Fig. 7.12) and physical space (see Fig. 7.13) confirms the cartoon displayed in Fig. 7.5, but is restricted to the sole toroidal component of the flow.

It is worth noting that in actual flows internal waves would not be expected to be isotropic. The reason why is that some of them are generated by the adjustment of

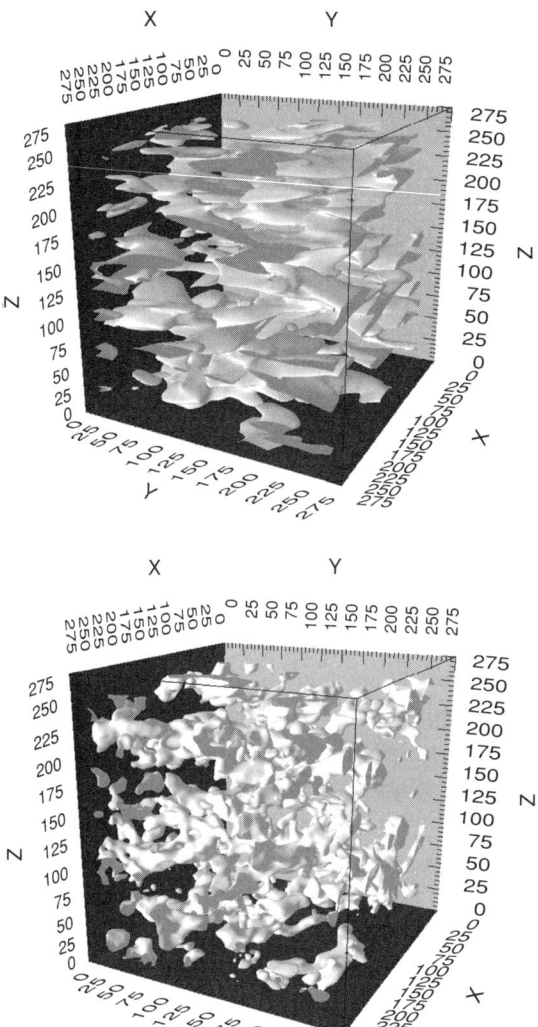

Figure 7.13. Isoenstrophy surfaces (snapshot), using only the toroidal (top) and the poloidal (bottom) contributions from the fluctuating-velocity field. High-resolution DNS of strongly stratified flow. Courtesy of L. Liechtenstein.

the toroidal modes (cyclostrophic adjustment), some can be affected by the toroidal shear, and others can undergo the resonant-wave–vortex interaction (see Lelong and Riley, 1991). The latter was reported to be important in numerical simulations in Bartello (1995).

Another type of more specific structure results from the zig-zag instability. Such instabilities were first identified in the presence of vertical columnar structures moving horizontally in a stratified tank. A typical tangling motion develops in the horizontal direction perpendicularly to the main motion of eddies, with typical velocity U breaking their vertical coherence with a typical length scale U/N. In addition to the case of a pair of counterrotating eddies, advancing with almost constant velocity because of mutual induction, the case of corotating eddies was investigated (Otheguy, Billant, and Chomaz, 2006). In the latter case, mutual induction results in circular motion, and tangling zig-zag motion develops in the radial direction. This instability was proposed as a generic mechanism to create layering with a

7.6 Coherent Structures

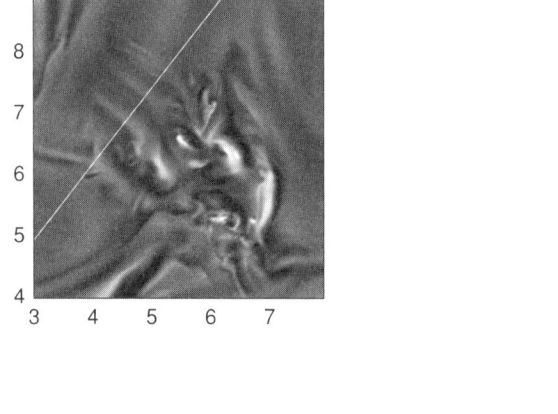

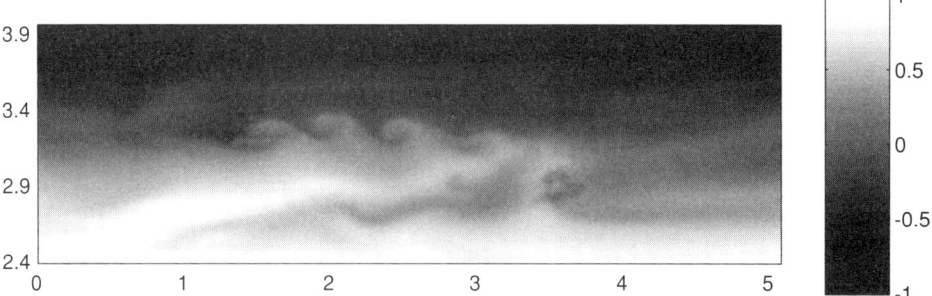

Figure 7.14. Top panel shows part of a horizontal slice through the w field. The white dashed line gives the orientation of a vertical slice through the horizontal plane. The bottom panel shows ρ_T on that vertical slice. Reproduced from Riley and deBruynKops (2003) with permission of AIP.

universal scaling U/N in a strongly turbulent stratified flow (Billant and Chomaz, 2001). This assumption is probably too simple. On the one hand, the zig-zag instability requires the presence of prexisting coherent vortices with vertical lengths much larger than U/N: In some experiments, in which a moving rake favors 2D structuring, the zig-zag motion is recovered (Praud, Fincham, and Sommeria, 2005), but it is not found in other ones with smaller dimensions (Peter Davidson, experimental study in progress), in which turbulence is generated by a grid with not a too large mesh. On the other hand, even in the presence of an array of vertical 2D vortices, significant horizontal velocity U must result from translational or rotational motion of eddies: For instance, the zig-zag instability is inhibited if the base flow is the 2D Taylor–Green flow, or Taylor's "four rollers mill," in which all degrees of freedom are blocked by mutual induction of vortices.

A more promising type of instability is of the Kelvin–Helmholtz type and can result from the intense vertical shearing between pancake layers. Such structures do not appear in the fully 3D DNS (in cubic boxes) of decaying stratified turbulence by Liechtenstein, Godeferd, and Cambon (2005) because the moderate Reynolds number probably limits the shearing process. For a different reason, because of the insufficient vertical resolution, they hardly appear in the DNS/LES with flattened computational domains (Lindborg, 2006).

Only in the DNS by Riley and co-workers is there a significant occurrence found for such Kelvin–Helmholtz instabilities, as shown in Fig. 7.14.

The resolution of these DNSs (which can be considered as really 3D) is comparable with the ones by Liechtenstein, Godeferd, and Cambon (2005), but the largest structures are initialized by a network of 3D Taylor–Green vortices, allowing for a much larger Reynolds number.

It has been suggested for a long time that a stability criterion, such as the one of Miles (1961), i.e., $R_i = N^2/S^2 \sim 1/4$, can control the "efficient" local shear S. More generally, Riley and deBruynKops (2003) introduced the nondimensional number

$$R_b = F_r^2 Re,$$

referred to as the *buoyancy efficiency parameter*, to identify a regime of strongly turbulent and strongly stratified flows characterized by $R_b > 1$, capable of developing strong interlayer shearing. The latter threshold was recently rediscovered by Brethouwer et al. (2007), with application to their underresolved DNS. The parameter R_b is almost equivalent to the parameter $\epsilon/(\nu N^2)$, which is called the *activity parameter* or the *buoyancy Reynolds number* by oceanographers. Bill Smyth and Jim Riley (private communication) suggest that this parameter must be greater than about 20 for the flow to sustain turbulence.

Bibliography

BARTELLO, P. (1995). Geostrophic adjustment and inverse cascades in rotating stratified turbulence, *J. Atmos. Sci.* **52**, 4410–4428.

BILLANT, P. AND CHOMAZ, J. M. (2000). Experimental evidence for a new instability of a vertical columnar vortex pair in a strongly stratified fluid, *J. Fluid Mech.* **418**, 167–188.

BILLANT, P. AND CHOMAZ, J. M. (2001). Self-similarity of strongly stratified inviscid flows, *Phys. Fluids* **13**, 1645–1651.

BOUSSINESQ, J. (1876). *Théorie analytique de la chaleur, mise en harmonie avec la thermodynamique et la théorie mécanique de la lumière*, Paris, Gauthier-Villars (two vols.)

BRETHOUWER, G., BILLANT, P., LINDBORG, E., AND CHOMAZ, J. M. (2007). Scaling analysis and simulation of strongly stratified turbulent flows, *J. Fluid Mech.*, **585**, 343–368.

CAMBON, C. (2001). Turbulence and vortex structures in rotating and stratified flows, *Eur. J. Mech. B (Fluids)* **20**, 489–510.

CAMBON, C., GODEFERD, F. S., AND KANEDA, Y. (2007). Phase-mixing and toroidal cascade in rotating and stratified flows, presented at the Congrès Français de Mécanique, Grenoble, August 27–31, 2007.

CHANDRASEKHAR, S. (1981) *Hydrodynamic and Hydromagnetic Stability*, Dover.

CHARNEY, J. G. (1971). Geostrophic turbulence, *J. Atmos. Sci.* **28**, 1087–1095.

CHO, J. Y. N. AND LINDBORG, E. (2001). Horizontal velocity structure functions in the upper troposphere and lower stratosphere 1. Observations. *J. Geophys. Res.* **106**, 10223–10232.

CRAFT, T. J. AND LAUNDER, B. E. (2002). Application of TCL modelling to stratified flows, in *Closure Strategies for Turbulent and Transitional Flows*, Launder, B. and Sandham, N., eds., Cambridge University Press.

FJORTOFT, R. (1953). On the changes in the spectral distribution of kinetic energy for two-dimensional, non-divergent flows, *Tellus* **5**, 225–230.

GODEFERD, F. S. AND CAMBON, C. (1994). Detailed investigation of energy transfers in homogeneous stratified turbulence, *Phys. Fluids* **6**, 2084–2100.

GODEFERD, F. S. AND STAQUET, C. (2003). Statistical modelling and direct numerical simulations of decaying stably stratified turbulence. Part 2. Large-scale and small-scale anisotropy, *J. Fluid Mech.* **486**, 115–159.

HERRING, J. R. AND MÉTAIS, O. (1989). Numerical experiments in forced stably-stratified turbulence, *J. Fluid Mech.* **25**, 505–534.

KIMURA, Y. AND HERRING, J. R. (1996). *J. Fluid Mech.* **328**, 253–269.

LELONG, M. P. AND RILEY, J. J. (1991). Internal wave-vortical mode interactions in strongly stratified flows, *J. Fluid Mech.* **232**, 1–19.

LIECHTENSTEIN, L., GODEFERD, F. S., AND CAMBON, C. (2005). Nonlinear formation of structures in rotating stratified turbulence, *J. Turbulence* **6**, 1–18.

LILLY, D. K. (1983). Stratified turbulence and the mesoscale variability of the atmosphere, *J. Atmos. Sci.* **40**, 749–761.

LINDBORG, E. (2006). The energy cascade in a strongly stratified fluid, *J. Fluid Mech.* **550**, 207–242.

LINDBORG, E. AND CHO, J. Y. N. (2001). Horizontal velocity structure functions in the upper troposphere and lower stratosphere 2. Theoretical considerations, *J. Geophys. Res.* **106**, 10233–10241.

MILES, J. W. (1961). On the stability of heterogeneous shear flows, *J. Fluid Mech.* **10**, 496–508.

MOWBRAY, D. E. AND RARITY, B. S. H. (1967). A theoretical and experimental investigation of the phase configuration of internal waves of small amplitude in a density stratified liquid, *J. Fluid Mech.* **28**, 1–16.

NASTROM, G. D. AND GAGE, K. S. (1985). A climatology of atmospheric wavenumber spectra of wind and temperature observed by commercial aircraft, *J. Atmos. Sci.* **42**, 950–960.

ORSZAG, S. A. (1970). Analytical theories of turbulence, *J. Fluid Mech.* **41**, 363–386.

OTHEGUY, P., BILLANT, P., AND CHOMAZ, J. M. (2006). Effect of the planetary rotation on the zig-zag instability of co-rotating vortices in a stratified fluid, *J. Fluid Mech.* **553**, 273–281.

PEDLOWSKY, J. (1987). *Geophysical Fluid Dynamics*, Springer-Verlag.

PRAUD, O., FINCHAM, A. M., AND SOMMERIA, J. (2005). Decaying grid turbulence in a strongly stratified fluid, *J. Fluid Mech.* **522**, 1–33.

PRAUD, O., SOMMERIA, J., AND FINCHAM, A. M. (2006). Decaying grid turbulence in a rotating stratified fluid, *J. Fluid Mech.* **547**, 389–412.

RILEY, J. J. AND DEBRUYNKOPS, S. M. (2003). Dynamics of turbulence strongly influenced by buoyancy, *Phys. Fluids* **15**, 2047–2059.

RILEY, J. J., METCALFE, R. W., AND WEISSMAN, M. A. (1981). In *Proceedings of the AIP Conference on Nonlinear Properties of Internal Waves*, West, B. J., ed., AIP, pp. 72–112.

SALHI, A. AND CAMBON, C. (2007). Anisotropic phase-mixing in homogeneous turbulence in a rapidly rotating or in a strongly stratified fluid: An analytical study, *Phys. Fluids* **19**, 055102.

SMITH, L. M. AND WALEFFE, F. (2002). Generation of slow large scales in forced rotating stratified turbulence, *J. Fluid Mech.* **451**, 145–168.

WALEFFE, F. (1992). The nature of triad interactions in homogeneous turbulence, *Phys. Fluids* A **4**, 350–363.

8 Coupled Effects: Rotation, Stratification, Strain, and Shear

A combination of system rotation and stable stratification is essential for geophysical applications, even if the former effect is significantly smaller than the latter in 3D flows, e.g., for scales much smaller than the *synoptic* ones in the atmosphere. As for pure rotating turbulence in Chapter 4 and purely stratified turbulence in Chapter 7, linear analysis, i.e., RDT, describes only neutral stability and will lead to the definition of both the wave-vortex eigenmode decomposition and dispersion frequencies of inertia–gravity waves in the present chapter. Nonlinear dynamics is essential, and allows us to revisit a *quasi-geostrophic (QG) cascade*, which generalizes the toroidal cascade discussed in Chapter 7 with additional Coriolis effects.

Other coupled effects investigated in this chapter can create linear instabilities that can be analyzed within the RDT framework. These instabilities are associated with turbulence-production mechanisms, which are the main striking new physical phenomena when compared with other flows discussed in this book. Therefore only the linear approach will be emphasized in these cases. In the presence of mean shear, *barotropic instabilities* occur, with a strong analogy between the rotating-shear-flow case and the stratified shear flow. A special case combining the three ingredients, namely the mean shear, system rotation, and stable stratification, is shown to give new insight into the baroclinic instability. Finally, the very important *elliptical-flow instability* is investigated, which results from a coupled effect of mean vorticity with weak additional mean strain. This instability can trigger nonlinear cascade and turbulence in physical systems with large strained vortices.

A general conclusion about linear stability theory is provided at the end of this chapter in order to delineate the domain of relevance of homogeneous RDT and of its natural extension by means of WKB RDT.

8.1 Rotating Stratified Turbulence

In the absence of mean shear, it is possible to consider that both the system angular velocity and the mean-buoyancy gradient are vertical. The vertical mean vorticity can be replaced with the Coriolis parameter

$$f = 2\Omega \cos \lambda, \qquad (8.1)$$

where λ is the *colatitude* (see Fig. 8.1).[*]

---

[*] Projecting the angular velocity onto the local vertical axis, which yields the angle-dependent factor $\cos \lambda$, is analogous to projecting the angular velocity on the wave-vector direction, which yields the

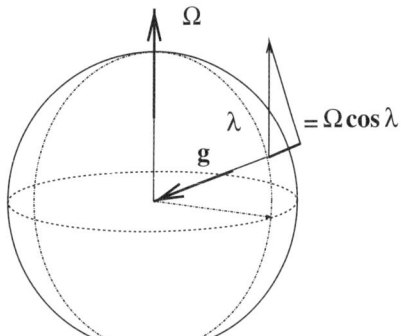

Figure 8.1. Definition and meaning of the Coriolis parameter f in geophysics.

Only neutral stability can be described by use of RDT in this case. The linear inviscid solution for velocity–buoyancy fluctuations consists of superimposed steady and oscillating modes, as for the purely stratified-flow case. The former corresponds to the *QG mode* (which is equal to the toroidal mode in the absence of rotation) and the latter, referred to as the *ageostrophic (AG) mode*, is related to inertia–gravity waves. This new decomposition is commonly referred to as QG/AG decomposition. RDT yields very simple behavior, in which single-time velocity–buoyancy correlations are possibly affected by damped oscillations, the damping resulting from the dispersivity of wave motion. The poloidal–toroidal-like decomposition, closely related to the QG/AG one, was shown to simplify the RDT prediction. For instance, the total turbulent kinetic energy is conserved in pure rotation, and damped oscillations yield an asymptotic equidistribution of poloidal and toroidal turbulent kinetic energy. In pure stratification, both total (kinetic + potential) and toroidal kinetic energy are conserved, whereas poloidal kinetic and potential energies asymptotically equilibrate after damped oscillations (see Fig. 7.4). In these cases, linear dynamics – if restricted to single-time, two- or single-point correlations – is of little interest, as relevant structuring effects result from only the nonlinear terms. In the present case, RDT suggests building a full nonlinear model in terms of the slowly varying amplitudes of QG/AG modes, which are constant in the pure linear inviscid limit.

As for the purely stratified case [see Eq. (7.11)], a single vector w can gather both velocity and buoyancy fluctuations, with the following decomposition:

$$\hat{w} = \sum_{s=0,\pm 1} \underbrace{a^{(s)}(k,t)\exp(s\iota\sigma_k t)}_{\xi^{(s)}(k,t)} N^{(s)}(k), \quad s = 0, \pm 1. \quad (8.2)$$

It is important to note that both the eigenmodes $N^{(s)}, s = 0, \pm 1$ and the (unsigned) dispersion frequency of inertia–gravity waves,

$$\sigma_k = \sqrt{N^2 \sin^2\theta + f^2 \cos^2\theta}, \quad (8.3)$$

angle-dependent dispersion frequency of inertial waves with $\cos\theta$. This projection reflects a dominant role of the Coriolis force in the tangent plane of the rotating spheroid (our Earth), whereas it reflects pure solenoidal motion looking at a sphere in 3D Fourier space. In the geophysical community, the angle of latitude is commonly used and therefore a sine often appears instead of a cosine.

8.1 Rotating Stratified Turbulence

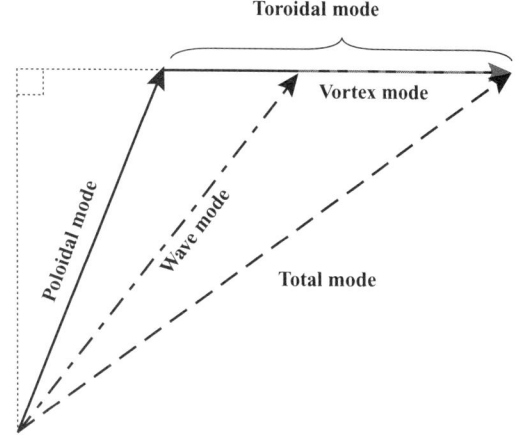

Figure 8.2. Cartoon of the "wave-vortex" (or linear QG-AG) decomposition from the Craya–Herring decomposition or related toroidal–poloidal one in physical space. The toroidal mode is the part of the horizontal velocity fluctuation with zero divergence in terms of horizontal coordinates; it corresponds to (velocity contribution to) the QG mode only without rotation; (velocity contribution to) the QG mode is only a fraction of this toroidal mode at a given f/N ratio and identically vanishes at $N = 0$ (pure rotation).

are explicit functions of the ratio f/N (Cambon, 2001). Here, $s = 0$ is related to the QG mode, which is also referred to as the nonpropagating mode, whereas $s = \pm 1$ corresponds to the AG modes, which are propagating inertia–gravity waves.

As shown in Fig. 8.2, the QG/AG (or wave-vortex) decomposition is dependent on f/N: the inertia–gravity-wave contribution to the velocity field is the poloidal part of the flow without rotation and includes an increasing part of the toroidal component when the rotation (by means of the ratio f/N) increases. In the case of pure rotation, the inertial wave mode includes the entire velocity field, and the nonpropagating (QG) part of the flow identically vanishes.

Because the QG mode, with $a^{(0)}$ amplitude, makes a significant contribution, the cascade related to triadic interactions including QG modes can be expected to be dominant only with respect to the other resonant or quasi-resonant wave interactions. It is recalled that, using the nomenclature used in other chapters, these pure QG triadic interactions are denoted as $(0, 0, 0)$. This purely QG cascade is discussed in the next subsection.

Discarding *a priori* the QG mode, it can be shown that resonant inertia–gravity waves can play a role in concentrating energy toward larger scales and quasi-VSHF modes, at least if f/N is small. This was found by Smith and Waleffe (2002), for instance, by forcing isotropically the small scales without putting energy at large scale at the initial time.

At least two radically different mechanisms of concentration of energy toward the VSHF mode are thus possibly present:

- the *toroidal cascade*, already investigated in Chapter 7 for the pure stratification case; the toroidal mode is the limit of the QG mode in the absence of system rotation. This mechanism plays a dominant role and leads to a rapid layering of the flow if a significant large-scale toroidal part of the flow exists initially and f/N is not too large.
- the *transfers induced by the resonant inertia–gravity waves*.

The relevance of the QG model was expected by Smith and Waleffe (2002) to prevail when $1/2 < N/f < 2$, as triadic "wave" resonances are forbidden in this range of parameters, as seen from looking at dispersion law (8.3).

8.1.1 Basic Triadic Interaction for Quasi-Geostrophic Cascade

If the flow is seen in a rotating frame of reference with angular velocity $(f/2)\mathbf{n}$, f being the Coriolis parameter for geophysical applications, only the basic momentum equation for $\mathbf{u}$ is affected by the additional Coriolis force $f\mathbf{n} \times \mathbf{u}$. The associated evolution equation for b is identical to the one obtained in the pure stratification case; see Eq. (7.3). In agreement with Eq. (8.2) and with the cartoon displayed in Fig. 8.2, the nonpropagating mode now combines toroidal kinetic energy and potential energy,

$$\xi^{(0)} = a^{(0)} = \frac{\sigma_s}{\sigma} u^{(1)} + \frac{\sigma_r}{\sigma} u^{(3)}, \tag{8.4}$$

with

$$\sigma_s = N \frac{k_\perp}{k}, \quad \sigma_r = f \frac{k_\parallel}{k}, \quad \sigma = \sqrt{\sigma_r^2 + \sigma_s^2}, \tag{8.5}$$

and where $\xi^{(0)}$ is related to the *QG energy*. The two other terms $\xi^{(\pm 1)}$ deal with inertia–gravity waves whose dispersion frequency σ_k is given by Eq. (8.3) (Cambon, 2001).

Let us now investigate detailed properties of the interactions within a single isolated triad. Following the same procedure as for the toroidal cascade in Chapter 7, i.e., removing all nonlinear interactions involving wave modes in order to retain only the QG ones, one obtains a set of equations similar to (7.18):

$$\dot{\xi}_k^{(0)} = (p^2 \sigma_p^2 - q^2 \sigma_q^2) G' \xi_p^{(0)*} \xi_q^{(0)*}, \tag{8.6}$$

$$\dot{\xi}_p^{(0)} = (q^2 \sigma_q^2 - k^2 \sigma_k^2) G' \xi_q^{(0)*} \xi_k^{(0)*}, \tag{8.7}$$

$$\dot{\xi}_q^{(0)} = (k^2 \sigma_k^2 - p^2 \sigma_p^2) G' \xi_k^{(0)*} \xi_p^{(0)*}, \tag{8.8}$$

which involves a modified factor $G'(k, p, q)$. This new geometrical factor is fully symmetric in terms of (k, p, q) (Cambon, Godeferd, and Kaneda, 2007). One finds from this system of equations that detailed conservation holds for both QG energy and potential enstrophy, i.e.,

$$\dot{\xi}_k^{(0)} \xi_k^{(0)*} + \dot{\xi}_p^{(0)} \xi_p^{(0)*} + \dot{\xi}_q^{(0)} \xi_q^{(0)*} = 0$$

and

$$k^2 \sigma_k^2 \dot{\xi}_k^{(0)} \xi_k^{(0)*} + p^2 \sigma_p^2 \dot{\xi}_p^{(0)} \xi_p^{(0)*} + q^2 \sigma_q^2 \dot{\xi}_q^{(0)} \xi_q^{(0)*} = 0.$$

These two detailed conservation laws are analogous to those for toroidal energy and vertical enstrophy associated with the pure toroidal cascade for purely stratified

8.1 Rotating Stratified Turbulence

flows and to those for the total energy and total enstrophy in 2D isotropic turbulence. The linkage to *potential enstrophy* comes from

$$\frac{k^2 \sigma_k^2}{N^2} \xi^{(0)} \xi^{(0)*} = \kappa_\perp^2 u^{(1)} u^{(1)*} + \left(\frac{f}{N} k_\parallel\right)^2 u^{(3)} u^{(3)*}, \tag{8.9}$$

using Eq. (8.4). This result is to be compared with vertical enstrophy, which is recovered if $f = 0$. These two conservations laws were already quoted by Bartello (1995), his eigenmode decomposition being essentially the same[†] as the one deduced from Craya–Herring decomposition, and its geometric factor $N^{(000)}$ is similar to the preceding one (but it does not display the factor $p^2 \sigma_p^2 - q^2 \sigma_q^2$).

An important remark must be made here. Equation (8.9) gives the spectral density of the variance of the *linearized absolute potential vorticity* (APV). The fluctuating linearized APV can be defined as

$$\boldsymbol{n} \cdot \boldsymbol{\omega} + \frac{f}{N^2} \frac{\partial b}{\partial x_\parallel} \tag{8.10}$$

in physical space and as

$$\iota k_\perp u^{(1)} + \iota \frac{f}{N} k_\parallel u^{(3)} \tag{8.11}$$

in Fourier space. Its fully nonlinear counterpart is defined as the scalar product of the absolute vorticity (mean $f\boldsymbol{n}$ + fluctuating $\boldsymbol{\omega}$) by the gradient of buoyancy (mean $N^2 \boldsymbol{n}$ + fluctuating ∇b), and is eventually divided by N^2 for keeping the dimension of vorticity:

$$\frac{1}{N^2} (f\boldsymbol{n} + \boldsymbol{\omega}) \cdot (N^2 \boldsymbol{n} + \nabla b). \tag{8.12}$$

In short, from the preceding equation, one can derive an additive threefold decomposition in terms of

- a "mean" contribution f,
- the linearized fluctuation given by Eq. (8.10),
- a *nonlinear quadratic* contribution $\boldsymbol{\omega} \cdot (\nabla b / N^2)$.

8.1.2 About the Case With Small but Nonnegligible f/N Ratio

Especially in oceanography, one encounters at mesoscales f/N ratios of the order of 10^{-1}–10^{-2}. Nevertheless, even with a weak f/N ratio, the effect of the Coriolis force cannot be ignored. The dominant structures are pancake structures, as in strongly stratified turbulence without rotation, but their scaling and dynamics are influenced by the Coriolis force. For instance, the contribution from gravity-wave dispersion

---

[†] The geometrical decomposition used in Godeferd and Cambon (1994) and Cambon (2001) has the only advantage in that a true pseudo-compressible vector can be defined in physical space, by inverse Fourier transformation of the single "true" vector $\hat{\boldsymbol{w}} = \hat{\boldsymbol{u}} + \iota(\boldsymbol{k}/k)(\hat{b}/N) = \sum_{i=1,3} u^{(i)} \boldsymbol{e}^{(i)} = \sum_{s=0,\pm 1} \xi^{(s)} \boldsymbol{N}^{(s)}$ with all related simplifications that are due to orthonormality. Mixing a true vector $\hat{\boldsymbol{u}}$ and a scalar $\hat{b}$ is less tractable in general.

frequency ($\sigma_s = N \sin \theta$) is assumed to be of the same order as the contribution from the inertial-wave frequency, or

$$N \frac{k_\perp}{k} \sim f \frac{k_\parallel}{k}. \tag{8.13}$$

The latter relation is consistent with a Burgers' number,

$$Bu = \left(\frac{f}{N} \frac{L_h}{L_v} \right)^2,$$

close to one, if the ratio of typical vertical L_v and horizontal L_h length scales is related to the inverse of the ratio $k_\parallel / k_\perp$.

8.1.3 The QG Model Revisited. Discussion

The QG model discussed by Charney (1971) was already touched on in Chapter 7. Conservation of full nonlinear APV defined by Eq. (8.12) is a direct consequence of the Ertel theorem (see also Staquet and Riley, 1989). Roughly speaking, this amounts to replacing the vertical direction with the local normal to isopycnal surfaces in the toroidal–poloidal decomposition, the QG motion being along these surfaces and AG motion being across them.

Of course, if these surfaces are weakly undulated and therefore close to horizontal surfaces, linearization of APV is physically justified.

Too strong properties are often attributed to QG motion, in an analogous way to the unjustified use of the Proudman theorem to explain the transition toward a 2D state, yielding some wrong statements:

- QG motion is analogous to 2D motion, with dual energy cascades, so that a direct cascade is expected for APV and an inverse cascade is for QG energy,
- In spite of the strong anisotropy of QG motion, a simple rescaling in terms of f/N can restore an apparent isotropy.

The recent experimental study by Praud, Sommeria, and Fincham (2006), which made use of the very large Coriolis platform in Grenoble, contributed to support this point of view, at least for values of f/N ranging from 0.8 to 1.2. Beautiful cigar vortices are created just behind a moving rake, looking very similar to 2D structures, and then evolve toward less elongated structures (Fig. 8.3) in a rotating tank full of brine. The use of the PIV technique allowed almost 3D velocity and vorticity measurements.

In principle, vertical variability is captured in this experimental study. Even if the horizontal dimension greatly exceeds the vertical dimension of the layer, this does not reduces to the shallow-water case in which "2D," "QG," and "horizontal" motion concepts collapse. On the other hand, the use of a rake with only vertical sticks can favor pure 2D motion with respect to the use of a grid.

In a DNS without any forcing, it is possible to switch from "pancake" to "cigar" structures by increasing the ratio f/N, as shown in Fig. 8.4, for instance. Incidentally,

8.1 Rotating Stratified Turbulence

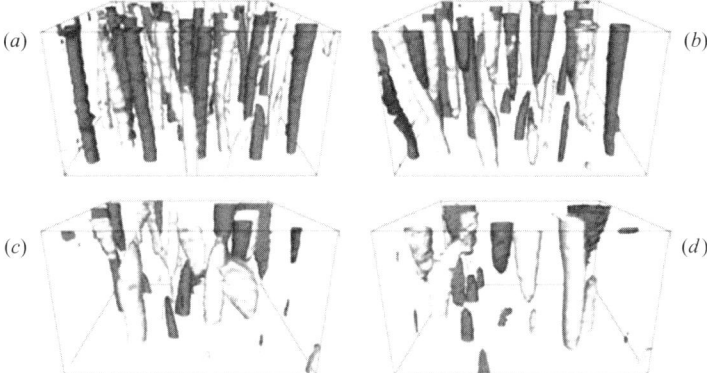

Figure 8.3. Visualization by PIV of vortex structures, $N/f = 1.2$, from Praud, Sommeria, and Fincham (2006) with permission of CUP.

the fact that an apparently isotropic structure is found for $f/N = 1$ is possibly misleading: This case is very different from pure isotropic turbulence. It is also worthwhile to note that "structures," identified by isovorticity surfaces in high-resolution DNS, are much more scrambled than the ones visualized by PIV techniques in physical experiments: Cigars are very different from smooth Taylor columns in such DNSs! A bit paradoxically, underresolved DNS and LES can show apparently smoother structures, but this is a numerical artefact.

Neglecting wall effects and anisotropic forcing, one could write a more subtle QG model for the new Lin equation,

$$\left(\frac{\partial}{\partial t} + 2\nu k^2\right) e^{(\mathrm{QG})}(k_\parallel, k_\perp, t) = T^{(\mathrm{QG})}(k_\parallel, k_\perp, t), \tag{8.14}$$

with

$$\frac{1}{2}\langle \xi^{(0)*}(\boldsymbol{p}, t)\xi^{(0)}(\boldsymbol{k}, t)\rangle = e^{(\mathrm{QG})}(\boldsymbol{k}, t)\delta^3(\boldsymbol{k} - \boldsymbol{p}), \tag{8.15}$$

extending Eqs. (7.22)–(7.24) for the toroidal cascade in purely stratified flows to various f/N ratios. The generalized transfer term $T^{(\mathrm{QG})}(k_\parallel, k_\perp, t)$ can be constructed from triadic contributions strictly preserving detailed conservation of both QG energy and (linearized) APV. Depending on f/N, a very complex and multiform anisotropic cascade is expected in terms of the full distribution of the transfer term expressed as a function of $k_\perp$ and $k_\parallel$. As for the model of toroidal cascade in

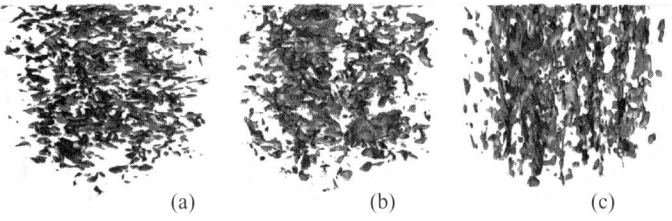

Figure 8.4. Enstrophy isosurfaces; (a) $f/N = 0.1$, (b) $f/N = 1$, (c) $f/N = 10$. 256^3 unforced DNS results from Liechtenstein, Godeferd, and Cambon (2005).

Chapter 7 for $f/N = 0$, the strong dependence of the term G' in Eq. (8.6) with respect to $k_\parallel$ can induce a dynamical cascade process very different from the one observed in 2D turbulence, in spite of the conservation of APV. Only the case of pure rotation cannot be recovered this way, as the QG mode vanishes in this case and only inertial-wave turbulence is relevant. Nevertheless, the case of pretty large f/N ratios – as the highest ratios addressed by Praud, Sommeria, and Fincham (2006) – remains of interest. Note that Herring (1980) proposed a first QG model based on a Lin equation with a transfer term closed by an EDQNM technique, but its relevance was limited by additional quasi-isotropic assumptions.

A very different angle of attack is to try to generalize the Kolmogorov 4/5 law using both velocity and APV increments for building third-order structure functions (Kurian, Smith, and Wingate, 2005). The advantage of working on structure functions in physical space is avoiding a too strict limitation to quadratic nonlinearity (a fully nonlinear APV variable can be used), whereas cubic and quartic nonlinearities are very difficult to handle in Fourier space (even the classical convolution product mediated by quadratic nonlinearity is not so simple!). A drawback is that a complete solenoidal description, with exact removal of explicit pressure terms from the very beginning, is very complicated in physical space; removal of mixed pressure–velocity terms in the 4/5 Kolmogorov law is a very marginal case, for instance.

Of course, a more complete analysis may be developed on the grounds of the most general set of spectra and cospectra consistent with axisymmetry without mirror symmetry. In this more general case, another Lin equation is written for the spectrum of total (kinetic + potential) inertia–gravity-wave energy, $e^{(w)}$, defining another true transfer term[‡] $T^{(w)}$. The system must be supplemented by two additional terms similar to Z in Eq. (7.24) characterizing the imbalance between potential and kinetic energy of waves, together with poloidal and toroidal buoyancy fluxes exchanged between them. Only these Z terms are affected by linear (rapid) factors $\iota\sigma t$, and they are not associated with global conservation laws. This is a complete generalization of the (e, Z) system for pure rotation and the system $[e^{(\text{tor})}, e^{(\text{pol+pot})}, Z']$ for pure stratification.

It is important to note that the vanishing integral for $T^{(\text{QG})}$ and $T^{(w)}$ reflects global conservation laws, whereas the *detailed conservation of both QG energy and potential vorticity is found only when inertia–gravity-wave modes are discarded in triads*. Such a removal can be made *a priori*, or invoking the physical context of very small Rossby and Froude numbers.

8.2 Rotation or Stratification With Mean Shear

In contrast to the cases without mean motion (in addition to the sole solid-body rotation), it is better to reintroduce specific notation for the fluctuation u' in the following discussion. A similar distinction is made for the buoyancy scalar, introducing $\overline{b}$ and b'.

[‡] Let us recall that a nonlinear term in a Lin-type equation is referred to as a *true* transfer term if its integral over all wavenumbers is zero, i.e., if it is associated with a global conservation law.

8.2 Rotation or Stratification With Mean Shear

Within the Boussinesq approximation framework, the equation for the solenoidal velocity field in a rotating frame displays two additional terms, reflecting both Coriolis and buoyancy forces, the intensity of the latter being denoted by b (the fluctuation being denoted by b' from now on):

$$\boldsymbol{F}_b = 2\boldsymbol{\Omega} \times \boldsymbol{u} - b\boldsymbol{n}. \tag{8.16}$$

The use of b allows us to get rid of the different formulations in terms of temperature or density. Let us just recall that b is related to the density ρ by the relation $b = \rho g/\rho_0$ in a liquid, where g is the gravitational acceleration and ρ_0 is the constant density of reference. The vector $\boldsymbol{n}$ is the vertical unit upward vector such that $\boldsymbol{g} = -g\boldsymbol{n}$. Before the field is split into mean and fluctuating components, the momentum equation is recast as follows,

$$\frac{\partial u_i}{\partial t} + u_j \frac{\partial u_i}{\partial x_j} + 2\Omega_n \epsilon_{inj} u_j = -\frac{1}{\rho_0} \frac{\partial p}{\partial x_i} + b u_i \tag{8.17}$$

and the mass conservation equation $\dot{\rho} + \rho \frac{\partial u_i}{\partial x_i} = 0$ yields

$$\frac{\partial b}{\partial t} + u_j \frac{\partial b}{\partial x_j} = 0, \quad \frac{\partial u_i}{\partial x_i} = 0. \tag{8.18}$$

The mean flow consists of a vertical mean shear such as $A_{ij} = S\delta_{i1}\delta_{j3}$ and of a vertical density gradient, i.e.,

$$\frac{\partial \overline{u}_i}{\partial x_j} = A_{ij} = S\delta_{i1}\delta_{j3}, \quad \frac{\partial \overline{b}}{\partial x_j} = N^2 \delta_{j3},$$

so that the Brunt–Väisälä frequency N appears as the characteristic frequency of buoyancy–stratification. From now on, the indices 1, 2, 3 refer to streamwise, spanwise, and vertical directions of the pure plane shear mean motion, respectively, with $n_i = \delta_{i3}$, as in Fig. 8.5(c).

This mean flow is an exact solution of the preceding equations *if the system rotation is in the spanwise direction*, i.e., $\Omega_i = \Omega \delta_{i2}$. In this case, the fluctuating buoyancy b' is governed by the equation

$$\frac{\partial b'}{\partial t} + Sx_3 \frac{\partial b'}{\partial x_1} + u'_j \frac{\partial b'}{\partial x_j} - \nu P_r \nabla^2 b' = -N^2 u'_3,$$

reintroducing the diffusivity of the stratifying agent for generality. The equation for $\boldsymbol{u}'$ is

$$\frac{\partial u'_i}{\partial t} + Sx_3 \frac{\partial u'_i}{\partial x_2} + S\delta_{i1} u'_3 + u_j \frac{\partial u_i}{\partial x_j} + 2\Omega^s_n \epsilon_{inj} u_j + \nu \nabla^2 \boldsymbol{u} = -\frac{1}{\rho_0} \frac{\partial p}{\partial x_i} + b'\boldsymbol{n}. \tag{8.19}$$

In the absence of the mean-density gradient, i.e., if $N = 0$, one can get rid of b' and the rotating mean-shear flow is a particular solution of the Euler equations in the rotating frame.

The main RDT results are now revisited in the presence of pure plane mean shear, system rotation, and/or mean-density stratification–buoyancy.

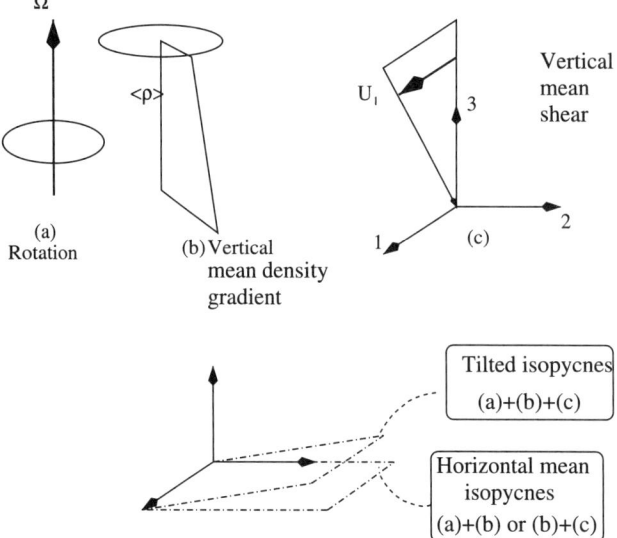

Figure 8.5. Sketch of the mean flow, including (a) system vorticity, (b) vertical stable stratification, and (c) mean shear. Tilting of isopycnal surfaces can trigger the baroclinic instability, if (a), (b), (c) are simultaneously present.

Equations for the fluctuating fields are written in Fourier space. The corresponding variables are denoted by $\hat{u}$ and $\hat{b}$, as before, the "prime" being omitted because there is no ambiguity with the mean flow in Fourier space.

In all cases subsequently addressed, including the baroclinic context, the linear advection process is induced only by a mean shear, so that the time dependency of the wave vector is always

$$k_1 = K_1, k_2 = K_2, k_3 = K_3 + K_1 St, \qquad (8.20)$$

choosing $S = \partial \bar{u}_1 / \partial x_3$. In a polar-spherical system of coordinates for the initial wave vector $\mathbf{K}$ in Eq. (8.20), the angles θ (polar) and φ (azimutal) are defined by

$$\sin\theta = \sqrt{k_1^2 + k_2^2}/K, \quad \cos\varphi = k_1/\sqrt{k_1^2 + k_2^2}, \qquad (8.21)$$

with the usual relations

$$k_1 = K_1 = K\sin\theta\cos\phi, \quad k_2 = K_2 = K\sin\theta\sin\phi, \quad K_3 = K\cos\theta.$$

Inviscid RDT governing equations are

$$\frac{d\hat{u}_i}{dt} = -\underbrace{\left[S\delta_{j3}\left(\delta_{i1} - 2\frac{k_i k_1}{k^2}\right) + 2\Omega P_{in}\epsilon_{n2j}\right]}_{M_{ij}}\hat{u}_j + P_{i3}\hat{b}, \qquad (8.22)$$

in which contributions from pure shear, Coriolis force, and buoyancy effects are taken into account, and

$$\frac{d\hat{b}}{dt} = -N^2 \hat{u}_3. \qquad (8.23)$$

8.2 Rotation or Stratification With Mean Shear

As in previous RDT studies, a system of equations is derived in the Craya–Herring frame of reference:

$$\begin{pmatrix} \dot{u}^{(1)} \\ \dot{u}^{(2)} \\ \dot{u}^{(3)} \end{pmatrix} = \begin{bmatrix} 0 & (2\Omega + S)\frac{k_2}{k} & 0 \\ -2\Omega\frac{k_2}{k} & S\frac{k_1 k_3}{k^2} & -N\frac{k_\perp}{k} \\ 0 & N\frac{k_\perp}{k} & 0 \end{bmatrix} \begin{pmatrix} u^{(1)} \\ u^{(2)} \\ u^{(3)} \end{pmatrix}, \qquad (8.24)$$

with $k_\perp = \sqrt{k_1^2 + k_2^2}$. One can recover easily the terms identified in the pure shear (with S factor), pure rotation (with the dispersion frequency $2\Omega k_2/k$), and in pure stratification (with the dispersion frequency $Nk_\perp/k$) cases. But the main difficulty remains the time dependency of k_3, also reflected in k, that is induced by the pure shear according to Eq. (8.20).

8.2.1 The Rotating-Shear-Flow Case

The rotating shear flow, with angular velocity aligned with the spanwise direction ($i = 2$ here) and without stratification ($N = 0$) is well documented. The homogeneous case was studied by single-point closure methods, RDT and DNS/LES. The mechanisms of stabilization and destabilization by rotation, identified in the homogeneous case, can explain what happens in rotating channels and blade cascades in turbomachinery. As a general result, the production of turbulence increases near the pressure-driven (or intrados) wall, in agreement with an anticyclonic rotation with respect to the rotation induced by the mean shear. On the opposite side, the turbulence is damped (relaminarization is even possible) near the suction side (or extrados) wall, in agreement with a cyclonic rotation.

The asymmetry in terms of cyclonic and anticyclonic spanwise system rotation is not reproduced by a basic $k - \epsilon$ model, which completely misses the Coriolis force effects. It could be recovered, however, by some more sophisticated "nonlinear" versions, which are close to algebraic stress models. Any RSM can work satisfactorily, the key point being to have an exact "production" tensor. The basic effect of production can be understood thanks to the very simple analysis based on a particle-displacement argument introduced by Bradshaw (1969) and revisited in a slightly different way by Tritton (1992).

The starting point is the following system of equations for the planar fluctuating flow, which corresponds to an oversimplified *pressure-released* inviscid RDT:

$$\frac{d}{dt}u'_1 + (S + 2\Omega)u'_3 = 0, \quad \frac{d}{dt}u'_3 + 2\Omega u'_1 = 0, \qquad (8.25)$$

the spanwise component u'_2 being constant in the same conditions. The dynamical behavior of this system is governed by the *Bradshaw number* (or *rotational Richardson number*, in order to avoid confusion with the "true" Richardson number introduced in the stably stratified shear case):

$$B = \frac{2\Omega}{S}\left(1 + \frac{2\Omega}{S}\right). \qquad (8.26)$$

An exponential growth is obtained for $B < 0$, an exponential damping for $B > 0$, and a neutral behavior is recovered for $B = 0$.

This effect of the Bradshaw number is confirmed by much more sophisticated analyses, such as RDT and even "nonhomogeneous" stability analyses in more complex flows. One of the most popular is the one by Pedley (1969), which deals with rapidly rotating pipes.

As pointed out by J. Riley (private communication; see also Yanase et al., 1992), it was almost fortuitous that this oversimplified analysis based on Eq. (8.25) gave the same criterion as the rigorous stability analysis performed by Pedley. A relevant explanation was given by Leblanc and Cambon (1997): In simplified system (8.25), u'_1 and u'_3 must be interpreted as the amplitudes of *solenoidal* disturbance modes with very high spanwise wavenumbers, which are naturally pressureless, and not as the primitive fluctuating horizontal velocity components. The term "very high spanwise wavenumber" is related to modes such that $k_2 \gg 1/L$, where L is a typical length scale of the horizontal motion and k_2 is the wavenumber of the disturbance in the direction normal to the plane of the 2D base flow.

Such modes, like $v = A(x)e^{\sigma t}e^{\imath k_2}$, with vectors A and x lying in the plane of the base flow, have dominant contributions to exponential instability with respect to all other modes. This result is valid for the stability of any 2D base flow in a rotating frame subjected to 3D disturbances. System (8.24) gives a simple illustration in "homogeneous" RDT, where $u^{(1)}$ (toroidal mode) and $u^{(2)}$ (poloidal mode) satisfy the same system as the two-component pressureless one if $k_1 = k_3 = 0, k_2 = k$. But the pure 2D contribution (no variability in the spanwise direction) is recovered at vanishing k_2, for which the Coriolis force makes no contribution.[§] A related point is that only exponential instability is governed by the Bradshaw number B whereas different dynamics are given by the parameters $Ro^{-1} = 2\Omega/S$ and $-(1 + Ro^{-1})$, even in cases having the same B, as investigated by Salhi and Cambon (1997). The simplest example is the case without rotation and the case with zero absolute vorticity ($Ro = -1$), which are very different, despite the fact that they both correspond to $B = 0$.

Looking more closely at RDT solutions, an Ince equation (1956) can be written for both the poloidal and toroidal velocity components. The simplest way is to start from the first two equations in (8.24), so that

$$\frac{1}{S^2}\frac{d}{dt}\left[k^2(t)\frac{du^{(1)}}{dt}\right] + k_2^2 B u^{(1)} = 0, \tag{8.27}$$

$$\frac{1}{S}\frac{du^{(1)}}{dt} = \left(1 + \frac{2\Omega}{S}\right)\frac{k_2}{k(t)}u^{(2)}. \tag{8.28}$$

[§] The mode of planar motion, which is relevant for explaining the stabilizing–destabilizing effect of rotation, is very close to the VSHF mode emphasized in stably stratified turbulence, replacing the vertical direction with the spanwise direction; as for the VSHF mode, it is completely different from a 2D mode.

8.2 Rotation or Stratification With Mean Shear

The first one gives the typical Ince equation and displays only the Bradshaw (or rotational Richardson) number.

Typical RDT results for the turbulent kinetic energy are shown in Fig. 8.10 and rediscussed in Section 8.4.

8.2.2 The Stratified-Shear-Flow Case

This case is much less documented than the previous one, but a good survey of available results can be found in Hanazaki and Hunt (2004) along with RDT analyses. The analogy between the two cases, rotating and stratified shear flows, was discussed by Bradshaw (1969), but only on the grounds of very simple arguments. This analysis suggested a quasi-complete analogy between the number in Eq. (8.26) (called the Richardson number in Bradshaw, 1969!) and the *true Richardson number*, which is defined by

$$Ri = \frac{N^2}{S^2} \tag{8.29}$$

in the stratified-shear-flow case.

In this case, RDT equations yield the following Ince equation for b, found by Hanazaki and Hunt (2004):

$$k^2(t)\frac{d^2\hat{b}}{dt^2} - 2Stk_1k_3(t)\frac{d\hat{b}}{dt} + (k_1^2 + k_2^2)N^2\hat{b} = 0, \tag{8.30}$$

which suggested new analytical solutions based on Legendre functions of complex order. Without shear, the time dependency of the coefficients vanishes in Eq. (8.30), and the periodic solutions are immediately recovered, with a frequency equal to the dispersion frequency of gravity waves: $\sigma = \pm N \sin\theta$. Of course, because this equation corresponds to the last one in Eq. (8.24), it can be rewritten as

$$\frac{1}{S^2}\frac{d}{dt}\left[k^2(t)\frac{d\hat{b}}{dt}\right] + (k_1^2 + k_2^2)Ri\hat{b} = 0, \tag{8.31}$$

which displays the Richardson number.

8.2.3 Analogies and Differences Between the Two Cases

When $k_1 = 0$ (this mode corresponds to an infinite streamwise wavelength) the time dependency of the wave vector vanishes, and RDT solutions with exponential growth (if $B < 0$ or $Ri < 0$) or periodic behavior (if $B > 0$ or $Ri > 0$) are immediately recovered. In the latter case, the typical frequency is the dispersion frequency of gravity waves in the stratified shear case, $\sigma = \pm N \sin\theta$, and the dispersion frequency of inertial waves in the rotating shear case, $\sigma = \pm 2\Omega \cos\theta$. In the general case ($k_1 \neq 0$), the solutions of the g_{ij} equations were found in terms of hypergeometric functions (Salhi and Cambon, 1997), and then in terms of Legendre functions of complex order (Salhi, 2002), generalizing the solutions given by Hanazaki and Hunt.

This complex order was denoted μ in the rotating case and γ in the stratified shear case (Salhi, 2002), where

$$\mu = \frac{1}{2}(-1 + \sqrt{1 - 4B \tan^2 \varphi}),$$
$$\gamma = \frac{1}{2}(-1 + \sqrt{1 - 4Ri/\cos^2 \varphi}).$$
(8.32)

According to these new RDT solutions, it can be confirmed that the exponential instability is governed only by the Bradshaw (or rotational Richardson) number, or by the Richardson number, with detailed analogy between rotating and buoyant–stratified cases. Algebraic instability, however, does not scale with either the Bradshaw or the Richardson number alone, as previously stated by comparing the rotating shear cases at Ro^{-1} and at $-(1 + Ro^{-1})$.

The general RDT solutions can be easily generalized to the shear case with both spanwise system rotation and vertical stratification (A. Salhi, private communication), and at least the solutions at $k_1 = 0$ were addressed by Kassinos, Akylas, and Langer (2006). Nevertheless, this case with three external parameters ($S, N, 2\Omega$) is less interesting than the one addressed in the next section, which gives new insight into baroclinic instability.

8.3 Shear, Rotation, and Stratification. RDT Approach to Baroclinic Instability

8.3.1 Physical Context, the Mean Flow

We now consider the same homogeneous turbulent shear flow having mean velocity in the x_1 direction in a Cartesian reference frame, but rotating with angular velocity Ω about the x_3 (vertical) axis (see Fig 8.5). The absolute mean vorticity is

$$\Omega_i^a = S\delta_{i2} + f\delta_{i3},$$
(8.33)

with $f = 2\Omega$. As in the previous section, the mean flow is subject to vertical stratification, with uniform density gradient

$$\overline{\rho} = \rho_0 - S_\rho x_3,$$
(8.34)

where ρ_0 is a constant reference density; S_ρ is chosen positive in the stabilizing case. Equivalently, one has

$$\overline{b} = -N^2 x_3.$$
(8.35)

This mean flow is not an exact solution of the Euler equations, which reduces to

$$\frac{\partial \overline{u}_i}{\partial t} + \overline{u}_j \frac{\partial \overline{u}_i}{\partial x_j} + f\epsilon_{i3j}\overline{u}_j = -\frac{1}{\rho_0}\frac{\partial \overline{p}}{\partial x_i} + \overline{b}\delta_{i3}.$$
(8.36)

8.3 Shear, Rotation, and Stratification

The corresponding equation for the vorticity is found by taking the curl of this equation, leading to

$$\frac{dW_i}{dt} + \bar{u}_j \frac{\partial W_i}{\partial x_j} - \frac{\partial \bar{u}_i}{\partial x_j}(W_j + f\delta_{j3}) = -\epsilon_{ij3}\frac{\partial \bar{b}}{\partial x_j}. \qquad (8.37)$$

This equation is satisfied by Eq. (8.33) in both vertical and spanwise directions, but not in the streamwise direction, for which

$$\frac{dW_1}{dt} - Sf = -\frac{\partial \bar{b}}{\partial x_2}. \qquad (8.38)$$

Without an additional spanwise component of the mean-density (or buoyancy) gradient, mean vorticity is created in the streamwise direction. Accordingly, to remove W_1, the following density-gradient component ought to be accounted for:

$$\frac{\partial \bar{b}}{\partial x_2} = Sf = -\underbrace{\frac{Sf}{N^2}}_{\epsilon}\frac{\partial \bar{b}}{\partial x_3} \quad \text{with} \quad N^2 = S_\rho \frac{g}{\rho_0}. \qquad (8.39)$$

In other words, the tendency for the horizontal density gradient $\frac{\partial \bar{b}}{\partial x_2}$ to generate vorticity [Eq. (8.38)] in the streamwise direction is exactly balanced by twisting the background vorticity (Sf term). This is often called the *geostrophic adjustment* in the geophysical community (Drazin and Reid, 1981).

Finally, the linearization of mass and momentum conservation equations yields

$$\frac{\partial u'_i}{\partial t} + \underbrace{Sx_3\frac{\partial u'_i}{\partial x_1} + S\delta_{i1}u_3}_{\text{shear}} + f\epsilon_{i3j}u'_j + \frac{1}{\rho_0}\frac{\partial p}{\partial x_i} = b'\delta_{i3}; \qquad \frac{\partial u'_i}{\partial x_i} = 0, \qquad (8.40)$$

$$\frac{\partial b'}{\partial t} + \underbrace{Sx_3\frac{\partial b'}{\partial x_1}}_{\text{shear}} = -N^2\left(u'_3 - \underbrace{\epsilon u'_2}_{\text{HDG}}\right). \qquad (8.41)$$

Viscous–diffusive terms are ommitted for the sake of brevity: In the b equation, the diffusivity $\kappa \nabla^2 b$ would be related to the kinematic diffusivity κ of the *stratifying agent*, salt or temperature, in an experimental or observational case (of course, the diffusion of ρ has no sense in the mass conservation equation). New terms induced by the shear are underlined. They consist of direct distortion terms (shear) and horizontal-density-gradient (HDG) effects. Equation (8.38) is a consequence of the basic flow admissibility constraint, which requires that the mean flow must be a particular solution of Euler or Helmholtz equations, as the admissibility condition ($d\mathbf{A}/dt + \mathbf{A}^2$ symmetric) in the pure kinematic nonbuoyant case with arbitrary $A_{ij}(t)$. The slope $\epsilon = \bar{b}_{,2}/\bar{b}_{,3}$ of the mean isopycnal (constant-density) surfaces with respect to the horizontal direction is due to the coupling between shear and rotation because $\epsilon = 0$ without shear or without the Coriolis force (the latter cases were addressed in the previous section).

The mean flow with three parameters S, f, and N is shown in Fig. 8.5. Two independent nondimensional numbers can be chosen among the Rossby number, the Richardson number, and the baroclinic coefficient:

$$Ro = \frac{S}{f}, \quad Ri = \frac{N^2}{S^2}, \quad \epsilon = \frac{Sf}{N^2}. \tag{8.42}$$

Let us emphasize that the present case with nonzero ϵ, which corresponds to the baroclinic instability, can be considered as a model for an important problem in meteorology, that is, the large-scale instability of the westerly winds in midlatitudes (Drazin and Reid, 1981).

8.3.2 RDT Equations

The equation for the velocity Fourier mode is derived from (8.22), only changing $2\Omega\epsilon_{n3j}$ into $f\epsilon_{n2j}$, whereas the new equation for $\hat{b}$ is found as

$$\frac{d\hat{b}}{dt} = -N^2 \left(-\epsilon \hat{u}_2 + \hat{u}_3\right). \tag{8.43}$$

For the sake of convenience, a new scaling is used to define the third buoyancy-related mode, keeping unchanged the two solenoidal modes: $u^{(3)} = (S/N^2)\hat{b}$. After some tedious algebra, a new system of equations, very similar to (8.24), is obtained:

$$\begin{pmatrix} \dot{u}^{(1)} \\ \dot{u}^{(2)} \\ \dot{u}^{(3)} \end{pmatrix} = \begin{bmatrix} 0 & \frac{k_2 + \epsilon R i k_3}{k} & 0 \\ -\epsilon R i \frac{k_3}{k} & \frac{k_1 k_3}{k^2} & -Ri \frac{k_\perp}{k} \\ -\epsilon \frac{k_1}{k_\perp} & \frac{k_\perp}{k} + \epsilon \frac{k_2 k_3}{k k_\perp} & 0 \end{bmatrix} \begin{pmatrix} u^{(1)} \\ u^{(2)} \\ u^{(3)} \end{pmatrix}. \tag{8.44}$$

A possible viscous factor modified only by mean shear at $Pr = 1$ (Prandtl number) is not recalled for the sake of brevity (Salhi and Cambon, 1997; Hanazaki and Hunt, 2004; Salhi, 2002).

Simple analytical solutions of system (8.44) are obtained when considering the $k_1 = 0$ mode that corresponds to an infinite streamwise wavelength. In this case, the wave vector is no longer time dependent, because the shear advection vanishes. Accordingly, the coefficients of the system of RDT are constant, and analytical solutions are easily found.

For $k_1 = 0$, one obtains

$$\dot{u}^{(1)} = \frac{k_2 + \epsilon R i k_3}{k} u^{(2)}, \tag{8.45}$$

$$\dot{u}^{(2)} = Ri \left[\frac{\epsilon k_3}{k} u^{(1)} - \frac{k_\perp}{k} u^{(3)} \right], \tag{8.46}$$

$$\dot{u}^{(3)} = \left(\frac{k_\perp}{k} + \epsilon \frac{k_2 k_3}{k k_\perp} \right) u^{(2)}, \tag{8.47}$$

8.4 Elliptical Flow Instability From "Homogeneous" RDT

leading to the following equation for the poloidal component:

$$\ddot{u}^2 + \underbrace{Ri\left[\left(\epsilon\frac{k_3}{k} + \frac{k_2}{k}\right)^2 - (1-Ri)\epsilon^2\frac{k_3^2}{k^2}\right]}_{\omega_0^2} u^{(2)} = 0. \tag{8.48}$$

These solutions exhibit an oscillating behavior (stable case) when $\omega_0^2 > 0$, an exponential growth (unstable case) whenever $\omega_0^2 < 0$, and a linear (algebraic) growth if $\omega_0 = 0$, where

$$\omega_0^2 = \left[\left(\frac{\cos\theta}{R_0} + \sin\theta\right)^2 - (1-Ri)\sin^2\theta\right]. \tag{8.49}$$

Neutral curves drawn in the $[Ri, \theta = \widehat{(\mathbf{k}, \mathbf{n})}]$ plane for $k_1 = 0$ for different values of Ro (left) and ϵ (right) are displayed in Fig. 8.6. For the latter case, a zoom is made on small values of ϵ, which are more relevant for geophysical applications.

It is shown that the threefold coupling among shear, rotation, and stratification allows one to extend the band of instability until $Ri = 1$. Without system rotation, the instability essentially concerns negative values of the Richardson number and is limited by rather small positive values of Ri: $Ri \sim 0.1$ from RDT, DNS, and LES studies, and $Ri = 1/4$ is recovered from the analysis of Miles (1961).¶

About the occurrence of baroclininic instability in the geophysical context, the pioneering approach by Eady (1949) seems radically different at first glance, but numerical solutions of the general RDT equations at $k_1 \neq 0$ yield amplification rates that are comparable with those found by Eady for small values of the parameter ϵ (such values are illustrated in the bottom part of Fig. 8.6). Typical DNS results, carried out for extending the RDT results, are shown in Fig. 8.7.

8.4 Elliptical Flow Instability From "Homogeneous" RDT

This instability is very generic and occurs in many flow configurations. The reader is referred to Kerswell (2002) for a detailed review. A sudden interest arose when Pierrehumbert discovered its characteristic properties by a conventional normal-mode analysis approach, whereas at the same time Bayly (1986) found the same results using a much simpler and more elegant method, which is essentially equivalent to RDT.

Ellipticity in the core of large vortices is very general. It can originate in the mutual interaction of adjacent vortices, whereas an isolated vortex can remain circular. As proposed by the authors previously mentioned, it is not necessary to study the stability of a pair of vortices, but rather one should study the stability of a single vortex, getting rid of the mutual-interaction origin of the ellipticity.** One just has

---

¶ The stability analysis of Miles, however, is different, because it accounts for a possible inflexion point of the mean-shear profile for a horizontal slab limited by two horizontal walls.

** A similar reasoning is made when the mean shear is considered *a priori*, getting rid of its origin, like solid-wall effects.

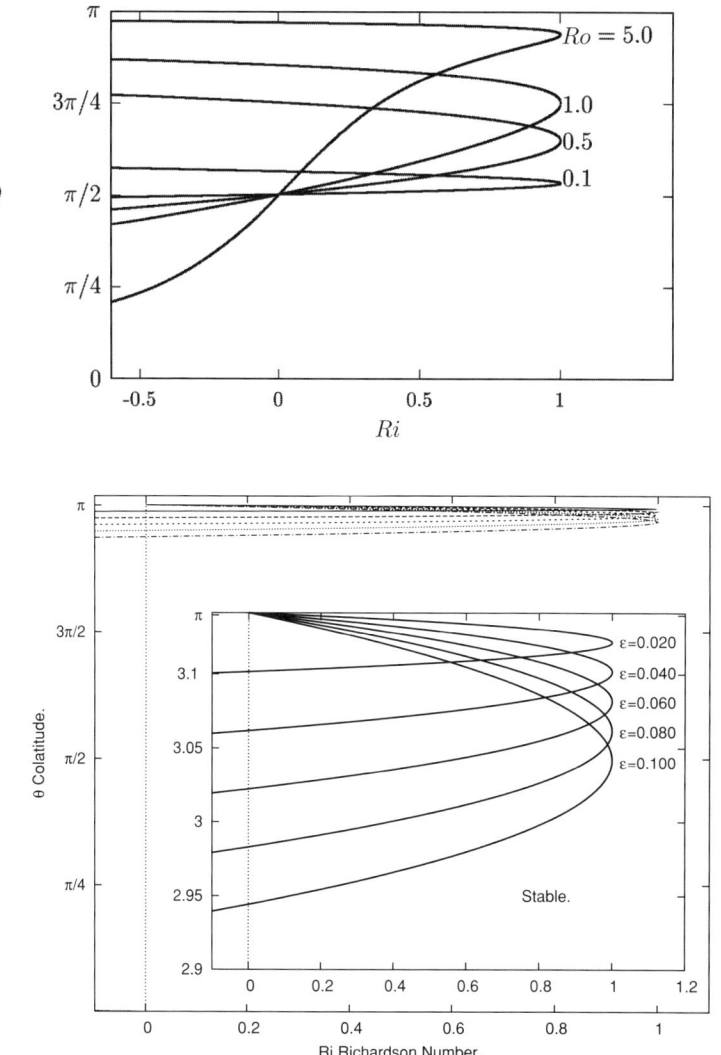

Figure 8.6. Neutral curves (exponentially instable zone are in the part of the surface delineated by the concave side of the curve), or $\omega_0 = 0$, for $k_1 = 0$, at different values of ϵ (bottom) and Ro (top).

to assume, in addition, that the typical wavelength of the instability is small with respect to the dimension of the core of the vortex. In this sense, elliptical instability is a local instability in actual flows, in contrast to cooperative instabilities (e.g., the Crow instability in a vortex pair) that involve the whole pattern of adjacent eddies.

Going back to RDT, one has to imagine that the mean flow given by Eq. (5.7) with $S < \Omega$ represents an infinite elliptical eddy. In this case, no mean length scale may appear, and dynamics of initial disturbances depends only on the orientation, but not on the modulus, of their wave vector. The effect of a viscous cutoff in viscous RDT can be easily accounted for, but it is not discussed here for the sake of brevity. Of course, it is more realistic to consider that the elliptic core has a finite size and that RDT (or WKB RDT, or short-wave asymptotics) is valid only for

8.4 Elliptical Flow Instability From "Homogeneous" RDT

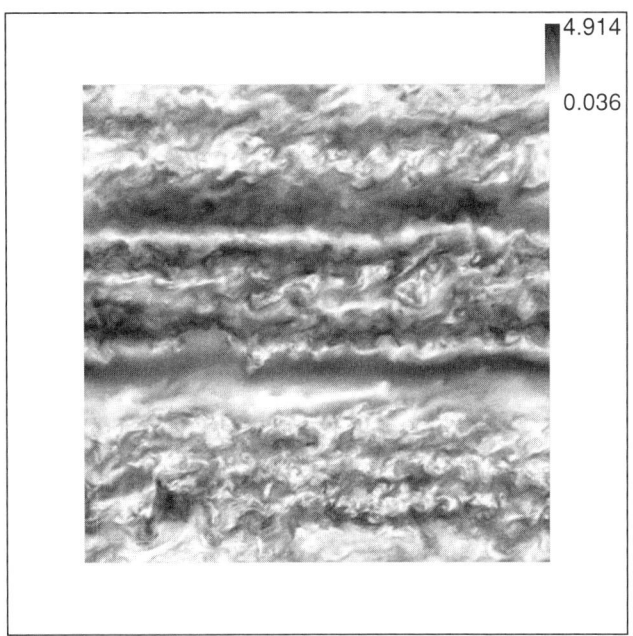

Figure 8.7. Isosurfaces for the buoyancy fluctuation in the zonal spanwise-vertical plane, from DNS. $\epsilon = 0.2$, $Ri = 0.99$, $Re_\lambda(t=0) = 66$. Courtesy of G. Simon.

disturbances with wavelengths much smaller than this size, but equations are essentially the same.

RDT calculations were carried out by Cambon (1982) and Cambon, Teissèdre, and Jeandel (1985) for $S < \Omega$, and the results foreshadowed the ones by Bayly. This study contributed to motivating the experimental study by Leuchter and coworkers, with the design of a very complex distorting duct capable of reproducing an elliptical-flow case with $S = \Omega/2$, as discussed in Chapter 5. Observation of a clear elliptical-flow instability was problematic, given the limited length of the duct. The emphasis was put on the complex evolution of Reynolds stress components, related spectra, and integral length scales, for statistical modeling purposes. To avoid confusion with the notation used in this chapter, the strain rate is given as D (and not S, kept for the shear rate only) and the vorticity of the elliptical eddy is given as W (and not 2Ω, kept for the system vorticity in the rotating frame). Expressed in terms of the solenoidal modes $u^{(1)}$ and $u^{(2)}$, the general RDT equations are

$$\begin{pmatrix} \dot{u}^{(1)} \\ \dot{u}^{(2)} \end{pmatrix} = \begin{bmatrix} 2aD & -\frac{k_3}{k}(W+2\Omega) \\ 2bD + \frac{k_3}{k}(W+2\Omega) & \frac{k}{k} - aD \end{bmatrix} \begin{pmatrix} u^{(1)} \\ u^{(2)} \end{pmatrix}, \quad (8.50)$$

with $a = e_1^{(1)} e_1^{(1)}$ and $b = e_1^{(2)} e_2^{(1)} + e_2^{(2)} e_1^{(1)}$ (Cambon, 1982; Cambon et al., 1994), choosing the axial vector $\boldsymbol{n}$ along the direction of mean vorticity ($\boldsymbol{n}$ normal to the plane of the 2D mean flow here). Choosing $\boldsymbol{n}$ in the (cross-gradient) direction of the shear is interesting too, but not discussed here (see Salhi, Cambon, and Speziale, 1997). For the sake of convenience, an additional Coriolis effect is accounted for in the previous equation, the case of the basic elliptical instability in a Galilean frame

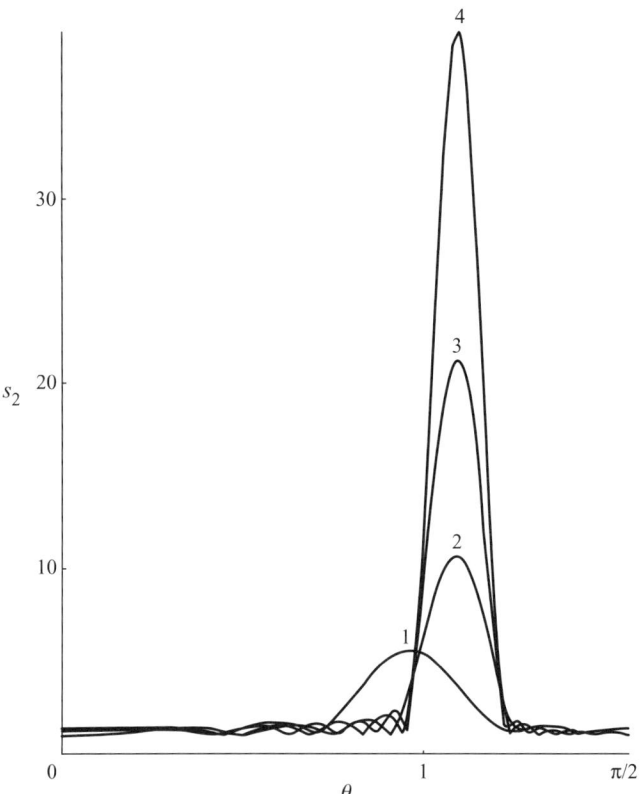

Figure 8.8. Figure reproduced from Cambon's thesis (1982) in Godeferd, Cambon, and Leblanc (2001): Maximum eigenvalue of the symmetric Green's function $\mathbf{G}\tilde{\mathbf{G}}$ as a function of the direction θ of the wave vector measured from the polar axis (taken at a period $t = 2\pi/\sigma_0$, this eigenvalue differs from the actual Floquet parameter only because of the use of a symmetric form of the Green's matrix). Curves labeled 1, 2, 3, 4, are obtained at times $t/T = 1.3, 2, 2.5, 3$ in terms of the period $T = 2\pi/\sigma_0$, $W/(2D) = 3$.

being recovered for $\Omega = 0$ and $D < W/2$. As a very simple term, *the matrix exhibits the projection of the absolute vorticity $2\mathbf{\Omega} + \mathbf{W}$ onto the wave vector, as the unique contribution from rotational terms.*

Analytically solving this system of two equations is difficult because $\mathbf{k}$ is periodic in time according to Eq. (5.27), so that all coefficients a, b, and $\dot{k}/k$ are periodic too. Computation of the Green's functions g and $\mathbf{G}$ must therefore be performed through numerical integration. The main result is displayed in Fig. 8.8.

This figure has the merit of suggesting the mechanism of *resonant amplification* by periodic forcing, which is more accurately captured by Bayly (1986), who used a Floquet analysis.

The Floquet analysis takes advantage of the fact that the coefficients in the linear system of equations are periodic with a frequency of

$$\sigma_0 = \sqrt{\left(\frac{W}{2}\right)^2 - D^2}. \tag{8.51}$$

8.4 Elliptical Flow Instability From "Homogeneous" RDT

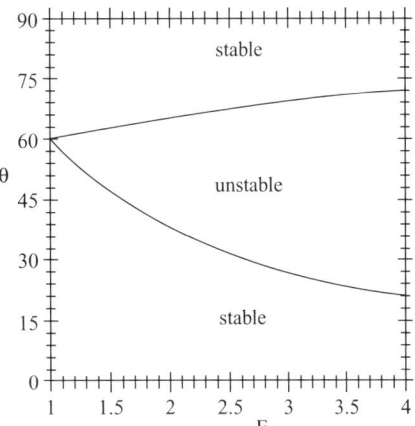

Figure 8.9. Typical instability band in the θ–E plane, $E = (W/2 - D)/(W/2 + D)$ being the ellipticity; from Salhi, Cambon, and Speziale (1997) with permission of the American Institute of Physics.

It is therefore possible to compute the Green's function for only a single period $T = 2\pi/\sigma_0$ and to extract its eigenvalues once and for all. The Green's function given by an arbitrary number p of periods is obtained by simply calculating the power p of the one-period matrix. Exponentiation of the one-period matrix amounts to exponentiation of its eigenvalues, so that the amplification rate is easily calculated from a single-period run.

The instability band was found to correspond to

$$W\frac{k_3}{k} \sim \pm W/2$$

at very small D: One recovers on the left-hand side the intrinsic frequency of inertial waves, which is also the dominant term in Eq. (8.50) at small D, whereas the right-hand side is the frequency of the external forcing following mean elliptical streamlines at small D. One can imagine a scheme in which the wave-vector direction describes a periodic trajectory (with frequency $W/2$ at vanishing D), whereas the Fourier component rotates in the plane normal to it with frequency $W \cos\theta = W k_3/k$: Resonance is found when $\cos\theta = \pm 1/2$. The subharmonic conditions with $\cos\theta$ given by a rational number other than $\pm 1/2$ yield no significant amplification here. The location of the instability peak near $(k_3/k) = \cos\theta = \pm 1/2$ in Fig. 8.8 is therefore explained. The maximum growth rate at leading order in terms of D/W is found as

$$\sigma = \frac{9}{16}D$$

for the particular wavenumber orientation $\cos\theta = 1/2$. On time average, this means that the vorticity $\hat{\omega}$ aligns itself with the underlying stretching direction.

Using the rigorous Floquet analysis, it is shown how the instability band, which emanates from the point $\theta = \pi/3$ at vanishing ellipticity, expands at larger ellipticities (see Fig. 8.9).

Among a lot of results not given here for the sake of brevity, one can mention the analytical study of an Ince equation by Waleffe (1990). One can obtain such an equation by deriving a single second-order ODE from system (8.50), as we have

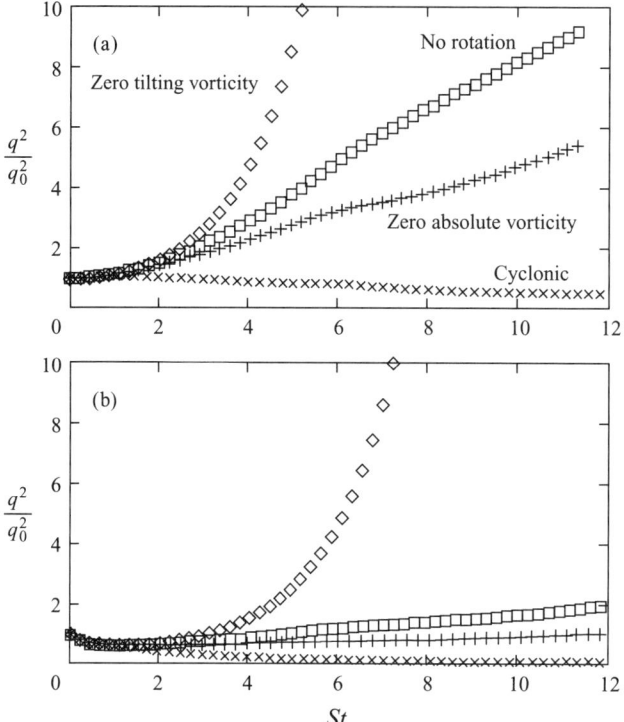

Figure 8.10. (a) Inviscid and (b) viscous RDT results for the four cases $R = 2\Omega/W$, pure shear flow in the rotating frame. Reproduced from Cambon et al. (1994) with permission of CUP.

seen in other examples in the previous sections. Transition to turbulence was further investigated using LES/DNS by Lundgren and Mansour (1996).

Let us also mention the shift of the instability band when the elliptical flow is seen in a reference frame rotating at angular velocity Ω (Craik, 1989): It is simply found by using

$$(2\Omega + W)\frac{k_3}{k} \sim \pm W/2.$$

This illustrates that the system vorticity and the relative (mean) vorticity do not act in the same way: They are simply added in the left-hand side, displaying the absolute vorticity in the dispersion frequency, but W is kept on the right-hand side. Typical results from inviscid and viscous RDTs for the pure shear flow are shown in Fig. 8.10 and for the rotating elliptical flow in Fig. 8.11. As shown by Cambon et al. (1994) and Salhi, Cambon, and Speziale (1997), four flow cases are very relevant among all possible 2D mean flows with rectilinear $D = W/2$, hyperbolical $D > W/2$, or elliptical $D < W/2$ streamlines in the frame rotating with angular velocity Ω:

- the reference case without system rotation,
- the case with zero-mean *tilting vorticity*, $2\Omega + W/2 = 0$, which gives always the most destabilizing one,

8.5 Axisymmetric Strain With Rotation

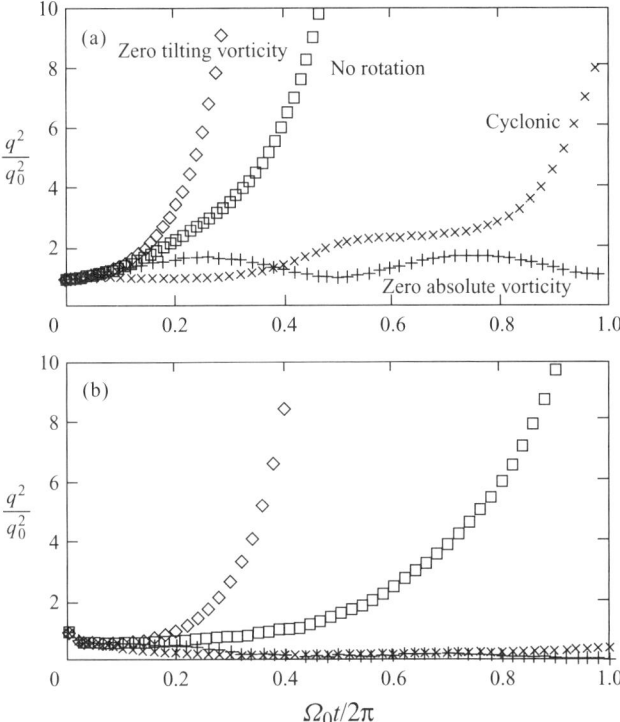

Figure 8.11. (a) Inviscid and (b) viscous RDT results for the four cases, $R = 2\Omega/W$, elliptical-flow case with $D = W/4$ in the rotating frame. Reproduced from Cambon et al. (1994) with permission of CUP.

- the case with zero-mean absolute vorticity, $2\Omega + W = 0$, which is always the only unconditionally stable one, even if subject to some algebraic growth (see also Craik, 1989),
- a cyclonic case, with $2\Omega = W/2$.

Viscous RDT is given here only as a reference, in order to show the effect of a viscous cutoff, and therefore to select only the most robust exponential growths.

8.5 Axisymmetric Strain With Rotation

Axial strain, such as the one obtained near the centerline of a convergent axisymmetric duct, was addressed in Chapter 5. An interesting case was obtained by Leuchter and Dupeuble (1993), when adding an axisymmetric convergent duct after the generator of rotation illustrated in Fig. 4.3. In close connection with what was observed when rapid rotation is suddenly applied to axisymmetric initial data (see Fig. 4.9), the anisotropy of the flow is dramatically modified. This effect is completely missed by any RSM, as it is linked to the selective rapid damping of polarization anisotropy by phase mixing, the directional anisotropy being conserved. This linear

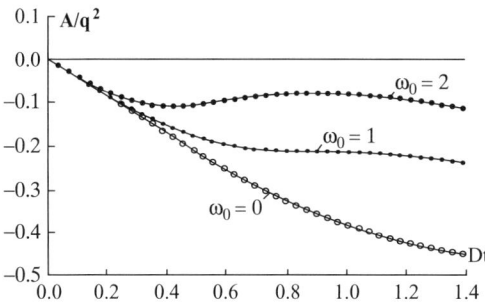

Figure 8.12. A-parameter (which characterizes the RST anisotropy in the case of axial symmetry) with $q^2 = 2\mathcal{K}$. ω_0 is the rotation-to-strain-rate ratio. RDT calculation, data taken from Leuchter and Cambon (1997). Courtesy of O. Leuchter.

effect can be reproduced by the structure-based model of Kassinos and Reynolds, as discussed in Cambon and Scott (1999).[††]

A more sophisticated study was carried out by Leuchter and Cambon (1997), with access to both linear and subtle nonlinear effects of rotation in the straining geometry, using RDT, DNS, and EDQNM2. The anisotropy parameter $A/(2\mathcal{K})$, with

$$A = b_{33} - (b_{11} + b_{22})/2,$$

is plotted in Figs. 8.12 and 8.13. This parameter is the unique one needed to characterize b_{ij} with axial symmetry, because b_{ij} is diagonal with $b_{11} = b_{22} = -b_{33}/2$. A RSM can reproduce the rise of A with negative value because of the axisymmetric strain, reflecting the rise of a "pancake-type" RST $b_{11} = b_{22} > 0$, but not the effect of additional rotation, which partly counterbalances this production.

The case of axial compression with rotation can be considered as another case of compressed turbulence, as discussed in Section 10.3. If the compression is periodic – this could be illustrated in a reciprocating engine with swirl – specific instabilities can be shown by use of homogeneous RDT, very close to the elliptical-flow case. After a first numerical RDT computation by C. Cambon, a complete Floquet's analysis was performed by Mansour and Lundgren (1990), showing different bands of instability similar to the one displayed in Fig. 8.9. A mechanism of parametric resonance of inertial waves by the external periodic compression is displayed. This mechanism was further investigated in the more realistic configuration of an axially rotating cylindrical vessel, with small-amplitude periodic compression (Duguet, Scott, and Le Penven, 2005).

8.6 Relevance of RDT and WKB RDT Variants for Analysis of Classical Instabilities

The short-wave asymptotic theory introduced by Lifschitz and Hameiri (1991) is presented in Chapter 13. It can be seen as a WKB variant of RDT. It is used in the following discussion for identifying localized elliptic, centrifugal, and hyperbolic instabilities (Godeferd, Cambon, and Leblanc, 2001).

[††] The damping of polarization anisotropy was referred to as "rotational randomization" in the structure-based model.

8.6 Relevance of RDT and WKB RDT Variants

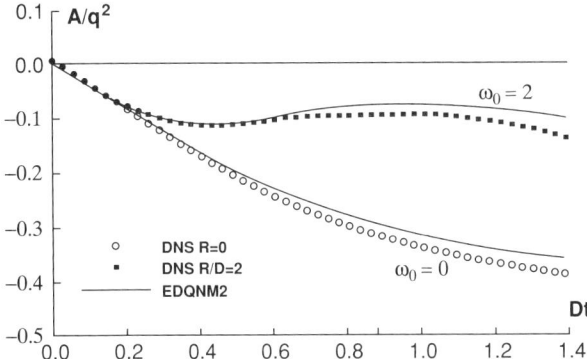

Figure 8.13. Same quantities and parameters as in Fig. 8.12. DNS and EDQNM2 results, data taken from Leuchter and Cambon (1997). Courtesy of O. Leuchter.

The Coriolis force alters the stability of 2D vortex flows subjected to 3D disturbances. As an illustration, it is possible to consider 2D base flows more complex than those of homogeneous RDTs illustrated in Fig. 2.1. For instance, the Taylor–Green flow in a rectangular cell (see Fig. 8.14) has an elliptic point in the core of each eddy and a hyperbolic point in the corner of the four cells. The Stuart flow (see Fig. 8.15) is elliptic in the core region with hyperbolic points inserted between adjacent vortices (only a single vortex is shown, but one has to consider periodicity in the horizontal direction). The stability of these flows can be revisited in a rotating frame, using the short-wave WKB theory developed by Lifschitz and Hameiri (1991), which amounts to a zonal RDT analysis. Such an analysis allows for the identification of the role of elliptic and hyperbolic points in 3D instabilities altered by system rotation, but also to capture the centrifugal instability that may affect anticyclonic vortices (Sipp, Lauga, and Jacquin, 1999; Godeferd, Cambon, and Leblanc, 2001). The three kinds of instability and their possible competition were studied by solving the Townsend (or Kelvin–Townsend) equations along different trajectories. For each closed trajectory, a temporal Floquet parameter can be calculated from the zonal RDT Green's function. This parameter, denoted by $\sigma(x_0, \theta)$, depends on both the space coordinate x_0, which labels the trajectory, and on the angle θ, which gives the orientation of the wave vector. A typical pattern of $\sigma(x_0, \theta)$ in the case of the rotating Stuart flow is shown in Fig. 8.16. The dominant instability is the centrifugal

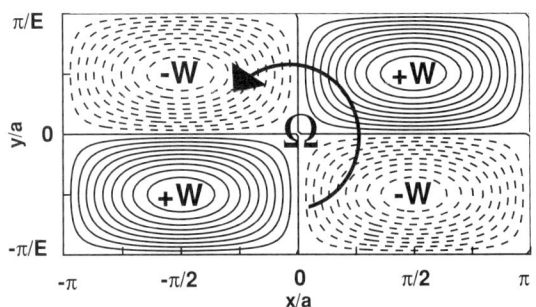

Figure 8.14. The Taylor–Green flow: isovalues of the vorticity. Case $E = 2$. Reproduced from Sipp, Lauga, and Jacquin (1999) with permission of AIP.

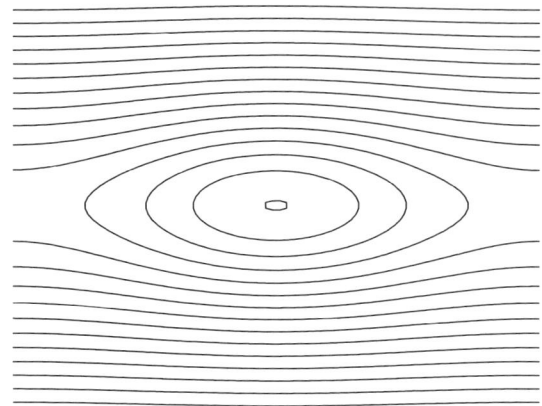

Figure 8.15. The Stuart flow. Isovalues of the stream function. Core ellipticity parameter $\rho = 1/3$.

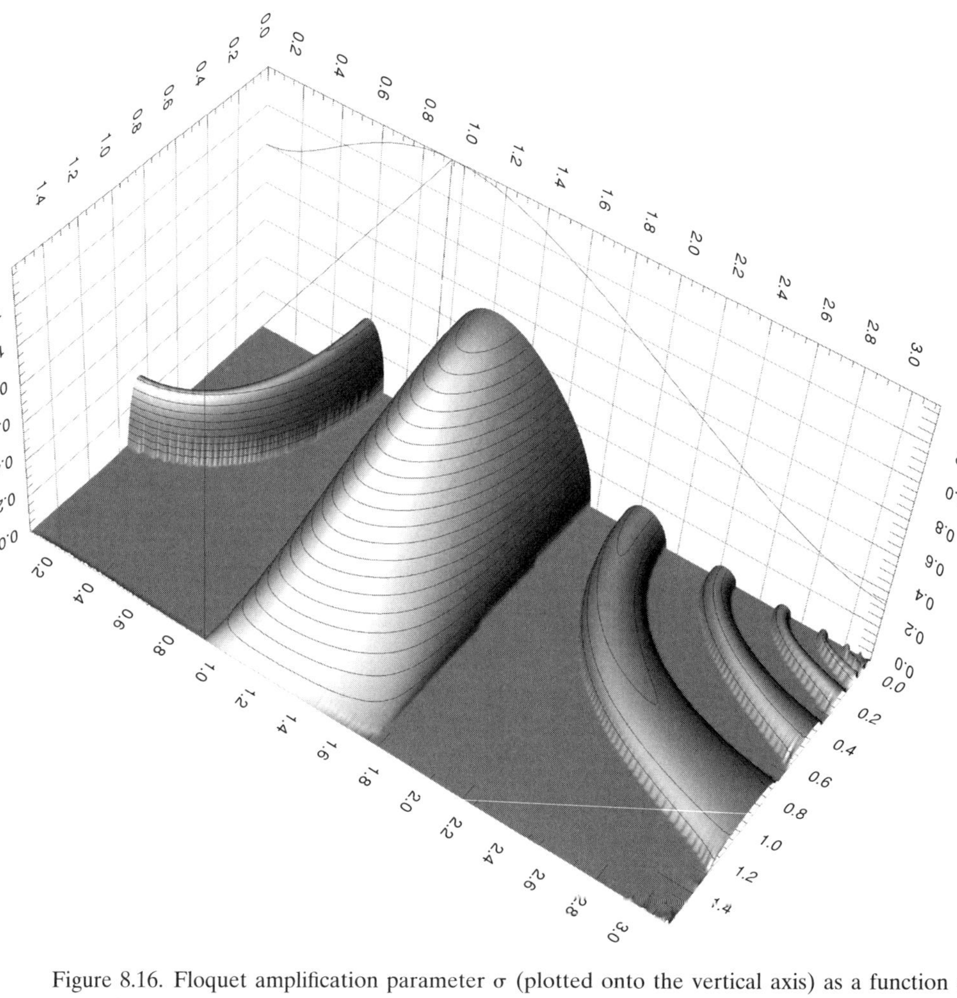

Figure 8.16. Floquet amplification parameter σ (plotted onto the vertical axis) as a function of the trajectory, indexed by the position x_0 (on the left), and the orientation θ (on the right) of the local wave vector to the spanwise (normal to the plane of the base flow) axis. x_0 varies from 0 (core) to π (periphery), and θ varies from 0 (pure spanwise modes) to $\pi/2$ (pure 2D modes). Anticyclonic system rotation: The dimensionless vorticity at the core ($x_0 = 0$) of a Stuart cell is -7 and the related Rossby number is -5. Reproduced from Godeferd, Cambon, and Leblanc (2001) with permission of CUP.

8.6 Relevance of RDT and WKB RDT Variants

one in the particular anticyclonic case illustrated here. In addition, a typical elliptical instability branch emanates from the core (left-hand part of the figure). It should be pointed out that the local WKB RDT for short-wave disturbances can provide real insight into the nature of instabilities – e.g., elliptical, hyperbolic, and centrifugal – that occur in nonparallel flow with and without system rotation. Classical massive eigenvalue problems provide little or no such insight. In such studies, for instance by Peltier and co-workers, different kinds of instabilities, called "core," "braid," and "edge," were identified, but local analysis allowed us to substitute "elliptical," "hyperbolical," and "centrifugal" into this terminology.

Finally, it is perhaps worthwhile to mention the instabilities that cannot be captured by RDT (possibly extended toward WKB RDT). A very important one is the inflexional shear instability. Even if every point of a linear profile is an inflexional point, the case of change of concavity cannot be recognized by RDT as such. The reason is that only the gradient of the mean velocity is accounted for in the local theory, but not the curvature. A related point is that the inflexional instability in actual flows is not a short-wave, very local, one.

The case of the baroclinic instability is more subtle and surprising. On the one hand, the baroclinic instability is a long-wave instability in many geophysical contexts; on the other hand, we hope that the specific section in this chapter could show a possible relevance of homogeneous and WKB RDT, and at least will generate a debate in the geophysical community.

Bibliography

BARTELLO, P. (1995). Geostrophic adjustment and inverse cascades in rotating stratified turbulence, *J. Atmos. Sci.* **52**, 4410–4428.

BAYLY, B. J. (1986). Three-dimensional instability of elliptical flow, *Phys. Rev. Lett.* **57**, 2160–2163.

BAYLY, B. J., HOLM, D. D., AND LIFSCHITZ, A. (1996). Three-dimensional stability of elliptical vortex columns in external strain flows, *Philos. Trans. R. Soc. London Ser. A* **354**, 895–926.

BRADSHAW, P. (1969). The analogy between streamline curvature and buoyancy in turbulent shear flow, *J. Fluid Mech.* **36**, 177–191.

CAMBON, C. (1982). Etude spectrale d'un champ turbulent incompressible soumis à des effets couplés de déformation et rotation imposés extérieurement, Thèse de Doctorat d'Etat, Université Lyon I, France.

CAMBON, C. (2001). Turbulence and vortex structures in rotating and stratified flows, *Eur. J. Mech. B (Fluids)* **20**, 489–510.

CAMBON, C., BENOIT, J. P., SHAO, L., JACQUIN, L. (1994). Stability analysis and LES of rotating turbulence with organized eddies, *J. Fluid Mech.* **278**, 175–200.

CAMBON, C., GODEFERD, F. S., AND KANEDA, Y. (2007). Phase-mixing and toroidal cascade in rotating and stratified flows, presented at the *18th Congrès Français de Mécanique*, Special International Session on stratified rotating flows, Grenoble, August 27–31. AFM Symp., Grenoble, Int. session on rotating stratified flows, August 2007.

CAMBON, C., GODEFERD, F. S., NICOLLEAU, F., AND VASSILICOS, J. C. (2004). Turbulent diffusion in rapidly rotating flows with and without stable stratification, *J. Fluid Mech.* **499**, 231–255.

CAMBON, C. AND SCOTT, J. F. (1999). Linear and nonlinear models of anisotropic turbulence, *Annu. Rev. Fluid Mech.* **31**, 1 53.

CAMBON, C., TEISSÈDRE, C., AND JEANDEL, D. (1985). Etude d' ettets couplés de rotation et de déformation sur une turbulence homogène, *J. Méc. Théor. Appl.* **5**, 629–657.

CHARNEY, J. G. (1971). Geostrophic turbulence, *J. Atmos. Sci.* **28**, 1087–1095.

CRAIK, A. D. D. (1989). The stability of unbounded two- and three-dimensional flows subject to body forces: some exact solutions, *J. Fluid Mech.* **198**, 275–292.

DRAZIN, P. G. AND REID, W. H. (1981). *Hydrodynamic Stability*, Cambridge University Press.

DUGUET, Y., SCOTT, J. F., AND LE PENVEN, L. (2005). Instability inside a rotating gas cylinder subject to axial periodic strain, *Phys. Fluids* **17**, 114103.

EADY, E. T. (1949). Long waves and cyclonic waves, *Tellus* **1**, 33–52.

GODEFERD, F. S. AND CAMBON, C. (1994). Detailed investigation of energy transfers in homogeneous stratified turbulence, *Phys. Fluids* **6**, 2084–2100.

GODEFERD, F. S., CAMBON, C., AND LEBLANC, S. (2001). Zonal approach to centrifugal, elliptic and hyperbolic instabilities in Suart vortices with external rotation, *J. Fluid Mech.* **449**, 1–37.

GOLDSTEIN, M. E. (1978). Unsteady vortical and entropic distortions of potential flows round arbitrary obstacles, *J. Fluid Mech.* **89**, 433–468.

HANAZAKI, H. AND HUNT, J. C. R. (2004). Linear processes in unsteady stably stratified sheared turbulence with mean shear, *J. Fluid Mech.* **507**, 1–42.

HERRING, J. R. (1980). Statistical theory of quasi-geostrophic turbulence, *J. Atmos. Sci.* **37**, 969–977.

KANEDA, Y. (1981). Renormalized expansions in the theory of turbulence with the use of the Lagrangian position function, *J. Fluid Mech.* **107**, 131–145.

KANEDA, Y. (2000). Single-particle diffusion in strongly stratified and/or rapidly rotating turbulence, *J. Phys. Soc. Jpn.* **69**, 3847–3852.

KASSINOS, S. C., AKYLAS, E., AND LANGER, C. A. (2007). Rapidly sheared homogeneous stratified turbulence in a rotating frame, *Phys. Fluids* **19**, 021701.

KASSINOS, S. C., REYNOLDS, W. C., AND ROGERS, M. M. (2001). One-point turbulence structure tensors, *J. Fluid Mech.* **428**, 213–248.

KERSWELL, R. R. (2002). Elliptical instability, *Annu. Rev. Fluid Mech.* **34**, 83–113.

KURIAN, S., SMITH, L., AND WINGATE, B. (2006). On the two-point correlation of potential vorticity in rotating and stratified turbulence, *J. Fluid Mech.* **555**, 131–140.

LEBLANC, S. AND CAMBON, C. (1997). On the three-dimensional instabilities of plane flows subjected to Coriolis force, *Phys. Fluids* **9**, 1307–1316.

LEUCHTER, O. AND CAMBON, C. (1997). EDQNM and DNS predictions of rotation effects in strained axisymmetric turbulence, presented at the *11th International Symposium on Turbulent Shear Flow*, Grenoble, September 8–10, 1997.

LEUCHTER, O. AND DUPEUBLE, A. (1993). Rotating homogeneous turbulence subjected to an axisymmetric contraction, presented at the *9th International Symposium on Turbulent Shear Flows*, Kyoto, Japan, August 16–18, 1993.

LIECHTENSTEIN, L., GODEFERD, F. S., AND CAMBON, C. (2005). Nonlinear formation of structures in rotating stratified turbulence, *J. Turbulence* **6**, 1–18.

LIFSCHITZ, A. AND HAMEIRI, E. (1991). Local stability conditions in fluid dynamics, *Phys. Fluids A* **3**, 2644–2641.

LUNDGREN, T. S. AND MANSOUR, N. N. (1996). Transition to turbulence in an elliptic vortex, *J. Fluid Mech.* **307**, 43–62.

MANSOUR, N. N. AND LUNDGREN, T. S. (1990). Three-dimensional instability of rotating flows with oscillating axial strain, *Phys. Fluids A* **2**, 2089–2091.

MILES, J. W. (1961). On the stability of heterogeneous shear flows, *J. Fluid Mech.* **10**, 496–508.

PEDLEY, T. J. (1969). On the stability of viscous flow in a rapidly rotating pipe, *J. Fluid Mech.* **35**, 97–115.

PEDLOWSKY, J. (1987). *Geophysical Fluid Dynamics*, Springer-Verlag.

PRAUD, O., SOMMERIA, J., AND FINCHAM, A. (2006). Decaying grid turbulence in a rotating stratified fluid, *J. Fluid Mech.* **547**, 389–412.

SALHI, A. (2002). Similarities between rotation and stratification effects on homogeneous shear flow, *Theor. Comput. Fluid Dyn.* **15**, 339–358.

SALHI, A. AND CAMBON, C. (1997). An analysis of rotating shear flow using linear theory and DNS and LES results, *J. Fluid Mech.*, **347**, 171–195.

SALHI, A. AND CAMBON, C. (2006). Advance in rapid distortion theory: From rotating shear to the baroclinic instability, *J. Appl. Mech.* **73**, 449–460.

SALHI, A., CAMBON C., AND SPEZIALE, C. G. (1997). Linear stability analysis of plane quadratic flows in a rotating frame, *Phys. Fluids* **9**, 2300–2309.

SIPP, D., LAUGA, E., AND JACQUIN, L. (1999). Vortices in rotating systems: Centrifugal, elliptic and hyperbolic type instabilities, *Phys. Fluids* **11**, 3716–3728.

SMITH, L. M. AND WALEFFE, F. (2002). Generation of slow, large scales in forced rotating, stratified turbulence, *J. Fluid Mech.* **451**, 145–168.

STAQUET, C. AND RILEY, J. J. (1989). On the velocity field associated to potential vorticity, *Dyn. Atmos. Ocean* **14**, 93–123.

TRITTON, D. J. (1992). Stabilization and destabilization of turbulent shear flow in a rotating fluid, *J. Fluid Mech.* **241**, 503–523.

WALEFFE, F. (1990). On the three-dimensional instability of strained vortices, *Phys. Fluid A*, **2**, 76–80.

YANASE, S., FLORES, C., MÉTAIS, O., AND RILEY, J. J. (1992). Rotating free shear flow I: Linear stability analysis, *Phys. Fluids A* **5**, 2725–2737.

9 Compressible Homogeneous Isotropic Turbulence

9.1 Introduction to Modal Decomposition of Turbulent Fluctuations

9.1.1 Statement of the Problem

A natural question that arises when dealing with compressible turbulent flows is this: How does one characterize the compressibility effects on turbulence? Or, in an equivalent way, what are the differences between the compressible turbulent fluctuations and the incompressible ones? To answer this question, it is first important to remark that in incompressible flows the full solution is contained in the sole velocity field because the pressure is nothing but an enslaved Lagrange multiplier. In compressible turbulence, this is no longer true because pressure is now an autonomous variable and at least one additional physical variable is required for describing the solution.* The basic governing equations for such flows are

$$\frac{\partial \rho}{\partial t} + \nabla \cdot (\rho \boldsymbol{u}) = m, \tag{9.1}$$

$$\rho \left(\frac{\partial \boldsymbol{u}}{\partial t} + \boldsymbol{u} \nabla \boldsymbol{u} \right) = -\nabla p - \frac{2}{3} \nabla (\mu \nabla \cdot \boldsymbol{u}) + \nabla \cdot (\mu \mathbf{S}) + \rho \boldsymbol{f}, \tag{9.2}$$

$$\frac{p}{R} \left(\frac{\partial s}{\partial t} + \boldsymbol{u} \cdot \nabla s \right) = \nabla \cdot (\kappa \nabla T) + \mu \left[\frac{1}{2} \mathbf{S} : \mathbf{S} - \frac{2}{3} (\nabla \cdot \boldsymbol{u})^2 \right] + Q, \tag{9.3}$$

where R, p, ρ, T, $\boldsymbol{u}$, s, μ, and κ denote the perfect gas constant, pressure, density, temperature, velocity, entropy, coefficients of viscosity, and heat conduction, respectively. The additional variables m, $\boldsymbol{f}$, and Q are related to the rate of mass injection per unit volume, the body force per unit mass, and the rate of heat addition per unit volume, respectively. Both the viscosity and the heat conduction are assumed to be monotonic functions of the temperature, i.e., $\mu = \mu(T)$ and $\kappa = \kappa(T)$. The system is supplemented by the perfect gas law

$$p = \rho RT \tag{9.4}$$

---

* The discussion in the present book is restricted to the case of single-phase, nonreactive, single-species flows of divariant fluids.

and the definition of the entropy

$$s - s_r = c_v \log\left[\left(\frac{p}{p_r}\right)\left(\frac{\rho_r}{\rho}\right)\right], \quad (9.5)$$

where s_r, p_r, and ρ_r are related to a reference state.

A common way to solve this problem is to try to decompose the observed fluctuations as the sum of a compressible part and an incompressible one, the latter being very often understood as the part of the solution that fulfills the incompressible Navier–Stokes equations, the former being defined as the difference between the full solution and the incompressible part. Unfortunately, no fully general decomposition based on this approach leading to tractable and useful analysis has been proposed up to now. A main reason for that is that such a decomposition does not explicitly distinguish between acoustic waves and other compressible phenomena.

To remedy this problem and to provide a meaningful and powerful decomposition of compressible fluctuations into physical modes, Kovasznay proposed a small-parameter expansion discussed in the next subsection, which is based on the assumption that the turbulent fluctuations will be small in some sense with respect to a uniform mean flow. As will be subsequently seen, this decomposition provides us with much meaningful information, but its validity is restricted because it relies on a linearized theory. To handle flows in which nonlinear mechanisms are dominant, another approach consists of using the exact Helmholtz decomposition of the compressible velocity field. Because this decomposition does not rely on any assumption dealing with the amplitude of the turbulent fluctuations, it is valid in all flows. But its weakness is that it does not allow a direct splitting of other flow variables like density, pressure, or entropy. Therefore it must be supplemented with arbitrary splitting procedures for these variables (Subsection 9.1.4).

9.1.2 Kovasznay's Linear Decomposition

The first step in Kovasznay's approach (Kovasznay, 1953) consists of expanding the turbulent field as

$$\boldsymbol{u} = \boldsymbol{u}_0 + \epsilon \boldsymbol{u}_1 + \epsilon^2 \boldsymbol{u}_2 + \cdots, \quad (9.6)$$

$$\rho = \rho_0 + \epsilon \rho_1 + \epsilon^2 \rho_2 + \cdots, \quad (9.7)$$

$$p = p_0 + \epsilon p_1 + \epsilon^2 p_2 + \cdots, \quad (9.8)$$

$$s = s_0 + \epsilon s_1 + \epsilon^2 s_2 + \cdots, \quad (9.9)$$

where ϵ is a small parameter related to the amplitude of the perturbation field and $(\boldsymbol{u}_0, \rho_0, p_0, s_0)$ are related to a uniform mean field. It is worth noting that the leading fluctuating terms in the pressure and the density fields are assumed to have the same scaling order with respect to ϵ. Different scaling laws can also be considered (e.g., Zank and Matthaeus, 1990, 1991; Bayly, Levermore, and Passot, 1992). The mean velocity $\boldsymbol{u}_0$ can be set to zero by changing the frame of reference.

9.1 Introduction to Modal Decomposition of Turbulent Fluctuations

Assuming that the source terms on the right-hand sides of Eqs. (9.1)–(9.3) scale like ϵ and inserting the preceding expansions into these equations, one obtains the following linearized set of equations for the first-order fluctuating field $(\boldsymbol{u}_1, \rho_1, p_1, s_1)$ (the subscript 1 will be omitted hereafter for the sake of clarity):

$$\nabla \cdot \boldsymbol{u} + \frac{\partial p}{\partial t} - \frac{\partial s}{\partial t} = \frac{m}{\rho_0}, \tag{9.10}$$

$$\frac{\partial \boldsymbol{u}}{\partial t} + a_0^2 \nabla p - \nu_0 \nabla^2 \boldsymbol{u} - \frac{1}{3}\nu_0 \nabla(\nabla \cdot \boldsymbol{u}) = \boldsymbol{f}, \tag{9.11}$$

$$\frac{\partial s}{\partial t} - \frac{4}{3}\nu_0 \nabla^2 s - \frac{4}{3}(\gamma - 1)\nu_0 \nabla^2 p = \frac{Q}{\rho_0 c_p T_0}, \tag{9.12}$$

where $\nu_0 = \mu_0/\rho_0$, $\gamma = c_p/c_v$, c_p being the specific heat at constant pressure and c_v that at constant volume. It is to be noted that the pressure and the entropy have been normalized by γp_0 and c_p, respectively (the notations have not been changed for the sake of simplicity). The speed of sound in the undisturbed medium, a_0, is computed as $a_0 = \sqrt{\gamma p_0/\rho_0}$. The Prandtl number $\mu c_p/\kappa$ has been taken equal to 3/4 for the sole purpose of simplifying the algebra. This linear system can be rewritten introducing the fluctuating vorticity $\boldsymbol{\Omega} = \nabla \times \boldsymbol{u}$. By a slight manipulation, one obtains

$$\frac{\partial \boldsymbol{\Omega}}{\partial t} - \nu_0 \nabla^2 \boldsymbol{\Omega} = \nabla \times \boldsymbol{f}, \tag{9.13}$$

$$\frac{\partial s}{\partial t} - \frac{4}{3}\nu_0 \nabla^2 s = \frac{4}{3}(\gamma - 1)\nu_0 \nabla^2 p + \frac{Q}{\rho_0 c_p T_0}, \tag{9.14}$$

$$\frac{\partial^2 p}{\partial t^2} - a_0^2 \nabla^2 p - \frac{4}{3}\gamma \nu_0 \frac{\partial}{\partial t}(\nabla^2 p) = \left[\left(\frac{\partial}{\partial t} - \frac{4}{3}\nu_0 \nabla^2\right)\frac{m}{\rho_0} - \nabla \cdot \boldsymbol{f} + \frac{\partial}{\partial t}\left(\frac{Q}{\rho_0 c_p T_0}\right)\right]. \tag{9.15}$$

This set of equations is supplemented by additional relations obtained by linearizing perfect gas law (9.4) and entropy definition (9.5):

$$\gamma p - \frac{\rho}{\rho_0} - \frac{T}{T_0} = 0, \tag{9.16}$$

$$p + \frac{1}{\gamma - 1}\left(s - \frac{T}{T_0}\right) = 0. \tag{9.17}$$

Using these equations, Kovasznay proposes to define three *physical modes*, each mode corresponding to the solution of a subsystem extracted from (9.13)–(9.15):

- The *vorticity mode*, whose fluctuating field is denoted by $(\boldsymbol{\Omega}_\Omega, p_\Omega, s_\Omega, \boldsymbol{u}_\Omega)$, is defined as follows:

$$\frac{\partial \boldsymbol{\Omega}_\Omega}{\partial t} - \nu_0 \nabla^2 \boldsymbol{\Omega}_\Omega = \nabla \times \boldsymbol{f}, \tag{9.18}$$

$$p_\Omega = 0, \quad s_\Omega = 0, \quad \nabla \times \boldsymbol{u}_\Omega = \boldsymbol{\Omega}_\Omega, \quad \nabla \cdot \boldsymbol{u}_\Omega = 0. \tag{9.19}$$

The vorticity mode is associated with a solenoidal rotational velocity field, and it can be interpreted as the "incompressible" part of the solution. But it is worth noting that there is no corresponding pressure disturbance because it is expected to be of the order of ϵ^2. If the source term is set equal to zero, an exact wavelike solution is

$$\boldsymbol{\Omega}_\Omega = \boldsymbol{Z}_\Omega \exp\left(\imath \boldsymbol{k}_\Omega \cdot \boldsymbol{x} - \nu_0 k_\Omega^2 t\right), \tag{9.20}$$

$$\boldsymbol{u}_\Omega = \imath \frac{\boldsymbol{k}_\Omega \times \boldsymbol{Z}_\Omega}{k_\Omega^2} \exp\left(\imath \boldsymbol{k}_\Omega \cdot \boldsymbol{x} - \nu_0 k_\Omega^2 t\right), \tag{9.21}$$

where the wave vector associated with the vorticity mode, $\boldsymbol{k}_\Omega$, and the complex amplitude fluctuation vector $\boldsymbol{Z}_\Omega$ are such that $\boldsymbol{Z}_\Omega \cdot \boldsymbol{k}_\Omega = 0$, i.e., the associated velocity field corresponds to a transverse wave.

- The *entropy mode* whose corresponding perturbation field is $(\boldsymbol{\Omega}_e, p_e, s_e, \boldsymbol{u}_e)$, is defined as

$$\frac{\partial s_e}{\partial t} - \frac{4}{3}\nu_0 \nabla^2 s_e = \frac{4}{3}(\gamma - 1)\nu_0 \nabla^2 p_e + \frac{Q}{\rho_0 c_p T_0}, \tag{9.22}$$

$$\boldsymbol{\Omega}_e = 0, \quad p_e = 0, \quad \nabla \times \boldsymbol{u}_e = 0, \quad \nabla \cdot \boldsymbol{u}_e = \frac{\partial s_e}{\partial t}. \tag{9.23}$$

The corresponding wavelike solution for the source-free problem is

$$s_e = S \exp\left(\imath \boldsymbol{k}_e \cdot \boldsymbol{x} - \frac{4}{3}\nu_0 k_e^2 t\right), \tag{9.24}$$

$$\boldsymbol{u}_e = \imath S \frac{4}{3}\nu_0 \boldsymbol{k}_e \exp\left(\imath \boldsymbol{k}_e \cdot \boldsymbol{x} - \frac{4}{3}\nu_0 k_e^2 t\right), \tag{9.25}$$

where S and $\boldsymbol{k}_e$ are the complex amplitude and the wave vector, respectively. The velocity perturbation is purely dilatational and is induced by the sole viscous effects, and therefore $\boldsymbol{u}_e = 0$ in the inviscid case.

- The *acoustic mode*, which is characterized by $(\boldsymbol{\Omega}_p, p_p, s_p, \boldsymbol{u}_p)$. The governing relations for this mode are

$$\frac{\partial^2 p_p}{\partial t^2} - a_0^2 \nabla^2 p_p - \frac{4}{3}\gamma \nu_0 \frac{\partial}{\partial t}\left(\nabla^2 p_p\right) = \left[\left(\frac{\partial}{\partial t} - \frac{4}{3}\nu_0 \nabla^2\right)\frac{m}{\rho_0} - \nabla \cdot \boldsymbol{f} + \frac{\partial}{\partial t}\left(\frac{Q}{\rho_0 c_p T_0}\right)\right], \tag{9.26}$$

$$\frac{\partial s_p}{\partial t} - \frac{4}{3}\nu_0 \nabla^2 s_p = \frac{4}{3}(\gamma - 1)\nu_0 \nabla^2 p_p, \tag{9.27}$$

$$\nabla \times \boldsymbol{u}_p = 0, \quad \nabla \cdot \boldsymbol{u}_p = \frac{\partial s_p}{\partial t} - \frac{\partial p_p}{\partial t} + \frac{m}{\rho_0}. \tag{9.28}$$

9.1 Introduction to Modal Decomposition of Turbulent Fluctuations

The wave solution is

$$p_p = P \exp\left[\imath(\mathbf{k}_p \cdot \mathbf{x} - \sigma t)\right], \tag{9.29}$$

$$s_p = \frac{4}{3} P \frac{(\gamma-1)\nu_0 k_p^2}{c - \frac{4}{3}\nu_0 k_p} \exp\left[\imath(\mathbf{k}_p \cdot \mathbf{x} - \sigma t)\right], \tag{9.30}$$

$$\mathbf{u}_p = \imath P \mathbf{k}_p \frac{a_0^2}{c - \frac{4}{3}\nu_0 k_p} \left[\imath(\mathbf{k}_p \cdot \mathbf{x} - \sigma t)\right], \tag{9.31}$$

where P and $\mathbf{k}_p$ are the complex amplitude and the wave vector, respectively. The complex propagation frequency σ is defined as

$$\sigma = -a_0 k_p \left(\sqrt{1 - \frac{4\gamma^2 \nu_0^2 k_p^2}{9 a_0^2}} - \imath \frac{2}{3} \frac{\gamma \nu_0 k_p}{a_0}\right). \tag{9.32}$$

The imaginary part of σ gives the damping rate of the acoustic waves whereas the real part is related to the frequency of oscillations. It is observed that the viscous effects lead to the existence of a dispersive solution. The existence of an acoustic–entropy fluctuation originates in the viscous dissipation of the pressure waves. In the inviscid case, one recovers $s_p = 0$ and waves travel at the speed a_0 (i.e., $\sigma = -a_0 k_p$).

It is seen that disturbances associated with the entropy mode and the vorticity mode are passively advected by the mean field (velocity $\mathbf{u}_0$ in a reference frame at rest), whereas acoustic disturbances travel at the speed of sound relative to the mean flow.

Using this three-mode decomposition, all turbulent fluctuations can be decomposed as

$$p = p_\Omega + p_e + p_p, \tag{9.33}$$

$$s = s_\Omega + s_e + s_p, \tag{9.34}$$

$$\mathbf{u} = \mathbf{u}_\Omega + \mathbf{u}_e + \mathbf{u}_p, \tag{9.35}$$

$$\mathbf{\Omega} = \mathbf{\Omega}_\Omega + \mathbf{\Omega}_e + \mathbf{\Omega}_p. \tag{9.36}$$

Nonvanishing contributions in both viscous and inviscid cases are summarized in Table 9.1.

The analysis of governing equations for each mode also gives some information on the role of the forcing terms m, f, and Q.

The mass-addition term m leads to a production of the acoustic mode [see Eqs. (9.26) and (9.28)]. The effect is twofold: Mass addition induces the rise of a nonzero velocity perturbation, and it also generates pressure waves. But it is worth noting that if m obeys the diffusion equation,

$$\left(\frac{\partial}{\partial t} - \frac{4}{3}\nu_0 \nabla^2\right) \frac{m}{\rho_0} = 0, \tag{9.37}$$

Table 9.1. *Nonvanishing Kovasznay mode contribution to the fluctuating field*

Mode/fluctuation	s	p	u	Ω
Acoustic	X	○	○	
Entropy	○		X	
Vorticity			○	○

Note: Symbols are related to nonvanishing contributions. X: nonzero contribution in the viscous case only; ○: nonzero contribution in both viscous and inviscid cases.

then no pressure wave is created. In this case, the generated velocity field is a potential field whose potential ϕ_p satisfies the following Poisson equation:

$$\nabla^2 \phi_p = \frac{m}{\rho_0}. \qquad (9.38)$$

The forcing term f produces both the vorticity mode and the acoustic mode. A closer examination of the governing equations shows that the irrotational (resp. solenoidal) part of f generates the acoustic (resp. vorticity) mode and cannot generate the vorticity (resp. acoustic) mode. If f is a harmonic force field (i.e., it is both solenoidal and irrotational) no fluctuating vorticity and pressure fields are generated. The only effect is the creation of a harmonic velocity field u_H given by

$$\frac{\partial u_H}{\partial t} = f. \qquad (9.39)$$

The effect of heat addition (term Q) is to create both an entropy mode and an acoustic mode. Adding heat obviously yields an increase of the entropy (creation of the entropy mode) and leads to a local dilatation of the medium and a local disturbance in the pressure field (creation of the acoustic mode).

9.1.3 Weakly Nonlinear Corrected Kovasznay Decomposition

The linear decomposition previously presented makes it possible to define the three physical modes, but it does not provide any insight into the interactions between them because the modes evolve independently. Information dealing with the creation–destruction of fluctuations that are due to the modal interactions is recovered by looking at terms of the order of ϵ^2 resulting from bilateral interactions (Chu and Kovasznay, 1957). The full analysis brings in 18 terms and also involves a second nondimensional parameter[†] $\epsilon' = \nu_0 k/a_0$ that measures the ratio of the characteristic length scale of the perturbation, $1/k$, and the intrinsic scale of the medium ν_0/a_0. Second-order terms scale as ϵ^2 or $\epsilon^2 \epsilon'$. Because for turbulent flows at atmospheric pressure and density one has $\epsilon' \ll \epsilon$,[‡] it is chosen to neglect terms of the order

---

[†] It is recalled that the first nondimensional parameter ϵ is related to the amplitude of the perturbations.
[‡] Considering $\nu_0 = 0.15 \times 10^{-4}\,\text{m}^2\,\text{s}^{-1}$ and $a_0 = 300\,\text{ms}^{-1}$, one obtains $\nu_0/a_0 = 5 \times 10^{-8}$ m, which is of the order of the mean-free path of the molecules in the gas. For a frequency equal to 1 Hz, one

9.1 Introduction to Modal Decomposition of Turbulent Fluctuations

Table 9.2. *Source terms associated with second-order bilateral modal interactions according to Kovasznay decomposition*

Modal interaction	Acoustic source	Vorticity source	Entropy source
Acoustic–acoustic	Steepening and self-scattering $\nabla \cdot \nabla \cdot (\boldsymbol{u}_p \boldsymbol{u}_p) + a_0^2 \nabla^2 p_p^2$ $+ \frac{\gamma-1}{2} \frac{\partial^2 p_p^2}{\partial t^2}$	$O(\epsilon^2 \epsilon')$	$O(\epsilon^2 \epsilon')$
Vorticity–vorticity	Generation $2\nabla \cdot \nabla \cdot (\boldsymbol{u}_\Omega \boldsymbol{u}_\Omega)$	Self-convection and stretching $-\boldsymbol{u}_\Omega \nabla \boldsymbol{\Omega}_\Omega + \boldsymbol{\Omega}_\Omega \nabla \boldsymbol{u}_\Omega$	$O(\epsilon^2 \epsilon')$
Entropy–entropy	$O(\epsilon^2 \epsilon')$	$O(\epsilon^2 \epsilon')$	$O(\epsilon^2 \epsilon')$
Acoustic–vorticity	Scattering $2\nabla \cdot \nabla \cdot (\boldsymbol{u}_\Omega \boldsymbol{u}_p)$	Vorticity convection and stretching $-\boldsymbol{u}_p \nabla \boldsymbol{\Omega}_\Omega + \boldsymbol{\Omega}_\Omega \nabla \boldsymbol{u}_p - \boldsymbol{\Omega}_\Omega \nabla \cdot \boldsymbol{u}_p$	$O(\epsilon^2 \epsilon')$
Acoustic–entropy	Scattering $\frac{\partial}{\partial t} \nabla \cdot (s_e \boldsymbol{u}_p)$	Baroclinic source $-a_0^2 (\nabla s_e) \times (\nabla p_p)$	Heat convection $-\boldsymbol{u}_p \cdot \nabla s_e$
Vorticity–entropy	$O(\epsilon^2 \epsilon')$	$O(\epsilon^2 \epsilon')$	Heat convection $-\boldsymbol{u}_\Omega \cdot \nabla s_e$

of $\epsilon^2 \epsilon'$. Remaining terms and associated production mechanisms are displayed in Table 9.2.

It is important to note that these second-order corrections make all the modes fully coupled, because even self-interactions yield the growth of the other modes. Therefore the Kovasznay decomposition strictly holds for fully linear approximations only.

9.1.4 Helmholtz Decomposition and Its Extension

Let us recall that the Helmholtz decomposition of an arbitrary vector field $\boldsymbol{v}$ takes the form[§]

$$\boldsymbol{v} = \boldsymbol{v}_s + \boldsymbol{v}_d + \boldsymbol{v}_H, \tag{9.40}$$

where $\boldsymbol{v}_s, \boldsymbol{v}_d,$ and $\boldsymbol{v}_H$ are the solenoidal (i.e., incompressible), the dilatational (i.e., compressible), and harmonic components, respectively. This decomposition is shown to be useful throughout this book, even when strict incompressibility is assumed (it is not applied to the velocity but to its time variations, e.g., see Subsection 2.5.4). It can be generated by two terms, a potential vector such that $\boldsymbol{v}_s = \nabla \times \boldsymbol{\psi}$, yielding $\nabla \cdot \boldsymbol{v}_s = 0$, and a scalar potential $\boldsymbol{v}_d = \nabla \phi$, yielding $\nabla \times \boldsymbol{v}_d = 0$. The intersection of solenoidal and dilatational subspaces, $\boldsymbol{v}_H$, is generated by both curl($\nabla \times$) and div($\nabla \cdot$) operators, and is usually determined by the following harmonic problem: $\nabla^2 \boldsymbol{v}_H = 0$. The "vortical-divergent" nature of this decomposition also derives from $\boldsymbol{\omega} = \nabla \times \boldsymbol{u} = \nabla \times \boldsymbol{u}_s$ and $d = \nabla \cdot \boldsymbol{u} = \nabla \cdot \boldsymbol{u}_d$. Let us just emphasize that the

---

has $1/k = 300$ m and $\epsilon' = 1.66 \times 10^{-10}$. For 1 kHz one has $\epsilon' = 1.66 \times 10^{-7}$ and $\epsilon' = 1.66 \times 10^{-4}$ at 1 MHz. Even at 1 GHz, one obtains $\epsilon' = 1.66 \times 10^{-1} < 1$!

[§] The mathematical conditions for the existence and the uniqueness of such a decomposition are not discussed here.

toroidal–poloidal decomposition in physical space offers a useful and unique way to construct the potential vector $\boldsymbol{\psi}$ [see Eqs. (2.65), (2.66), and (2.70)]. This potential is not unique in general, and therefore it must be subjected to additional gauge conditions to be completely determined. As a last point, the poloidal–toroidal decomposition can be itself tied to a kind of Helmholtz decomposition, but restricted to the transverse $\boldsymbol{u} \times \boldsymbol{n}$ part of the flow, yielding a peculiar vortical-divergent decomposition for the horizontal (or transverse, $\perp \boldsymbol{n}$) velocity field only (see Chapter 7).

The Helmholtz decomposition is exact and does not rely on any assumption about the physical nature of v. In the absence of relevant boundary effects, or in the case of periodic boundaries (as used in DNS for approaching realizations of a homogeneous flow), the harmonic component is space uniform and can therefore be taken equal to zero, so that

$$\boldsymbol{u} = \boldsymbol{u}_s + \boldsymbol{u}_d. \tag{9.41}$$

More generally, even in the presence of solid boundaries, the harmonic term can be absorbed in, e.g., the second "dilatational-irrotational" term [the RDT solution for an irrotational straining process in the presence of a wall provides an example, based on the Helmholtz decomposition of linearized Weber equation (5.19)].

Generally, this decomposition is static, in the sense that it does not rely on the evolution equations that govern the dynamics, up to some exceptions mentioned throughout this book.

Helmholtz decomposition holds for the velocity, but does not provide any help in the task of splitting the pressure and the density (or the entropy) between compressible and incompressible components. Therefore the splitting given by Eq. (9.41) must be supplemented by some arbitrary definition of compressible and compressible parts of other physical variables.

The pressure is usually split as

$$p = p_s + p_d, \tag{9.42}$$

where p_s is *defined* as the part of the pressure field that satisfies the Poisson equation found for the pressure in the incompressible case, leading to

$$\nabla^2 p_s = -\nabla \cdot \nabla \cdot (\boldsymbol{u}_s \boldsymbol{u}_s). \tag{9.43}$$

The general decomposition discussed in Chapter 2 suggests a more general decomposition, applying the Helmholtz decomposition to any term $\boldsymbol{V} = \boldsymbol{V}_s + \boldsymbol{V}_d$ given by a relation of the form $\partial \boldsymbol{u}/\partial t + \boldsymbol{V} + \frac{1}{\rho}\nabla p = 0$. The classical previously mentioned Poisson equation is recovered assuming $\boldsymbol{u}_d = 0$, $\rho = \text{constant}$, and $\boldsymbol{V}_d = (\boldsymbol{u}_s \nabla \boldsymbol{u}_s)_d$. Another variant would consist of applying the Helmholtz decomposition to the alternative evolution equation for $\sqrt{\rho}\boldsymbol{u}$ subsequently discussed.

The component p_d is then defined as the remainder: $p_d = p - p_s$ from Eq. (9.43).

An important remark is that the solenoidal field $(\boldsymbol{u}_s, p_s)$ does not include acoustic waves, but that the residual field $(\boldsymbol{u}_d, p_d)$ is not restricted to acoustic phenomena. The characteristic velocity scale associated with $(\boldsymbol{u}_s, p_s)$ is the fluid velocity, whereas $(\boldsymbol{u}_d, p_d)$ can have two characteristic scales in the most general case: the fluid velocity and the speed of sound. The former will hold if $(\boldsymbol{u}_d, p_d)$ is dominated by heat transfer (convective phenomenon), whereas the latter will be relevant in acoustics-governed cases.

It is worth noting that, following Kida and Orszag (1990a, 1990b, 1992), Miura and Kida (1995) extend the usual Helmholtz decomposition by applying it to the vector field $\boldsymbol{w} = \sqrt{\rho}\boldsymbol{u}$. Using this definition, the authors enforce the positive definiteness of the spectra of the compressive and rotational kinetic energies.

9.1.5 Bridging Between Kovasznay and Helmholtz Decomposition

Kovasznay decomposition can be related to Helmholtz decomposition for the velocity field.

In the case in which $\boldsymbol{u}_H = 0$, one obtains

$$\boldsymbol{u}_s = \boldsymbol{u}_\Omega, \quad \boldsymbol{u}_d = \boldsymbol{u}_p + \boldsymbol{u}_e. \tag{9.44}$$

It is worth noting that the Helmholtz approach does not rely on a small-parameter expansion and is therefore exact, whereas the Kovasznay decomposition is nothing but a first-order approximation.

9.1.6 On the Feasibility of a Fully General Modal Decomposition

The second-order correction of the linear Kovasznay decomposition provides meaningful qualitative insight into bilateral interactions. Because of the small-amplitude hypothesis, it is not able to give correct quantitative predictions in fully developed turbulent flows. Therefore, more sophisticated decompositions must be found to deal with genuinely nonlinear dynamics.

The idea of defining nearly independent physical modes that can serve as a basis to decompose turbulent compressible fluctuations is very appealing, but unfortunately it cannot be extended to arbitrary mean fields. The very reason why is that in the general case the mean-field gradients bring in new terms in linearized equations that do not allow us to decouple the different fluctuating fields. The search for such a system of equations for the acoustic mode developing about an arbitrary mean field is the Holy Grail of aeroacoustics, and a large number of surrogate evolution equations have been proposed that will not be discussed here. The interested reader is invited to refer to reference books on acoustics.

Other chapters of this book deal with the coupling between the modes induced by the mean-field nonuniformity: Chapter 10 is devoted to the interactions induced by a nonuniform smooth mean-velocity field, and the case of the interaction with a planar normal shock wave is detailed in Chapter 11.

9.2 Mean-Flow Equations, Reynolds Stress Tensor, and Energy Balance in Compressible Flows

9.2.1 Arbitrary Flows

We first address the derivation of the governing equations for the mean field and the associated second-order turbulent stresses. The usual density-weighted average, referred to as the Favre averaging technique,¶ is retained here. For a dummy variable φ (either a scalar or a vectorial one), the mean part, $\tilde{\phi}$, and the fluctuating part, ϕ'', are defined as

$$\tilde{\phi} \equiv \frac{\overline{\rho \phi}}{\bar{\rho}}, \quad \phi'' \equiv \phi - \tilde{\phi}, \tag{9.45}$$

where the *bar* symbol is related to the usual statistical average. Let us consider the Navier–Stokes equations written in the strong conservation form:

$$\frac{\partial \rho}{\partial t} + \nabla \cdot (\rho \boldsymbol{u}) = 0, \tag{9.46}$$

$$\frac{\partial \rho \boldsymbol{u}}{\partial t} + \nabla \cdot (\rho \boldsymbol{u}\boldsymbol{u}) = -\nabla p + \nabla \cdot \boldsymbol{\tau}, \tag{9.47}$$

$$\frac{\partial \rho e}{\partial t} + \nabla \cdot (\rho \boldsymbol{u} e) = -p(\nabla \cdot \boldsymbol{u}) + \boldsymbol{\tau} : (\nabla \boldsymbol{u}) + \nabla \boldsymbol{q}, \tag{9.48}$$

where $e = c_v T$ is the internal energy. The viscous stress tensor $\boldsymbol{\tau}$ and the heat conduction flux vector $\boldsymbol{q}$ are defined as

$$\boldsymbol{\tau} = \mu \left[2\mathbf{S} - \frac{2}{3}(\nabla \cdot \boldsymbol{u})\mathbf{I} \right], \tag{9.49}$$

$$\boldsymbol{q} = -\kappa \nabla T. \tag{9.50}$$

For the sake of completeness, the evolution equation for the enthalpy, $h = c_p T$, the vorticity $\boldsymbol{\Omega} = \nabla \times \boldsymbol{u}$, and the pressure are

$$\frac{\partial \rho h}{\partial t} + \nabla \cdot (\rho \boldsymbol{u} h) = \left(\frac{\partial p}{\partial t} + \boldsymbol{u} \cdot \nabla p \right) + \boldsymbol{\tau} : (\nabla \boldsymbol{u}) + \nabla \boldsymbol{q}, \tag{9.51}$$

$$\frac{\partial \boldsymbol{\Omega}}{\partial t} + (\nabla \boldsymbol{\Omega})\boldsymbol{u} = (\nabla \boldsymbol{u})\boldsymbol{\Omega} + \nu \nabla^2 \boldsymbol{\Omega} - (\nabla \cdot \boldsymbol{u})\boldsymbol{\Omega}$$
$$+ \frac{1}{\rho^2}(\nabla \rho) \times \left\{ \nabla p - \frac{4}{3} \nabla[\mu(\nabla \cdot \boldsymbol{u})] + \mu \nabla \times \boldsymbol{\Omega} \right\}$$
$$+ \nabla \times \left[\frac{2}{\rho} \mathbf{S}(\nabla \mu) - \frac{2}{3} \frac{1}{\rho}(\nabla \cdot \boldsymbol{u})(\nabla \mu) \right]$$
$$- \frac{1}{\rho}(\nabla \mu) \times (\nabla \times \boldsymbol{\Omega}), \tag{9.52}$$

$$\frac{\partial p}{\partial t} + \boldsymbol{u} \cdot \nabla p = \frac{\gamma p}{c_p} \left(\frac{\partial s}{\partial t} + \boldsymbol{u} \cdot \nabla s \right) - \gamma p(\nabla \cdot \boldsymbol{u}). \tag{9.53}$$

¶ But let us notice that the density-weighted average was introduced by Osborne Reynolds in his seminal paper in 1884!

9.2 Mean-Flow Equations, RST, and Energy Balance in Compressible Flows

Considering constant molecular viscosity and diffusivity, and taking the inner product of (9.52) with $\mathbf{\Omega}$, one obtains an evolution equation for the enstrophy $\Theta = \mathbf{\Omega} \cdot \mathbf{\Omega}/2$:

$$\frac{\partial \Theta}{\partial t} + \nabla \cdot (\boldsymbol{u}\Theta) = -\Theta(\nabla \cdot \boldsymbol{u}) + \mathbf{\Omega} \cdot \mathbf{S} \cdot \mathbf{\Omega} - \frac{1}{\rho^2}\mathbf{\Omega} \cdot (\nabla p \times \nabla \rho)$$
$$+ \frac{\nu}{\rho}\mathbf{\Omega} \cdot \nabla^2 \mathbf{\Omega} - \nu\mathbf{\Omega} \cdot \left\{\frac{1}{\rho^2}\nabla\rho \times \left[\nabla^2 \boldsymbol{u} + \frac{1}{3}\nabla(\nabla \cdot \boldsymbol{u})\right]\right\}. \quad (9.54)$$

Applying mass-weighted averaging procedure (9.45) to system (9.46)–(9.48) and using the binary regrouping** approach for the convective terms,

$$\rho\phi u_i = \bar{\rho}\tilde{\phi}\tilde{u}_i + \overline{\rho\phi''u_i''} = \bar{\rho}\tilde{\phi}\tilde{u}_i + \bar{\rho}\widetilde{\phi''u_i''}, \quad (9.55)$$

one obtains the following equations for the mean-flow variables:

$$\frac{\partial \bar{\rho}}{\partial t} + \frac{\partial (\bar{\rho}\tilde{u}_j)}{\partial x_j} = 0, \quad (9.56)$$

$$\frac{\partial \bar{\rho}\tilde{u}_i}{\partial t} + \frac{\partial (\bar{\rho}\tilde{u}_i\tilde{u}_j)}{\partial x_i} = -\frac{\partial \bar{p}}{\partial x_i} + \frac{\partial \bar{\tau}_{ij}}{\partial x_j} - \frac{\partial \bar{\rho}R_{ij}}{\partial x_j}, \quad (9.57)$$

$$\frac{\partial \bar{\rho}\tilde{e}}{\partial t} + \frac{\partial (\bar{\rho}\tilde{e}\tilde{u}_j)}{\partial x_j} = \underbrace{-\bar{p}\frac{\partial \tilde{u}_i}{\partial x_i}}_{\text{I}} \underbrace{- \overline{p\frac{\partial u_i''}{\partial x_i}}}_{\text{II}} + \underbrace{\bar{\tau}_{ij}\frac{\partial \tilde{u}_i}{\partial x_j}}_{\text{III}} + \underbrace{\overline{\tau_{ij}\frac{\partial u_i''}{\partial x_j}}}_{\text{IV}} + \underbrace{\frac{\partial \bar{q}_i}{\partial x_i}}_{\text{V}} - \underbrace{\frac{\partial (\bar{\rho}\widetilde{e''u_j''})}{\partial x_j}}_{\text{VI}}, \quad (9.58)$$

where the RST is now defined as

$$R_{ij} \equiv \widetilde{u_i''u_j''}. \quad (9.59)$$

An additional equation for the mean kinetic energy $\tilde{K} \equiv \bar{\rho}\tilde{u}_i\tilde{u}_i/2$ is obtained by taking the inner product of mean-momentum equation (9.57) by the mean-density-weighted velocity vector $\tilde{\boldsymbol{u}}$:

$$\frac{\partial \tilde{K}}{\partial t} + \frac{\partial (\tilde{K}\tilde{u}_j)}{\partial x_j} = \underbrace{-\frac{\partial (\bar{p}\tilde{u}_i)}{\partial x_i}}_{\text{VII}} + \underbrace{\bar{p}\frac{\partial \tilde{u}_i}{\partial x_i}}_{\text{I}} - \underbrace{\frac{\partial (\bar{\rho}R_{ij}\tilde{u}_i)}{\partial x_j}}_{\text{VIII}} + \underbrace{\bar{\rho}R_{ij}\frac{\partial \tilde{u}_i}{\partial x_j}}_{\text{IX}} + \underbrace{\frac{\partial (\bar{\tau}_{ij}\tilde{u}_i)}{\partial x_j}}_{\text{X}} - \underbrace{\bar{\tau}_{ij}\frac{\partial \tilde{u}_i}{\partial x_j}}_{\text{III}}.$$

(9.60)

The physical meaning of source terms in the mean-internal-energy equation and the mean kinetic energy are as follows:

- I: mean pressure-dilatation energy transfer, which is strictly null if the mean-velocity field $\tilde{\boldsymbol{u}}$ is solenoidal.
- II: pressure-fluctuation dilatation correlation, which vanishes if the fluctuating-velocity field $\boldsymbol{u}''$ is solenoidal.
- III: viscous heat production associated with mechanical dissipation by the mean flow.

** This term was coined by Chassaing and co-workers (see Chassaing et al., 2002), who developed the alternative ternary regrouping approach.

- IV: viscous heat production associated with mechanical dissipation by the fluctuating flow (turbulent dissipation of turbulence kinetic energy).
- V: mean external heat source by conduction.
- VI: turbulent diffusion of internal energy.
- VII: power of the mean external pressure forces in the mean motion.
- VIII: power of the Reynolds stresses.
- IX: energy exchange with the turbulence kinetic energy (shear production). This term is null if the mean shear is zero.
- X: power of the mean external viscous stresses in the mean motion.

The evolution equation for the fluctuating velocity is

$$\frac{\partial(\bar{\rho}\overline{u_i''})}{\partial t} + \frac{\partial(\bar{\rho}\overline{u_i''}\tilde{u}_j)}{\partial x_j} = -\frac{\partial(\bar{\rho}R_{ij})}{\partial x_j} - \bar{\rho}\overline{u_j''}\frac{\partial \tilde{u}_i}{\partial x_j} - \bar{\rho}\overline{u_i''\frac{\partial u_i''}{\partial x_j}} + \overline{\frac{\rho'}{\rho}\frac{\partial p}{\partial x_i}} - \overline{\frac{\rho'}{\rho}\frac{\partial \tau_{ij}}{\partial x_j}}.$$

(9.61)

In the same way as in the incompressible flow case, evolution equations for the Reynolds stresses can be deduced from the Navier–Stokes equations. Still considering the binary regrouping, one obtains

$$\frac{\partial \bar{\rho}R_{ij}}{\partial t} + \frac{\partial(\bar{\rho}R_{ik}\tilde{u}_k)}{\partial x_k} = -\bar{\rho}\left(R_{ik}\frac{\partial \tilde{u}_j}{\partial x_k} + R_{jk}\frac{\partial \tilde{u}_i}{\partial x_k}\right) + \overline{p'\left(\frac{\partial u_i''}{\partial x_j} + \frac{\partial u_j''}{\partial x_i}\right)}$$
$$- \frac{\partial}{\partial x_k}\left(\overline{\rho u_i'' u_j'' u_k''} + \overline{p' u_i''}\delta_{jk} + \overline{p' u_j''}\delta_{ik} - \overline{\tau_{ik}' u_j''} - \overline{\tau_{jk}' u_i''}\right)$$
$$+ \overline{u_i''}\left(\frac{\partial \bar{\tau}_{jk}}{\partial x_k} - \frac{\partial \bar{p}}{\partial x_j}\right) + \overline{u_j''}\left(\frac{\partial \bar{\tau}_{ik}}{\partial x_k} - \frac{\partial \bar{p}}{\partial x_i}\right) - \overline{\tau_{ik}'\frac{\partial u_j''}{\partial x_k}} - \overline{\tau_{jk}'\frac{\partial u_i''}{\partial x_k}}.$$

(9.62)

Defining the instantaneous turbulence kinetic energy as $k = u_i'' u_i''/2$, one deduces from the Reynolds stress equations the following evolution equation for the mean-density-weighted turbulence kinetic energy $\widetilde{\mathcal{K}}$:

$$\frac{\partial \widetilde{\mathcal{K}}}{\partial t} + \frac{\partial(\widetilde{\mathcal{K}}\tilde{u}_j)}{\partial x_j} = -\underbrace{\frac{\partial(\bar{\rho}\widetilde{\mathcal{K}u_j''})}{\partial x_j}}_{\text{XI}} - \underbrace{\bar{\rho}R_{ij}\frac{\partial \tilde{u}_i}{\partial x_j}}_{\text{IX}} - \underbrace{\frac{\partial(\bar{p}\overline{u_i''})}{\partial x_i}}_{\text{XII}} - \underbrace{\frac{\partial \overline{p' u_i''}}{\partial x_i}}_{\text{XIII}} + \underbrace{\overline{p\frac{\partial u_i''}{\partial x_i}}}_{\text{II}}$$
$$+ \underbrace{\frac{\partial(\overline{\tau_{ij} u_i''})}{\partial x_j}}_{\text{XIV}} - \underbrace{\overline{\tau_{ij}\frac{\partial u_i''}{\partial x_j}}}_{\text{IV}},$$

(9.63)

where the physical mechanisms at play are as follows:

- XI: turbulent diffusion.
- XII: external power of mean-pressure forces acting through the fluctuating motion.

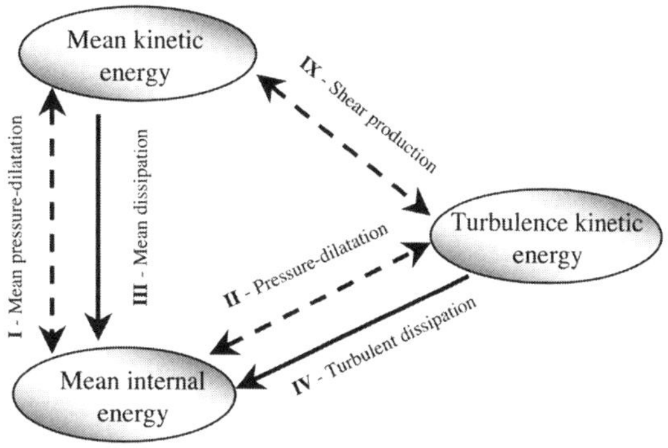

9.2 Mean-Flow Equations, RST, and Energy Balance in Compressible Flows

Figure 9.1. Schematic view of mean-energy exchanges in compressible turbulence.

- XIII: external power of pressure fluctuations in the fluctuating motion.
- XIV: external power of fluctuating viscous forces in the fluctuating motion.

Direct energy exchanges among the mean-flow kinetic energy, the mean internal energy, and the mean turbulence kinetic energy are due to common terms appearing in Eqs. (9.58), (9.60), and (9.63), namely terms I, II, III, IV, and IX. A schematic view of this dynamical scheme is displayed in Fig. 9.1.

9.2.2 Simplifications in the Isotropic Case

The dynamical scheme just presented simplifies dramatically in isotropic turbulence because of the absence of the mean-flow gradient and the symmetry properties of statistical moments of turbulent fluctuations. In this case, system (9.58)–(9.60)–(9.63) becomes

$$\frac{\partial \bar{\rho} \tilde{e}}{\partial t} = -\underbrace{\overline{p \frac{\partial u_i''}{\partial x_i}}}_{\text{II}} + \underbrace{\overline{\tau_{ij} \frac{\partial u_i''}{\partial x_j}}}_{\text{IV}}, \qquad (9.64)$$

$$\frac{\partial \tilde{K}}{\partial t} = 0, \qquad (9.65)$$

$$\frac{\partial \tilde{\mathcal{K}}}{\partial t} = \underbrace{\overline{p \frac{\partial u_i''}{\partial x_i}}}_{\text{II}} - \underbrace{\overline{\tau_{ij} \frac{\partial u_i''}{\partial x_j}}}_{\text{IV}}. \qquad (9.66)$$

One observes that, as in the case of incompressible flow, the mean kinetic energy is constant because the turbulent force in the mean-momentum equation vanishes. The remaining coupling terms, II and IV, correspond to energy exchanges between the mean internal energy and the mean turbulence kinetic energy (see Fig. 9.2). It is worth noting that term II depends on the sole dilatational part of the fluctuating-velocity field. Using the Kovasznay decomposition, one can see that this term is associated with the acoustic mode and the entropy mode. In the general case in which temperature-dependent viscosity is considered, term IV also accounts for turbulent fluctuations of the molecular viscosity. This term is also present in the

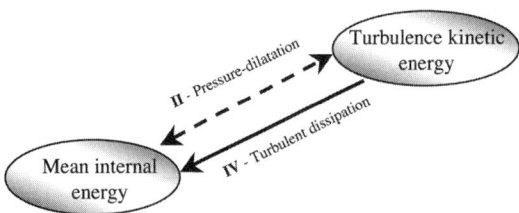

Figure 9.2. Schematic view of mean-energy exchanges in compressible isotropic turbulence.

incompressible case and therefore is associated with the three modes of the Kovasznay decomposition.

The full dynamical scheme in isotropic turbulence consists of energy exchanges at constant total mean energy between the mean turbulence kinetic and the mean internal energy because

$$\frac{\partial}{\partial t}(\tilde{e} + \tilde{K} + \widetilde{\mathcal{K}}) = \frac{\partial \tilde{e}}{\partial t} + \frac{\partial \widetilde{\mathcal{K}}}{\partial t} = 0. \tag{9.67}$$

Therefore the whole dynamics is governed by terms II and IV, and most studies dealing with compressible isotropic turbulence have been devoted to the analysis of these two terms and the underlying physical mechanisms.

To get a deeper insight into the contributions of each physical mode, it is useful to decompose terms II and IV.

The pressure-dilatation correlation (term II) can be rewritten as

$$\overline{p\frac{\partial u_i''}{\partial x_i}} = \overline{p'\frac{\partial u_i''}{\partial x_i}} = \overline{p'\frac{\partial u_i'}{\partial x_i}}, \tag{9.68}$$

where it is important to note that $u' \equiv (u - \bar{u})$ is defined using the usual statistical average and not the density-weighted one. This change is possible because, as pointed out by Feiereisen and co-workers (Feiereisen, Reynolds, and Ferziger, 1981), *the density-weighted average and the usual average are equivalent in homogeneous flows*. One observes this by writing the following exact decomposition of the density-weighted momentum:

$$\bar{\rho}\tilde{u}_i \equiv \bar{\rho}\bar{u}_i + \overline{\rho u_i'} = \bar{\rho}\bar{u}_i + \overline{\rho' u_i'}, \tag{9.69}$$

from which it follows that

$$\tilde{u}_i = \bar{u}_i + \frac{\overline{\rho' u_i'}}{\bar{\rho}}. \tag{9.70}$$

Because the momentum is conserved in homogeneous turbulence (and more generally in all periodic domains) and the statistical average can be interpreted as a volume average by invoking the ergodic theorem, the last term on the right-hand

9.2 Mean-Flow Equations, RST, and Energy Balance in Compressible Flows

side of Eq. (9.70) is constant in space and time. The two velocity fields $\tilde{u}$ and $\bar{u}$ are related by an additive constant, which can be chosen to be zero by selecting the ad hoc frame of reference.

Introducing the Helmholtz decomposition $u' = u'_s + u'_d$ and $p' = p'_s + p'_d$, the pressure-dilatation term is rewritten as

$$\overline{p' \frac{\partial u'_i}{\partial x_i}} = \overline{p'_s \frac{\partial u'_{di}}{\partial x_i}} + \overline{p'_d \frac{\partial u'_{di}}{\partial x_i}}. \tag{9.71}$$

This new expression emphasizes that this term is strictly null in incompressible flows, but also that the solenoidal field has a contribution associated with the correlation between the solenoidal pressure fluctuations and the divergence of the fluctuating-velocity field. Because the entropy mode has no pressure fluctuation (at least in the first-order Kovasznay approximation), it is seen that term II is null if there is no acoustic mode. In the true solution of nonlinear problems, it is expected to be very small if no acoustic waves are present or if no very intense entropy source is present.

Neglecting the molecular viscosity fluctuations,[††] the dilatation-dissipation term (term IV) can be decomposed in homogeneous turbulence as[‡‡]

$$\overline{\tau_{ij} \frac{\partial u''_i}{\partial x_j}} = \overline{\tau'_{ij} \frac{\partial u''_i}{\partial x_j}} = \bar{\rho}\bar{\varepsilon}_s + \bar{\rho}\bar{\varepsilon}_d, \tag{9.72}$$

where $\bar{\varepsilon}_s$ and $\bar{\varepsilon}_d$, which are respectively referred to as the solenoidal and the dilatational dissipation rate, are defined as

$$\bar{\varepsilon}_s = 2\frac{\bar{\mu}}{\bar{\rho}} \overline{W'_{ij} W'_{ij}} = \frac{\bar{\mu}}{\bar{\rho}} \overline{\Omega'_i \Omega'_i}, \quad W'_{ij} = \frac{1}{2}\left(\frac{\partial u'_i}{\partial x_j} - \frac{\partial u'_j}{\partial x_i}\right), \tag{9.73}$$

$$\bar{\varepsilon}_d = \frac{4}{3}\frac{\bar{\mu}}{\bar{\rho}} \overline{\left(\frac{\partial u'_i}{\partial x_i}\right)^2} = \frac{4}{3}\frac{\bar{\mu}}{\bar{\rho}} \overline{\left(\frac{\partial u'_{di}}{\partial x_i}\right)^2}, \tag{9.74}$$

where $\Omega \equiv \nabla \times u' = \nabla \times u'_s$. It is observed that $\bar{\varepsilon}_s$ (resp. $\bar{\varepsilon}_d$) does not depend on the dilatational (resp. solenoidal) field at all, and will therefore be exactly zero if the solenoidal (resp. dilatational) field is not present in the flow. In high-speed flows without a strong external entropy source, and restricting the analysis to the linear Kovasznay splitting, it is seen that the solenoidal dissipation $\bar{\varepsilon}_s$ is associated with the sole vorticity mode, whereas the dilatational dissipation is mainly due to the acoustic mode.

[††] This assumption is relevant for most high-speed nonreactive flows.
[‡‡] In nonhomogeneous flows a third contribution must be taken into account, which is defined as

$$\bar{\varepsilon}_n = 2\frac{\bar{\mu}}{\bar{\rho}}\left[\frac{\partial^2}{\partial x_i \partial x_j}\overline{u'_i u'_j} - 2\frac{\partial}{\partial x_j}\left(\overline{u'_j \frac{\partial u'_i}{\partial x_i}}\right)\right].$$

9.2.3 Quasi-Isentropic Isotropic Turbulence: Physical and Spectral Descriptions

A further simplified model is obtained assuming that turbulent fluctuations are isentropic. The resulting model is widely used to analyze the properties of turbulence in the compressible regime in the absence of significant thermal effects.

The associated set of governing equations is

$$\frac{\partial u'_i}{\partial t} + \frac{1}{\bar{\rho}} \frac{\partial p'}{\partial x_i} - \nu \frac{\partial u'_i}{\partial x_k \partial x_k} - \frac{\nu}{3} \frac{\partial}{\partial x_i} \left(\frac{\partial u'_k}{\partial x_k} \right) = -u'_j \frac{\partial u'_i}{\partial x_j}, \quad (9.75)$$

$$\frac{\partial}{\partial t} \left(\frac{p'}{\gamma P} \right) + \frac{\partial u'_i}{\partial x_i} = -u'_j \frac{\partial}{\partial x_j} \left(\frac{p'}{\gamma P} \right), \quad (9.76)$$

in which all nonlinear terms have been put on the right-hand side, and where

$$\frac{p'}{\gamma P} = \frac{p'}{\bar{\rho} a_0^2} \quad \text{with} \quad a_0^2 = \gamma \frac{P}{\bar{\rho}}, \quad (9.77)$$

in which P and $\bar{\rho}$, which can possibly be time-dependent variables (see Chapter 10), are chosen constant together with the speed of sound in this section. More general quasi-isentropic equations can be derived, as discussed in Chapter 10, but additional assumptions are very useful, such as the low-Mach-number assumption, which leads to $\gamma p/P \ll 1$ and the possible removal of the nonlinear term in the last equation for the pressure p.

Viscous terms may be omitted, in agreement with the isentropic assumption, but they have been kept here because they are used in some closure approaches and/or for numerical convenience (hence the term quasi-isentropic used for this subsection). The second viscous term, which involves the divergence of the velocity, is consistent with Eq. (2.14) supplemented with the Stokes law $3\lambda + 2\mu = 0$, as well as with Eq. (9.2). Let us also note that the role of viscosity in Kovasznay mode coupling, which has already been introduced in Subsection 9.1.2, will be subsequently rediscussed in a simpler way.

The problem can be recast in a much simpler and useful way by use of the local Craya basis in Fourier space.[§§] The solenoidal and dilatational parts of the velocity field, denoted by $\hat{u}_s$ and $\hat{u}_d$, can be decomposed as follows:

$$\hat{u}_s(k) = u^{(1)}(k) e^{(1)}(k) + u^{(2)}(k) e^{(2)}(k), \quad \hat{u}_d(k) = u^{(3)}(k) e^{(3)}(k), \quad (9.78)$$

where $e^{(i)}(k), i = 1, 3$ are defined as in the incompressible case [see Eq. (2.67)]. Their counterparts in terms of vorticity $[\boldsymbol{\omega}' = \mathrm{curl}(\boldsymbol{u}') = \mathrm{curl}(\boldsymbol{u}_s)]$ and divergence

---

[§§] This projection onto the local reference frame is valid without any assumption dealing with statistical symmetries, such as isotropy. Isotropy allows us to use the projectors $P_{ij}^\perp$ and $P_{ij}^\parallel$, instead of, or in addition to, the Craya–Herring modes, as extensively discussed in Subsection 2.5.2. This is true only because of the equipartition in terms of poloidal and toroidal modes that is imposed by 3D isotropy.

9.2 Mean-Flow Equations, RST, and Energy Balance in Compressible Flows

$(d = \nabla \cdot \boldsymbol{u} = \nabla \cdot \boldsymbol{u}_s)$ are immediately found:

$$\hat{\boldsymbol{\omega}} = \imath k \left[u^{(2)} \boldsymbol{e}^{(1)} - u^{(1)} \boldsymbol{e}^{(2)} \right] \quad \text{and} \quad \hat{d} = \imath k u^{(3)}. \tag{9.79}$$

The definition of vorticity in terms of the Craya modes is the same as the one used in the incompressible case. These three velocity modes must be supplemented by a fourth mode, which accounts for the remaining independent thermodynamic quantity.¶¶ To have a problem with homogeneous dimension, the pressure fluctuation can be scaled as a velocity, and considered as a fourth component (Simone, Coleman, and Cambon, 1997):

$$u^{(4)} = \imath \frac{\hat{p}}{\bar{\rho} a_0}. \tag{9.80}$$

This scaling is similar to the one used in Eckhoff and Storesletten (1978).

Therefore governing equations (9.75) and (9.76) are rewritten in terms of the four variables in Fourier space as follows:

$$\frac{d}{dt} \begin{pmatrix} u^{(1)} \\ u^{(2)} \\ u^{(3)} \\ u^{(4)} \end{pmatrix} + \begin{bmatrix} \nu k^2 & 0 & 0 & 0 \\ 0 & \nu k^2 & 0 & 0 \\ 0 & 0 & \frac{4}{3}\nu k^2 & -a_0 k \\ 0 & 0 & a_0 k & 0 \end{bmatrix} \begin{pmatrix} u^{(1)} \\ u^{(2)} \\ u^{(3)} \\ u^{(4)} \end{pmatrix} = \begin{pmatrix} T_{\text{NL}}^{(1)} \\ T_{\text{NL}}^{(2)} \\ T_{\text{NL}}^{(3)} \\ T_{\text{NL}}^{(4)} \end{pmatrix}, \tag{9.81}$$

where all linear terms have been grouped on the left-hand side. The nonlinear terms are defined as follows:

$$\begin{pmatrix} T_{\text{NL}}^{(1)} \\ T_{\text{NL}}^{(2)} \\ T_{\text{NL}}^{(3)} \\ T_{\text{NL}}^{(4)} \end{pmatrix} = \begin{pmatrix} -\boldsymbol{e}^{(1)} \cdot \widehat{(\boldsymbol{\omega}' \times \boldsymbol{u}')} \\ -\boldsymbol{e}^{(2)} \cdot \widehat{(\boldsymbol{\omega}' \times \boldsymbol{u}')} \\ -\boldsymbol{e}^{(3)} \cdot \widehat{\boldsymbol{\omega}' \times \boldsymbol{u}'} - \frac{1}{2} \imath k \widehat{u'_j u'_j} \\ \imath \widehat{u'_j \frac{\partial (p'/(\bar{\rho} a_0))}{\partial x_j}} \end{pmatrix}, \tag{9.82}$$

where the *hat* symbol denotes the Fourier transform. A more advanced closed form of the nonlinear terms of Eq. (9.81) in terms of $u^{(1)}$, $u^{(2)}$, $u^{(3)}$, and $u^{(4)}$ is obtained by injecting (9.78) and (9.79) into Eq. (9.82)(Cambon and Sagaut, 2007).

It is worth noting that because a simplified isentropic model is used, the computed solenoidal and dilatational fields are not identical to those obtained by projecting the solution of the full compressible Navier–Stokes equations. In the present simplified model, only Kovasznay's vortical and acoustic modes [characterized by $(u^{(1)}, u^{(2)})$ and $(u^{(3)}, u^{(4)})$, respectively] are retained, and the entropic mode is discarded.

Two-point statistical moments are now considered. Because of 3D isotropy, two-point second-order statistics are generated by three independent spectra, namely the spectrum of the solenoidal kinetic energy $E_{ss}(k)$, the kinetic-energy

---

¶¶ A single additional degree of freedom is enough thanks to the isentropy assumption.

spectrum of the dilatational component $E_{dd}(k)$, and the pressure spectrum $E_{pp}(k)$:

$$\langle u^{(1)*}(\boldsymbol{p},t)u^{(1)}(\boldsymbol{k},t)\rangle = \langle u^{(2)*}(\boldsymbol{p},t)u^{(2)}(\boldsymbol{k},t)\rangle = \frac{E_{ss}(k,t)}{8\pi k^2}\delta^3(\boldsymbol{k}-\boldsymbol{p}), \quad (9.83)$$

$$\langle u^{(3)*}(\boldsymbol{p},t)u^{(3)}(\boldsymbol{k},t)\rangle = \frac{E_{dd}(k,t)}{4\pi k^2}\delta^3(\boldsymbol{k}-\boldsymbol{p}), \quad (9.84)$$

$$\langle u^{(4)*}(\boldsymbol{p},t)u^{(4)}(\boldsymbol{k},t)\rangle = \frac{E_{pp}(k,t)}{4\pi k^2}\delta^3(\boldsymbol{k}-\boldsymbol{p}), \quad (9.85)$$

and the cross spectrum

$$\langle u^{(3)*}(\boldsymbol{p},t)u^{(4)}(\boldsymbol{k},t)\rangle = \frac{E_{dp}(k,t)}{4\pi k^2}\delta^3(\boldsymbol{k}-\boldsymbol{p}), \quad (9.86)$$

whose imaginary part is neglected, consistent with 3D isotropy with mirror symmetry. The *solenoidal* and *dilatational kinetic energies*, respectively denoted $\mathcal{K}_s$ and $\mathcal{K}_d$, are computed as follows:

$$\mathcal{K}_s(t) = \int_0^{+\infty} E_{ss}(k,t)dk, \quad \mathcal{K}_d(t) = \int_0^{+\infty} E_{dd}(k,t)dk, \quad (9.87)$$

and the corresponding dissipations are defined as

$$\bar{\varepsilon}_s(t) = 2\nu\int_0^{+\infty} k^2 E_{ss}(k,t)dk, \quad \bar{\varepsilon}_d(t) = 2\frac{4}{3}\nu\int_0^{+\infty} k^2 E_{dd}(k,t)dk. \quad (9.88)$$

The pressure variance is recovered as follows:

$$\overline{p'p'}(t) = \rho_0^2 a_0^2 \int_0^{+\infty} E_{pp}(k,t)dk. \quad (9.89)$$

The evolution equations associated with the four spectra are similar to the original Lin equation (3.88) derived in the incompressible case for the kinetic-energy spectrum $E(k)$:

$$\frac{d}{dt}\begin{pmatrix} E_{ss} \\ E_{dd} \\ E_{dp} \\ E_{pp} \end{pmatrix} + \begin{bmatrix} 2\nu k^2 & 0 & 0 & 0 \\ 0 & 2\frac{4}{3}\nu k^2 & 2a_0 k & 0 \\ 0 & -a_0 k & \frac{4}{3}\nu k^2 & a_0 k \\ 0 & 0 & -2a_0 k & 0 \end{bmatrix} \begin{pmatrix} E_{ss} \\ E_{dd} \\ E_{dp} \\ E_{pp} \end{pmatrix} = \begin{pmatrix} T_{ss} \\ T_{dd} \\ T_{dp} \\ T_{pp} \end{pmatrix}, \quad (9.90)$$

where nonlinear terms have been grouped on the right-hand side. As for the cases of rotating and/or stratified turbulence, only the transfer terms related to true energy spectra must have a zero integral, i.e., exhibit a global conservation property. This constraint is fulfilled by T_{ss} (solenoidal energy transfer) and $T_{pp} + T_{dd}$ (transfer of total acoustic-wave energy) but not by $T_{pp} - T_{dd}$ and T_{dp}, which therefore are not true transfer terms. The nonlinear terms are related to the physical mechanisms mentioned in Table 9.2. Neglecting viscous terms, this system can be recast in the following compact form (see also rotating and/or stratified incompressible flow cases):

$$\frac{dE_{ss}(k)}{dt} = T_{ss}(k), \quad (9.91)$$

$$\frac{dE_w(k)}{dt} = T_w(k), \quad (9.92)$$

$$\frac{dZ(k)}{dt} + 2\imath(a_0 k)Z(k) = T_z(k), \quad (9.93)$$

in which

$$E_w(k) = E_{dd}(k) + E_{pp}(k) \quad (9.94)$$

is the *total turbulent acoustic-energy spectrum*,[*] and

$$Z(k) = E_{dd}(k) - E_{pp}(k) + 2\imath E_{dp}(k) \quad (9.95)$$

characterizes the imbalance between kinetic and potential energies of waves.

9.3 Different Regimes in Compressible Turbulence

Numerical experiments and theoretical analyses show that several dynamical regimes exist in isotropic compressible turbulence, even in the free-decay case in which no external forcing is present. This is a noticeable difference with incompressible decaying turbulence that exhibits a single behavior. A major difficulty is that these regimes are very sensitive to a large number of parameters, such as the turbulent Mach number[†] and the initial conditions (i.e., the relative energy of each mode in the Kovasznay or Helmholtz decomposition). That can be intuitively understood by looking at the corrected Kovasnay analysis, which reveals that each physical mode has a very specific dynamics: Changing the initial condition might therefore have a strong influence on the development of the flow.

Four main regimes have been identified, according to the influence of compressibility effects on the turbulence dynamics[‡]:

- The *low-Mach-number quasi-isentropic regime*, in which the turbulent Mach number is low and the interactions between the solenoidal and dilatational components are weak. Moreover, the dilatational component is assumed to obey

---

[*] It is recalled that the *acoustic energy* is usually defined as

$$\rho_0 \frac{u^2}{2} + \frac{p^2}{2\rho_0 a_0^2}$$

in the framework of linear acoustics, where u and p denote acoustic velocity and pressure disturbances, respectively.

[†] Let us recall that the turbulent Mach number is defined as $M_t = \sqrt{\mathcal{K}}/a_0$.

[‡] A very low-Mach-number regime of compressed – but not really compressible – turbulence is discussed in Chapter 10. Despite the fact that isotropy is assumed for this flow, it is more convenient to include the corresponding discussion in the next chapter, as the isotropic fluctuating flow is subjected to an external mechanism of mean spherical compression–dilatation.

quasi-linear acoustic dynamics. A vast majority of available studies are devoted to the case in which the dilatational mode is restricted to the acoustic mode. Two kinds of theories are subsequently emphasized: a purely linear one, which basically predicts that *acoustic equilibrium* holds at all scales, and a more powerful nonlinear one, which shows that *acoustic equilibrium* is restricted at very small wavenumbers only, whereas another regime, referred to as *pseudo-sound*, is observed at large wavenumbers.

- The *weakly nonlinear thermal regime*, in which the dilatational component includes thermal effects that are not governed by acoustic phenomena.
- The *nonlinear subsonic regime*, in which the turbulent Mach number is still less than one, but the fluctuations of the dilatational mode are strong enough to make nonlinear phenomena arise. In this case, some turbulence-induced very small shocks (referred to as *shocklets* or *eddy shocklets*) are detected.
- The *supersonic regime*, in which the turbulent Mach number is larger than one. In this case, the dilatational mode is of great importance and shocklets have a large impact on the full field.

9.3.1 Quasi-Isentropic Turbulent Regime

Theoretical developments for this regime are pivotal from our viewpoint as they all rely on all the basic exact equations of Subsection 9.2.3. The meaning of the term "exact" must be taken here in the same sense as for exactness of Eq. (9.76). Pure linear theory allows us to recover the essentials of the acoustic equilibrium, which is possibly altered by laminar viscous terms. Regarding nonlinear theories, no fewer than three versions are presented, giving very different results even if they rely on the same "exact" Lin equations and use similar QN closures that have been improved for extradissipative terms.

9.3.1.1 Linear Theory

The basic equations of linear theory for compressible turbulence are obtained in a trivial way, dropping the right-hand sides in Eqs. (9.81) and (9.90). It is observed that the solenoidal and the wavy components of the solution are totally decoupled, in agreement with the usual linear acoustic theory. The dilational and the pressure modes are coupled through acoustic-wave dynamics, which induces some exchanges between the dilatational kinetic energy and the turbulent potential energy.

Let us first consider the linear viscous regime. The solenoidal kinetic energy decays exactly as in the incompressible case (see Chapter 3), leading to

$$u^{(\alpha)}(k,t) = e^{-\nu k^2 t} u^{(\alpha)}(k,0), \alpha = 1,2 \quad E_{ss}(k,t) = e^{-2\nu k^2 t} E_{ss}(k,0). \quad (9.96)$$

The linear system related to acoustic waves simplifies to

$$\frac{d}{dt}\begin{pmatrix} u^{(3)} \\ u^{(4)} \end{pmatrix} + \begin{bmatrix} \frac{4}{3}\nu k^2 & -a_0 k \\ a_0 k & 0 \end{bmatrix} \begin{pmatrix} u^{(3)} \\ u^{(4)} \end{pmatrix} = 0, \quad (9.97)$$

9.3 Different Regimes in Compressible Turbulence

whose solution is governed by the sign of the discriminant:

$$D = [(2/3)\nu k^2]^2 - [a_0 k]^2.$$

A cutoff value $k_d = 3a_0/(2\nu)$ is therefore introduced: For $k < k_d$ (i.e., $D < 0$) the eigenvalues of the preceding linear system of equations are complex conjugates, yielding damped oscillating solutions, whereas only damping without oscillations is found for $k > k_d$ (i.e., $D > 0$) in relation with real eigenvalues. At low Mach number and large Reynolds number, the cutoff value k_d is very large,[§] so that the domain $k > k_d$ is irrelevant, but a renormalized version of this system can be useful, with drastic modification of k_d: This issue is subsequently discussed.

The linear inviscid theory has received much more attention (e.g., Erlebacher et al., 1990; Sarkar et al., 1991; Erlebacher and Sarkar, 1993) because it leads to the prediction of possible equilibrium states. It is handled in a very simple way (Cambon, Coleman, and Mansour, 1993; Simone, Coleman, and Cambon, 1997) using both the local Craya–Herring decomposition and the pressure rescaling given by Eq. (9.80).[¶] Dropping all nonlinear and viscous terms, one can see that the solenoidal component is frozen, whereas the following conservation relations hold at all wavenumbers:

$$u^{(3)}(k,t) + s\iota u^{(4)}(k,t) = e^{\iota s a_0 k t}\left[u^{(3)}(k,0) + s\iota u^{(4)}(k,0)\right], s = \pm 1, \quad (9.98)$$

$$E_w(k,t) = E_{ss}(k,t) + E_{dd}(k,t) = E_{ss}(0,t) + E_{dd}(0,t) = E_w(k,0), \quad (9.99)$$

along with

$$E_{dd}(k,t) = \frac{1}{2}E_w(k,0) + \frac{1}{2}[E_{dd}(k,0) - E_{pp}(k,0)]\cos(2a_0 k t)$$
$$- E_{dp}(k,0)\sin(2a_0 k t), \quad (9.100)$$

$$E_{pp}(k,t) = \frac{1}{2}E_w(k,0) - \frac{1}{2}[E_{dd}(k,0) - E_{pp}(k,0)]\cos(2a_0 k t)$$
$$+ E_{dp}(k,0)\sin(2a_0 k t), \quad (9.101)$$

$$E_{dp}(k,t) = \frac{1}{2}[E_{dd}(k,0) - E_{pp}(k,0)]\sin(2a_0 k t)$$
$$+ E_{dp}(k,0)\cos(2a_0 k t). \quad (9.102)$$

The *acoustic-equilibrium state* is defined as an equilibrium state in which *the kinetic energy of the dilatational mode is equal to the potential energy of the pressure mode*. One has to distinguish between two variants of the acoustic-equilibrium assumption.

[§] Considering air at common pressure and temperature, one has $a_0 \simeq 340\,\mathrm{m\,s^{-1}}$ and $\nu \simeq 10^{-5}$ m^2 s^{-1}, yielding $k_d \sim 5.10^7$ m^{-1}.

[¶] It is worth noting that a large number of works dealing with the inviscid linear theory have been carried out in the physical space, using system (9.75)–(9.76). In this case, a multiple scale expansion is needed to operate the splitting between the solenoidal and the dilatational component, whereas it is trivial in the local spectral frame of reference. It can also yield to a premature occurrence of nondimensional parameters and to an artificial unnecessary complication of the problem.

The first one, referred here to as the *strong acoustic-equilibrium hypothesis*, assumes that this equilibrium holds at all wavenumbers, yielding

$$E_{dd}(k) = \frac{1}{\rho_0^2 a_0^2} E_{pp}(k), \quad \forall k \quad \text{(strong acoustic-equilibrium hypothesis)}. \quad (9.103)$$

The second variant, namely the *weak acoustic-equilibrium hypothesis*, deals with the asymptotic values of global quantities, such as $\mathcal{K}_d(t)$ and $\overline{p'p'}(t)$ at large time. Using the analytical solutions previously given, one has

$$\lim_{t \to +\infty} \mathcal{K}_d(t) = \lim_{t \to +\infty} \int_0^{+\infty} E_{dd}(k,t) dk$$

$$= \frac{1}{2} \int_0^{+\infty} E_w(k,0) dk = \frac{1}{2} \left[\mathcal{K}_d(0) + \frac{\overline{p'p'}(0)}{\rho^2 a_0^2} \right]$$

$$= (\mathcal{K}_d)_\infty, \quad (9.104)$$

$$\lim_{t \to +\infty} \overline{p'p'}(t) = \rho_0^2 a_0^2 \lim_{t \to +\infty} \int_0^{+\infty} E_{pp}(k,t) dk$$

$$= \frac{\rho_0^2 a_0^2}{2} \left[\mathcal{K}_d(0) + \frac{\overline{p'p'}(0)}{\rho^2 a_0^2} \right] = (\overline{p'p'})_\infty, \quad (9.105)$$

along with

$$\lim_{t \to +\infty} \int_0^{+\infty} E_{dp}(k,t) dk = \lim_{t \to +\infty} \int_0^{+\infty} (E_{pp} - E_{dd}) dk = 0. \quad (9.106)$$

The latter relation comes from relation (9.95), which leads to

$$Z(k,t) = e^{\iota a_0 k t} Z(k,0), \quad \lim_{t \to +\infty} \int_0^{+\infty} Z(k,t) dk = 0. \quad (9.107)$$

This last result is seen to be a consequence of the phase-mixing phenomenon. Let us recall that phase mixing was induced by an angle-dependent factor in the dispersion law for inertia and/or gravity waves, whereas it results from the presence of the factor k in the integrand in the present case.

An important conclusion is that, whatever initial condition is considered, the solution converges toward the following equilibrium state:

$$(\mathcal{K}_d)_\infty = \frac{1}{\rho_0^2 a_0^2} (\overline{p'p'})_\infty \quad \text{(weak acoustic-equilibrium hypothesis)}. \quad (9.108)$$

It is of course easily seen that strong acoustic equilibrium is a sufficient but not necessary condition for the weak acoustic equilibrium to be satisfied.

The weak acoustic-equilibrium solutions can be represented in a very simple and elegant way by use of the ratio of the mean compressible kinetic energy to the total mean turbulent kinetic energy, $\chi(t)$, and the function $\mathcal{F}(t)$ introduced by Sarkar:

$$\mathcal{F}(t) \equiv \rho_0^2 a_0^2 \frac{\mathcal{K}_d(t)}{\overline{p'p'}(t)} = \rho_0^2 a_0^4 M_t^2 \frac{\chi(t)}{\overline{p'p'}(t)}, \quad (9.109)$$

9.3 Different Regimes in Compressible Turbulence

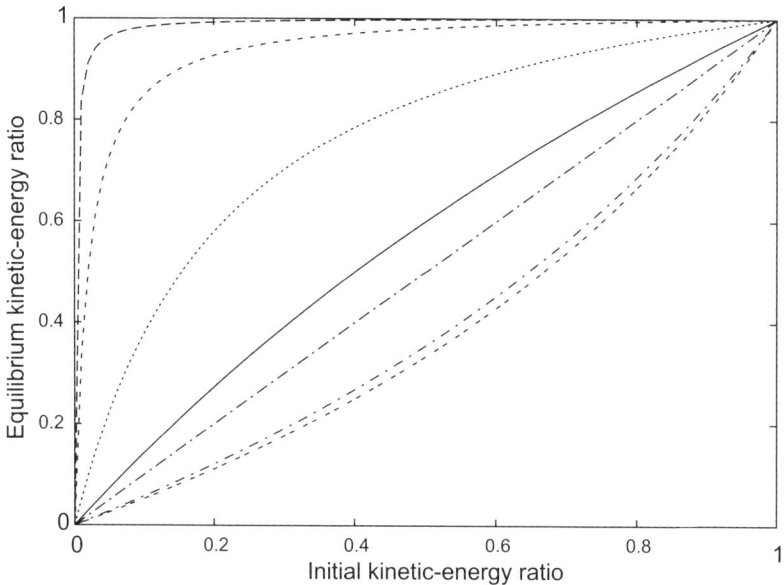

Figure 9.3. Values of the equilibrium turbulent kinetic-energy ratio χ_∞ as functions of $\chi(0)$ and $\mathcal{F}(0)$. Increasing values of $\mathcal{F}(0)$ correspond to decreasing values of χ_∞ at fixed $\chi(0)$. Lines plotted correspond to (from left to right) $\mathcal{F}(0) = 10^{-3}, 10^{-2}, 10^{-1}, 0.5, 1, 10, 10^5$.

with $\mathcal{K} = \mathcal{K}_s + \mathcal{K}_d$, $M_t = \sqrt{\mathcal{K}}/a_0$ and

$$\chi(t) = \frac{\mathcal{K}_d(t)}{\mathcal{K}_s(t) + \mathcal{K}_d(t)} = \frac{\mathcal{K}_d(t)}{\mathcal{K}_s(0) + \mathcal{K}_d(t)}. \tag{9.110}$$

The equilibrium values can be rewritten as

$$\overline{(p'p')}_\infty = \frac{1}{2}\overline{p'p'}(0)\left[1 + \mathcal{F}(0)\right], \tag{9.111}$$

$$(\mathcal{K}_d)_\infty = \frac{1}{2}\mathcal{K}_d(0)\left[1 + \frac{1}{\mathcal{F}(0)}\right]. \tag{9.112}$$

A very interesting result obtained by inserting equilibrium values into (9.109) is that the acoustic-equilibrium value of Sarkar's function, $\mathcal{F}_\infty$, is equal to unity:

$$\mathcal{F}_\infty = \lim_{t \to \infty} \mathcal{F}(t) = 1. \tag{9.113}$$

This result indicates that at acoustic equilibrium there is an equipartition between the kinetic [numerator of Eq. (9.109)] and the potential component [half the denominator of Eq. (9.109)] of the compressible energy. One obtains the relative weights of the incompressible and compressible parts of the kinetic energy by evaluating the acoustic-equilibrium value of the parameter χ:

$$\chi_\infty = \chi(0)\frac{1 + \mathcal{F}(0)}{2\mathcal{F}(0) + \chi(0)[1 - \mathcal{F}(0)]}. \tag{9.114}$$

Main features of the weak acoustic-equilibrium state are illustrated in Figs. 9.3 and 9.4. It is clearly seen that a low value of $\mathcal{F}(0)$ yields a very rapid increase of

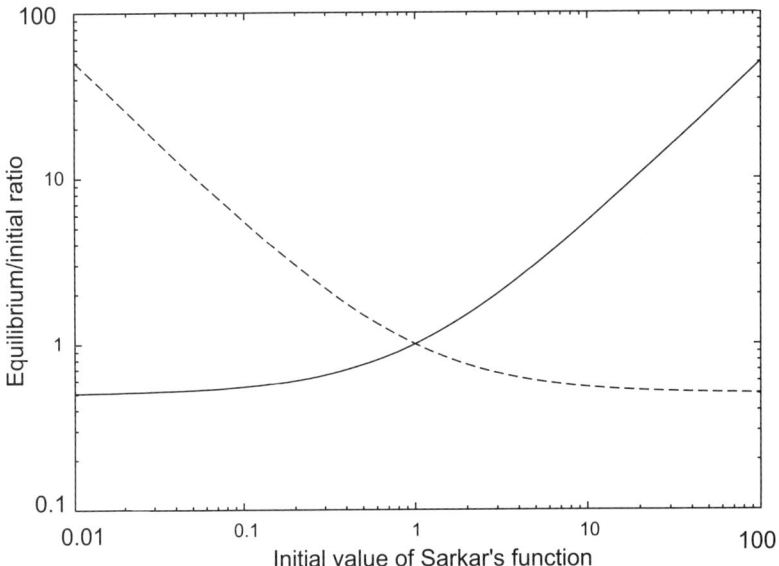

Figure 9.4. Acoustic-equilibrium value referred to the initial value as a function of $\mathcal{F}(0)$. Dashed curve, dilatational velocity field; solid curve, dilatational pressure.

the kinetic-energy ratio in terms of $\chi(0)$, meaning that initial strong nonequilibrium leads to very rapid changes in the solution for $\mathcal{F}(0) \ll 1$. This relaxation is also seen to yield a dramatic change in the repartition of the acoustic energy. An interesting conclusion is that the linear dynamics that corresponds to the pseudo-acoustic regime is compatible with very important changes in the dilatational field.

9.3.1.2 The Relevant Incompressible Limit for Both Spectra of Solenoidal Energy and Pressure Variance

Before the spectra and cospectra related to dilatational and pressure components are modeled, a definition of the incompressible reference is the first mandatory task.

At low Mach number, it is possible to neglect the feedback from dilatational and pressure modes in Eq. (9.91), so that the spectrum of the solenoidal mode, $E_{ss}(k)$, is given as in strictly incompressible turbulence. Therefore a closed form for $T_{ss}(k)$ is obtained by a classical isotropic incompressible EDQNM model, and $E_{ss}(k)$ exhibits a classical Kolmogorov inertial range at high Reynolds number. This is consistent with a selection among all resonant triads, neglecting all resonant triads involving waves with respect to the pure solenoidal ones (more details are given Chapter 15). In a similar way, neglecting all triads involving at least one wave mode versus pure vortex interactions yields selecting the toroidal cascade in Chapter 7 and the QG cascade in Chapter 8.

The calculation of the spectrum of the pure incompressible part of the pressure fluctuation, denoted by E_{pp}^{inc}, is performed starting from the Poisson equation and

9.3 Different Regimes in Compressible Turbulence

using a QN approximation as in Batchelor's approach.** As a result, the following robust model of $E_{pp}(k)$ as a function of $E_{ss}(k)$ is obtained:

$$E_{pp}^{inc}(k) = \frac{1}{\bar{\rho}^2 a_0^2} \cdot \frac{\bar{\rho}^2}{2} k \iint_{\Delta_k} (1-x^2)(1-y^2) E_{ss}(p) E_{ss}(q) \frac{dpdq}{pq}, \quad (9.115)$$

using the same notations as for the isotropic incompressible EDQNM transfer term (see Subsection 3.5.7).

The first prefactor $1/(\bar{\rho}^2 a_0^2)$ originates in pressure scaling (9.77) and can be ommitted in order to interpret $E_{pp}^{inc}(k)$ as the spectrum of the pressure variance with its original dimension. Even if the preceding integral cannot be analytically solved in general, this equation is consistent with a pressure spectrum shape at small k and the scaling $kE^2(k) \sim k^{-7/3}$, at larger k, for a Kolmogorov energy spectrum.

9.3.1.3 Quasi-Inviscid Limit: Toward an Extended Wave-Turbulence Model

In the low-Mach-number case, acoustic perturbations travel at a much higher speed than hydrodynamic fluctuation, and a two-time-scale problem can be defined. Associating the fast time scale with acoustic perturbations and the slow time scale with hydrodynamic fluctuations, a wave-turbulence-type problem is obtained whose governing equations can be expressed in terms of the slow amplitudes $a_\alpha^{(0)}$, $\alpha = 1, 2$ and $a^{(s)}$, which are defined as

$$u^{(\alpha)} = a_\alpha^{(0)}(\mathbf{k}, t), \quad \alpha = 1, 2, \quad (9.116)$$

$$u^{(3)} + s u^{(4)} = e^{i s a_0 k t} a^{(s)}(\mathbf{k}, t), \quad s = \pm 1. \quad (9.117)$$

Here, $\mathbf{a}^{(0)}$ denotes the amplitude of the vortical nonpropagating mode and $a^{(s)}, s = \pm 1$ are related to the amplitudes of wavy acoustic modes. General properties of the associated nonlinear system of equations are discussed in Chapter 14. Nonlinear terms, once expressed as functions of these new variables, still involve convolution products inherited from their quadratic nature. The exact inviscid equation can be written as

$$\dot{a}^{(s)} = \sum_{s'=0,\pm 1, s''=0,\pm 1} \iiint_{\mathbf{k}+\mathbf{p}+\mathbf{q}=0} N_{ss's''}(\mathbf{k}, \mathbf{p}) e^{i a_0 t (sk + s'p + s''q)} a^{(s')*}(\mathbf{p}, t) a^{(s'')*}(\mathbf{q}, t) d^3\mathbf{p},$$

(9.118)

up to some formal difficulties: Because $\mathbf{a}^{(0)}$ is a two-component *vector*, the index s no longer refers to the solenoidal part in this particular subsubsection, and only the value $s = 0$ is related to the solenoidal mode. The influence matrix $N_{ss's''}$ is derived from (9.82) in a straightforward – but tedious – way. As in similar cases of rotating (which do not involve any $s = 0$ mode), stratified (in which $s = 0$ would correspond to the toroidal + VSHF mode), and rotating stratified (in which

---

** The pressure variance is linked to fourth-order velocity correlations by means of the Poisson equation, and fourth-order correlations are factorized in term of products of second-order ones by means of a QN approximation.

$s = 0$ would correspond to the QG mode), an interesting feature is that all products $a^{(s')*}(\boldsymbol{p}, t) a^{(s'')*}(\boldsymbol{q}, t)$ present in the convolution integral governing $a^{(s)}(\boldsymbol{k}, t)$ are weighted by the following resonance operator:

$$\exp[\imath a_0 t (sk + s'p + s''q)], \, s, s', s'' = 0, \pm 1, \quad \boldsymbol{k} + \boldsymbol{p} + \boldsymbol{q} = 0. \qquad (9.119)$$

Different interactions are characterized only by the set (s, s', s''):

- *pure vortex* (solenoidal here) triadic interactions associated with $(0, 0, 0)$,
- *pure wavy* triadic interactions corresponding to $(\pm 1, \pm 1, \pm 1)$,
- *mixed* triadic interactions with $(0, \pm 1, 0)$ or $(0, \pm 1, \pm 1)$. This last class is assumed to be very weak with respect to pure vortex interactions when the nonlinear transfer term T_{ss} is modeled.

The generation of all transfer terms using asymptotic QNM theory readily follows, but the absence of a relevant ED term (with vanishing eddy viscosity) would generate an inertial range with a k^{-2} slope and not a $k^{-5/3}$ one for T_{ss}. The optimal compromise between "strong" turbulence and "weak" wave-turbulence theory, is to introduce an ED correction in generating the typical Green's function (or Kraichnan's response function) as follows:

$$\frac{\partial}{\partial t} \begin{pmatrix} u^{(1)} \\ u^{(2)} \\ u^{(3)} \\ u^{(4)} \end{pmatrix} + \begin{bmatrix} \eta & 0 & 0 & 0 \\ 0 & \eta & 0 & 0 \\ 0 & 0 & \epsilon_{\text{acous}} & -a_0 k \\ 0 & 0 & a_0 k & \epsilon_{\text{acous}} \end{bmatrix} \begin{pmatrix} u^{(1)} \\ u^{(2)} \\ u^{(3)} \\ u^{(4)} \end{pmatrix} = \begin{pmatrix} X^{(1)} \\ X^{(2)} \\ X^{(3)} \\ X^{(4)} \end{pmatrix}. \qquad (9.120)$$

The arbitrary vector $(X^{(1)}, X^{(2)}, X^{(3)}, X^{(4)})^T$ can be replaced with a Dirac term (impulsional response) in the most general definition of the Green's function, but the same result (identifying the response function once and for all) is most easily obtained from the general initial-value problem with $(X^{(1)}, X^{(2)}, X^{(3)}, X^{(4)})^T = (0, 0, 0, 0)^T$ (Cambon and Scott, 1999). *It is very important to stress that the preceding system of equations is used for generating only the nonlinear Green's function, and for solving only corresponding equations for triple correlations needed in the integrands of T_{dd}, T_{pp}, and T_{pp}, or equivalently for $T_{pp} + T_{dd}$ and $T^{(z)}$.* The ED term η can be chosen as in Subsection 3.5.7, and ϵ_{acous} is a formal small parameter, used only for the sake of mathematical regularization of the resonance operator. Of course, the very high-Reynolds-number limit allows us to get rid of details for the laminar viscous terms, which are displayed in the next subsubsection.

9.3.1.4 Introducing Relevant Eddy-Damping. Main Results

A first application of the EDQNM procedure was performed by Marion and coworkers [Marion et al. (1998a, 1988b)], with some inaccuracies corrected by Bataille (1994), and new numerical results given in Bataille and Zhou (1999) and Bertoglio, Bataille, and Marion (2001). This procedure follows the one proposed by Leslie (1973) and invokes Kraichnan's DIA theory as an intermediate step, using two-point

9.3 Different Regimes in Compressible Turbulence

spectral tensors of the form $\hat{R}_{ij}(\mathbf{k}, t, t')$, before deriving EDQNM-type equations.[††] One can reinterpret the system generating the nonlinear response function as

$$\frac{d}{dt}\begin{pmatrix} u^{(1)} \\ u^{(2)} \\ u^{(3)} \\ u^{(4)} \end{pmatrix} + \begin{bmatrix} \eta & 0 & 0 & 0 \\ 0 & \eta & 0 & 0 \\ 0 & 0 & (4/3)\nu k^2 + \eta & -a_0 k \\ 0 & 0 & a_0 k & \eta \end{bmatrix} \begin{pmatrix} u^{(1)} \\ u^{(2)} \\ u^{(3)} \\ u^{(4)} \end{pmatrix} = \begin{pmatrix} X^{(1)} \\ X^{(2)} \\ X^{(3)} \\ X^{(4)} \end{pmatrix}. \quad (9.121)$$

The choice of the ED term $\eta(k)$ for the compressible nonlinear terms associated with $E_{dd}(k)$, $E_{pp}(k)$, and $E_{dp}(k)$ has a dramatic influence on the results. The choice of the same $\eta(k)$ for all modes amounts to introducing the very simple and unique decorrelation function $e^{-\eta(t-t')}$. This procedure is not theoretically grounded and must therefore be considered as an empirical closure, as it relies on the direct use of a damping term built for solenoidal modes for the nonlinear interactions involving dilatational modes, which obey very different physics.

Two main results are obtained: First, the acoustic equilibrium is recovered in a strong sense; second, a typical slope for the pressure spectrum is found, as $-7/2$ (Marion et al., 1998a), then $-11/3$ (Bataille, 1994; Bataille and Zhou, 1999; Bertoglio, Bataille, and Marion, 2001), but in any case that of Batchelor, $-7/3$, is not recovered in the incompressible limit $M_t \to 0$. Another reported problem is that strong acoustic equilibrium, together with the high level of pressure spectrum in the inertial range, yields overestimating the level of E_{dd}.

These results, considered at least as puzzling and somehow unphysical, motivated Fauchet (1998) and Fauchet and Bertoglio (1999a, 1999b) to choose a new decorrelation function. Mentioning some informal proposal made by Kraichnan, Fauchet and Bertoglio proposed replacing the usual ED factor $-\eta(t-t')$ with $\eta^2(t-t')^2$ for the dilatational and pressure modes. This result can be recast in a more general way. Instead of renormalizing the pure dissipative laminar term, one may try to renormalize the dispersion frequency of acoustic waves, so that the relevant response function would be generated by

$$\frac{\partial}{\partial t}\begin{pmatrix} u^{(1)} \\ u^{(2)} \\ u^{(3)} \\ u^{(4)} \end{pmatrix} + \begin{bmatrix} \eta & 0 & 0 & 0 \\ 0 & \eta & 0 & 0 \\ 0 & 0 & (4/3)\nu k^2 + \epsilon_{acoust} & -a_0 k - r \\ 0 & 0 & a_0 k + r & \epsilon_{acoust} \end{bmatrix} \begin{pmatrix} u^{(1)} \\ u^{(2)} \\ u^{(3)} \\ u^{(4)} \end{pmatrix} = \begin{pmatrix} X^{(1)} \\ X^{(2)} \\ X^{(3)} \\ X^{(4)} \end{pmatrix}. \quad (9.122)$$

In other words, η is not used as an additional nonlinear dissipative effect for dilatational pressure and dilatational modes. It is replaced with a nonlinear correction $r(t)$ for the purely linear acoustic dispersion frequency $a_0 k$. The too simple choice $r \sim \eta$ is not correct, because the decorrelation effect cannot be obtained with a deterministic r, and only a modified resonance operator would be generated in Eq. (9.119), changing $a_0(t-t')$ into $(a_0+r)(t-t')$. As a more subtle interpretation,

---

[††] A direct procedure for solving linear operators at the level of triple corelations, however, would probably be more general and more convenient for mathematical analysis, like the one for deriving EDQNM1 to EDQNM3, the latter giving wave turbulence in the limit of inviscid wave propagator.

Table 9.3. *Two-point closure prediction dealing with inertial range in the low-Mach-number regime ($M_t < 0.1$)*

Decorrelation function	$\sim \exp[-\eta(k)(t-t')]$	$\sim \exp[-\eta^2(k)(t-t')^2]$
$E_{dd}(k)$	$\propto M_t^2 Re_L^1 k^{-11/3}$	$\propto M_t^4 Re_L^0 k^{-3}$
$E_{pp}(k)$	$\propto M_t^2 Re_L^1 k^{-11/3}$	$\propto M_t^2 Re_L^0 k^{-7/3}$
$E_{p'p'}^{\text{acous}}(k)$	$\sim E_{dd}(k)$	$\propto M_t^6 Re_L^0 k^{-11/3}$
$\lim_{M_t \to 0} E_{pp}(k)$	$\neq E_{pp}^{\text{inc}}(k)$	$= E_{pp}^{\text{inc}}(k)$
$\mathcal{K}_d/\mathcal{K}_s$	$\propto M_t^2 Re_L^1$	$\propto M_t^4 Re_L^0$
$\bar{\varepsilon}_d/\bar{\varepsilon}_s$	$\propto M_t^2 Re_L^0$	$\propto M_t^4 Re_L^{-1} \ln(Re_L)$

Notes: Left and right columns display results given by EDQNM and the improved Fauchet–Bertoglio model, respectively. $E_{p'p'}^{\text{acous}}(k)$ denotes the spectrum of the acoustic pressure fluctuation defined as $p' = p - p_s$, where p_s is the pressure field associated with the solenoidal velocity field $\mathbf{u}_s$. The last two lines summarize results dealing with the ratio of solenoidal-dilatational kinetic energy and dissipation, respectively. Adapted from Fauchet (1998).

r is really a random factor (hence our coining r for random), changing from a realization of $[u^{(3)}, u^{(4)}]$ to another one. One must assume that r is a Gaussian process, having zero mean and variance η. This point will be further discussed in the next section using the simple example of Kraichnan's random oscillator explained in Kaneda (2007).

The main results obtained with this improved model are the following (predictions related to the inertial range are gathered in Table 9.3, and a simplified analytical model will be exhaustively discussed in Subsection 9.3.1.6):

- The *strong acoustic-equilibrium hypothesis*, which states that $E_{pp}(k) = E_{dd}(k)$, is violated, showing the importance of nonlinear effects. This is seen in Fig. 9.5, which displays computed spectra in the nonlinear equilibrium state.
- Strong acoustic equilibrium is now observed at very small k only (and not over the entire inertial range), at scales really corresponding to acoustic wavelengths that are much larger than the integral velocity length scale given by $E_{ss}(k)$.
- In the inertial range, the $E_{pp}(k)(k)$ spectrum almost collapses with its incompressible counterpart $E_{pp}^{\text{inc}}(k)$, with a related $-7/3$ slope, but $E_{dd}(k)$ is found far below, with a -3 slope. This last result is consistent with a much smaller magnitude of the dilatational motion with respect to the evaluation consistent with strong acoustic equilibrium [which yields the total collapse $E_{pp}^{\text{inc}}(k) \sim E_{pp}(k) \sim E_{dd}(k)$]. The scaling is now $E_{dd} \sim M_t^4 k^{-3}$ in the inertial range. The corresponding behavior of the compressibility ratio χ defined by Eq. (9.110) as a function of the turbulent Mach number is therefore $\chi \sim M_t^4$, and not $\chi \sim M_t^2$, as suggested by Bataille (1994). The fact that the potential energy of waves, with an $E_{pp}(k)$ spectrum, can so greatly exceed their kinetic energy, with an $E_{dd}(k)$ spectrum, even questions the very concept of acoustic waves. This observation led many authors, including historical specialists of aeroacoustics, such as Lighthill, to refer to this state as the *pseudo-sound regime* instead of real acoustics.

9.3 Different Regimes in Compressible Turbulence

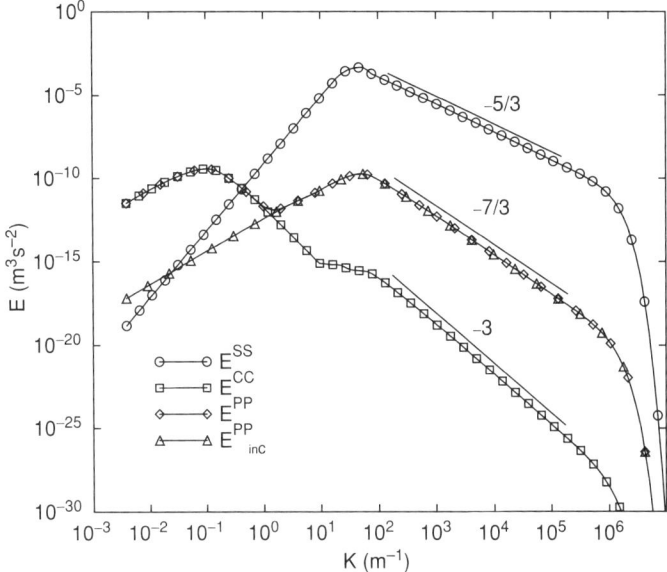

Figure 9.5. Spectra in the nonlinear equilibrium state predicted using an extended EDQNM-type closure for compressible flows (the dilational energy spectrum is denoted by E^{cc} instead of E_{dd}). Courtesy of G. Fauchet and J. P. Bertoglio.

- It is worth noting that previous results hold in the low-Mach-number regime only. For $M_t > 0.1$, the pseudo-sound regime disappears and the k^{-3} inertial range is not observed anymore on $E_{dd}(k)$. A $k^{-5/3}$ is recovered for M_t close to 1 according to the improved two-point closure, but this result must be considered with care because several assumptions that underlie this theory are not satisfied anymore.
- The ratio $\bar{\varepsilon}_d/\bar{\varepsilon}_s$ is observed to scale as M_t^5 for $M_t > 0.2$ instead of M_t^4 at lower Mach numbers (see Fig. 9.6).

9.3.1.5 Additional Discussion About the Modified Decorrelation Function

The use of the modified decorrelation function $e^{-\eta^2(t-t')^2}$ is the essential improvement brought by Fauchet with respect to earlier developments. Now we show that the Gaussian form is suggested by the "random oscillator," Kraichnan's toy model, rediscussed by Leslie (1973) and Orszag (1977), with an excellent survey given by Kaneda (2007). The starting point is the following single-mode model:

$$\frac{d}{dt}g(t) = -\{\imath[b_0 + b(t)] + \nu_0\}g(t), \quad g(0) = 1,$$

in which b_0 holds for the acoustic frequency $a_0 k$, $b(t)$ for a time-dependent, *possibly random*, contribution to the former, and ν_0 for a viscous parameter, e.g., proportional to νk^2. The exact solution of this single-mode equation is

$$g(t) = \exp[-(\nu_0 - \imath b_0)t - \imath \int_0^t b(s)ds],$$

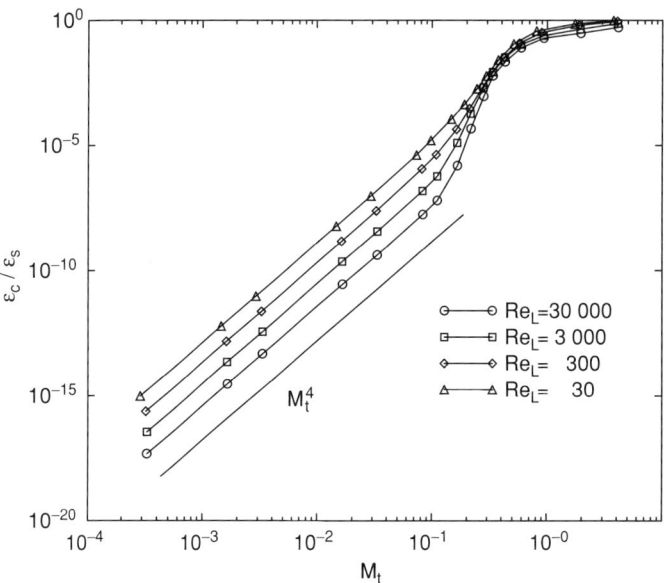

Figure 9.6. Ratio of the dilatational dissipation (denoted by ε^c here) over the solenoidal dissipation (denoted by ε^s here) in the nonlinear equilibrium state predicted with an extended EDQNM-type closure for compressible flows, vs. the turbulent Mach number for different values of the turbulent Reynolds number. Courtesy of G. Fauchet and J. P. Bertoglio

whereas the solution for its ensemble average is

$$\langle g(t) \rangle = \exp\left[-(\nu_0 - \iota b_0)t - \frac{1}{2}\iint_0^t \langle b(s)b(s')\rangle ds'\right].$$

In particular, the latter equation reduces to

$$\langle g(t) \rangle = \exp\left[-(\nu_0 - \iota b_0)t - \frac{1}{2}\sigma^2 t^2\right],$$

with $\sigma^2 = \langle b^2 \rangle$, if b is a real time-independent Gaussian process with zero mean. It is recalled that $\langle \exp(\iota s b) \rangle = \exp(-\sigma^2 t^2/2)$ for such a process.

Of course, $g(t)$ is a scalar function here, but the analogy between its equation and the subsystem $[u^{(3)} - u^{(4)}]$, with $X^{(3)} = X^{(4)} = 0$ in (9.122) is obvious, considering $u^{(3)} \pm \iota u^{(4)}$ and removing the laminar factor $(4/3)\nu k^2$, with $b(t) = r(t)$.

This simple analysis shows that a Gaussian part of the renormalized response function can be generated by a random contribution added to the mean frequency of linear acoustic waves. Going back to our specific problem of weakly nonlinear compressible turbulence, this suggests that a renormalization of the viscous terms is likely less crucial – and even less adequate – than a renormalization of the dispersion frequency. Main conclusions and semi-open questions are as follows:

- Pure acoustic-wave turbulence, corresponding to triadic interactions without contribution of solenoidal modes, i.e., with $(s = \pm 1, s' = \pm 1, s'' = \pm 1)$, is a marginally relevant model. It probably preserves the acoustic equilibrium, without

9.3 Different Regimes in Compressible Turbulence

the need to renormalize either the viscous factors or the acoustic frequency. One can expect a strong analogy with inertial-wave rotating turbulence from this point of view, as damping terms are needed only to regularize the resonance operators in the high-Reynolds-number limit.

- Modeling issues not present in pure wave turbulence occur when strong turbulence (e.g., usual solenoidal turbulence) interacts with wave turbulence. The concept of waves, which is associated with a balance between potential and kinetic energy, may even become irrelevant because the compressible state can be very far from acoustic equilibrium if k is not too small, yielding E_{pp}(potential) $\gg$ E_{dd}(kinetic).
- Breakdown of acoustic equilibrium in the closure model seems to be linked to the introduction of a Gaussian factor in the response tensor, possibly resulting from the renormalization of the acoustic-wave frequency, rather than from renormalized viscosity. The Gaussian decorrelation factor can inhibit the time memory of triple correlations in a more efficient way than the classical exponential term does. As a result, typical oscillations in resonance operator (9.119) are inhibited too.
- There is no physical explanation for choosing the order of magnitude of the renormalized dispersion frequency as $\langle r^2 \rangle = \eta^2$ in Eq. (9.122). It is also suggested that the extended wave-turbulence models using (9.120) can be of interest too. The breakdown of the "strong" acoustic equilibrium is also possible in such a model. This is evidenced by the linearly sheared (with shear rate S) flow model discussed in Chapter 10, in which there is a source term [$\sim Su^{(2)}$ in the shear case, whereas T_{dd} plays this role in the present case] induced by the coupling with the solenoidal mode in the equation for $u^{(3)}$. Because this source term has no counterpart in the equation for $u^{(4)}$, it can break the acoustic equilibrium at sufficiently large values of the $S/(a_0 k)$ parameter.

9.3.1.6 Analytical Fauchet–Bertoglio Model

In the absence of analytical results from the extended wave-turbulence model just discussed, let us give more details provided by the model proposed in Fauchet (1998). Carrying out an asymptotic analysis in the limit of very low Mach numbers and very high Reynolds numbers, and considering the following solenoidal kinetic-energy spectrum model

$$E_{ss}(k) = \begin{cases} Bk^\sigma & k < k_L \\ K_0 \bar{\varepsilon}_s^{2/3} k^{-5/3} & k_L \leq k \leq k_\eta, \\ 0 & k > k_\eta \end{cases} \quad (9.123)$$

where

$$B = K_0 \bar{\varepsilon}_s^{2/3} k_L^{-5/3-\sigma}, \quad (9.124)$$

Fauchet and Bertoglio obtained analytical models for both nonlinear transfer terms and related spectra. The analysis is restricted to $T_{dd}(k)$ and $T_{dp}(k)$, because $T_{ss}(k)$

is not modified by compressiblity effects and there is no feedback of the the dilatational part of the solution on u_s. The leading-order terms for the local and nonlocal tranfers in $T_{dp}(k)$ are

$$T_{dp}(k) \simeq \begin{cases} \underbrace{\dfrac{k^{2(1+\sigma)}}{a_0}}_{\text{I}} + \underbrace{\dfrac{8}{65} \dfrac{K_0^2}{a_0} \bar{\varepsilon}_s^{4/3} \dfrac{k^3}{k_L^{13/3}}}_{\text{II}} & (k < k_L) \\ \underbrace{a_0 k E_{pp}^{\text{inc}}(k)}_{\text{III}} + \underbrace{\left(\dfrac{k}{k_L}\right)^{4/3} a_0^2 k^2 E_{pp}^{\text{inc}}(k)}_{\text{IV}} & (k \geq k_L) \end{cases}, \qquad (9.125)$$

in which terms I and III are related to local interactions, whereas nonlocal intercations are grouped in contributions II and IV. The exact form of the incompressible pressure spectrum $E_{pp}^{\text{inc}}(k)$ is not known at this stage. One just has to know that $E_{pp}^{\text{inc}}(k) \propto k^{-7/3}$ in the inertial range. A careful analysis of the relative amplitude of these terms shows that nonlocal transfer term IV is dominant at small wavenumbers such that $k < k_P$, whereas the local interaction term III is dominant at higher wavenumbers $k \geq k_P$. The threshold wavenumber is evaluated by the following formula:

$$k_P = \left(\dfrac{65}{16} C_G\right)^{3/13} k_L \simeq 1.47 k_L, \quad C_G \simeq 1.32. \qquad (9.126)$$

Now using the relation

$$T_{dp}(k) = a_0 k E_{pp}^{\text{inc}}(k), \qquad (9.127)$$

one obtains the following expression for the incompressible pressure spectrum.

Taking $\sigma = 4$, the corresponding expression for the incompressible pressure spectrum is

$$E_{pp}^{\text{inc}}(k) = \begin{cases} \dfrac{8}{65} K_0^2 \bar{\varepsilon}_s^{4/3} \dfrac{k^2}{a_0^2 k_L^{13/3}} & k < k_P \\ \dfrac{C_G}{2a_0^2} K_0^2 \bar{\varepsilon}_s^{4/3} k^{-7/3} & k_P \leq k \leq k_\eta \\ 0 & k > k_\eta \end{cases}. \qquad (9.128)$$

Therefore it is seen that the incompressible pressure spectrum and the solenoidal kinetic-energy spectrum do not have their maxima at the same wavenumbers, because $k_P > k_L$. It is worth noting that the exact expression for $E_{pp}^{\text{inc}}(k)$ in the inertial range, i.e., $k \geq k_P$, is not a direct output of the asymptotic analysis that is

9.3 Different Regimes in Compressible Turbulence

supplemented by an auxiliary model. The dilatational nonlinear term can be expanded as

$$T_{dd}(k) \simeq \begin{cases} \underbrace{\frac{4}{3}\nu\frac{k^{(4+3\sigma)}}{a_0^4}}_{\text{V}} + \underbrace{\frac{64}{5}\beta^{3/2}\sqrt{2\pi}K_0^{7/2}\frac{\bar{\varepsilon}_s^{7/3}}{a_0^4 k_L^{7/3}}F_p\left(\frac{k}{k_{\text{acous}}}\right)}_{\text{VI}} \\ \quad + \underbrace{\frac{64}{15}\beta\frac{4}{3}\nu K_0^3 \frac{\bar{\varepsilon}_s^2}{a_0^4 k_L^3}k^2}_{\text{VII}} \qquad (k < k_L) \\ \underbrace{12\frac{4}{3}\nu\frac{\beta}{a_0}k^3 E_{ss}(k)E_{pp}^{\text{inc}}(k)}_{\text{VIII}} + \underbrace{k^{-7/3}\exp(-\beta a_0^2 k^{2/3})}_{\text{IX}} \\ \quad + \underbrace{\frac{4}{3}\nu\frac{k^{-7/3}}{a_0^4}}_{\text{X}} \qquad (k \geq k_L) \end{cases} \quad , (9.129)$$

with $\beta \simeq 0.2$, where the function F_p in term VI is defined as

$$F_p(x) = \frac{\Gamma(15/4) - \Gamma(15/4, x^2)}{x^7/2}. \tag{9.130}$$

The parameter k_{acous} denotes the wavenumber associated with the peak of the acoustic spectrum. It is evaluated as

$$k_{\text{acous}} = \frac{2\sqrt{2\beta}}{C_b} M_t k_L, \quad C_b = \sqrt{\frac{3\sigma+5}{3(\sigma+1)}}. \tag{9.131}$$

Here, local interactions are represented by terms V and VIII, whereas other terms are related to nonlocal contributions.

The analysis of the relative amplitudes reveals the existence of three different spectral zones:

- The *acoustic region*, which corresponds to very small wavenumbers such that $k < k_{r1}$, with

$$k_{r1} = \frac{3}{4}(10C_G)^{1/3}k_L \simeq 1.8k_L. \tag{9.132}$$

In this region, nonlocal term VI is dominant. It is therefore interpreted as the production of an acoustic wave. The spectrum of acoustic production, $\mathcal{P}_{\text{acous}}(k)$, can therefore be rewritten as

$$\mathcal{P}_{\text{acous}}(k) = (\text{VI}) = \frac{256}{15}\frac{\beta}{C_b^7}\sqrt{\frac{\pi\beta}{3}}\mathcal{K}_s^{3/2}M_t^4 F_p\left(\frac{k}{k_{\text{acous}}}\right). \tag{9.133}$$

The total radiated acoustic power, P_{tot}, is then equal to

$$P_{\text{tot}} = \int_0^{k_\eta} \mathcal{P}_{\text{acous}}(k)dk = 4.2 C_b^{-5}\bar{\varepsilon}_s M_t^5, \tag{9.134}$$

which is in very good agreement with the usual estimates found by using the acoustic Lighthill analogy. Relation (9.133) does not yield the right estimate for both the value and the location of the peak of the normalized acoustic production spectrum $\mathcal{P}^*_{\text{acous}}(k) \equiv \mathcal{P}_{\text{acous}}(k)/P$. Using the normalized frequency $\omega^* = a_0 k u'^2/\bar{\varepsilon}_s$, one recovers the correct prediction $\max[\mathcal{P}^*_{\text{acous}}(\omega^*)] \simeq 0.1$ for $\omega^* = \omega^*_{\max} \simeq 3.5$ by using a realistic spectrum shape for $E_{ss}(k)$ instead of (9.123). Doing so, term VI yields a normalized production spectrum that is very close to the one proposed by Lilley in 1993:

$$\mathcal{P}^*_{\text{acous}}(\omega^*) = \frac{8}{3\pi S_t} \frac{\omega^*/(2S_t)^4}{(1+(\omega^*/2S_t)^2)^3}, \quad S_t = 1.24. \quad (9.135)$$

- A *transition region*, in which both nonlocal transfer terms IX and X play an important role, term X having the largest amplitude. This region is defined as $k_{r1} < k < k_{r2}$, where

$$k_{r2} \simeq \frac{2.5}{C_b^{5/11} C_a^{2/11}}, M_t^{7/11} Re_L^{2/11} k_L, \quad C_a = \frac{3\sigma + 5}{3\sigma}. \quad (9.136)$$

- An *inertial-range region* for $k > k_{r2}$, in which local transfer term VIII is now dominant. This inertial range is associated with the *pseudo-sound regime*, in which the dilatational velocity field is in equilibrium with the solenoidal pressure.

This simplified form of $T_{dd}(k)$ enables a detailed analysis of the dilatational energy spectrum $E_{dd}(k)$. To this end, the following expression is derived from Eq. (9.90):

$$2\left[\frac{4}{3}\nu + \nu_t^{\text{acous}}(k)\right] k^2 E_{dd}(k) = T_{dd}(k), \quad (9.137)$$

where the ED term $\nu_t^{\text{acous}}(k)$ accounts for cumulative effects of higher-order terms neglected during the derivation of the simplified form of $T_{dd}(k)$, previously given. The improved two-point closure suggests that

$$\nu_t^{\text{acous}}(k) = \frac{1}{30}\sqrt{\frac{2\pi K_0}{\beta}} \left(\frac{\bar{\varepsilon}_s}{k_L^4}\right)^{1/3} F_\nu\left(\frac{k}{k_{\text{acous}}}\right), \quad (9.138)$$

with

$$F_\nu(x) = 5\frac{1-e^{-x^2}}{x^2} - e^{-x^2}. \quad (9.139)$$

The effect of this damping term is to prevent the occurrence of an *acoustic catastrophe* in the limit of infinite Reynolds number, i.e., $E_{dd}(k)$ remains bounded. A similar term was introduced by Crow in 1967. This damping term is very small for wavenumbers larger than k_{acous}, and therefore is neglected in both the transition region and the inertial range because it is much smaller than the molecular viscosity at these scales. Combining equilibrium relation (9.137) with the simplified form of

9.3 Different Regimes in Compressible Turbulence

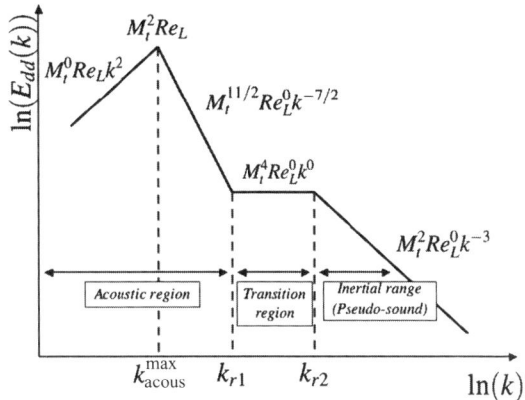

Figure 9.7. Sketch of the dilatational kinetic-energy spectrum $E_{dd}(k)$ in the nonlinear equilibrium state, according to the simplified Fauchet–Bertoglio analytical model derived from the extended two-point closure for compressible flows.

the nonlinear transfer term, one obtains (see Fig. 9.7)

$$E_{dd}(k) = \begin{cases} 32 \dfrac{\sqrt{2}\beta^{3/2}}{C_b^5 k_{\text{acous}}} M_t^3 \mathcal{K}_s F_{\text{acous}}\left(\dfrac{k}{k_{\text{acous}}}\right), & k < k_{r1} \\ \dfrac{64}{135} \dfrac{\beta}{C_b^6} M_t^4 Re_L^0 \mathcal{K}_s \dfrac{k^0}{k_L}, & k_{r1} < k < k_{r2}, \\ 2a \dfrac{C_G}{C_b^6} M_t^4 Re_L^0 \mathcal{K}_s k_L^{-1} \left(\dfrac{k}{k_L}\right)^{-3}, & k \geq k_{r2} \end{cases} \quad (9.140)$$

where

$$F_{\text{acous}}(x) = \frac{1}{x^2} \frac{F_p(x)}{F_\nu(x)}. \quad (9.141)$$

The peak of $E_{dd}(k)$ in the acoustic region observed at $k = k_{\text{acous}}^{\max} \simeq 1.32 k_{\text{acous}}$ is

$$E_{dd}(k_{\text{acous}}^{\max}) \simeq \frac{0.046}{C_b^7} \frac{\mathcal{K}_s^{5/2}}{\bar{\varepsilon}_s} M_t^2 Re_L^0. \quad (9.142)$$

In the acoustic region, further analyses show that

$$E_{dd}(k) \propto \begin{cases} M_t^0 Re_L^0 k^2, & k < k_{r1}, k \ll k_{\text{acous}}^{\max} \\ M_t^{11/2} Re_L^0 k^{-7/2}, & k < k_{r1}, k \gg k_{\text{acous}}^{\max} \end{cases}. \quad (9.143)$$

This analytical expression for $E_{dd}(k)$ enables a straightforward evaluation of the turbulent acoustic kinetic energy:

$$\mathcal{K}_{\text{acous}} = \int_0^{+\infty} 32 \frac{\sqrt{2}\beta^{3/2}}{C_b^5 k_{\text{acous}}} M_t^3 \mathcal{K}_s F_{\text{acous}}\left(\frac{k}{k_{\text{acous}}}\right) dk \propto 32 \frac{\sqrt{2}\beta^{3/2}}{C_b^5} M_t^3 \mathcal{K}_s, \quad (9.144)$$

which is in perfect agreement with Crow's scaling law for this quantity.

The compressible turbulent kinetic energy contained in the inertial range and the corresponding dilatational dissipation are found to be equal to

$$\mathcal{K}_d = \int_{k \geq k_{r2}} E_{dd}(k) dk = 0.25 C_b^{-6} \mathcal{K}_s M_t^4, \quad (9.145)$$

$$\bar{\varepsilon}_d = \frac{4}{3} \nu \int_{k \geq k_{r2}} k^2 E_{dd}(k) dk = 1.65 C_a C_b^{-5} \bar{\varepsilon}_s M_t^4 \frac{\ln(Re_L)}{Re_L}. \quad (9.146)$$

9.3.1.7 Numerical Experiments

Isotropic compressible turbulence has been investigated by several research groups by means of DNS of the full compressible Navier–Stokes equations. In most cases, the low-Reynolds-number free-decay regime is considered. The main results are as follows:

1. A statistical equilibrium is observed that corresponds to the *weak acoustic-equilibrium hypothesis* with $\mathcal{F}(t) \approx 1 = \mathcal{F}_\infty$. The function $\mathcal{F}$ fluctuates almost periodically around unity. This result is remarkable, as all couplings with internal energy are neglected in the theoretical derivation of the linear model for the pseudo-acoustic regime and the *strong acoustic-equilibrium hypothesis* is shown to be violated by both improved two-point closures and numerical simulations. These slight fluctuations might be explained by an almost periodic energy exchange between the acoustic mode and the entropy mode (i.e., the internal energy). This point is further discussed.
2. This weak equilibrium state is very robust: It is has been observed for a very wide range of turbulent Mach numbers and initial conditions. As a matter of fact, statistical equilibrium states with $\mathcal{F} \approx 1$ have been found for turbulent Mach numbers as high as 0.5 (Sarkar et al., 1991).
3. The theoretical prediction (9.114) for χ_∞ was found to be accurate for a large number of cases simulated in Erlebacher et al. (1990).

The energy balance associated with the statistical equilibrium states has been very finely analyzed in both decaying and forced isotropic turbulence (Kida and Orszag, 1990a, 1992; Miura and Kida, 1995). In these studies, the transfers between the dilatational turbulent kinetic energy $\mathcal{K}_d$, the solenoidal turbulent kinetic energy $\mathcal{K}_s$, and the fluctuating internal energy $\tilde{e}$ have been investigated. The main conclusions dealing with the global energy transfers at the equilibrium state are as follows:

1. In the acoustic-equilibrium state, both $\mathcal{K}_d(t)$ and $\tilde{e}(t)$ fluctuate sinusoidally about a constant mean value (see Fig. 9.8). The two signals are in exact phase opposition and have similar amplitude, leading to $\mathcal{K}_d(t) + \tilde{e}(t) \simeq$ constant. This is consistent with the finding that $\mathcal{F}(t)$ is nearly constant.
2. The solenoidal kinetic energy $\mathcal{K}_s$ varies slowly with irregular fluctuations of small amplitude and does not exhibit phase locking with either $\mathcal{K}_d(t)$ or $\tilde{e}(t)$.
3. The interactions between the solenoidal and compressive components of the turbulent kinetic energy are weaker than self-interactions of the respective components.
4. The pressure–dilatation term [term II in Eqs. (9.64) and (9.66)] governs the coupling between $\mathcal{K}_d(t)$ and $\tilde{e}(t)$ (see Fig. 9.9). It is also observed to overwhelm other terms that appear in the evolution equations for $\mathcal{K}_d(t)$, the total mean turbulent kinetic energy $\mathcal{K} = \mathcal{K}_d + \mathcal{K}_s$, and the internal energy $\tilde{e}$. It exhibits a periodic behavior with the same period as $\mathcal{K}_d(t)$ and $\tilde{e}(t)$, and it is due to acoustic pressure fluctuations.

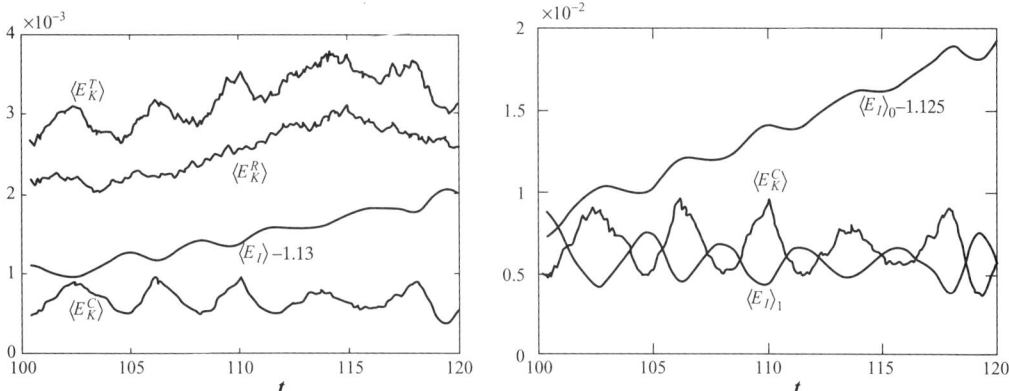

Figure 9.8. Computed time history of volume-averaged energies in forced compressible isotropic turbulence. $\langle E_K^T \rangle$, $\langle E_K^R \rangle$, $\langle E_K^C \rangle$, and $\langle E_I \rangle$ are the full turbulent kinetic energy, solenoidal kinetic energy, dilatational kinetic energy, and internal energy, respectively. Because a source term is present, the mean internal energy undergoes a constant growth and is split as the sum of a uniform part $\langle E_I \rangle_0$ and a turbulent part $\langle E_I \rangle_1$. Reproduced from Miura and Kida (1995) with permission of AIP.

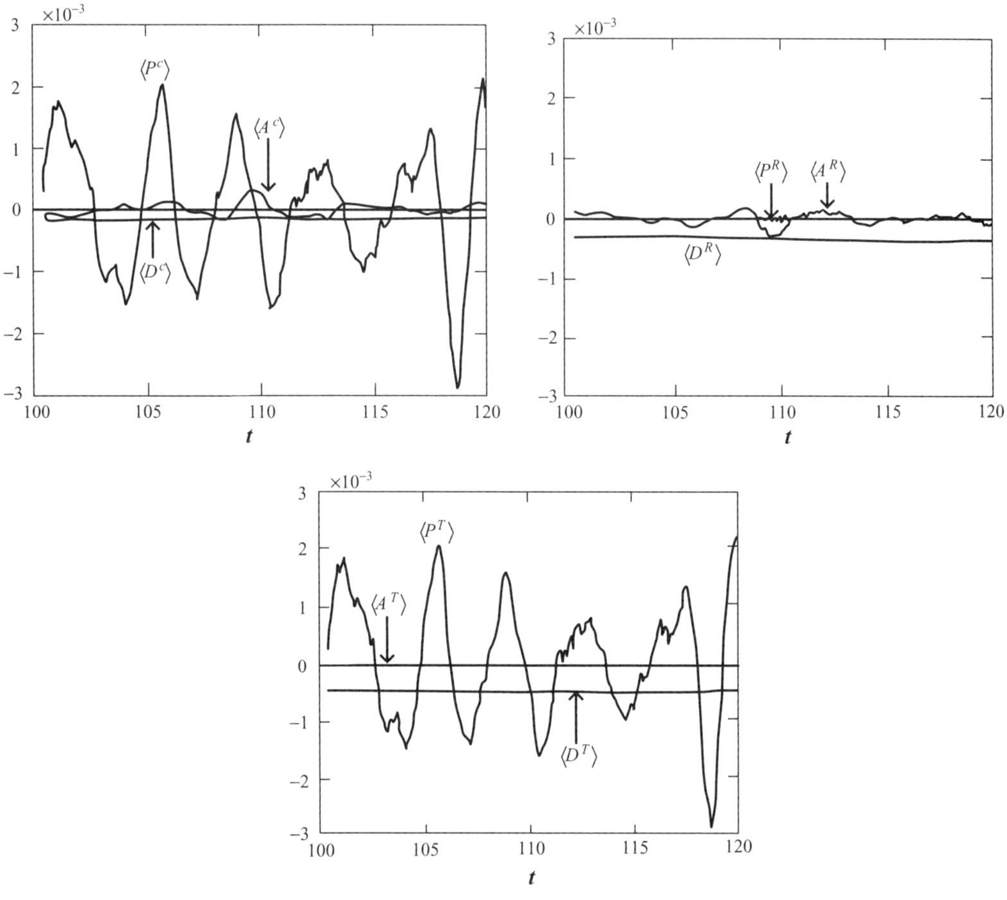

Figure 9.9. Time histories of the volume-averaged budget terms for the dilatational turbulent kinetic energy (left), the solenoidal turbulent kinetic energy (right), and the internal energy (bottom). Terms A, D, and P denote advection, viscous diffusion, and pressure terms, respectively. Reproduced from Miura and Kida (1995) with permission of AIP.

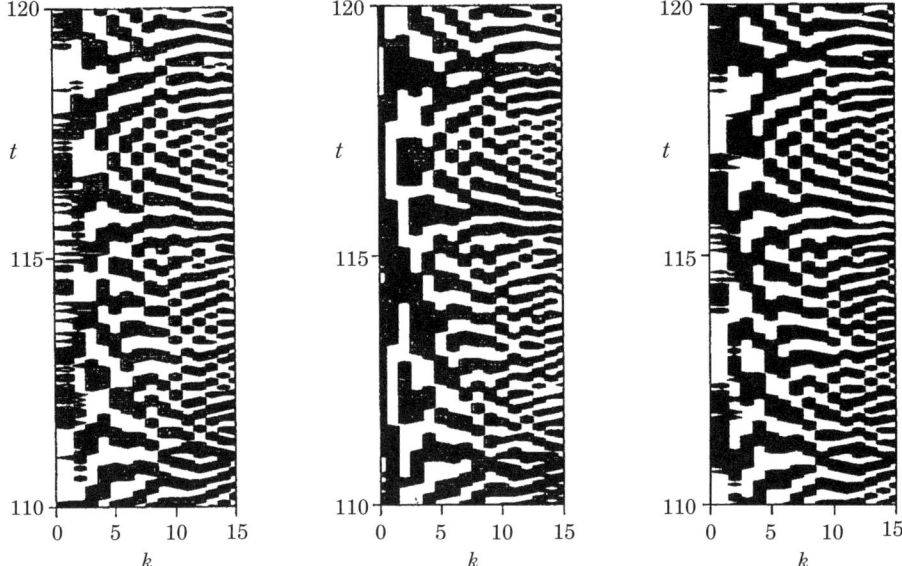

Figure 9.10. Time histories of the wavenumber spectra of dilatational turbulent kinetic energy (left), internal energy (middle), and dilatational pressure (right). White (resp. dark) regions are regions where the instantaneous spectrum coefficients are decreasing (resp. increasing) in time. Reproduced from Miura and Kida (1995) with permission of AIP.

This dynamical picture can be further refined by looking at energy exchanges at individual wavenumbers. The main findings of Miura and Kida are as follows:

1. The periodic behavior of the compressible kinetic energy and the internal energy is observed at each wavenumber in the spectra associated with these quantities, $E_{dd}(k,t)$ and $E_e(k,t)$, respectively (see Fig. 9.10). The same observation holds for the compressible pressure spectrum $E_{pp}(k,t)$.
2. The period of oscillation $\tau(k)$ depends on the wavenumber and is the same for the three spectra at each wavenumber. The measured period corresponds almost exactly to the one associated with acoustic waves:

$$\tau(k) = \frac{\pi}{\omega(k)}, \quad \omega(k) \sim \pm a_0 k. \tag{9.147}$$

At every wavenumber, it is found that the phase of oscillation of $E_{pp}(k,t)$ is in advance of those of $E_{dd}(k,t)$ and behind those of $E_e(k,t)$ by a quarter of a period.

A schematic view of the energy transfers associated with this regime is displayed in Fig. 9.11.

9.3.2 Weakly Compressible Thermal Regime

The analysis of the pseudo-acoustic regime previously presented relies on the assumption that the density and the temperature fluctuations are governed by acoustic

9.3 Different Regimes in Compressible Turbulence

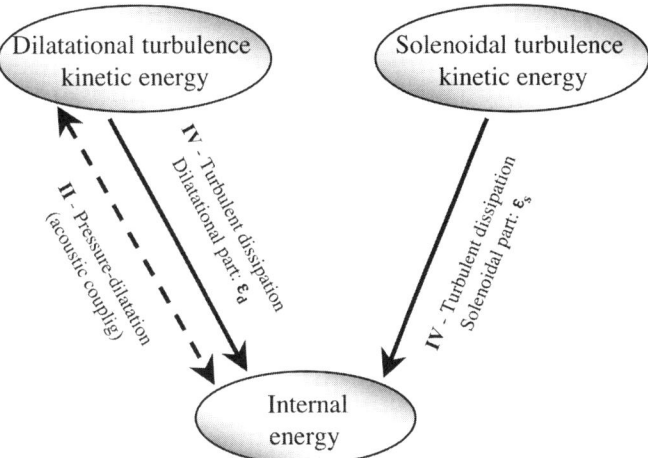

Figure 9.11. Detailed schematic view of mean-energy exchanges in compressible isotropic turbulence in the pseudo-acoustic-equilibrium state.

waves. This analysis can be extended by considering flows in which the density and temperature fluctuations are much larger than those induced by the acoustic fluctuations. In such flows, the asymptotic analysis presented in the preceding section is no longer valid and must be extended to describe the weakly compressible thermal regime.

9.3.2.1 Asymptotic Analysis and Possible Thermal Regimes

The weakly compressible thermal regime has been investigated by several authors, who proposed leading-order compressible corrections to the true incompressible Navier–Stokes equations. The following discussion puts the emphasis on the results of Bayly and co-workers (Bayly, Levermore, and Passot, 1992) and Zank and Matthaeus (1990, 1991).

The complexity of the problem is easily understood by recalling that the incompressible Navier–Stokes dynamics is recovered as the limit of the compressible Navier–Stokes equations when two small parameters are taken equal to zero: a first one, δ, related to the ratio of the fluid velocity about the speed of sound (i.e., a characteristic Mach number) and a second one, δ', related to ratio of thermal-energy scales. Therefore the problem of the relative size of these two small parameters arises when the leading-order correction to the incompressible Navier–Stokes equations is sought. Let us anticipate the following discussion to say that several regimes can be obtained, depending on the ratio of these two control parameters.

The first step consists of nondimensionalizing the full compressible Navier–Stokes equations and introducing the two small parameters. The resulting system is (Bayly, Levermore, and Passot, 1992)

$$\frac{\partial \rho}{\partial t} + \nabla \cdot (\rho \boldsymbol{u}) = 0, \tag{9.148}$$

$$\rho \left(\frac{\partial \boldsymbol{u}}{\partial t} + \boldsymbol{u} \cdot \nabla \boldsymbol{u} \right) = -\frac{1}{\delta^2} \nabla p + \frac{1}{Re} \nabla \cdot \tau + \rho \boldsymbol{f}, \tag{9.149}$$

$$\rho c_p \left(\frac{\partial T}{\partial t} + \boldsymbol{u} \cdot \nabla T \right) = \sigma \alpha T \left(\frac{\partial p}{\partial t} + \boldsymbol{u} \cdot \nabla p \right) + \frac{1}{2} \frac{\delta^2}{Re} \tau : \tau$$

$$+ \frac{1}{Re Pr} \nabla \cdot (\kappa \nabla T) + \delta' \rho q, \tag{9.150}$$

where $Re = L_r u_r \rho_r / \mu_r$ and $Pr = \mu_r c_p / \kappa_r$ are the Reynolds number and Prandtl number, respectively, α denotes the thermal expansion parameter, and $\sigma \equiv T_r \alpha_r$. The subscript r is related to reference scales. The system is supplemented by the perfect gas law (9.4). The characteristic speed u_r is associated with fluid velocity, and a reference thermal speed $v_r = \sqrt{c_p T_r}$ is introduced. The reference pressure is defined as $p_r = \rho_r v_r^2$. Here, q is a nondimensional thermal forcing term that accounts for the presence of heat sources–sinks in the flow (e.g., reactive flows). The two small parameters are defined as follows:

$$\delta \equiv \frac{u_r}{v_r} = \sqrt{\gamma - 1} M_r, \quad \delta' \equiv \frac{q_r}{v_r^2}, \tag{9.151}$$

where $M_r = u_r / a_r$ is the usual reference Mach number and q_r is a characteristic heat production scale.

The existence of these two scaling parameters yields a formal double expansion problem, whose treatment is cumbersome. To avoid such a complex development, Bayly and co-workers set $\delta' = \delta^{2/l}$, where l is a positive integer. By use of this relationship, all dynamic quantities are expanded in asymptotic series of the form (here expressed for a dummy variable ϕ):

$$\phi = \phi^{(0)} + \delta^{2/l} \phi^{(1)} + \delta^{4/l} \phi^{(2)} + \cdots . \tag{9.152}$$

Assuming that the fluid fluctuates close to the reference state, the zeroth-order terms for density and temperature must have their values for that state, yielding $\rho^{(0)} = T^{(0)} = 1$. An immediate consequence of the perfect gas law is that the zeroth-order pressure term $p^{(0)}$ also corresponds to a uniform field. Therefore the leading-order fluctuating field that accounts for the weak compressible thermal turbulence is made up of $\boldsymbol{u}^{(0)}, \rho^{(1)}, T^{(1)}$, and $p^{(1)}$.

Two cases can be defined that correspond to different thermal regimes:

1. *Relatively small external heating* with respect to both the viscous heating and the pressure-induced temperature fluctuations: $l = 1$. In this case, the lowest-order nontrivial equations are

$$\nabla \cdot \boldsymbol{u}^{(0)} = 0, \tag{9.153}$$

9.3 Different Regimes in Compressible Turbulence

$$\frac{\partial \boldsymbol{u}^{(0)}}{\partial t} + \boldsymbol{u}^{(0)} \cdot \nabla \boldsymbol{u}^{(0)} = -\nabla p^{(1)} + \frac{1}{Re}\nabla^2 \boldsymbol{u}^{(0)}, \qquad (9.154)$$

$$\frac{\partial T^{(1)}}{\partial t} + \boldsymbol{u}^{(0)} \cdot \nabla T^{(1)} = \sigma \left[\frac{\partial p^{(1)}}{\partial t} + \boldsymbol{u}^{(0)} \cdot \nabla p^{(1)} \right] + \frac{1}{2Re}\tau^{(0)} : \tau^{(0)}$$

$$+ \frac{1}{RePr}\nabla^2 T^{(1)} + q, \qquad (9.155)$$

supplemented by the linearized equation of state:

$$p^{(1)} = \frac{\gamma - 1}{\gamma}\left[\rho^{(1)} + T^{(1)}\right]. \qquad (9.156)$$

These equations must be interpreted as the incompressible Navier–Stokes equations for $\boldsymbol{u}^{(0)}$ and $p^{(1)}$ supplemented by a passive scalar equation for $T^{(1)}$ with several source terms. Therefore the fluctuating pressure is completely determined up to an additive function of time through the relation

$$\nabla^2 p^{(1)} = -\nabla \cdot \nabla \cdot [\boldsymbol{u}^{(0)}\boldsymbol{u}^{(0)}] + \nabla \cdot \boldsymbol{f}. \qquad (9.157)$$

The time evolution of the density perturbation is deduced from Eqs. (9.155)–(9.157).

2. *Strong external heating*: $l \geq 2$. In this case, the lowest-order nontrivial system is

$$\nabla \cdot \boldsymbol{u}^{(0)} = 0, \qquad (9.158)$$

$$\frac{\partial \boldsymbol{u}^{(0)}}{\partial t} + \boldsymbol{u}^{(0)} \cdot \nabla \boldsymbol{u}^{(0)} = -\nabla p^{(l)} + \frac{1}{Re}\nabla^2 \boldsymbol{u}^{(0)}, \qquad (9.159)$$

$$\frac{\partial T^{(1)}}{\partial t} + \boldsymbol{u}^{(0)} \cdot \nabla T^{(1)} = \frac{1}{RePr}\nabla^2 T^{(1)} + q, \qquad (9.160)$$

and

$$\rho^{(1)} + T^{(1)} = 0. \qquad (9.161)$$

The leading-order pressure fluctuation is given by

$$\nabla^2 p^{(l)} = -\nabla \cdot \nabla \cdot [\boldsymbol{u}^{(0)}\boldsymbol{u}^{(0)}]. \qquad (9.162)$$

Here again the system appears to be composed of the incompressible Navier–Stokes equations supplemented by a passive scalar equation. The latter is simpler than in the weak heating case, because pressure-induced and viscous-dissipation-induced temperature fluctuations are now negligible. An important difference with the previous case is that the density fluctuations are now totally enslaved to the temperature fluctuations and are anticorrelated.

These developments are consistent with the presence of acoustic modes of the order of δ^2. The first dilatational correction to the velocity field is $\boldsymbol{u}^{(1)}$.

9.3.2.2 Statistical Equilibrium States

We now discuss the features of the statistical equilibrium states associated with the two weakly compressible thermal models previously discussed. For these models, no exact analytical solutions can be found, and the analysis will be restricted to the properties of the inertial ranges of the spectra of the fluctuating quantities in a fully developed turbulent isotropic flow. In both cases, the kinetic-energy spectrum and the pressure spectrum are expected to be the same as in incompressible isotropic turbulent flows. Corresponding inertial-range scalings are $E_{ss}(k) \propto k^{-5/3}$ for the former and $E_{pp}(k) \propto k^{-7/3}$ for the latter.

In the strong heating case, the temperature fluctuations obey the passive scalar equation. Assuming that the external heating acts at relatively large scales and neglecting conduction effects, one obtains the usual scaling law from the temperature spectrum $E_{TT}(k) \propto k^{-5/3}$. Because the density and temperature fluctuations are anticorrelated, they have the same spectrum, yielding $E_{\rho\rho}(k) \propto k^{-5/3}$.

In the weak heating case, it must be remembered that entropy behaves as a passive scalar and therefore exhibits the usual scaling law in the inertial range $E_s(k) \propto k^{-5/3}$. The leading-order entropy fluctuation is given by

$$s^{(1)} = T^{(1)} - p^{(1)}. \tag{9.163}$$

Comparing the spectral slopes of the entropy spectrum and the pressure spectrum, one can see that the temperature fluctuations must overwhelm the pressure fluctuations at small scales to recover a $-5/3$ slope for the entropy spectrum, leading to $E_{TT}(k) \propto k^{-5/3}$. As a consequence, in the inertial range, density fluctuations will also be governed by temperature fluctuations, leading to $E_{\rho\rho}(k) \propto k^{-5/3}$.

It is seen that both regimes lead to the same scaling laws for the inertial-range spectra. But it is worth noting that these scaling laws differ from those obtained for a quasi-isentropic flow, in which $E_{\rho\rho}(k) \propto k^{-7/3}$.

9.3.2.3 Numerical Observations

The existence of the different regimes and the related turbulent statistical equilibrium states predicted by the theoretical analysis have been checked through numerical experiments (Bayly, Levermore, and Passot, 1992; Cai, O'Brien, and Ladeinde, 1997).

The main observations are summarized as follows:

1. Both weak ($l = 1$) and strong ($l = 2$) external heating regimes can be reproduced in numerical simulations and are stable if consistent initial conditions are prescribed.
2. The anticorrelation between density and pressure fluctuations is observed for almost incompressible initial conditions [$\chi(0) = 0$] at relatively low initial turbulent Mach numbers ($M_t \leq 0.3$). In other cases, the growth of the acoustic mode scrambles the correlation.

3. In freely decaying turbulence with consistent initial conditions, the asymptotic regime, i.e., the value of l at the final stage of the simulation, depends on the Prandtl number. For low values of the Prandtl number ($Pr \leq 1$) the density fluctuations are observed to decay until the $l = 1$ regime is encountered. For larger values of the Prandtl number, states with $l \geq 2$ are observed.
4. If initial conditions are not fully consistent with the governing equations, the pressure fluctuations are observed to grow very quickly, corresponding to a transfer of internal energy toward acoustic energy. The weakly compressible thermal regimes are then observed to bifurcate toward the pseudo-acoustic or the nonlinear subsonic regimes.
5. Despite the fact that it was derived neglecting the heat conduction effect, the weak equilibrium relation $\mathcal{F}(t) = 1$ [in which $\mathcal{F}$ is defined according to Eq. (9.109)] is observed to hold in simulations with large initial temperature fluctuations after a short transient phase. But the oscillations of $\mathcal{F}$ about 1 are much larger than in simulations with pseudo-acoustic initial conditions.

9.3.3 Nonlinear Subsonic Regime

The two regimes discussed in Subsections 9.3.1 and 9.3.2 are expected to occur in the limit of nearly incompressible turbulence, i.e., at turbulent Mach numbers: $M_t \ll 1$. For turbulent Mach numbers less than unity but not negligible, M_t can no longer be used as a small parameter. Therefore the asymptotic analyses previously presented are theoretically no longer valid, as one expects that the nonlinearities arising from the convective terms will play a major role. As a matter of fact, numerical simulations show that very small shocks, referred to as *shocklets* or *eddy shocklets*, develop.

9.3.3.1 Conditions for Occurrence of Shocklets

Before discussing the properties of these shocklets and analyzing their influence on the dynamics and the statistical properties of compressible isotropic turbulence, it is worth noting that they can occur at nominally very low Mach numbers depending on the initial condition. In the case in which initial acoustic pressure perturbations are very strong, the linear theory is no longer relevant to describe the dynamics. Considering such an initial condition, convective terms will play an important role, leading to the propensity for the occurrence of nonlinear acoustic phenomena such as wave steepening and focusing and shock formation. Therefore a strong nonlinear transient phase will be present during which shocklets can form.

In a similar manner, the analyses dealing with the weakly compressible thermal regime were based on the assumption that both δ and δ' as defined in Eq. (9.151) are very small. Because $\delta \propto M_r$, the asymptotic series expansion is no longer valid at higher turbulent Mach numbers. Moreover, it has been observed in numerical experiments that the thermal regimes are very sensitive to the initial conditions, and that, if the initial perturbations are not consistent with the governing equations and are strong enough, acoustic waves grow and the dynamics bifurcates toward the nonlinear regime in which shocklets are present.

No exact threshold value for the turbulent Mach number M_t associated with the occurrence of shocklets is known, as this phenomenon also depends on other parameters. It seems that all simulations carried out for $M_t \geq 0.4$ exhibit shocklets. But it is important to note that shocklets can appear at lower Mach numbers, depending on the initial condition.

9.3.3.2 Energy Budget and Shocklet Influence

Shocklets are small bow shocks that have been observed to satisfy the Rankine–Hugoniot jump conditions. Therefore these events exhibit all the properties of usual shocks. In particular, they induce sharp pressure and density gradients and, because they are associated with a compression, a negative value of the divergence of the velocity. As a consequence, one can reasonably expect that they should have a non-negligible influence on the energy balance. We subsequently summarize the observations retrieved from DNSs. Because all these simulations were carried out at small Reynolds numbers, viscous effects are important and they damp the effect of the shocklets. Therefore exact values of the quantities subsequently given must be interpreted as a qualitative description of high-Reynolds-number flows rather than as an accurate quantitative one.

Numerical experiments show that the pdf of the dilatation (i.e., the divergence of the velocity field) is strongly skewed: About 2/3 of the volume is associated with an expansion ($\nabla \cdot \boldsymbol{u} > 0$), whereas only 1/3 corresponds to compression. On average, the expansion regions are responsible for 80%–90% of the solenoidal dissipation $\bar{\varepsilon}_s$ and 50%–60% of the total dissipation ($\bar{\varepsilon}_d + \bar{\varepsilon}_s$). The global dilatational dissipation $\bar{\varepsilon}_d$ is found to be small with respect to the global solenoidal dissipation $\bar{\varepsilon}_s$: Lee and co-workers (Lee, Lele, and Moin, 1991) report that $\bar{\varepsilon}_d$ is less than or equal to 10% of the total dissipation for M_t up to 0.6.

The shocklets fill only a few percent of the total volume: Pirrozoli and Grasso (2004) found that they represent only 1.4% of the volume at $M_t = 0.8$ whereas Samtaney et al. (2001) report a fraction smaller than 2% in their set of numerical experiments. Nevertheless, the shocklets strongly modify the local relative importance of the physical mechanisms: Near shocklets, the dilatational dissipation is up to 10 times larger than the solenoidal dissipation. Despite the fact that they fill only a very small part of the fluid domain, shocklets are responsible for about 20% of the global dilatational dissipation.

The shocklets perturb the dilatation field. This perturbation can be roughly estimated by looking at the jump condition for the dilatation for a bow shock moving into a 2D inviscid steady flow provided by Kida and Orszag (1990a, 1992):

$$[[\nabla \cdot \boldsymbol{u}]] \simeq \frac{2}{R(\gamma+1)} \left(\frac{(\gamma-1)M_s^2 + 2}{(\gamma+1)M_s^2} - \frac{3M_s^2 + 1}{M_s^2 - 1} \tan^2 \theta \right) u_n, \quad (9.164)$$

9.3 Different Regimes in Compressible Turbulence

where R, u_n, M_s, and θ are the radius of curvature of the shock, relative velocity normal to the shock, shocklet Mach number defined by the ratio of u_n about the upstream speed of sound, and the angle between the fluid velocity and the shock normal, respectively. It is seen that the sign of the induced dilatation depends on both R and θ, and that its amplitude is a function of the square of the normal Mach number. The use of a 2D simplified model was proved to be qualitatively relevant by Kida and Orszag, as the 3D curved shock can be locally projected on a 2D space. Lee and co-workers observed that the correlation between pressure fluctuations and dilatation fluctuations is large near shocklets, leading to a local enhancement of the transfers between the internal energy and the turbulence kinetic energy. These authors also report that the overall effect of the pressure–dilatation term $\overline{p'd}$ on the evolution of kinetic energy in freely decaying isotropic turbulence is comparable with the overall dilatational dissipation $\bar{\varepsilon}_d$. This effect is typical of the presence of the shocklets, as this term is theoretically and experimentally found to be negligible in the pseudo-acoustic regime.

9.3.3.3 Enstrophy Budget and Shocklet Influence

The shocklets also have a large impact on the dynamics of vorticity and enstrophy. The general jump relation for vorticity derived from Rankine–Hugoniot will be discussed in the chapter devoted to the shock–turbulence interaction (Chapter 11), and the interested reader can refer to it. But it is very important to emphasize that the main trends and the relative importance of the different physical mechanisms are not the same in the shocklet case as in the large-scale shock case discussed in Chapter 11. The main reason for this is a scale effect: Shocklets are small shock waves that form when turbulent eddies allow for the local steepening of pressure waves, and their size is therefore comparable with those of the turbulent eddies, whereas large-scale shock size is much greater than that of the turbulent vortical structures.

We first recall some estimates related to vorticity creation by a bow shock moving in a steady, inviscid 2D flow (Kida and Orszag, 1990b):

$$[[\mathbf{\Omega}]] \simeq \frac{4(M_s^2 - 1)\sin\theta}{R(\gamma + 1)M_s^2((\gamma - 1)M_s^2 + 2)}|\mathbf{u}|, \tag{9.165}$$

where the nomenclature is the same as in the previous paragraph. The sign of the created vorticity is seen to depend on the local shock curvature and the angle of incidence. The created vorticity is zero for normal shocks ($\theta = 0$) and Mach waves $[\theta = \pm\cos^{-1}(1/M_s)]$. The effect of the sole baroclinic term $-(\nabla p \times \nabla \rho)/\rho^2$ is evaluated as

$$[[u_n\mathbf{\Omega}]] \simeq \frac{4(M_s^2 - 1)^2}{R(\gamma + 1)^2 M_s^4} u_n^2 \tan\theta \tag{9.166}$$

and is observed to depend on M_s^4 instead of M_s^2 for the global vorticity creation.

Numerical experiments show that the volume-averaged enstrophy budget is governed by the vortex-stretching term $\mathbf{\Omega} \cdot \mathbf{S} \cdot \mathbf{\Omega}$ and the viscous dissipation. The former is positive and creates some vorticity, whereas the latter is strictly negative. The baroclinic term is negligible whereas the compression term $\Theta(\nabla \cdot u)$ exhibits an oscillatory behavior, with a period very similar to those of the compressive kinetic energy and the internal energy. Therefore this phenomenon is interpreted as a coupling between acoustic waves and the vorticity.

A finer analysis can be achieved that distinguishes between regions of negative dilatation and regions of positive dilatation. The compression term is observed to be dominant in shocklet areas, whereas the stretching term is the most important in the expansion region. A careful look at DNS data reveals that vorticity is created on shocklets through the baroclinic interaction and is enhanced in expansion regions by the vortex-stretching phenomenon. The baroclinic production is relatively small because there is a clear trend for the pressure gradient and the density gradient to be aligned against each other: The global pdf of the angle between these vectors exhibits a peak near 4° and is almost null for angles higher than 10°, even for values of M_t as high as 0.74. It is also found that increasing the turbulent Mach number yields a stronger alignment of these vectors. Because of the weakness of the baroclinic production, Kida and Orszag observed that the barotropic relation

$$\left(\frac{p}{\bar{p}}\right) = \left(\frac{\rho}{\bar{\rho}}\right)^{\gamma} \tag{9.167}$$

is valid overall.

A last observation is that the vorticity has a statistical preference to align with the density gradient $\nabla\rho$ near the shocklets and to be orthogonal to it outside shocklet aeras. Because the shocklets fill a very small fraction of the fluid domain, the overall pdf of the angle between Ω and $\nabla\rho$ has a peak at 90°. As in the incompressible case, the vorticity is observed to be aligned overall with the intermediate eigendirection of the velocity-gradient tensor.

9.3.3.4 Statistical Equilibrium State

Numerical simulations show that statistical equilibrium is reached in the nonlinear subsonic case after a short transient phase.

This statistical equilibrium state is very similar to the one observed in the pseudo-acoustic regime: Sarkar's function fluctuates almost periodically about unity because of energy exchanges between internal energy and the dilatational field. The period of oscillation corresponds to the characteristic acoustic time scale, and the amplitude is an increasing function of the Mach number.

9.3.4 Supersonic Regime

The supersonic regime, in which the turbulent Mach number is greater than 1, is much less known than the other regimes. The main reasons why so little attention has been paid to this configuration are that it is encountered in astrophysics only and that it escapes most theoretical tools because it does not allow small-parameter expansion.

Only very few numerical experiments are available (Porter, Pouquet, and Woodward, 1992a, 1992b, 1994), which all reveal the existence of two distinct quasi-equilibrium phases separated by a short transition phase:

- The *quasi-supersonic phase*, whose typical duration is of the order of a few acoustic time scales. During this initial period, nonlinear phenomena yield the formation of a myriad of small but intense shock waves. No vortical structures are observed during this period. Then the shocks interact, leading to the existence of vortex sheets that roll up because of Kelvin–Helmholtz-type instabilities, yielding the existence of vortex tubes. These vortex tubes then experience vortex stretching, leading to the appearance of the usual kinetic-energy cascade phenomenon. During this phase, which is dominated by shock formation and shock interaction, the evolution of vorticity is governed by the baroclinic production and the linear terms (vortex stretching and dilatation terms), which are of equal amplitude. At the end of the quasi-supersonic phase, both dilatational velocity spectrum and solenoidal velocity spectrum exhibit an inertial range with a -2 slope:

$$E_{dd}(k) \propto k^{-2}, \quad E_{ss}(k) \propto k^{-2}. \tag{9.168}$$

It is worth noting that most of turbulent kinetic energy is contained in the solenoidal mode once the vortical structures have been created.

- The *immediate postsupersonic phase* that is governed by vortex interaction and vortical decay. The main processes involved in subsonic vortex dynamics are present, but shocks are still present and very active. As a consequence, the following inertial-range scalings are observed:

$$E_{dd}(k) \propto k^{-2}, \quad E_{ss}(k) \propto k^{-1}. \tag{9.169}$$

The vorticity dynamics is dominated by the vortex stretching and the dilation term during this phase, the baroclinic production being now much weaker because of the decrease of the turbulent Mach number.

At much longer times, an equilibrium state similar to the subsonic regime is recovered in which the shocks are much weaker and the solenoidal velocity dynamics is decoupled (at the leading-order approximation) from the acoustic field. The measured inertial range behaviors are $E_{dd}(k) \propto k^{-2}$ and $E_{ss}(k) \propto k^{-5/3}$.

9.4 Structures in the Physical Space

Compressible isotropic turbulence, like incompressible isotropic turbulence, exhibits coherent events that can be classified according to some criteria. In the pseudo-acoustic regime at very low Mach numbers, the solenoidal field is nearly decoupled from the dilatational field and they evolve almost independently. The kinetic energy being concentrated in the solenoidal velocity component (the ratio of the two components of the turbulent kinetic energy scales like M_t^4), the velocity field is almost identical to the one observed in purely incompressible flows, and all the results dealing with the velocity-field topology given in Section 3.6 hold.

The analysis must be modified to account for new configurations in the nonlinear subsonic regime. This was achieved by Kevlahan, Mahesh, and Lee (1992), who proposed a topological analysis of the velocity field that accounts for compressibility effects. Main elements of this classification are displayed in Subsection 9.4.1. It has also been observed that shocklets form in this regime. The main known characteristics of these structures are discussed in Subsection 9.4.2.

9.4.1 Turbulent Structures in Compressible Turbulence

The analysis carried out by Kevlahan and co-workers (Kevlahan, Mahesh, and Lee, 1992) relies on the local analysis of the topology of the velocity field. Introducing the anisotropic part of the instantaneous strain tensor $\mathbf{S}^* = \mathbf{S} - (S_{kk}/3)\mathbf{I}$, it is possible to define three region types:

- *Eddy-dominated regions*:

$$\mathbf{W} : \mathbf{W} > 2\mathbf{S}^* : \mathbf{S}^*. \tag{9.170}$$

- *Shear zones*:

$$\frac{1}{2}\mathbf{S}^* : \mathbf{S}^* \leq \mathbf{W} : \mathbf{W} \leq 2\mathbf{S}^* : \mathbf{S}^*. \tag{9.171}$$

- *Convergence zones*:

$$\mathbf{W} : \mathbf{W} < \frac{1}{2}\mathbf{S}^* : \mathbf{S}^*. \tag{9.172}$$

This decomposition is supplemented by a criterion related to the local degree of compressibility. Defining the sensor $\mathcal{C}$ as

$$\mathcal{C} = \frac{(\nabla \cdot \boldsymbol{u})^2}{\mathbf{S}^* : \mathbf{S}^* + \mathbf{W} : \mathbf{W}}, \tag{9.173}$$

it is proposed that

- $\mathcal{C} \leq 0.05$: structures behave as incompressible ones,
- $\mathcal{C} > 0.05$: structures are compressible.

The analysis can be further refined by retrieving some information about the structure shape. Denoting by λ_1^*, λ_2^*, and λ_3^* the three eigenvalues of $\mathbf{S}^*$,[‡‡] the structure shapes can be classified according to the sign of the third invariant of $\mathbf{S}^*$, $III^* = -\lambda_1^* \lambda_2^* \lambda_3^*$:

- $III^* < 0$: *cigar*-type structures,
- $III^* > 0$: *pancake*-type structures.

[‡‡] One can easily observe that

$$\lambda_i^* = \lambda_i - \frac{(\nabla \cdot \boldsymbol{u})}{3}, \tag{9.174}$$

where λ's are the eigenvalues of $\mathbf{S}$. A direct consequence is

$$\sum_{i=1,3} \lambda_i^* = 0, \tag{9.175}$$

as in incompressible turbulence.

9.4 Structures in the Physical Space

Regions can also be grouped as *focal* or compression regions and *nonfocal* or expansion regions according to the sign of the determinant of $\mathbf{S}^*$, $\mathcal{D}^* = \det(\mathbf{S}^*)$:

- $\mathcal{D}^* > 0$: focal/compression region,
- $\mathcal{D}^* < 0$: nonfocal/expansion region.

It is worth noting that $\mathcal{D}^*$ can be computed from the second and third invariants of $\mathbf{S}^*$ like

$$\mathcal{D}^* = \frac{27}{4}(\mathrm{III}^*)^2 + (\mathrm{II}^*)^3, \quad \mathrm{II}^* = \lambda_1^* \lambda_2^* + \lambda_1^* \lambda_3^* + \lambda_2^* \lambda_3^*. \qquad (9.176)$$

Using a wide database including flows with turbulent Mach numbers up to 0.8, Pirrozoli and Grasso (2004) observed that several features of isotropic turbulence are not sensitive to the Mach number and are therefore similar to those of perfectly incompressible isotropic turbulence. They are listed as follows:

1. The eigenvalues of the strain tensor $\mathbf{S}^*$ are in the ratio $-4:3:1$.
2. The number of pancake and cigar structures is in the ratio 3:1.
3. Vorticity has a statistical preference to align with the intermediate eigendirection of $\mathbf{S}^*$, being either parallel or antiparallel (probabilities are equal).
4. The joint pdf of II^* and III^* does not depend on the Mach number.

Their analyses also show that focal regions are of great importance for the solenoidal field: These regions fill about 2/3 of the total volume and account for 80%–90% of the enstrophy and 50%–60% of the solenoidal dissipation $\bar{\varepsilon}_s$. More precisely:

1. At low M_t, incompressible structures dominate and the fractions of the volume filled by shear zones, convergence zones, and eddies are 44%, 35% and 21%, respectively.
2. Shear regions account for 45% of the enstrophy regardless of the turbulent Mach number.
3. At low M_t, kinetic energy is dissipated nearly equally in focal and nonfocal structures, whereas at high M_t focal structures are more active than the nonfocal ones.
4. At high M_t, dilatational dissipation mainly takes place in shear and convergence zones.
5. Shocklets are rare (less than 2% of the volume) but represent up to 20% of the global dilatational dissipation.

9.4.2 A Probabilistic Model for Shocklets

We now present the main features of the probabilistic model for shocklets derived by Samtaney and co-workers (Samtaney, Pullin, and Kosovic, 2001) on the grounds of DNS data.

The initial step consists of parameterizing the pdf of the longitudinal velocity increment, which will serve as a basis to evaluate the shocklet-based Mach number.

The pdf is observed to be very similar to those measured in incompressible isotropic turbulence, leading to the following exponential expression:

$$P(\Delta u) \simeq \frac{1}{\sigma_{\Delta u}} \exp\left[-b(r)\left|\frac{\Delta u}{\sigma_{\Delta u}}\right|\right], \quad (9.177)$$

where Δu is the velocity increment along the direction of u between two points separated by a distance r and $\sigma_{\Delta u} = \sqrt{\overline{(\Delta u)^2}}$. The function b can be written as $b(r) = \alpha(r/\eta)^\beta$, where η is the Kolmogorov length scale and $\alpha = 1.5$ and $\beta = 0.16$ are constant parameters.[§§]

The second step deals with the derivation of a model pdf for the shocklet strength, the shocklet being modeled as a weak shock (i.e., $M_s - 1 \ll 1$).

Let Δu be the normal velocity difference across the shocklet. Usual jump conditions yield the exact relation:

$$\frac{\Delta u}{a} = -\frac{2}{\gamma + 1}\left(M_s - \frac{1}{M_s}\right), \quad (9.178)$$

where a is the speed of sound in the fluid upstream of the shocklet. The shocklet thickness δ_s is evaluated with the classical weak-shock theory:

$$\delta_s \simeq \frac{\nu}{a}\frac{3}{(M_s - 1)} \quad (9.179)$$

where ν is related to the viscosity upstream of the shock. To get a reliable model, one must evaluate all quantities using turbulence-related variables. To this end, it is assumed that the following expression for the dissipation derived in the incompressible case holds,

$$\varepsilon = 15\nu \overline{\left(\frac{\partial u}{\partial x}\right)^2} \simeq 15\nu \overline{\left(\frac{\Delta u}{r}\right)^2}, \quad (9.180)$$

from which the following leading-order estimates in terms of $(M_s - 1)$ are derived:

$$\Delta u \simeq \frac{4a}{\gamma + 1}(M_s - 1), \quad (9.181)$$

$$\sigma_{\Delta u} \simeq \frac{3}{\sqrt{15a}}\sqrt{\nu\varepsilon}\frac{1}{(M_s - 1)}, \quad (9.182)$$

$$\left|\frac{\Delta u}{\sigma_{\Delta u}}\right| \simeq \frac{4\sqrt{15}a^2}{3(\gamma + 1)\sqrt{\nu\varepsilon}}(M_s - 1)^2. \quad (9.183)$$

The last expression can be further refined by introduction of the mean turbulent Mach number M_t and the Taylor–Reynolds number Re_λ:

$$\left|\frac{\Delta u}{\sigma_{\Delta u}}\right| \simeq \frac{4}{(\gamma + 1)}\frac{Re_\lambda}{M_t^2}(M_s - 1)^2, \quad (9.184)$$

[§§] The value $\alpha = 1.5$ was measured in low-Reynolds-number simulations. High-Reynolds-number wind-tunnel experiments suggest $\alpha \simeq 0.5$.

9.4 Structures in the Physical Space

leading to the following expression for the pdf of the shocklet strength (the constant β is taken equal to zero for the sake of simplicity):

$$P(M_s - 1) \simeq \frac{8\alpha}{(\gamma+1)} \frac{Re_\lambda}{M_t^2} (M_s - 1) \exp\left[-\frac{4\alpha}{(\gamma+1)} \frac{Re_\lambda}{M_t^2} (M_s - 1)^2\right]. \quad (9.185)$$

This expression is observed to be in good agreement with a large set of experimental data, with $Re_\lambda = 50$–100 and $M_t = 0.1$–0.5. In the same manner, a model pdf can be found for the shocklet thickness:

$$P(\delta/\eta) \simeq \left(\frac{\delta}{\eta}\right)^3 \exp\left[-\frac{12\sqrt{15}\alpha}{\gamma+1} \left(\frac{\delta}{\eta}\right)^2\right]. \quad (9.186)$$

One can see from Eqs. (9.185) and (9.186) that the *most probable shocklet* corresponds to

$$M_s = 1 + M_t \sqrt{\frac{\gamma+1}{8\alpha Re_\lambda}}, \quad \delta = (15)^{1/4} \eta \sqrt{\frac{8\alpha}{\gamma+1}}. \quad (9.187)$$

Using the preceding value of α and considering air, one finds that the most probable shocklet thickness is about $5\,\eta$, which is much larger than the mean free path of the molecules.

Bibliography

ABBOTT, L. F. AND DESER, S. (1982). Stability of gravity with a cosmological constant, *Nucl. Phys.* **B195**, 76–96.

BATAILLE, F. (1994). Etude d'une turbulence faiblement compressible dans le cadre d'une modélisation quasi-normale avec Amortissement tourbillonnaire, *Ph.D. Thèse*, Ecole Centrale de Lyon.

BATAILLE, F. AND ZHOU, Y. (1999). Nature of the energy transfer process in compressible turbulence, *Phys. Rev. E* **59**, 5417–5426.

BAYLY, B. J., LEVERMORE, C. D., AND PASSOT, T. (1992). Density variations in weakly compressible flows, *Phys. Fluids* **4**, 945–954.

BERTOGLIO, J. P., BATAILLE, F., AND MARION, J. D. (2001). Two-point closures for weakly compressible turbulence, *Phys. Fluids* **13**, 290–310.

CAI, X. D., O'BRIEN, E. E., AND LADEINDE, F. (1997). Thermodynamic behavior in decaying, compressible turbulence with initially dominant temperature fluctuations, *Phys. Fluids* **9**, 1754–1763.

CAI, X. D., O'BRIEN, E. E., AND LADEINDE, F. (1998). Advection of mass fraction in forced, homogeneous, compressible turbulence, *Phys. Fluids* **10**, 2249–2259.

CAMBON, C., COLEMAN, G. N., AND MANSOUR, N. N. (1993). Rapid distortion analysis and direct simulation of compressible homogeneous turbulence at finite Mach number, *J. Fluid Mech.* **257**, 641–665.

CAMBON, C. AND SAGAUT, P. (2007). Internal report.

CAMBON, C. AND SCOTT, J. F. (1999). Linear and nonlinear models of anisotropic turbulence, *Annu. Rev. Fluid Mech.* **31**, 1–53.

CHASSAING, P., ANTONIA, R. A., ANSELMET, F., JOLY, L., AND SARKAR, S. (2002). *Variable Density Turbulence*, Springer-Verlag.

CHU, B. T. AND KOVASZNAY, L. S. G. (1957). Non-linear interactions in a viscous heat-conducting compressible gas, *J. Fluid Mech.* **3**, 494–514.

ECKHOFF, K.S. AND STORESLETTEN, L. (1978). A note on the stability of steady inviscid helical gas flows, *J. Fluid Mech.* **89**, 401–411.

ERLEBACHER, G., HUSSAINI, M. Y., KREISS, H. O., AND SARKAR, S. (1990). The analysis and simulation of compressible turbulence, *Theor. Comput. Fluid Dyn.* **2**, 73–95.

ERLEBACHER, G. AND SARKAR, S. (1993). Statistical analysis of the rate of strain tensor in compressible homogeneous turbulence, *Phys. Fluids* **5**, 3240–3254.

FAUCHET, G. (1998). Modélisation en deux points de la turbulence isotrope compressible et validation à l'aide de simulations numériques, *Thèse de Doctorat, Ecole Centrale de Lyon*.

FAUCHET, G. AND BERTOGLIO, J. P. (1999a). A two-point closure for compressible turbulence. *C. R. Acad. Sci. Paris Série IIb* **327**, 665–671.

FAUCHET, G. AND BERTOGLIO, J. P. (1999b). Pseudo-sound and acoustic régimes compressible turbulence. *C. R. Acad. Sci. Paris Série IIb* **327**, 665–671.

FEIEREISEN, W. J., REYNOLDS, W. C., AND FERZIGER, J. H. (1981). Numerical simulation of compressible homogeneous turbulent shear flow, *Report No. TF 13*, Stanford University.

KANEDA, Y. (2007). Lagrangian renormalized approximation of turbulence, *Fluid Dyn. Res.* **39**, 526–551.

KEVLAHAN, N., MAHESH, K., AND LEE, S. (1992). Evolution of the shock front and turbulence structures in the shock/turbulence interaction, in *Proceedings of the Summer Program, CTR*, Stanford University.

KIDA, S. AND ORSZAG, S. A. (1990a). Energy and spectral dynamics in forced compressible turbulence, *J. Sci. Comput.* **5**, 85–125.

KIDA, S. AND ORSZAG, S. A. (1990b). Enstrophy budget in decaying compressible turbulence, *J. Sci. Comput.* **5**, 1–34.

KIDA, S. AND ORSZAG, S. A. (1992). Energy and spectral dynamics in decaying compressible turbulence, *J. Sci. Comput.* **7**, 1–34.

KOVASZNAY, L. S. G. (1953). Turbulence in supersonic flow, *J. Aeronaut. Sci.* **20**, 657–682.

LEE, S., LELE, S. AND MOIN, P. (1991). Eddy shocklets in decaying compressible turbulence, *Phys. Fluids* **3**, 657–664.

LELE, S. (1994). Compressibility effects on turbulence, *Annu. Rev. Fluid Mech.* **26**, 211–254.

LESLIE, D. C. (1973). *Developments in the Theory of Turbulence*, Clarendon.

MARION, J. D., BERTOGLIO, J. P., AND CAMBON, C. (1988). Two-point closures for compressible turbulence, presented at the 11th International Symposium on Turbulence, Rolla, MO, Oct. 17–19. Preprints (A-89-33402 13-14), University of Missouri-Rolla, B22-1–B22-8.

MARION, J. D., BERTOGLIO, J. P., CAMBON, C., AND MATHIEU, J. (1988). Spectral study of weakly compressible turbulence. Part II: EDQNM, *C. R. Acad. Sci. Paris Série II* **307**, 1601–1606.

MIURA, H. AND KIDA, S. (1995). Acoustic energy exchange in compressible turbulence, *Phys. Fluids* **7**, 1732–1742.

ORSZAG, S. A. (1977). Lectures on the statistical theory of turbulence, in Balian, R. and Peube, J. L., eds., *Fluid Dynamics*, Gordon & Breach, pp. 235–374.

PIRROZOLI, S. AND GRASSO, F. (2004). Direct numerical simulations of isotropic compressible turbulence: Influence of compressibility on dynamics and structure, *Phys. Fluids* **16**, 4386–4407.

PORTER, D. H., POUQUET, A., AND WOODWARD, P. R. (1992a). Three-dimensional supersonic homogeneous turbulence: A numerical study, *Phys. Rev. Lett.* **68**, 3156–3159.

PORTER, D. H., POUQUET, A., AND WOODWARD, P. R. (1992b). A numerical study of supersonic turbulence, *Theor. Comput. Fluid Dyn.* **4**, 13–49.

PORTER, D. H., POUQUET, A., AND WOODWARD, P. R. (1994). Kolmogorov-like spectra in decaying three-dimensional supersonic flows, *Phys. Fluids* **6**, 2133–2142.

SAMTANEY, R., PULLIN, D. I., AND KOSOVIC, B. (2001). Direct numerical simulation of decaying compressible turbulence and shocklets statistics, *Phys. Fluids* **13**, 1415–1430.

SARKAR, S., ERLEBACHER, G., HUSSAINI, M. Y., AND KREISS, H. O. (1991). The analysis and modelling of dilatational terms in compressible turbulence, *J. Fluid Mech.* **227**, 473–493.

SIMONE, A., COLEMAN, G. N., AND CAMBON, C. (1997). The effect of compressibility on turbulent shear flow: A RDT and DNS study, *J. Fluid Mech.* **330**, 307–338.

ZANK, G. P. AND MATTHAEUS, W. H. (1990). Nearly incompressible hydrodynamics and heat conduction, *Phys. Rev. Lett.* **64**, 1243–1245.

ZANK, G. P. AND MATTHAEUS, W. H. (1991). The equations of nearly incompressible fluids. I. Hydrodynamics, turbulence and waves, *Phys. Fluids* **3**, 69–82.

10 Compressible Homogeneous Anisotropic Turbulence

10.1 Effects of Compressibility in Free-Shear Flows. Observations

To understand and model compressibility-induced effects on turbulence is an important topic, as these effects are significant in many engineering applications, particularly in the fields of propulsion and supersonic aerodynamics, which are concerned with jets or wakes subjected to large velocity and density gradients. The compressible plane mixing layer is a generic problem for these applications, and explaining and modeling how compressibility reduces turbulent mixing in a shear layer has motivated the large research effort devoted to this topic during the 1980s and 1990s. Mixing usually refers to interpenetration of two streams. It is characterized by two scales: a length scale δ and a velocity scale ΔU, which evaluate the thickness of the interface and intensity of the fluctuations, respectively. The reduction of mixing by compressibility is illustrated in Fig. 10.1 in which δ is the previously mentioned thickness and $M_c = \Delta U/a$ is the *convective Mach number*, with a the average speed of sound.

There is now a consensus in the literature that the "intrinsic compressibility" (nonzero-velocity divergence in Mach-number-dependent flows) of a turbulent velocity field tends to reduce the amplification rate of turbulent kinetic energy produced by mean-velocity gradients, with respect to the solenoidal case. These effects were particularly investigated in shear flows, including both experimental and numerical studies of the plane mixing layer and DNS of homogeneous shear flows. The reader is referred to the review by Lele (1994) and the references given therein, and to Sarkar (1995) and Simone, Coleman, and Cambon (1997) for more recent results.

Two preliminary questions immediately follow:

1. To what extent is homogeneous turbulence relevant to explain such mechanisms in inhomogeneous flows?
2. Is compressibility always stabilizing (i.e., leading to a decrease in the turbulent kinetic growth rate) in homogeneous turbulence?

The answer to the first question is somewhat difficult. The "stabilizing" effect of compressibility in the mixing layer, for instance, can be attributed to the inhibition of the Kelvin–Helmholtz instability at a convective Mach number larger than 0.6. Such an instability escapes homogeneous RDT. Nevertheless, it is expected that

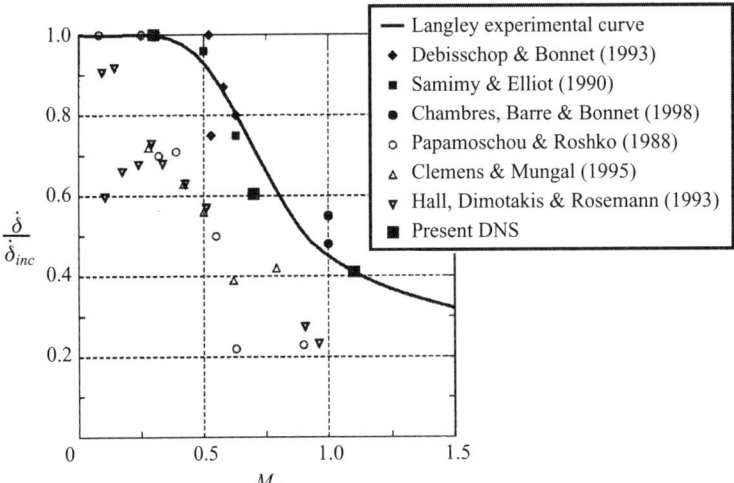

Figure 10.1. Dependence of shear-layer growth rate on convective Mach number. DNS and experimental results from Pantano and Sarkar (2002).

our analysis can exploit strong analogies between the homogeneous shear and the mixing layer.

The answer to the second question is simple: No.

10.1.1 RST Equations and Single-Point Modeling

Equations for the RST are not so different from the one in the homogeneous incompressible case, at least in the absence of specific additional production by mean-pressure, mean-density, or mean-temperature gradients. If we restrict our attention to flows in which the "production" term in the RST equations results only from mean-velocity gradients, we must consider that the pressure–strain-rate tensor is no longer trace free and that the dissipation tensor can display an explicit dilatational contribution. Using Eq. (9.62) and taking into account the homogeneity constraint, one obtains

$$\frac{\partial \bar{\rho} R_{ij}}{\partial t} = -\bar{\rho}\left(R_{ik}\frac{\partial \tilde{u}_j}{\partial x_k} + R_{jk}\frac{\partial \tilde{u}_i}{\partial x_k}\right) + \overline{p'\left(\frac{\partial u_i''}{\partial x_j} + \frac{\partial u_j''}{\partial x_i}\right)}$$
$$+ \overline{u_i''}\left(\frac{\partial \bar{\tau}_{jk}}{\partial x_k} - \frac{\partial \bar{p}}{\partial x_j}\right) + \overline{u_j''}\left(\frac{\partial \bar{\tau}_{ik}}{\partial x_k} - \frac{\partial \bar{p}}{\partial x_i}\right)$$
$$- \overline{\tau'_{ik}\frac{\partial u_j''}{\partial x_k}} - \overline{\tau'_{jk}\frac{\partial u_i''}{\partial x_k}}. \tag{10.1}$$

The corresponding evolution equation for the turbulent kinetic energy $\mathcal{K}$ is

$$\frac{d\mathcal{K}}{dt} = P + \Pi^{(d)} - \bar{\varepsilon}_s - \bar{\varepsilon}_d, \tag{10.2}$$

10.1 Effects of Compressibility in Free-Shear Flows

in which the production term is

$$P = -\bar{\rho} R_{ik} \frac{\partial \tilde{u}_i}{\partial x_k} \qquad (10.3)$$

and $\bar{\varepsilon}_s$ and $\bar{\varepsilon}_d$ are the solenoidal and dilatational dissipations introduced in Subsection 9.2.2. The pressure–dilatation correlation term, denoted by $\Pi^{(d)}$, is equal to half the trace of the pressure–strain-rate tensor.

Historically, two kinds of explanations were proposed to account for compressibility effects and used to derive specific turbulence models:

1. According to the first explanation, the main effect is attributed to the explicit terms in Eq. (10.2). For instance, a reduction of the growth rate of $\mathcal{K}$ is assumed to result from a significant negative value of $\Pi^{(d)}$ and the negative term $-\bar{\varepsilon}_d$.
2. According to the second approach, the main mechanism responsible for the decrease of the turbulent kinetic-energy growth rate is an alteration of the dynamics of pressure fluctuations. This modification results in a mollification of the pressure–strain-rate tensor in Eq. (10.1) and an associated depletion of the production term P in Eq. (10.2). This effect appears an implicit one, at least if one considers the evolution equation for $\mathcal{K}$.

An important issue is how to measure the compressibility. At least two Mach numbers are relevant in shear flows: the conventional *turbulent Mach number* $M_t = u'/a$, introduced in the previous chapter, and the *gradient Mach number*,

$$M_g = \frac{SL}{a_0}. \qquad (10.4)$$

The gradient Mach number compares the velocity scale SL with the speed of sound a_0, where S and L are a mean-velocity-gradient scale and a typical length scale of the largest turbulent eddies, respectively. A similar parameter that accounts for the change of a mean-flow Mach number across an eddy, denoted by Δm, was introduced by Durbin and Zeman (1992). A more general meaning and interpretation was then introduced by Jacquin, Cambon, and Blin (1993) and Sarkar (1995), with slightly different terminologies ("distortion" and "gradient" Mach number). Despite Jacquin's precedence, we adopt here the terminology gradient Mach number because it is the most popular. We think, however, that the RDT equations first investigated by Jacquin and co-workers gave the best interpretation of this parameter, as its counterpart at a fixed wavenumber, $S/(a_0 k)$, is the pivotal parameter for separating different flow regimes.

Looking at DNS results, one can see that the implicit alteration of the production is linked to a significant change in the Reynolds stress anisotropy. To illustrate this point, some values of b_{ij} are proposed for the compressible homogeneous shear-flow case and compared with their counterparts in the incompressible case (see Section 6) in Table 10.1. Estimates for the compressible shear case presented here were proposed in Heinz (2004) on the grounds of relatively low Reynolds DNS performed by Sarkar and co-workers.

Table 10.1. *Reynolds stress anisotropy in compressible and incompressible homogeneous shear flows at large St*

Quantity	Incompressible case	Compressible case	Pressure-released RDT
b_{11}	0.203	$2/3 - 0.4e^{-0.3M_g} \pm 0.01$	$2/3$
b_{22}	-0.143	$-1/3 + 0.17e^{-0.3M_g} \pm 0.01$	$-1/3$
b_{33}	-0.06	$-1/3 + 0.23e^{-0.3M_g} \pm 0.01$	$-1/3$
b_{12}	$-0.15(0)$	$-0.17e^{-0.3M_g} \pm 0.005$	0

Note: Values given in the two first columns come from DNS and wind-tunnel experiments (if available).

The reduction of b_{12} in absolute value with respect to the incompressible case is directly connected to the reduction of production in this case, whereas the increase in b_{11} reflects a less efficient redistribution of the kinetic energy among the normal Reynolds stresses by the pressure–strain-rate tensor, which can be interpreted as an alteration of the so-called return-to-isotropy mechanism. Even if compressibility correction factors in terms of M_g essentially come from empirical fitting, their asymptotic values are of interest, as they are observed to differ from both the incompressible and pressure-released RDT case.

Even if the "implicit" compressibility effect previously mentioned is probably more relevant than the "explicit" one to explain the decrease in the growth rate of $\mathcal{K}$, rationales based on single-point statistics, which ignore the detailed consequences of the Helmholtz decomposition, cannot be fully satisfactory and universal. This is illustrated by the fact that some flows can be found in which the mollification of pressure–strain-rate correlations at increasing gradient Mach number is not the right explanation, because it results in an *increase* of the turbulent kinetic-energy growth rate. These flows include the irrotational strain case and even shear flows at moderate elapsed time, as shown by RDT and DNS results. As discussed in Chapter 6 in the pure incompressible case, the "slow" (nonlinear) and the "rapid" (linear) contributions to the pressure–strain-rate tensor may have opposite effects on the production of $\mathcal{K}$: Reducing the linear term yields increasing the production. The conventional stabilizing effect of compressibility is recovered in homogeneous shear flows at larger elapsed times, but the right explanation is different from the one based on the sole reduction of pressure–strain terms (Sarkar, 1995; Pantano and Sarkar, 2002). This point will be extensively addressed in this chapter.

10.1.2 Preliminary Linear Approach: Pressure-Released Limit and Irrotational Strain

Linearizing the Euler equations for velocity and pressure fluctuations (u'_i, p') around a mean flow with velocity $\bar{u}_i$ and discarding the pressure–fluctuation term yields the following pressure-released solution (e.g., Cambon, Teissèdre, and Jeandel, 1985):

$$u_i^{\prime(\text{pr})}(\boldsymbol{x}, t) = H_{ij}(\boldsymbol{X}, t, t_0) u'_j(\boldsymbol{X}, t_0), \tag{10.5}$$

10.1 Effects of Compressibility in Free-Shear Flows

where x is the position of a fluid particle at time t having the position X at time t_0, following a mean trajectory. The matrix H_{ij} is closely linked to the Cauchy matrix, or semi-Lagrangian displacement gradient matrix, $\mathbf{F}$, related to the mean flow (see Section 2.1), so that H_{ij} and F_{ij} reflect a time-accumulated effect of mean-velocity gradients. They are obtained by solving the following equations:

$$\dot{F}_{ij} = \frac{\partial U_i}{\partial x_m} F_{mj}, \quad F_{ij}(X, t_0, t_0) = \delta_{ij}, \tag{10.6}$$

$$\dot{H}_{ij} = -\frac{\partial U_i}{\partial x_m} H_{mj}, \quad H_{ij}(X, t_0, t_0) = \delta_{ij}. \tag{10.7}$$

One has $H_{ij} = F_{ji}^{-1}$ if and only if the mean-velocity-gradient matrix $\mathbf{A}$ is symmetric, i.e., if the mean, flow is irrotational. A transposed mean-velocity-gradient matrix must be used in Eq. (10.6) to connect H_{ij} to the modified $\mathbf{F}$ in the general rotational case.

More generally, the linear response to an *irrotational* mean flow can be expressed by a linearized form of Weber equation (2.22), leading to

$$u'_i(x, t) = F_{ji}^{-1}(X, t, t_0) u'_j(X, t_0) + \frac{\partial \phi}{\partial x_i}, \tag{10.8}$$

in which the scalar potential ϕ accounts for the effects of fluctuating pressure. The associated equation for the vorticity fluctuation is

$$\omega'_i(x, t) = \frac{1}{\text{Det } \mathbf{F}} F_{ij}(X, t, t_0) \omega'_j(X, t_0). \tag{10.9}$$

Both equations are valid not only in the solenoidal case ($u'_{i,i} = 0$), as used in Chapter 5, but also in various barotropic compressible cases. The vorticity equation is not valid in the presence of a linearized baroclinic torque, for instance. These equations were used by several authors, including Hunt (1973), Goldstein (1978), and Durbin and Zeman (1992).

As conjectured in Jacquin, Cambon, and Blin (1993) and rediscussed in Coleman, and Mansour (1993) and Simone, Coleman, and Cambon (1997), the solenoidal linear response u'_s obtained by applying the Helmholtz decomposition to Eq. (10.8) yields the minimum kinetic-energy growth rate, and its pressure-released counterpart leads to the maximum growth rate. This yields the following evolution equation:

$$u' = u'_s + \underbrace{f(M_g)\left[u'^{(\text{pr})} - u'_s\right]}_{u'_d}, \tag{10.10}$$

where $u'^{(\text{pr})}$ is given by Eq. (10.5) with $\mathbf{H} = {}^T\tilde{\mathbf{F}}^{-1} = e^{-\int_{t_0}^{t} \mathbf{A}(t, t_0) dt'}$. The following simple model can be derived (e.g., Cambon, Coleman, and Mansour, 1993):

$$\mathcal{K} = \mathcal{K}_s + \underbrace{f(M_g)\left[\mathcal{K}^{(\text{pr})} - \mathcal{K}_s\right]}_{\mathcal{K}_d} \tag{10.11}$$

for the linear history of the turbulent kinetic energy, where the weighting factor $f(M_g)$ is a monotonically increasing function of M_g. For a consistency reason, one has $f(0) = 0$. This decomposition was successfully assessed for homogeneous turbulence, using both isentropic RDT and full DNS, as subsequently discussed, and it is reasonable if the pressure-released limit is more energetic than the solenoidal limit, as it often is. Incidentally, one can mention that a simple model by Debiève et al. (1982) for the evolution of the RST in turbulence–shock-wave interaction can be derived from Eq. (10.5). The reader is referred to Jacquin, Cambon, and Blin (1993) for a detailed discussion of this approach. A short discussion is also given in Chapter 11, in the section devoted to the comparison between RDT and LIA.

This preliminary analysis has the advantage in that it does not involve detailed expressions for the Helmholtz decomposition in Fourier space. But the extraction of $u^{(s)}$ from the solution of Eq. (10.8) requires solving the Poisson equation,

$$\nabla^2 \phi = -\frac{\partial u_i'^{(\text{pr})}}{\partial x_i},$$

which comes from the dilatational balance $u_d'^{(\text{pr})} + \nabla \phi = 0$.

More generally, two aspects must be kept in mind:

- The linearized Weber equation is valid for *irrotational mean flows without baroclinic effects* only. Finding the final expression of the scalar potential ϕ in terms of initial velocity remains an additional task to do anyway.
- The pressure-released linear limit (10.5) is completely general and is valid for a rotational mean flow. But it appears as the limiting case of a linearized Weber equation for irrotational mean strains only.

10.2 A General Quasi-Isentropic Approach to Homogeneous Compressible Shear Flows

Generally, if the Mach number effect is significant, the effects of compressibility are complex, as both acoustic and entropy modes are called into play, as well as the vortical mode inherited from the incompressible case (see Section 9.1). Irrotational mean flows have been studied by Goldstein (1978), who used an inhomogeneous RDT formulation [which can be based on Eq. (10.8)], whereas homogeneous RDT has been extended to quasi-isentropic compressible turbulence at significant Mach numbers in the presence of either irrotational compression or mean-shear flows (Simone, Coleman, and Cambon, 1997). For high-speed compressible flows, it is no longer possible to consider the velocity field as divergence free. Accordingly, the pressure disturbance can recover its role of thermodynamical variable. It is no longer a Lagrange multiplier bound by the divergence-free constraint, which can be eliminated.

10.2 Quasi-Isentropic Approach to Homogeneous Compressible Shear Flows

10.2.1 Governing Equations and Admissible Mean Flows

Compressible isentropic equations are

$$\frac{\partial \rho}{\partial t} + u_i \frac{\partial \rho}{\partial x_i} + \rho \frac{\partial u_i}{\partial x_i} = 0, \tag{10.12}$$

$$\rho \left(\frac{\partial u_i}{\partial t} + u_j \frac{\partial u_i}{\partial x_j} \right) = -\frac{\partial p}{\partial x_i} = 0, \tag{10.13}$$

$$\frac{1}{p} \left(\frac{\partial p}{\partial t} + u_i \frac{\partial p}{\partial x_i} \right) - \gamma \frac{1}{\rho} \left(\frac{\partial \rho}{\partial t} + u_i \frac{\partial \rho}{\partial x_i} \right) = 0. \tag{10.14}$$

Extending the analysis performed for strictly incompressible flows (including the special case of buoyant flows addressed in Chapters 7 and 8) is not a easy task and can be done in several different ways.

One can at least try to obey the following rules or principles:

- To define a base flow, which could be identified with the mean flow, or $\bar{\rho}, \bar{u}, \bar{p}$, as a special solution of the governing equations.
- To derive evolution equations for a disturbance flow, $\rho' = \rho - \bar{\rho}$, $p' = p - \bar{p}$, $u' = u - \bar{u}$ by subtracting equations for the base flow from governing equations. The structure of these equations may satisfy some properties of invariance by translation, which are consistent with statistical homogeneity.
- To restrict the degree of nonlinearity to quadratic terms, neglecting higher-order nonlinear terms.

As in homogeneous incompressible turbulence, the first and second conditions can be considered independently of statistical assumptions and treatment. This is done by Craik and co-workers, who define the first condition as an *admissibility condition*. Accordingly, an admissible base flow is also compatible with a wavelike form for the disturbance flow, and the superposition of both is called "a class of exact solutions" for Euler equations. This is nothing other than a formal rediscovery of RDT, but one in which nonlinearity is rigorously excluded in the equations for the disturbance flow: Only single-mode perturbation is considered *and nonlinearity is zero for a single Fourier mode in the incompressible case*. This is no longer true in compressible turbulence: Nonlinearity does exist even for a single Fourier mode (monochromatic disturbance). We will try to define, however, a system of simplified equations in which only quadratic nonlinearities appear.

The condition that the mean-velocity gradient must be uniform in space is inherited from the incompressible case, leading to

$$\frac{\partial \bar{u}_i}{\partial x_j} = A_{ij}(t).$$

An admissibility condition for the mean density follows as

$$\frac{\partial \bar{\rho}}{\partial t} + A_{jm} x_m \frac{\partial \bar{\rho}}{\partial x_j} + \bar{\rho} A_{ii} = 0.$$

This condition is compatible with the existence of a mean-density gradient, or $\bar{\rho} = \rho_0(t) + C_i(t)x_i$, as in the buoyant flow case with the Boussinesq approximation addressed in Chapter 7. But this case is too complicated if the velocity field is not solenoidal, so that only

$$\bar{\rho} = \rho(t), \quad \rho_0 = \bar{\rho}(0) \tag{10.15}$$

is considered.

As a consequence, the momentum equation for the base flow is very similar to its incompressible counterpart:

$$\bar{\rho}(t)\left(\frac{dA_{ij}}{dt} + A_{im}A_{mj}\right)x_j = -\frac{\partial \bar{p}}{\partial x_i}.$$

One recovers the condition that $d\mathbf{A}/dt + \mathbf{A}^2$ must be a symmetric tensor, taking the curl of the preceding equation, and that $\bar{\rho}(t)(dA_{ii}/dt + A_{im}A_{mi}) = -\nabla^2 \bar{p}$, taking its divergence. Looking now at the linearized momentum equation for the disturbance flow,

$$\rho'\left(\frac{dA_{ij}}{dt} + A_{im}A_{mj}\right)x_j + \bar{\rho}(t)\left[\frac{\partial u'_i}{\partial t} + A_{jm}x_m\frac{\partial u'_i}{\partial x_j} + A_{ij}(t)u'_j\right]x_j = -\frac{\partial \bar{p}}{\partial x_i}, \tag{10.16}$$

one observes that the contribution from the left-hand side is twofold, as it combines both the mean acceleration weighted by the fluctuating density and the fluctuating acceleration weighted by the mean density. Violation of translational invariance by the first term is probable, but difficult to prove.* A simplified class of mean flows is finally proposed as

$$\frac{d\mathbf{A}}{dt} + \mathbf{A}^2 = 0, \quad \bar{p} = P(t). \tag{10.17}$$

Relaxing the assumption of irrotational mean strain is possible for developing at least linear RDT solutions, but the assumption of statistical homogeneity must be enforced. Even if this condition is much less stringent than is generally admitted, the fact that the mean flow is characterized only by its spatial gradient matrix $\mathbf{A}$ has important consequences. First, it is not possible to define a length scale and a velocity scale for the mean flow (such as δ and ΔU in the shear-layer case), but only a time scale. This explains why linearization is not justified, as in conventional linear-stability analysis, by a (small) ratio of disturbance-to-base-velocity scale, but by an assumption of small elapsed time. S, which has the dimension of the inverse of a time, is a typical scalar scale for $\mathbf{A}$; the linear solution is expected to hold for small St only. But the maximum St at which it is valid depends crucially on the initial *shear-rapidity* factor SL/u', where L and u' are typical scales for the disturbance flow, and other features of the flow. Practical experience shows that the validity of RDT cannot be predicted *a priori*, considering only St and the initial value of

---

* In this case, one could check if a wavelike disturbance form for all the disturbance terms, including also ρ', can be consistently defined, without canceling *a priori* the mean-acceleration term.

10.2 Quasi-Isentropic Approach to Homogeneous Compressible Shear Flows

SL/u'. In some extreme cases, like rotating turbulence with angular velocity Ω and low Rossby number $Ro = u'/(2\Omega L)$, the nonlinearity becomes significant only after a very long time, such that $\Omega t = Ro^{-2}$ (see Chapter 4). This is partly explained by the depletion of nonlinearity that is due to phase mixing by dispersive inertial waves at small Rossby numbers. This "rapid" adjective in RDT is even less relevant here because at least two "rapid" time scales exist, namely $1/S$ and L/a. In brief, short time is a sufficient condition to ensure the validity of the linear solution, but not a necessary one.

A background mean flow is defined by space-uniform density $\bar{\rho}(t)$, pressure $P(t)$, and mean-velocity gradients,

$$U_i = A_{ij} x_j,$$

along with Eq. (10.17), and it is possible at least to consider that the fluctuations of density and pressure are weak with respect to their mean reference values, i.e.,

$$\rho' \ll \bar{\rho}(t), \quad p' \ll P(t).$$

Finally, the following simplified system of two equations is found for the fluctuating flow:

$$\frac{D}{Dt} u'_i + A_{ij} u'_j + \frac{1}{\bar{\rho}} \frac{\partial p'}{\partial x_i} = -u'_j \frac{\partial u'_i}{\partial x_j}, \tag{10.18}$$

$$\frac{D}{Dt}\left(\frac{p'}{\gamma P}\right) + \frac{\partial u'_i}{\partial x_i} = -u'_j \frac{\partial}{\partial x_j}\left(\frac{p'}{\gamma P}\right), \tag{10.19}$$

where

$$a^2 = \gamma \frac{P}{\bar{\rho}} \tag{10.20}$$

is the square of the speed of sound. The symbol $\frac{D}{Dt}$ denotes the material derivative following the mean-flow streamlines. Viscous terms are omitted, in agreement with isentropic assumption, but they can be added for numerical convenience. These equations are the starting point for both the nonlinear statistically isotropic approach in the absence of a mean-velocity gradient (Fauchet et al., 1997) (see also Subsection 9.3.1), and the linear approaches in the presence of **A** (Jacquin, Cambon, and Blin, 1993; Cambon, Coleman, and Mansour, 1993; Simone, Coleman, and Cambon, 1997).

10.2.2 Properties of Admissible Mean Flows

A zero-mean acceleration in Eq. (10.17) corresponds to

$$\ddot{F}_{ij}(t,0) X_j = 0,$$

so that the general expression for F is

$$F_{ij}(t,0) = \delta_{ij} + S_{ij} t, \tag{10.21}$$

where S_{ij} is an arbitrary constant matrix. The mean-velocity-gradient matrix is readily derived, using $\mathbf{A} = \frac{d\mathbf{F}}{dt}\mathbf{F}^{-1}$, as

$$\mathbf{A}(t) = \mathbf{S}(\mathbf{I} + \mathbf{S})^{-1}. \tag{10.22}$$

Using the volumetric ratio

$$J(t, 0) = Det\mathbf{F}(t, t_0) = \exp\left[\int_{t_0}^{t} A_{ii}(t')dt'\right] \tag{10.23}$$

and the mean-isentropic equation $P(t)\rho_0^{-\gamma}(t) = $ constant, one obtains

$$\rho_0(t) = \frac{\rho_0(0)}{J(t, 0)}, \quad P(t) = \frac{P(0)}{J^{\gamma}(t, 0)}, \tag{10.24}$$

resulting in the definition of a time-dependent speed of sound $a(t)$ if $J(t, 0)$ is time varying.

These equations were given in Cambon, Coleman, and Mansour (1993) to extend the conditions proposed by Blaisdell, Mansour, and Reynolds (1991), and perhaps are not the most general solutions consistent with the three admissibility conditions mentioned in the previous section. The search for more complex admissible base flows (e.g., Craik and Allen, 1992) is more a mathematically skilled task than a physically relevant problem. One must mention a possible way to derive a hierarchy of model equations, using asymptotic expansions: The ambitious study by Rupat Klein (private communication) yielded a hierarchy of flow models including more and more complex effects of compressibility, but is especially relevant for geophysical turbulence.

10.2.3 Linear Response in Fourier Space. Governing Equations

Equations (10.18) and (10.19) are linearized around a mean flow with space-uniform gradient $\mathbf{A}$, discarding their right-hand sides. As usual in RDT and in related stability analyses, the equations are simplified by considering Fourier modes

$$(u'_i, p')(\mathbf{x}, t) = \int (\hat{u}_i, \hat{p})[\mathbf{k}(t), t]e^{\imath \mathbf{k}(t) \cdot \mathbf{x}} d^3\mathbf{k},$$

with

$$\dot{k}_i = -A_{ji}k_j.$$

The only difference with incompressible RDT for the treatment of advection is the occurrence of the term A_{nn}, because the Fourier counterpart of u'_i is

$$\frac{\partial \hat{u}_i}{\partial t} - A_{nn}\hat{u}_i - A_{jm}k_j\frac{\partial \hat{u}_i}{\partial k_m}.$$

Extending the notation with the overdot to derivatives in Fourier space, the latter equation can be recast as

$$\dot{\hat{u}}_i - A_{nn}\hat{u}_i = \frac{\partial \hat{u}_i}{\partial t} + \frac{dk_m}{dt}\frac{\partial \hat{u}_i}{\partial k_m} - A_{nn}\hat{u}_i. \tag{10.25}$$

10.2 Quasi-Isentropic Approach to Homogeneous Compressible Shear Flows

The system of Eqs. (10.18)–(10.19) yields

$$\dot{\hat{u}}_i - A_{nn}\hat{u}_i + A_{ij}\hat{u}_j = -\frac{\iota}{\rho_0}k_i\hat{p}, \quad (10.26)$$

$$\dot{\hat{p}} + (\gamma - 1)A_{nn}\hat{p} = -\gamma P \iota k_i \hat{u}_i. \quad (10.27)$$

An important step in investigating this system is to use the Helmholtz decomposition:

$$\hat{\mathbf{u}} = \underbrace{\left(\mathbf{I} - \frac{\mathbf{k}\mathbf{k}^T}{k^2}\right)\hat{\mathbf{u}}}_{\hat{\mathbf{u}}_s} + \underbrace{\frac{\mathbf{k}\mathbf{k}^T}{k^2}\hat{\mathbf{u}}}_{\hat{\mathbf{u}}_d}. \quad (10.28)$$

In addition, the Craya–Herring frame of reference can be used for specifying the two solenoidal modes: The third vector is nothing other than $\mathbf{k}/k$ and is used to define the dilatational mode, with a superscript 3, consistently with

$$\mathbf{e}^{(3)} = \frac{\mathbf{k}}{k}, \quad u^{(3)} = \hat{\mathbf{u}} \cdot \mathbf{e}^{(3)},$$

from which comes

$$\hat{\mathbf{u}} = \underbrace{u^{(1)}\mathbf{e}^{(1)} + u^{(2)}\mathbf{e}^{(2)}}_{\hat{\mathbf{u}}_s} + \underbrace{u^{(3)}\frac{\mathbf{k}}{k}}_{\hat{\mathbf{u}}_d}. \quad (10.29)$$

In agreement with the decomposition in physical space discussed in Subsection 9.1.4, the subscripts s and d denote solenoidal and dilatational modes, respectively. Because the Craya–Herring frame of reference is a direct orthonormal frame, vortical (applying the curl operator) and dilatational (applying the divergence) velocity contributions have simple counterparts in this frame:

$$\hat{\boldsymbol{\omega}}(\mathbf{k}, t) = \iota k \left[u^{(1)}\mathbf{e}^{(2)} - u^{(2)}\mathbf{e}^{(1)}\right], \quad (10.30)$$

$$\widehat{u_{i,i}}(\mathbf{k}, t) = \iota k u^{(3)}. \quad (10.31)$$

To recover a homogeneous problem, the pressure fluctuation is scaled as a velocity and considered as fourth component of the solution vector (Simone, Coleman, and Cambon, 1997):

$$u^{(4)} = \iota \frac{\hat{p}}{\rho_0 a}. \quad (10.32)$$

This scaling is similar to the one of Eckhoff and Storesletten (1978).

Linear solutions are therefore expressed in terms of $u^{(i)}, i = 1, 4$ components, solving the following linear system of ODEs:

$$\begin{pmatrix} \dot{u}^{(1)} \\ \dot{u}^{(2)} \\ \dot{u}^{(3)} \\ \dot{u}^{(4)} \end{pmatrix} + \begin{bmatrix} m_{11} - A_{nn} & m_{12} & m_{13} & 0 \\ m_{21} & m_{22} - A_{nn} & m_{23} & 0 \\ m_{31} & m_{32} & m_{33} - A_{nn} & +ak \\ 0 & 0 & -ak & \frac{3-\gamma}{2}A_{nn} \end{bmatrix} \begin{pmatrix} u^{(1)} \\ u^{(2)} \\ u^{(3)} \\ u^{(4)} \end{pmatrix} = 0, \quad (10.33)$$

or equivalently,

$$\dot{u}^{(i)} - A_{nn}u^{(i)} + m_{ij}u^{(j)} = 0, \tag{10.34}$$

where the coefficients of the RDT matrix **m** are defined as follows:

$$m_{\alpha\beta} = e_i^{(\alpha)} A_{ij} e_j^{(\beta)} - \dot{e}_i^{(\alpha)} e_i^{(\beta)}$$
$$= e_i^{(\alpha)} A_{ij} e_j^{(\beta)} + \epsilon_{\alpha 3\beta} e_j^{(2)} A_{ij} e_j^{(1)}, \tag{10.35}$$

$$m_{\alpha 3} = e_i^{(\alpha)} A_{ij} e_j^{(3)} - \dot{e}_i^{(\alpha)} e_i^{(3)} = e_i^{(\alpha)} (A_{ij} - A_{ji}) e_j^{(3)}, \tag{10.36}$$

$$m_{3\alpha} = e_i^{(3)} A_{ij} e_j^{(\alpha)} - \dot{e}_i^{(3)} e_i^{(\alpha)} = 2 e_i^{(3)} A_{ij} e_j^{(\alpha)}, \tag{10.37}$$

$$m_{33} = e_i^{(3)} A_{ij} e_j^{(3)}, \tag{10.38}$$

$$m_{34} = -m_{43} = -a_0 k. \tag{10.39}$$

As previously, Greek indices take the value 1 or 2 only and refer to solenoidal modes. The calculation of the "solenoidal block" is made assuming that the polar axis n of the Craya–Herring frame of reference is one of the eigenvectors of **A**.

Of course, the 4D problem in physical space (u_1, u_2, u_3, p) remains a four-component problem in Fourier space. When the Craya–Herring frame is used, no reduction of the number of variables is obtained as in solenoidal cases because the dilatational mode $u^{(3)}$ does not vanish, but the matrix $m_{ij}, i = 1, 4, j = 1, 4$ has some zero components and the role of each nonzero component is more easily understood. All the coefficients in the "velocity block" of the preceding matrix, or $m_{ij}, i = 1, 3, j = 1, 3$, depend on **A** and therefore scale with S, which is a norm of **A**. The acoustic terms m_{34} and m_{43} scale with the dispersion frequency of acoustic waves, ka_0. As a consequence, the parameter $S/(a_0 k)$ is immediately found to be the *pivotal* parameter; of course it is a spectral counterpart of the gradient Mach number M_g. The different couplings between the solenoidal modes $u^{(1)}$ and $u^{(2)}$, the dilatational mode $u^{(3)}$, and the pressure mode $u^{(4)}$ are illustrated on Fig. 10.2.

The solution of Eq. (10.33) is expressed as

$$u^{(i)}[\boldsymbol{k}(t), t] = J(t, 0) g_{ij}(\boldsymbol{k}, t, 0) u^{(j)}(\boldsymbol{K}, 0), \tag{10.40}$$

with J given by (10.23). As usual in RDT, the deterministic function $g_{ij}, i, j = 1, 4$, can be computed analytically or numerically, solving sequentially system (10.33) for a set of arbitrary simple initial data, such as $u^{(i)} = \delta_{i1}, \delta_{i2}, \delta_{i3}, \delta_{i4}$. Corresponding solutions for the statistical moments are obtained from the initial value of these statistical moments through g products. Simplified forms of these initial values usually come from isotropy and "strong" acoustic-equilibrium assumptions (Simone, Coleman, and Cambon, 1997).

In the general case with time-dependent J, new "divergence" and "pressure" terms can be used, introducing the following new integration variables:

$$y = J^{-1} \frac{u^{(3)}}{k}, \quad z = J^{-1} \frac{u^{(4)}}{a_0}, \tag{10.41}$$

10.2 Quasi-Isentropic Approach to Homogeneous Compressible Shear Flows

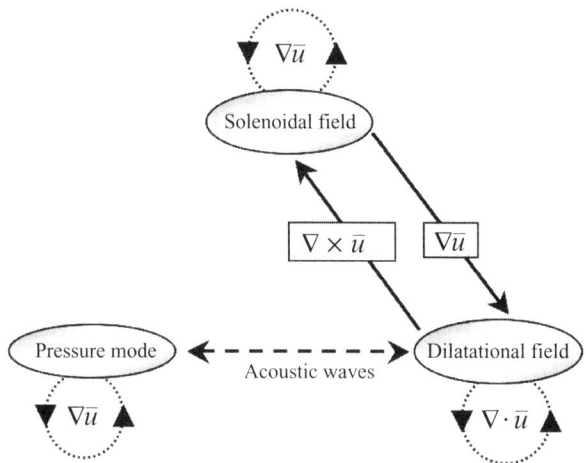

Figure 10.2. Schematic view of interactions between modes defined in the local Craya–Herring frame according to the RDT analysis of compressible homogeneous shear flows. Dotted lines denote self-interaction.

and by using

$$\frac{\dot{k}}{k} = -A_{ij}\frac{k_i k_j}{k^2}. \tag{10.42}$$

one finds a useful "pressure" equation:

$$\frac{D}{Dt}\left(\frac{\dot{z}}{k^2}\right) + a_0^2 z = a_0^2 z_s, \tag{10.43}$$

in which

$$z_s = J^{-1}\frac{\hat{p}_s}{\bar{\rho}a_0^2} = \iota \frac{J^{-1}}{ka_0^2} m_{3\alpha} u^{(\alpha)} \tag{10.44}$$

involves only the solenoidal velocity field and exactly corresponds to the solution of the Poisson equation found in the strictly incompressible case.

10.2.3.1 Recovering the Acoustic Regime

In the absence of mean flow, i.e., setting $\mathbf{A} = \mathbf{0}$ in Eq. (10.33), the solenoidal mode is strictly conserved, whereas pressure and dilatational velocity modes are governed by

$$\left[u^{(3)} \pm \iota u^{(4)}\right](t) = e^{\pm \iota a_0 k t}\left[u^{(3)} \pm \iota u^{(4)}\right](t=0), \tag{10.45}$$

which corresponds to the acoustic regime discussed in Chapter 9.

10.2.3.2 Recovering the Solenoidal Limit

The solenoidal limit, also known as incompressible RDT, is found by solving only the block $i = 1, 2$, $j = 1, 2$, as $u^{(3)} = 0$ gives the solenoidal limit [e.g., Eq. (10.31)].

The governing equation for the pressure mode is

$$u^{(4)} = u_s^{(4)} = \frac{m_{3\alpha}}{ka_0} u^{(\alpha)}. \tag{10.46}$$

This equation gives the counterpart of the solution for the Poisson equation satisfied by the fluctuating pressure in this limit, or

$$\hat{p} = \hat{p}_s = -\frac{\imath}{\rho_0} m_{3\alpha} u^{(\alpha)}, \quad m_{3\alpha} = 2\frac{k_i}{k} A_{ij} e_j^{(\alpha)}.$$

10.2.3.3 Irrotational Mean-Strain Case

In addition to pure solenoidal coupling terms $m_{\alpha\beta}$, which are the same as in solenoidal RDT, and to "acoustical" or "pseudo-sound" terms $m_{34} = -a_0 k$, $m_{43} = a_0 k$, previously discussed, some very interesting terms are

$$m_{\alpha 3} = e_i^{(\alpha)} (A_{ij} - A_{ji}) \frac{k_j}{k}.$$

These terms represent a feedback from the dilatational mode to the solenoidal modes, and they are generated by the *rotational part* of the mean flow.

As an immediate consequence, the solenoidal flow is decoupled in the presence of an irrotational straining process. Another less obvious consequence is that the kinetic-energy growth rate is larger in compressible RDT than in solenoidal RDT, as the kinetic energy of the dilatational mode, which is always positive, is just added to the kinetic energy of the solenoidal mode, which is independent of compressibility in this case. The effect of the fluctuating pressure in the solenoidal linear limit is just to kill this dilatational contribution. Accordingly, as first demonstrated by Jacquin, Cambon, and Blin (1993), the kinetic-energy growth rate increases monotonically with increasing gradient Mach number M_g, from solenoidal RDT to "pressure-released" RDT, in full agreement with Eq. (10.11). These results were revisited and confirmed by full DNS (Cambon, Coleman, and Mansour, 1993) (also quoted in Lele, 1994, and Simone, Coleman, and Cambon, 1997) for homogeneous axial compression [$A_{ij} = S(t)\delta_{i1}\delta_{j1}$], as shown on Fig. 10.3. For the sake of convenience, the time-advancement parameter is not $S(0)t$ but the inverse of the mean volumetric ratio $J(t)$, with $S(t) = -S(0)/(1 - S(0)t) = -S(0)/J(t)$. The contribution of the solenoidal mode corresponds to a quasi-linear growth of kinetic energy in terms of the mean-compression ratio J^{-1}, whereas the dilatational contribution leads to a quasi-parabolic growth. It is worth noticing that compressibility is always shown to have a destabilizing effect regarding RDT for irrotational mean flow. Of course, the general relevance of this result can be questioned because it relies on both the irrotational condition and the short time condition $S(0)t < 1$ in the case of axial compression.

Analytical solutions for solenoidal RDT and pressure-released limit are

$$\frac{\mathcal{K}(t)}{\mathcal{K}(0)} = \frac{1}{2}\left(1 + J^{-2}\frac{\tan^{-1}(\sqrt{J^{-1} - 1})}{\sqrt{J^{-1} - 1}}\right), \tag{10.47}$$

10.2 Quasi-Isentropic Approach to Homogeneous Compressible Shear Flows

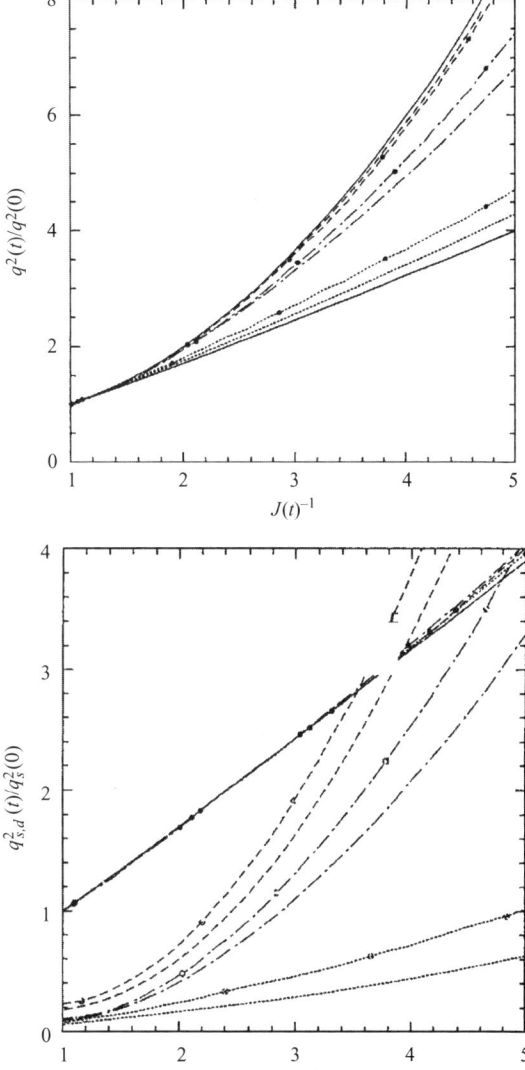

Figure 10.3. Turbulent kinetic-energy histories for different values of the gradient Mach number M_g, in the axial compression case. Top: Full DNS (dots) and linear theory (lines). Bottom: in addition, splitting into solenoidal and dilatational parts (Simone, Coleman, and Cambon, 1997).

yielding a quasi-linear growth in terms J^{-1}, and

$$\frac{\mathcal{K}(t)}{\mathcal{K}(0)} = \frac{2 + J^{-2}}{3}, \qquad (10.48)$$

yielding a parabolic growth if $J^{-1} \geq 1$, respectively. Despite the presence of the varying mean volumetric ratio, the solenoidal equation is very close to the classical equation for mean incompressible axial strain given by Batchelor [see also Eqs. (5.36) and (5.37) in Chapter 5], up to a $J^{4/3}$ factor. This result was also found by Ribner (1953). The framework of solenoidal turbulence subjected to mean strain with variable volume is also addressed in Cambon, Mao, and Jeandel (1992), with the particular flow case addressed in the next subsection.

10.3 Incompressible Turbulence With Compressible Mean-Flow Effects: Compressed Turbulence

An interesting class of solenoidal (i.e., with divergence-free velocity fluctuations) homogeneous turbulent flows can be considered in the presence of a mean flow with space-uniform gradients, which take into account a variation in the mean volume. Provided that the Mach number is small enough, this set of assumptions is self-consistent, and it is possible to extend solenoidal RDT to *compressed turbulence*, i.e., to a divergence-free fluctuating-velocity field in the presence of a mean dilatational flow, neglecting acoustics and thermal effects.

The mean flow is characterized by the volumetric ratio (10.23), which differs from 1 when the constraint $A_{ii} = 0$ is relaxed. For the sake of brevity t_0 is omitted in what follows, so that abridged notations $\mathbf{F}(t)$, $J(t)$ are now used in this section. Among different compressing mean flows, the case of isotropic compression deserves particular attention. In this case, the matrices $\mathbf{A}$ and $\mathbf{F}$, and the trajectory equations are written as

$$A_{ij}(t) = S(t)\delta_{ij}, \quad F_{ij}(t) = J^{1/3}(t)\delta_{ij}, \quad x_i = J^{1/3}(t)X_j, \qquad (10.49)$$

in which $S = \frac{1}{3}\frac{1}{J}\frac{dJ}{dt}$. The fluctuating field is governed by

$$\frac{\partial u'_i}{\partial t} + Sx_j\frac{\partial u'_i}{\partial x_j} + Su'_i + \frac{1}{\rho}\frac{\partial p'}{\partial x_i} = -u'_j\frac{\partial u'_i}{\partial x_j} + \nu\nabla^2 u'_i, \qquad (10.50)$$

in which explicit nonlinear terms and viscous terms are gathered in the right-hand side. Setting the right-hand side to zero, the RDT solution is directly found in physical space[†]:

$$\mathbf{u}'(\mathbf{x}, t) = J^{-1/3}(t)\mathbf{u}'(\mathbf{X}, 0).$$

More interesting is the possibility of deriving a rescaling for full nonlinear equation (10.50) in terms of spatial coordinates, velocity, and time. It is expressed as follows:

$$\mathbf{x}^* = J^{-1/3}\mathbf{x}, \quad \mathbf{u}^*(\mathbf{x}^*, t^*) = J^{1/3}\mathbf{u}'(\mathbf{x}, t) \quad dt^* = J^{-2/3}(t)dt. \qquad (10.51)$$

Such a dynamical rescaling can also be used in Boltzmann equations and applied to cosmological gas in order to account for the expansion of the universe. When substituting it in Eq. (10.50) that governs the primitive unscaled variables, the rescaled quantities are shown to satisfy the Navier–Stokes equations without the additional mean terms that depend on S on the left-hand side. For consistency reasons, the pressure is rescaled as $p^* = J^{5/3}p$, and the only difference with uncompressed freely decaying isotropic turbulence for the velocity field $\mathbf{u}^*(\mathbf{x}^*, t^*)$ is a possible influence

---

[†] This is a very special case, in which the nonlocal potential term is zero in Eq. (10.8).

10.3 Incompressible Turbulence With Compressible Mean-Flow Effects

of time variations of the viscosity $v^*(t)$. The variation in Reynolds number follows directly because $u'L = u^*L^*$. If the Reynolds number is high enough, however, it is reasonable to expect that all classical results dealing with spatiotemporal dynamics and statistics of isotropic freely decaying turbulence are still valid for (u^*, x^*, t^*), so that the corresponding laws for primitive variables (u', x, t) can be readily derived using Eq. (10.51). The reader is referred to Cambon, Mao, and Jeandel (1992) for various applications.

This scaling deserves attention for two reasons. First, it illustrates a particular "dynamical" version of the general scale invariance [see, e.g., Frisch, 1995, and Eq. (3.25)]:

$$x^* = \lambda x, \quad u^* = \lambda^h u, \quad t^* = t\lambda^{1-h}, \quad v^* = \lambda^{1+h}v, \quad (10.52)$$

so that λ corresponds to the time-dependent mean-density ratio $J^{-1/3}$, with $h = -1$. In the latter invariance group, the viscosity would be left unchanged if $h = -1$, but it should be borne in mind that the dynamical rescaling deals with a continuously time-varying parameter $J^{-1/3}(t)$ in contrast to λ. It is worth noting that, by taking $h = -1$, one recovers scaling law (3.20), whereas one finds transformation (3.21) by setting $h = 1$.

Second, it can be used to check the consistency of any model or theory, ranging from $\mathcal{K} - \varepsilon$ to elaborate EDQNM, DIA, or LRA versions.

As a simple example, let us start with a classical decay law such as

$$\mathcal{K}(t) = \mathcal{K}(0)\left(1 + \frac{t}{nt_0}\right)^{-n}, \quad L(t) = L(0)\left(1 + \frac{t}{nt_0}\right)^{1-n/2},$$

consistently obtained for the turbulent kinetic energy, its dissipation rate, and the single relevant integral length scale L, with $1/t_0 = -(1/\mathcal{K})d\mathcal{K}/dt$ at $t = 0$. Applying the rescaling, which amounts to rewriting the same equations in terms of "starred" variables, the following equations are derived for the "compressed" decay:

$$\mathcal{K}(t) = \mathcal{K}(0)e^{2Ct}\left(1 + \frac{e^{2Ct} - 1}{2nCt_0}\right)^{-n}, \quad L(t) = L(0)e^{-Ct}\left(1 + \frac{e^{2Ct} - 1}{2nCt_0}\right)^{1-n/2},$$

for a mean compression or dilatation at constant rate $S(t) = -C$. These equations show immediately that the domain of relevance of RDT in terms of elapsed time is more restricted as is usually conjectured, with a dominant nonlinearity having an effect opposite to the linear one. Choosing a spherical compression, i.e., $C > 0$, the RDT growth-rate factor for $\mathcal{K}$, e^{2Ct}, is always balanced and rapidly dominated by a nonlinear term given by e^{-2nCt}. This reflects the fact that, when the velocity u' is affected by a linear RDT factor e^{Ct}, the nonlinear term of dimension u'^2/l is affected by a factor e^{-3Ct}, the full nonlinear effect being finally accounted for by the time rescaling $dt = dt^*e^{2Ct}$.

This flow is particular in the sense that turbulence is not really compressible, but it offers a very simple way to exactly evaluate the impact of nonlinearity; this is an unique instance for comparing linear RDT with full nonlinear theory. In contrast, a depletion of nonlinearity is rather expected in true compressible turbulence, with respect to the incompressible flow case, but in the anisotropic case, as discussed further. It is also possible to study the spherical (isotropic) compression or dilatation applied to really compressible homogeneous turbulence. Very consistent results were found by Blaisdell, Coleman, and Mansour (1996) and Simone, Coleman, and Cambon (1997), using full DNS and isentropic RDT: As a particular result, the strong acoustic equilibrium can be sustained, as illustrated by Fig. 2 in Simone, Coleman, and Cambon (1997).

10.4 Compressible Turbulence in the Presence of Pure Plane Shear

The background velocity field of pure plane shear addressed in this section is identical to the one considered in the incompressible case (see Chapter 6). It is defined by

$$\mathbf{A} = \begin{bmatrix} 0 & S & 0 \\ 0 & 0 & 0 \\ 0 & 0 & 0 \end{bmatrix}, \quad \mathbf{F}(t) = \begin{bmatrix} 1 & St & 0 \\ 0 & 1 & 0 \\ 0 & 0 & 1 \end{bmatrix}. \tag{10.53}$$

The associated characteristic lines in the both Fourier and physical (trajectories) space are given by

$$k_1 = K_1, \quad k_2 = K_2 - K_1 St, \quad k_3 = K_3; \quad x_1 = X_1 + St X_2, \quad x_2 = X_2, \quad x_3 = X_3. \tag{10.54}$$

In this case $J \equiv 1$, so that $\bar{\rho}$, P, and a ($= a_0$) are constant.

10.4.1 Qualitative Results

Even in the pure shear case, the pressure-released limit is more energetic than the linear solenoidal limit. Accordingly, a reduction of pressure fluctuations, in the linear limit, would yield a monotonic increase of turbulent kinetic energy with increasing M_g, as for the case of irrotational mean straining! The fact that the pressure-released growth rate is higher than the solenoidal one, in the linear limit, results from Eqs. (10.5) and (10.53), which yield a quadratic growth rate for the kinetic energy, i.e., $\mathcal{K}(t) \propto (St)^2$.[‡] In the same conditions, the solenoidal RDT predicts only a linear growth rate: $\mathcal{K}(t) \propto St$. Once recast in a relevant nondimensional form, the kinetic-energy growth rate is characterized by

$$\Lambda = \frac{1}{S\mathcal{K}} \frac{d\mathcal{K}}{dt}, \tag{10.55}$$

which is equal to

$$\Lambda = -2 \left[b_{12} + \frac{\bar{\varepsilon}_s + \bar{\varepsilon}_d - \Pi^{(d)}}{S\mathcal{K}} \right] \tag{10.56}$$

[‡] The reader is referred to Chapter 6 for a detailed discussion of the incompressible shear case. Results are summarized in Table 6.1.

10.4 Compressible Turbulence in the Presence of Pure Plane Shear

according to Eq. (10.2). In the pressure-released linear limit with isotropic initial data, the kinetic-energy growth rate is equal to

$$\Lambda^{(\text{pr})} = \frac{2St}{3 + (St)^2}. \tag{10.57}$$

This equation gives the upper solid line in the bottom part of Fig. 10.4. The corresponding limit of solenoidal RDT is plotted in Fig. 10.4, as the lower solid line. Is there a simple explanation for this stabilizing effect of pressure in solenoidal RDT, without looking at RDT details? As in all shear-flow cases, the answer is given by the dynamics of the vertical velocity component: u'_2 is passively advected in the pressure-released linear limit, whereas it is its Laplacian $\nabla^2 u'_2$ that is advected in the RDT solenoidal limit. According to the corresponding RDT complete solution in Fourier space, $D(k^2 \hat{u}_2)/Dt = 0$, leading to $\hat{u}_2(\mathbf{k}, t) = \frac{K^2}{k^2} \hat{u}_2(\mathbf{K}, 0)$ and a decrease of u'_2 as $K/k < 1$ for $t > 0$. Using the Craya–Herring frame with the polar axis parallel to the cross-gradient direction of the mean shear, the complete RDT solution (see also Chapter 6) is much simpler than the one given by Townsend (1976) in the fixed frame. The corresponding equation dealing with $u'^{(2)}$ is

$$\frac{D}{Dt}(ku^{(2)}) = 0, \tag{10.58}$$

which is consistent with the pure advection of $\nabla^2 u_2$ in the physical space. Note that this analysis only confirms that the role of the so-called "rapid" pressure–strain-rate tensor in Reynolds stress equations is a stabilizing one. *This result is in qualitative agreement with crude single-point models, in which the "rapid" pressure–strain-rate tensor is modeled as reducing the production.* Using these simple considerations, a destabilizing effect of compressibilty is observed, as shown in Fig. 10.4 for $St < 4$ in both full DNS and quasi-isentropic compressible linear theory. This result is very similar to what happened in the irrotational mean-strain case, with a monotonic increase of Λ with increasing M_g, from the solenoidal to the pressure-released case.

10.4.2 Discussion of Results

As shown in Fig. 10.4, the conventional "stabilizing" behavior of compressibility is recovered at largest time $St > 4$. It is therefore clear that this stabilizing behavior is explained by the presence of the $m_{\alpha 3}$ coupling terms, at least in the linear limit. Figure 10.4 displays the main part of the turbulent kinetic-energy growth rate Λ defined in Eq. (10.55), which reduces to $-2b_{12}$, ignoring other terms, as also justified by Sarkar (1995).

Equations for the pure plane shear case are subsequently rewritten using Eqs. (10.35)–(10.37) and the three nontrivial components from the solenoidal Craya–Herring frame:

$$e_1^{(1)} = -\frac{K_3}{K_\perp}, \quad e_2^{(2)} = -\frac{K_\perp}{k}, \quad e_1^{(2)} = \frac{K_1 k_2}{K_\perp k}.$$

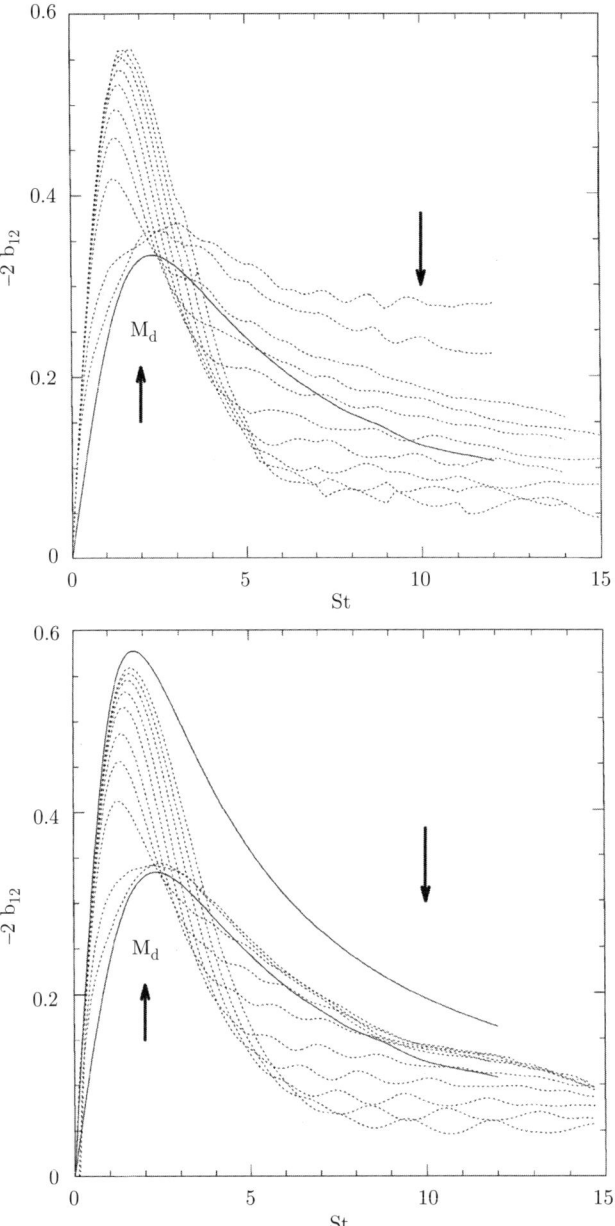

Figure 10.4. Histories of the nondimensional production term $-2b_{12}$, full DNS (top) and linear quasi-isentropic compressible theory (bottom), so-called (improperly) RDT. Upper and lower solid curves correspond to pressure-released limit and solenoidal limit, respectively. Initial M_g (called M_d in the figure) ranges from 4 to 67 for both DNS and RDT; arrows show trends with increasing M_g. From Simone, Coleman, and Cambon (1997).

10.4 Compressible Turbulence in the Presence of Pure Plane Shear

It is recalled that the optimal choice for the arbitrary vector $\boldsymbol{n}$ in the definition of the local frame (see Table 2.2) is to choose the polar axis in the cross-gradient shear direction: $n_i = \delta_{i2}$. Further simplifications are obtained with integrating factors to remove some diagonal coupling terms, in agreement with Eq. (10.42). After some algebra, one obtains

$$\dot{u}^{(1)} + S \frac{K_3}{k(t)} u^{(2)} = S \frac{K_3 k_2(t)}{K_\perp} \frac{u^{(3)}}{k(t)}, \tag{10.59}$$

$$\frac{D}{Dt}(k u^{(2)}) = -S \frac{K_1}{K_\perp} k(t) u^{(3)}, \tag{10.60}$$

$$\frac{D}{Dt}\left(\frac{u^{(3)}}{k(t)}\right) = 2S \frac{K_1 K_\perp}{k^4(t)} k(t) u^{(2)} - a_0 u^{(4)}, \tag{10.61}$$

$$\dot{u}^{(4)} = a_0 k(t) u^{(3)}, \tag{10.62}$$

with

$$K_\perp = \sqrt{K_1^2 + K_3^2}, \tag{10.63}$$

as a special case of Eq. (10.33). The counterpart of Eq. (10.43) is

$$\frac{D}{Dt}\left(\frac{\dot{\hat{p}}}{k^2(t)}\right) = a_0^2 \left(\underbrace{2\imath \rho_0 S \frac{K_1 K_\perp}{k^4}(k u^{(2)})}_{\hat{p}_s} - \hat{p} \right). \tag{10.64}$$

Going back to the (generally expected) stabilizing effect of compressibility, it is commonly accepted, following Sarkar (1995) and Pantano and Sarkar (2002), that the weakening of pressure correlation is the sole explanation. As a matter of fact, the weakening of pressure can be demonstrated from the solution of Eq. (10.64), considering the following scalar Green's function for pressure to velocity coupling:

$$\hat{p}(\mathbf{k}, t) = \int_{t_0}^{t} \mathcal{G}(\mathbf{k}, t, t') \hat{p}_s(\mathbf{k}, t') dt'. \tag{10.65}$$

Recently, Thacker, Sarkar, and Gatski (2006) proposed an analytical solution for a similar scalar Green's function in the pure shear case, generalizing the form $\mathcal{G} = \frac{\sin[a_0 k(t-t')]}{a_0 k}$ recovered in the shearless case (e.g., Pantano and Sarkar, 2002). One can point out that this scalar Green's function is generated by the equation

$$\ddot{\hat{p}} + a_0^2 k^2 \hat{p} = a_0^2 k^2 \hat{p}^{(s)} \tag{10.66}$$

(translating their results with our notations), which is simpler and less general than Eq. (10.64). Both Eqs. (10.64) and (10.66) account for the time dependency of k by means of Eq. (10.54), but the removal of the divergence term was not accurately obtained in Thacker, Sarkar, and Gatski (2006). In addition, the equation of type

(10.65) was used to express the "rapid" pressure–strain-rate tensor in terms of the velocity spectral tensor involved in $\overline{\hat{p}^*\hat{u}_i^{(s)}}$: A solenoidal spectral model for $\overline{\hat{u}_i^{(s)*}\hat{u}_j^{(s)}}$ was used by Thacker, Sarkar, and Gatski (2006) for this purpose (see also Cambon and Rubinstein, 2006, for a discussion of this model).

It is advocated here that the explanation based on Eq. (10.64) (weakening of pressure fluctuations) for stabilizing–destabilizing compressibility effects is only a partial one. This equation is also a by-product of the general study based on the full system of linear equations considered here. The conventional explanation is valid, for instance, to account for the difference between the less compressible case in the top part Fig. 10.4 (which corresponds to an almost constant production rate at largest St), which is also a fully nonlinear result, and the pressure released case (upper curve in solid line in the bottom part of Fig. 10.4). This "explanation" is irrelevant when the plot of compressible RDT results at large M_g and large elapsed time lies below the incompressible RDT limit curve (solid line in Fig. 10.4): In this case the sole argument of mollification of pressure would lead a destabilizing effect of compressibility. In contrast, the second explanation based on the feedback [in Eq. (10.60)] of the dilatational mode onto the relevant poloidal mode (which includes the whole vertical velocity component, a key component for the production by shear in any case) is valid.

As a final remark, let us recall that the argument dealing with the weakening of pressure is always relevant in the irrotational "mean" case, or at $St < 4$ in the shear case, but yields a *systematic destabilizing effect* because the pressure-released limit is always over the incompressible RDT limit! Of course, looking at the Reynolds stress equations, the weakening of the *nonlinear* (so-called slow) pressure–strain-rate tensor yields a stabilizing effect in the pure shear-flow case, but this reflects more a depletion of nonlinearity at increasing M_g than a stabilizing effect of compressibility: In addition, the ratio of gradient to turbulent Mach number M_g/M_t is nothing other than the shear-rapidity factor, and increasing M_g without increasing M_t in the same proportion means depleting the nonlinearity. The latter remark also holds for DNS results presented in Fig. 10.4, but not for pure linear theories.

10.4.3 Toward a Complete Linear Solution

As a direct continuation of the study by Simone, Coleman, and Cambon (1997), some work remains to be done to retrieve more information from the linear equations in the pure shear-flow case. The existence of the invariant quantity

$$\xi = ku^{(2)} + \frac{S}{a_0}\frac{K_1}{K_\perp}u^{(4)}, \tag{10.67}$$

which is passively advected [i.e., conserved along the characteristic lines (10.54)], as seen by combining Eqs. (10.60) and (10.62), offers new perspectives for analytical solutions. Analytical solutions by Thacker, Sarkar, and Gatski (2006) for the scalar pressure Green's function can be useful for this purpose too.

10.5 Perspectives and Open Issues

A single second-order equation is found at $K_1 \neq 0$ for $x = ku^{(2)}$:

$$\frac{D}{Dt}\left(\frac{\dot{x}}{k^2}\right) - K_1^2\left(2\frac{S^2}{k^4} + \frac{a_0^2}{K_\perp^2}\right)x = -\frac{K_1^2}{K_\perp^2}a_0^2\xi. \tag{10.68}$$

As in all RDT cases in the presence of pure plane shear, an analytical solution is found if $K_1 = 0$, because $\boldsymbol{k} = \boldsymbol{K}$. In this case acoustic solution (10.45) for $u^{(3)}$ and $u^{(4)}$ is valid, whereas the solution of (10.60) and (10.59) is

$$u^{(2)}(\boldsymbol{k}, t) = u^{(2)}(\boldsymbol{k}, 0), \tag{10.69}$$

$$u^{(1)}(\boldsymbol{k}, t) = u^{(1)}(\boldsymbol{k}, 0) - St\frac{k_3}{k}u^{(2)}(0)$$
$$+ S\frac{k_2}{k}\left[\frac{\sin(a_0 kt)}{a_0 k}u^{(3)}(\boldsymbol{k}, 0) - \frac{\cos(a_0 kt)}{a_0 k}u^{(4)}(\boldsymbol{k}, t)\right]. \tag{10.70}$$

The feedback from dilatational to toroidal mode is displayed in the latter equation. The probably more important (for global production) feedback from dilatational to poloidal is canceled at $k_1 = 0$, showing the need for a general solution at any K_1.

10.5 Perspectives and Open Issues

Perspectives for modeling nonhomogeneous and/or nonlinear effects can be briefly discussed. On the one hand, extending homogeneous RDT toward zonal (localized) RDT is possible, but the related assumption of short-wave disturbance can disconnect the acoustic modes in practice. A more promising case is found when disturbances are localized in the vicinity of rays (along which total energy, including the acoustic one, propagates), instead of being localized near mean trajectories. More details are given in Chapter 13.

On the other hand, in the absence of mean-velocity gradients, interactions among solenoidal, dilatational, and pressure modes are purely nonlinear and can be analyzed and modeled in pure isotropic homogeneous turbulence. In this context, the model by Fauchet et al. (1997) gave promising spectral information, as shown in the previous chapter. To reconcile both cases, i.e., taking into account both linear distortion by the mean flow and nonlinearity, is a formidable challenge. At least, the nonlinear model could be used for initializing in a better way the compressible RDT equations, replacing a questionable "strong" acoustic equilibrium with a more realistic "weak" one.

Regarding "strong" acoustic equilibrium, even the RDT solution can significantly break it, independently of initial data, if M_g is sufficiently large, or more precisely if $S/(a_0 k)$ is large and $K_1 \neq 0$. Linear equation (10.62) is probably always valid, even in the nonlinear case, and the forcing by the solenoidal term [poloidal mode in

Eq. (10.61)] can play a similar role here as the dominant part of the $T_{\text{NL}}^{(3)}$ term does in the nonlinear case.

10.5.1 Homogeneous Shear Flows

A critical survey of previous studies has shown that the alteration of pressure equation by compressibility, without significant change in its source term, is not the correct explanation for the "stabilizing" effect, at least for homogeneous shear flow. It is suggested that the alteration previously mentioned results from the depletion of nonlinearity, and that this is the nonlinear part of the pressure–strain rate and not the linear (so-called rapid) one that is concerned in this case. In contrast, the subtle coupling between solenoidal and dilatational velocity modes is essential for explaining the stabilizing effect in the linear limit, and especially the feedback from the dilatational mode induced by the rotational part of the mean flow, as in Eq. (10.60). Such an analysis escapes the description permitted by Reynolds stress modeling. A general linear solution such as (10.40) contains a lot of information, and it is a pity to derive from it only conventional single-point statistics: More information can be obtained about spectral distribution, because the ratio $S/(ak)$ that underlies the distortion Mach number is wavenumber dependent, and also about specific vortical and dilatational contributions.

10.5.2 Perspectives Toward Inhomogeneous Shear Flows

In an incompressible mixing layer, the velocity scale is unequally determined by the difference in the two stream velocities, $v_0 \sim \Delta U$, and variation in the length scale unequally depends on the velocity ratio. Compressibility changes this dimensional rule by making the speed of sound a relevant parameter with the consequence that the two preceding scales now possibly depend on a Mach number (the gradient Mach number $M_g \sim \Delta U/a$ or the turbulent Mach number $M_t \sim v_0/a$ with a an average of the two speeds of sound). As already mentioned, the consensus that emerged from DNS of compressible mixing layers is that compressibility stabilizes a mixing layer by decreasing its pressure fluctuations; see, e.g., Pantano and Sarkar (2002). This leads in particular to reduction of the pressure–strain terms that produce the turbulent shear stress through redistribution among the Reynolds stresses of the energy provided by the work of the mean shear. Indeed, these DNSs provide us with decisive results. But, according to our preceding analyses, the detailed sequence of mechanisms leading to mixing-layer stabilization still escapes our understanding. It is important to note that linear analyses of compressible flows are somewhat in contradiction with the proposed interpretations because, given a shear rate $\Delta U/\delta$, damped pressure fluctuations should make both the kinetic energy and the shear stress *increase* instead of decrease, through contribution of dilatational velocity fluctuations growing with the gradient Mach number. This should remain effective in the conditions that hold in a mixing layer because one does not expect a strong imbalance between linear and nonlinear time scales in such a free flow (actually $\delta/\Delta U$ does not depart so much from δ/v_0 and mildly rapid shear conditions must prevail). This indicates that nonlinear compressibility effects should be addressed, in particular the changes in the mechanism of "isotropization" of the fluctuations by pressure, which are essential for producing kinetic energy in a shear flow. Evaluation of the respective impact of compressibility on linear and nonlinear pressure terms is required for understanding and modeling correctly the compressible mixing layer. Note at last that in this inhomogeneous flow transport terms are also deeply modified by the drop of pressure fluctuations. Namely, one observes that the decrease in the production of turbulent kinetic energy is almost compensated for by an equivalent decrease in its transport, letting the dissipation rate ε remain almost unchanged.

This last result also deserves attention. With Favre averaging and normalization by ΔU and by the momentum thickness δ being used in the DNS, the result is that $\varepsilon\delta/(\Delta U)^3$ depends weakly on compressibility. If at the same time the normalized kinetic energy $v_0^2/(\Delta U)^2$ is reduced by compressibility, basic dimensional analysis leaves us with the paradox that the rate of dissipation of the turbulent kinetic energy [which is proportional to $(\Delta U)^3/\delta$] exceeds by a factor proportional to $(\Delta U/v_0)^3$ the rate of injection of the kinetic energy into the cascade (proportional to v_0^3/δ).

Indeed, this reasoning is very crude, but it indicates that the detailed mechanisms fixing energetic equilibrium in compressible mixing layers are not yet fully asserted. This was addressed for instance by the results of Jacquin et al. (1996), who observed that changing M_g thanks to variations in the total temperature of the interacting streams had almost no effect on mixing of the total pressure in their flow: Weak variations in the total pressure spreading rate with compressibility were obtained and were also observed in other experiments. This may be an indication that the dissipation processes, which set the losses (i.e., the transformation of mechanical energy into heat), weakly depend on compressibility in free compressible shear flows. If this were true, this should be integrated into the models. Although it is, there remains still much to make and to understand on the subject.

Finally, explanations based on the hydrodynamic stability, as the inhibition of Kelvin–Helmholtz instabilities by compressibility, cannot be ignored, even if our main theme here is developed turbulence. The reader is referred to Friedrich (2006) for the problem of compressibility in wall-bounded flows.

10.6 Topological Analysis, Coherent Events and Related Dynamics

Because it is nearly impossible to generate compressible homogeneous flows in wind tunnels, DNS is the main tool for coherent event eduction and analysis. The case of compressible homogeneous shear flows has received much less attention than the incompressible homogeneous shear flow and compressible isotropic turbulence. Only a very few papers address the issue of the dynamics of coherent events in the compressible homogeneous shear case, among which are Sarkar, Erlebacher, and Hussaini (1991), Blaisdell, Mansour, and Reynolds (1993), Erlebacher and Sarkar (1993), Simone, Coleman, and Cambon (1997) and Hamba (1999).

10.6.1 Nonlinear Dynamics in the Subsonic Regime

As in the case of isotropic compressible turbulence, several flow regimes can be identified, according to the level of compressibility, the relative importance of thermal versus acoustic mechanisms, etc. Only the subsonic case without strong thermal effects has been investigated, corresponding to the pseudo-acoustic regime described in Subsection 9.3.1 or the nonlinear subsonic regime discussed in Subsection 9.3.3, in which the main part of the turbulent kinetic energy is carried by the solenoidal component of the velocity field. In practice, initial turbulent Mach numbers M_t ranging from 0.1 to 0.5 have been considered in the references previously cited. No results dealing with the supersonic regime and homogeneous shear with strong thermal effects are available. Only relatively weak shear effects have been investigated, as the final-value nondimensional time reached in the simulations is typically $St \simeq 10$. A noticeable exception is found in Blaisdell, Mansour, and Reynolds (1993), in which simulations have been carried out up to $St = 24$. The main reason for that is that, because of the production mechanisms, the turbulent Mach number is monotonically increasing after a transient phase, leading to the occurrence of shocklets, which are poorly captured by the spectral methods used for this kind of simulation.

All simulations show that the flow converges toward a state that does not depend on the initial value of the compressibility ratio $\chi(0)$ defined in Eq. (9.110). After a transient state, the production effects associated with the mean shear seem to lead to nearly universal behavior, in which a solenoidal field and an acoustic field interact. This evolution is illustrated in Fig. 10.5, which displays the evolution of the balance of the terms in Eq. (10.1) as a function of St for two values of the initial Mach number. It is observed that the relative importance of each balance term does not depends on the turbulent Mach number and St (after the initial transient phase). In all cases, the dissipation is negligible. For low values of the turbulent Mach number, the solenoidal and acoustic fields are relatively decoupled in this growth regime. This is mainly due to the fact that the kinetic energy of the dilatational mode is very small compared with that of the solenoidal field. At higher values of the turbulent Mach number, shocklets are observed, as in the isotropic case. An interesting point is that this nonlinear subsonic regime is reached in all cases if the final value of St is high enough. When shocklets are present, the dilatational dissipation $\bar{\varepsilon}_d$ is enhanced. Even in the presence of shocklets, it is observed in Blaisdell, Mansour, and Reynolds (1993) that, after the initial transient phase, the relative weights of dilatational and solenoidal kinetic-energy dissipations reach a constant value, with $\bar{\varepsilon}_d/\bar{\varepsilon}_s = 0.1$. This ratio is reached for $M_t \geq 0.3$. In this regime, it is observed that 23% (resp. 58%) of the total dilatational dissipation is associated with the 1% (resp. 10%) volume of the flow with the most compressive dilatations.

10.6 Topological Analysis, Coherent Events and Related Dynamics

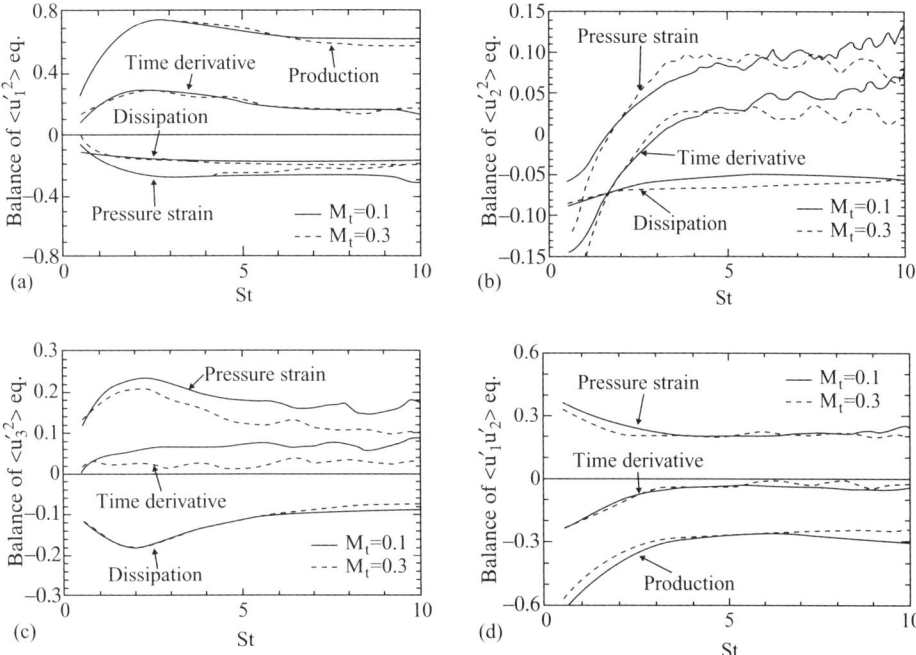

Figure 10.5. Evolution of the budget terms of the Reynolds stresses as a function of St in compressible homogeneous shear flows. Dissipation is not plotted when it is negligible. From Hamba (1999) with permission of the American Institute of Physics.

This relative decoupling was analyzed by Erlebacher and Sarkar (1993), who looked at the balance of budget terms in the equation for the dilatation variance $\overline{d^2}$ and the enstrophy variance $\overline{\omega^2}$. In the homogeneous shear case, the corresponding evolution equations are

$$\frac{1}{2}\frac{\partial \overline{d^2}}{\partial t} = -\frac{1}{6}\overline{d^3} - \overline{S_{ij}^I S_{ij}^I d} - \overline{S_{ij}^C S_{ij}^C} - 2\overline{S_{ij}^C S_{ij}^I d}$$
$$+ \frac{1}{2}\overline{\omega^2 d} - 2S\overline{dS_{12}^I} - 2S\overline{dS_{12}^C}$$
$$- S\overline{\omega_3 d} - \overline{d\frac{\partial}{\partial x_j}\left(\frac{1}{\rho}\frac{\partial p}{\partial x_j}\right)} - \frac{4}{3}\nu\overline{\frac{\partial d}{\partial x_j}\frac{\partial d}{\partial x_j}} \qquad (10.71)$$

and

$$\frac{1}{2}\frac{\partial \overline{\omega^2}}{\partial t} = \overline{\omega_i S_{ij}^I \omega_j} + \overline{\omega_i S_{ij}^C \omega_j} - \frac{2}{3}\overline{\omega^2 d} + S\overline{\omega_1 \omega_2}$$
$$+ \frac{2}{3}S\overline{\omega_3 d} - S\overline{S_{3j}^I \omega_j} + S\overline{S_{3j}^C \omega_j}$$
$$- \overline{\epsilon_{ijk}\omega_i\frac{\partial}{\partial x_j}\left(\frac{1}{\rho}\frac{\partial p}{\partial x_k}\right)} - \nu\overline{\frac{\partial \omega_i}{\partial x_j}\frac{\partial \omega_i}{\partial x_j}}, \qquad (10.72)$$

Table 10.2. *Values of the terms in the budget equations for dilatation and vorticity variance in compressible homogeneous shear flows at* $St = 9$ *and* $M_t = 0.27$, *according to Erlebacher and Sarkar (1993)*

Eq.	1	2	3	4	5	6	7	8	9	10
(10.71)	28	0	96	0	34	0	355	69	33	−292
(10.72)	4124	−26	−45	4555	−46	0	0	1	−6355	
(10.72), $M = 0$	6699			5301				0	−8453	

Note: The terms are sorted from the left to the right on the right-hand side of the evolution equation. The third line displays the value of the budget term for the vorticity variance in the incompressible case at the same value of St and the same Reynolds number as in the compressible case.

where $\mathbf{S}^C$ and $\mathbf{S}^I$ denote the the dilatational and solenoidal parts of the turbulent velocity-gradient tensor $\mathbf{S}$, respectively. Amplitudes of the balance terms that appear on the right-hand side of these two equations computed at $St = 9$ when $Re_\lambda = 23.4$, $M_t = 0.27$ and $S\mathcal{K}/\varepsilon = 6.05$ are displayed in Table 10.2. From looking at the dilatation variance balance, it is clear that interactions terms between solenoidal and dilatational modes are much smaller than between the dilatational components themselves. The main production term is $-2\overline{SdS^C_{12}}$, which is related to an interaction of the dilatational field with the background shear. In a similar way, it is seen that the dilatational mode has a very weak direct influence on the vorticity variance, but that the total enstrophy increase rate in the compressible case is 50% of the one found in the strictly incompressible case. Here, compressibility is observed to reduce both the enstrophy variance production by the solenoidal vortex-stretching term and the enstrophy variance dissipation. It is worth noting that the main production mechanism is the nonlinear vortex stretching, and that the direct production by the background shear is negligible at $St = 9$.

10.6.2 Topological Analysis of the Rate-of-Strain Tensor

The effect of compressibility on the statistical features of the rate-of-strain tensor was investigated by Erlebacher and Sarkar (1993). To this end, these authors split the rate-of-strain tensor $\mathbf{S}$ into a solenoidal component $\mathbf{S}^I$ and an irrotational component $\mathbf{S}^C$, which are computed by applying the Helmholtz decomposition to the global fluctuating-velocity field.

In the incompressible case, the rate-of-strain ellipsoid (based on the eigenvalues of $\mathbf{S}$) has the preferred shape (−4:1:3) in strongly dissipative regions. In the compressible case, with $St = 9$ and $M_t = 0.27$, the eigenvalue ratios of the solenoidal rate-of-strain tensor are almost identical to those of $\mathbf{S}$ in the incompressible case. The irrotational part exhibits a very different behavior, as pdfs of the eigenvalue ratios have two peaks. The main peak is associated with an ellipsoid of the shape (−2.2:1:1.2), whereas the secondary peak corresponds to (−1:−0.7:1.7). The former suggests that the structures that are associated with regions of high dilatation are sheetlike in the x–z plane, the strain rates being extensional in the plane of the

10.6 Topological Analysis, Coherent Events and Related Dynamics

sheet and strongly compressive normal to it. The latter shows that regions with one large expansion strain are also associated with high dilatation. The exact shape of the two preferential dilatational rate-of-strain ellipsoids are certainly Reynolds and Mach number dependent, but the finding that the two rate-of-strain components have very different features is trustworthy.

The same simulation also shows that compressibility (at least in this regime) has no influence on the relative orientation of the vorticity vector.

10.6.3 Vortices, Shocklets, and Dynamics

The vortical structures observed in available simulations are qualitatively the same as in the incompressible case discussed in Chapter 6. This is in agreement with the analysis of the balance of the budget terms of the vorticity variance previously presented. Therefore it can be concluded that, in the pseudo-acoustic regime and in the nonlinear subsonic regime, compressiblity does not result in a qualitative change in the vortical-structure dynamics, and most of the results presented for the incompressible case dealing with vortex dynamics still hold. But it should be mentioned that the existence of a self-sustaining process in compressible homogeneous shear flows has not been investigated.

The occurrence of shocklets in compressible homogeneous shear flows has been reported in several DNSs (Blaisdell, Mansour, and Reynolds, 1993; Sarkar, Erlebacher, and Hussaini, 1991). They appear as elongated ribbonlike structures lying at an angle about 15°–20° to the x axis in the $(x-y)$ plane. The most plausible scenario is that they are created by the upwash and downwash mechanisms induced by the streamwise streaky vortical structures in the direction of the mean-velocity gradient (Blaisdell, Mansour, and Reynolds, 1993). This entrainment effect causes high-speed and low-speed fluid pockets to come into contact, yielding a compression that causes a shocklet. DNS data show that shocklets do not contribute directly significantly to the dilatational dissipation rate $\bar{\varepsilon}_d$, but they play an important role in the dynamics of $\bar{\varepsilon}_d$.

The weak influence of shocklets on the global dynamics is also revealed by the fact that the thermodynamic fields follow a quasi-isentropic behavior, despite the occurrence of the shocklets, which are entropic phenomena. The conclusion is that, looking at the value of the polytropic coefficient n in the relation

$$\frac{\overline{p'p'}}{\bar{p}} = n \frac{\overline{\rho'\rho'}}{\bar{\rho}} = \frac{n}{n-1} \frac{\overline{T'T'}}{\bar{T}}, \tag{10.73}$$

$n = 1.35$ was found in Blaisdell, Mansour, and Reynolds (1993). It is recalled that $n = \gamma = 1.4$ corresponds to an isentropic flow, whereas $n = 0$ and $n = 1$ are related to isobaric and isothermal flows, respectively.

Bibliography

BLAISDELL, G. A., COLEMAN, G. N., AND MANSOUR, N. N. (1996). RDT for compressible homogeneous turbulence under isotropic mean strain, *Phys. Fluids* **8**, 2692–2708.

BLAISDELL, G. A., MANSOUR, N. N., AND REYNOLDS, W. C. (1991). Numerical simulation of compressible homogeneous turbulence. *Report TF-50*. Department of Mechanical Engineering, Stanford University.

BLAISDELL, G. A., MANSOUR, N. N., AND REYNOLDS, W. C. (1993). Compressibility effects on the growth and structure of homogeneous turbulent shear flow, *J. Fluid Mech.* **256**, 443–485.

CAMBON, C., COLEMAN, G. N., AND MANSOUR, N. N. (1993). Rapid distortion analysis and direct simulation of compressible homogeneous turbulence at finite Mach number, *J. Fluid Mech.* **257**, 641–665.

CAMBON, C., MAO, Y., AND JEANDEL, D. (1992). On the application of time dependent scaling to the modeling of turbulence undergoing compression, *Eur. J. Mech. B (Fluids)* **6**, 683–703.

CAMBON, C. AND RUBINSTEIN, R. (2006). Anisotropic developments for homogeneous shear flows, *Phys. Fluids* **18**, 085106.

CAMBON, C., TEISSÈDRE, C., AND JEANDEL, D. (1985). Etude d'effets couplés de déformation et de rotation sur une turbulence homogène, *J. Méc. Théor. Appl.* **4**, 629–657.

CRAIK, A. D. D. AND ALLEN, H. R. (1992). *J. Fluid Mech.* **234**, 613–627.

DEBIÈVE, J. F., GOUIN, H., AND GAVIGLIO, J. (1982). Evolution in the Reynolds stress tensor in a shock wave turbulence interaction, *Indian J. Technol.* **20**, 90–97.

DURBIN, P. A. AND ZEMAN, O. (1992). RDT for homogeneous compressed turbulence with application to modeling, *J. Fluid Mech.* **242**, 349–370.

ECKHOFF, K. S. AND STORESLETTEN, L. (1978). A note on the stability of steady inviscid helical gas flows, *J. Fluid Mech.* **89**, 401–411.

ERLEBACHER, G. AND SARKAR, S. (1993). Statistical analysis of the rate of strain tensor in compressible homogeneous turbulence, *Phys. Fluids* **5**, 3240–3254.

FABRE, D., JACQUIN, L., AND SESTERHENN, J. (2001). Linear interaction of a cylindrical entropy spot with a shock wave, *Phys. Fluid* **13**, 1–20.

FAUCHET, G., SHAO, L., WUNENBERGER, R., AND BERTOGLIO, J. P. (1997). An improved two-point closure for weakly compressible turbulence, presented at the *11th Symposium on Turbulent Shear Flow*, Grenoble, September 8–10, 1997.

FRIEDRICH, R. (2006). Effects of compressibility and heat release in turbulent wall-bounded and free shear flows, *SIG4 ERCOFTAC and GST13 workshop*, Porquerolles, France, June 1, 1997.

FRISCH, U. (1995). *Turbulence: The Legacy of A.N. Kolmogorov*, Cambridge University Press.

GOLDSTEIN M. E. (1978). Unsteady vortical and entropic distortions of potential flows round arbitrary obstacles, *J. Fluid Mech.* **89**, 433–468.

HAMBA, F. (1999). Effects of pressure fluctuations on turbulence growth in compressible homogeneous shear flow, *Phys. Fluids* **11**, 1623–1635.

HEINZ, S. (2004). *Statistical Mechanics of Turbulent Flows*, Springer.

HUNT, J. C. R. (1973). A theory of turbulent flow around two-dimensional bluff bodies, *J. Fluid Mech.* **61**, 625–706.

JACQUIN, L., CAMBON, C., AND BLIN, E. (1993). Turbulence amplification by a shock wave and Rapid Distortion Theory, *Phys. Fluids A* **10**, 2539–2550.

JACQUIN, L., MISTRAL, S., GEFFROY, P., AND CRUAUD, F. (1996). Mixing of a heated supersonic jet with a parallel stream, in *Advances in Turbulence V*, Springer-Verlag.

KOVASZNAY, L. S. G. (1953). Turbulence in supersonic flow, *J. Aeronaut. Sci.* **20**, 657–682.

LELE, S. K. (1994). Compressibility effects on turbulence, *Annu. Rev. Fluid Mech.* **26**, 211–254.

LIFSCHITZ, A. AND HAMEIRI, E. (1991). Local stability conditions in fluid dynamics, *Phys. Fluids A* **3**, 2644–2641.

LIGHTHILL, M. J. (1978). *Waves in Fluids*, Cambridge University Press.

PANTANO, C. AND SARKAR, S. (2002). A study of compressibility effects in the high-speed turbulent shear layer using DNS, *J. Fluid Mech.* **451**, 329–371.

RIBNER, H. S. AND TUCKER, M. (1953). Spectrum of turbulence in a contracting stream, *NACA Report*, No. 113.

SABEL'NIKOV, V. A. (1975). Pressure fluctuations generated by uniform distortion of homogeneous turbulence, *J. Mech. Sov. Res.* **4**, 46–56.

SARKAR S. (1995). The stabilizing effect of compressibility in turbulent shear flow, *J. Fluid Mech.* **282**, 163–286.

SARKAR S., ERLEBACHER, G., AND HUSSAINI, M. Y. (1991). Direct simulation of compressible turbulence in a shear flow, *Theor. Comput. Fluid Dyn.* **2**, 291–305.

SIMONE, A., COLEMAN G. N., AND CAMBON C. (1997). The effect of compressibility on turbulent shear flow: A RDT and DNS study, *J. Fluid Mech.* **330**, 307–338.

THACKER, W. D., SARKAR, S., AND GATSKI, T. B. (2006). Analyzing the influence of compressibility on the rapid pressure-strain rate correlation in turbulent shear flow, *Theor. Comput. Fluid Dyn.*, DOI 10.1007/s00162-007-0043-4.

TOWNSEND, A. A. (1976). *The Structure of Turbulent Shear Flow*, 2nd ed., Cambridge University Press.

11 Isotropic Turbulence–Shock Interaction

This chapter is devoted to the analysis of the interaction of an initially isotropic turbulence with a normal plane shock wave. Even though this case is very simple from a geometrical viewpoint, it will be seen that it involves most physical mechanisms observed in more complex configurations. It also makes it possible to carry out an extensive theoretical analysis, leading to a deep understanding of the underlying physics.

11.1 Brief Survey of Existing Interaction Regimes

Several interaction regimes exist, which can be grouped into two families. The first one, referred to as the *destructive interaction* family, encompasses all configurations in which the structure of the shock wave is deeply modified during the interaction in the sense that a single well-defined shock wave can no longer be identified, the limiting case being the shock destruction. The second family, i.e., the *nondestructive interaction* family, is made up of all cases in which the structure of the shock wave is preserved during the interaction. It is important to note that, in the latter case, the shock wave can be strongly corrugated by the incoming turbulence.

11.1.1 Destructive Interactions

The first case of destructive interactions is that of *unstable shocks*, in which any small disturbances will lead to the destruction of the shock wave because of instability mechanisms. In such a case, the destruction mechanism is tied to the shock itself and not to the turbulence dynamics (see Lubchich and Pudovkin, 2004, and references given therein).

According to Dyakov (1954), a shock is *absolutely unstable* if one of the two following conditions is fulfilled:

$$(\rho_2 u_2)^2 \left[\frac{\partial(1/\rho_2)}{\partial p_2} \right]_{\mathcal{H}} < -1 \qquad (11.1)$$

or

$$(\rho_2 u_2)^2 \left[\frac{\partial(1/\rho_2)}{\partial p_2} \right]_{\mathcal{H}} > 1 + 2M_2, \qquad (11.2)$$

11.1 Brief Survey of Existing Interaction Regimes

where M is the Mach number and subscripts 1 and 2 refer to the shock upstream and downstream states, respectively. The index $\mathcal{H}$ indicates that the derivative is calculated along the Hugoniot curve in the pressure-specific volume plane. The absolute instability regime corresponds to cases in which the solution of the jump conditions is not unique, and small perturbations trigger the bifurcation toward stable states made of combinations of discontinuities and simple waves. It can be shown that the absolute instability cannot occur for plane shocks in perfect gases. It can also be observed in perfect gases for curved shocks or plane shocks with viscous effects such as an interaction with a boundary layer.

Another shock instability, referred to as the *relative instability*, was identified by Kontorovich (1957). Here, a perturbation, once having emerged at the discontinuity, stands for arbitrarily long times, emitting acoustic, vorticity, and entropy waves without attenuation and amplification. The criterion for the occurrence of the relative instability is

$$\frac{1 - M_2^2[1 + (\rho_2/\rho_1)]}{1 - M_2^2[1 - (\rho_2/\rho_1)]} < (\rho_2 u_2)^2 \left[\frac{\partial(1/\rho_2)}{\partial p_2}\right]_{\mathcal{H}} < 1 + 2M_2. \tag{11.3}$$

It is observed that the range of the relative instability is adjacent to one of the two ranges of the absolute instability. In the relative instability regime the shock wave is not destroyed by infinitesimal initial perturbations but it cannot exist alone, as the downstream solution is made of the superposition of a uniform field and propagating perturbation waves. For initial perturbations of finite amplitude, the shock wave disintegrates to a shock wave of essentially different intensity and other elements.

Other cases of destructive interactions are associated with cases in which the turbulent fluctuations are strong enough to yield a local deep modification of the shock wave. These configurations escape the linear instability theory used to define the preceding destructive regimes and can therefore be classified as nonlinear destructive interactions. The first case is associated with the case in which a turbulent eddy is strong enough to render the flow locally subsonic. When it reaches the shock wave, the latter will be locally annihilated. In the second case, the upstream flow remains supersonic but the perturbation is strong enough to make some secondary shocks appear.

11.1.2 Nondestructive Interactions

The nondestructive interactions are trivially defined as all the interactions that are not destructive, meaning that a single well-defined continuous shock wave can be indentified all through the interaction process. Two subcases are identified:

1. The *linear interaction regime* (see Section 11.2), in which the perturbations are assumed to be weak in the sense that the mean-flow quantities obey the usual Rankine–Hugoniot jump conditions, whereas the turbulent fluctuations satisfy linearized jump relations.

2. The *nonlinear interaction regime* (see Section 11.3), in which the turbulent intensity is so high that the mean flow is modified by the turbulent fluctuations. In this case, turbulent fluxes must be taken into account when jump conditions are written for the mean flow.

11.2 Linear Nondestructive Interaction

11.2.1 Shock Modeling and Jump Relations

The available developments dealing with linear approximation theory in the nondestructive case do not take viscous effects into account. The rationale for that is that viscous effects are negligible compared with other physical mechanisms during the interaction (this will be proved *a posteriori* when we compare theoretical results with DNS and experimental results), and that relaxation times associated with vibrational, rotational, and translational energy modes of the molecules are very small with respect to macroscopic turbulent time scales. Therefore the shock is modeled as a surface discontinuity with zero thickness. An important consequence is that the shock has no intrinsic time or length scale, and its corrugation is entirely governed by incident fluctuations. Its effects are entirely captured by the Rankine–Hugoniot jump conditions for mass, momentum, and energy:

$$[[\rho u_n]] = 0, \tag{11.4}$$

$$[[\rho u_n^2 + p]] = 0, \tag{11.5}$$

$$[[u_t]] = 0, \tag{11.6}$$

$$\left[\left[e + \frac{p}{\rho} + u^2\right]\right] = [[H]] = 0, \tag{11.7}$$

where H is the stagnation enthalpy and u is the velocity in the reference frame tied to the shock wave, i.e., $u = v - u_s$ where v and u_s are the fluid velocity and the shock speed in the laboratory frame, respectively. Subscripts n and t are related to the normal and tangential components of vector fields with respect to the shock wave, respectively:

$$u_n \equiv u \cdot n, \quad u_t \equiv n \times (u \times n), \quad u = u_n n + u_t, \tag{11.8}$$

where n is the shock normal unit vector. An exact general jump condition for the vorticity can be derived from the relations just given (Hayes, 1957). First noting that the vorticity vector $\boldsymbol{\Omega} = \nabla \times u$ can be decomposed as $\boldsymbol{\Omega} = \Omega_n n + \boldsymbol{\Omega}_t$ with

$$\Omega_n = (\nabla \times u_t)_n \tag{11.9}$$

and

$$\boldsymbol{\Omega}_t = n \times \left(\frac{\partial u_t}{\partial n} + u_t \cdot \nabla n - \nabla_{\parallel} u_n\right) \tag{11.10}$$

11.2 Linear Nondestructive Interaction

where $\nabla_\|$ denotes the tangential (with respect to the shock surface) part of the nabla operator, one obtains the following vorticity jump conditions in unsteady flows in which the shock experiences deformations:

$$[[\Omega_n]] = 0, \qquad (11.11)$$

$$[[\Omega_t]] = \mathbf{n} \times \left(\nabla_\|(\rho u_n) \left[\left[\frac{1}{\rho}\right]\right] - \frac{1}{\rho u_n} [[\rho]] (D_\| \mathbf{u}_t + u_s D_\| \mathbf{n}) \right), \qquad (11.12)$$

with

$$D_\| \mathbf{u}_t = \left(\frac{d\mathbf{u}_t}{dt}\right)_t + \mathbf{u}_t \cdot \nabla_\| \mathbf{u}_t = \left(\frac{\partial \mathbf{u}_t}{\partial t} + u_s \frac{\partial \mathbf{u}_t}{\partial n}\right)_t + \mathbf{u}_t \cdot \nabla_\| \mathbf{u}_t, \qquad (11.13)$$

and

$$D_\| \mathbf{n} = \frac{d\mathbf{n}}{dt} + \mathbf{u}_t \cdot \nabla_\| \mathbf{n} = -\nabla_\| u_s + \mathbf{u}_t \cdot \nabla_\| \mathbf{n}. \qquad (11.14)$$

It is seen that the normal component of the vorticity is continuous across the shock, whereas the jump of the tangential component depends on the density jump, the tangential velocity, and the shock-wave deformation. In steady flows, the jump condition for the tangential vorticity simplifies as

$$[[\Omega_t]] = \mathbf{n} \times \left(\nabla_\|(\rho u_n) \left[\left[\frac{1}{\rho}\right]\right] - \frac{1}{\rho u_n} [[\rho]] \mathbf{u}_t \cdot \nabla_\| \mathbf{u}_t \right). \qquad (11.15)$$

11.2.2 Introduction to the Linear Interaction Approximation Theory

We now briefly introduce the linear interaction approximation (LIA), which is a very powerful tool pioneered in the 1950s (Ribner, 1953; Moore, 1954) to analyze the nondestructive linear interaction regime. Details of the LIA procedure are given in Chapter 12. It relies on the following simplified dynamic scheme:

1. The shock wave has no intrinsic scale, and therefore it is governed by incident perturbations. It will act only through the jump conditions.
2. Both mean and fluctuating parts of the upstream field (i.e., the field in the supersonic part of the flow) are arbitrarily fixed.
3. The downstream field is fully determined by the upstream field and the jump conditions. More precisely, it is assumed that the interaction process between turbulent fluctuations and the shock is mostly linear, so that
   a. the mean flow obeys the usual Rankine–Hugoniot conditions and
   b. the fluctuating field obeys linearized jump conditions.

This physical scheme is illustrated in Fig. 11.1.

Two conditions must be fulfilled to ensure that the linear approximation is relevant:

1. The fluctuations must be weak in the sense that the distorted shock wave must remain well defined. Numerical experiments led Lee and co-workers (Lee,

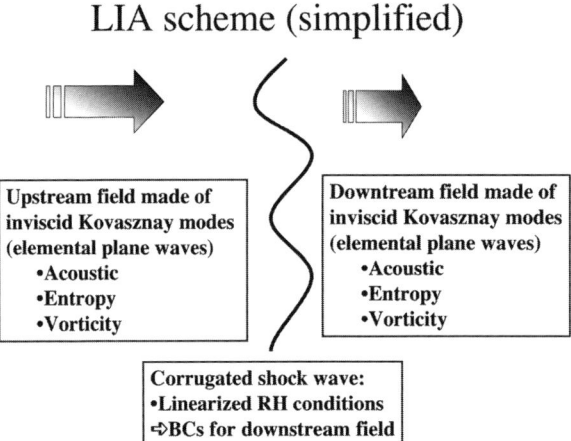

Figure 11.1. Schematic view of the LIA for shock–turbulence interaction.

Lele, and Moin, 1993) to propose the following empirical criterion for the linear regime:

$$M_t^2 < \alpha(M_1^2 - 1), \tag{11.16}$$

where M_t and M_1 are the upstream turbulent and mean Mach numbers, respectively, and $\alpha \approx 0.1$.

2. The time required for turbulent events to cross the shock must be small compared with the turbulence time scale $\mathcal{K}/\varepsilon$ (with $\mathcal{K}$ and ε the turbulent kinetic energy and the turbulent kinetic-energy dissipation rate, respectively), so that nonlinear mechanisms cannot have significant effects.

The LIA analysis is made more accurate by decomposing the fluctuating field using Kovasznay decomposition: Both the upstream and downstream fluctuating fields are split as sums of individual modes, each mode being characterized by its nature (acoustic, vorticity, or entropy mode) and wavenumber or frequency. Because linearized jump conditions are utilized, all cross interactions between modes are precluded, and the downstream fluctuating field is obtained by means of a simple superposition of the LIA results obtained for each upstream fluctuating mode. Let us emphasize here that, as will be subsequently demonstrated, the fact that interactions are precluded does not mean that an upstream perturbation wave is associated with an emitted downstream wave of the same nature (as a matter of fact, all physical modes are excited in the downstream region in the general case), but that the interaction process is not sensitive to shock deformations induced by other upstream fluctuations.

The resulting LIA scheme is as follows: One considers two semi-infinite domains separated by the shock wave. Both the mean and fluctuating fields in the upstream domain are arbitrarily prescribed. Because the flow is hyperbolic in this domain, it is not sensitive to the presence of the shock wave. The mean downstream field is computed with the mean upstream field and the usual Rankine–Hugoniot

11.2 Linear Nondestructive Interaction

jump relations (11.4)–(11.7). The emitted fluctuating field is then computed by the linearized jump relations as boundary conditions. Using results displayed in Chapter 12, it is important to note that the wave vectors of the emitted waves are computed with the dispersion relation associated with each physical mode, the frequency and the tangential component of the wave vector being the same as the upstream perturbation. The linearized jump conditions are used to compute only the amplitudes of the emitted waves.

11.2.3 Vortical Turbulence–Shock Interaction

We first address the case in which the incident turbulence is isotropic and composed of vorticity modes only. This case was investigated by several researchers (Lee, Lele, and Moin, 1993, 1997), who used LIA and DNS. The trends found by means of DNS and LIA are corroborated by wind-tunnel experiments, but a strict quantitative agreement is hopeless because the exact nature of the incident turbulence in experiments cannot be controlled because of technological limitations. The main observations are as follows:

1. *Velocity fluctuations.* The streamwise distributions of the kinetic energy of the three velocity components given by both DNS and LIA are displayed in Fig. 11.2. It is observed that all velocity components are amplified, leading to a global increase in the turbulent kinetic energy. The amplification rate is well recovered by the LIA calculation, showing that the amplification is mainly due to linear mechanisms. In agreement with LIA, the velocity field behind the shock wave is axisymmetric. Both LIA and DNS predict that the amplification is Mach number dependent. The amplification level is plotted as a function of the upstream Mach number, M_1, in Fig. 11.3. It is interesting to note that the amplification of the transverse velocity components is an increasing monotonic function, whereas the shock normal velocity component amplification exhibits a maximum near $M_1 = 2$. The transverse components are more amplified than the streamwise component for $M_1 > 2$, and the amplification of the total turbulent kinetic energy tends to saturate beyond $M_1 = 3$. The streamwise DNS profiles reveal that the velocity field experiences a rapid evolution downstream of the shock, leading to the definition of two different regions behind the shock wave. This observation is in full agreement with the LIA analysis, which predicts the existence of a near-field region where the evanescent acoustic waves emitted during the interaction are not negligible. Comparing the LIA and DNS profiles (see Fig. 11.2) once again leads to the conclusion that the process is mainly governed by linear mechanisms. The rapid evolution in the near-field region is due to the exponential decay of evanescent acoustic waves, which are responsible for the anticorrelation of the (acoustic) dilatational and (vortical) solenoidal field just downstream of the shock. The nature of the relaxation process that takes place in the near-field region is better understood when one recalls that the solution of linearized Euler equations about a 1D mean field is

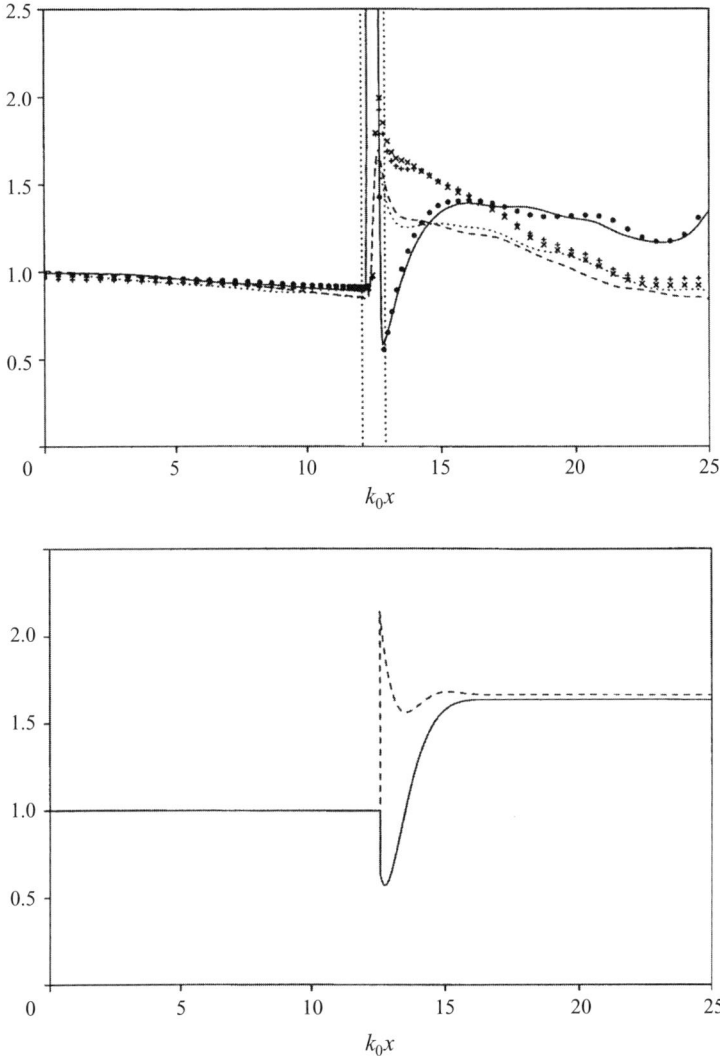

Figure 11.2. Streamwise evolution of normalized Reynolds stresses. Top: DNS results (lines for $M_1 = 2$, $M_t = 0.108$, $Re_\lambda = 19.0$ and symbols for $M_1 = 3$, $M_t = 0.110$, $Re_\lambda = 19.7$)); streamwise component $R_{11} = \overline{u'u'}$: solid line and dots; spanwise component $R_{22} = \overline{v'v'}$: dashed line and "x"; spanwise component $R_{33} = \overline{w'w'}$: dotted line and "+". Bottom: LIA results for $M_1 = 2$, $M_t = 0.108$; streamwise component R_{11}: solid line; spanwise components R_{22} and R_{33}: dashed line. Vertical dotted lines show the limit of the shock displacement region. From Lee, Lele, and Moin (1997) with permission of CUP.

such that the following acoustic-energy balance holds:

$$\frac{\partial}{\partial x}\left[M_2\left(\frac{\mathcal{K}}{a_2^2} + \frac{1}{2}\frac{\overline{\rho'^2}}{\bar{\rho}^2}\right) + \frac{1}{\gamma}\frac{\overline{p'u''}}{\bar{p}a_2}\right] = 0, \qquad (11.17)$$

where

$$\mathcal{K} = \frac{1}{2}\left(\overline{u''u''} + \overline{v''v''} + \overline{w''w''}\right) \qquad (11.18)$$

11.2 Linear Nondestructive Interaction

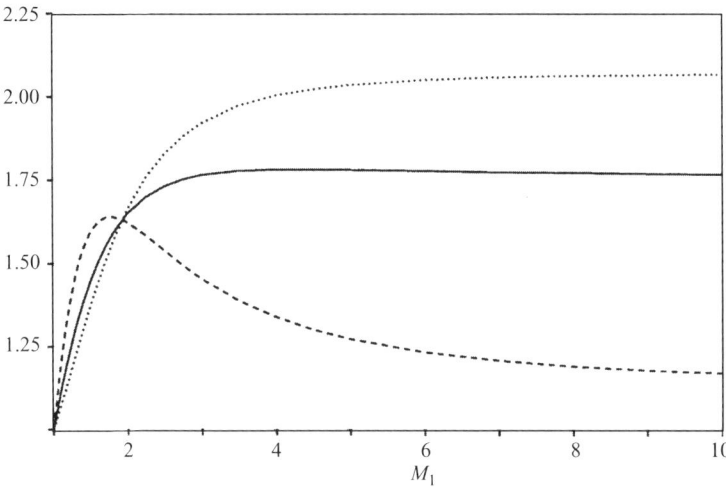

Figure 11.3. LIA prediction of far-field Reynolds stress amplification vs. the upstream Mach number. Solid line: turbulent kinetic energy; dashed line: streamwise Reynolds stress $R_{11} = \overline{u'u'}$; dotted line: spanwise Reynolds stresses $R_{22} = \overline{v'v'}$ and $R_{33} = \overline{w'w'}$. From Lee, Lele, and Moin (1997) with permission of CUP.

if viscous and entropy–dilatation correlation effects are neglected. A close examination of DNS data shows that these two contributions are small in the near-field region and that the near-field evolution is associated with an energy transfer from the acoustic potential energy in the form of density or pressure fluctuations to turbulent kinetic energy. This transfer is done by means of the pressure transport term $\nabla \cdot \overline{(p'\boldsymbol{u}'')}$. The pressure–dilatation term $\overline{p'\nabla \cdot \boldsymbol{u}''}$ is observed to be strictly positive in this region, corresponding to a reversible transfer from the mean internal energy to the turbulent kinetic energy.

Outside the near-field region, the global behavior results from the competition between the viscous decay and the return-to-isotropy process. In low-Reynolds-number DNS,* the viscous effect is dominant: The turbulent kinetic-energy balance simplifies as an equilibrium between the convection term and the viscous term, showing that the main effects are convection of turbulent velocity fluctuations by the mean field and their destruction by viscous effects.

2. *Vorticity field.* The vorticity is also strongly affected by the interaction with the shock wave. The streamwise evolution of the vorticity components computed in two different simulations are presented in Figs. 11.4 and 11.5. Several typical features are observed. First, the streamwise (i.e., shock normal) vorticity component is not affected, in agreement with the conclusion drawn from jump condition (11.11). The two other components are amplified, in a symmetric way, leading to the definition of a statistically axisymmetric vorticity field being the shock wave. This behavior was predicted by the LIA analysis [see

---

* The turbulent Reynolds numbers based on the Taylor microscale Re_λ at the inlet plane of DNS presented in Lee, Lele, and Moin (1993, 1997) range from 11.6 to 21.6.

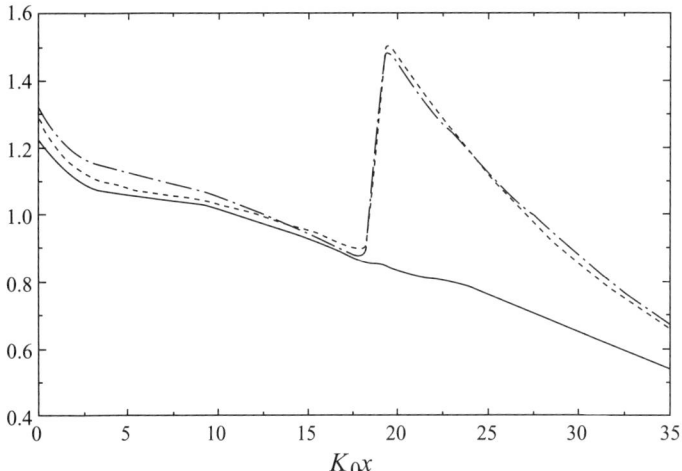

Figure 11.4. Vorticity amplification across shock (DNS data, $M_1 = 1.2$, $Re_t = 84.8$). Solid line: streamwise component $\overline{\Omega'_1 \Omega'_1}$; dashed line: spanwise component $\overline{\Omega'_2 \Omega'_2}$; dash–dot line: spanwise component $\overline{\Omega'_3 \Omega'_3}$. From Lee, Lele, and Moin (1993) with permission of CUP.

Eq. (12.95)]. The amplification of the transverse component is Mach number dependent, and the LIA analysis presented in Fig. 11.6 shows that it is a monotonically increasing function that tends to saturate at very high Mach numbers. Because the vorticity has no contribution from the acoustic modes, it does not exhibit a near field. But it is interesting to note that two different behaviors of the streamwise vorticity component are observed downstream of the shock: It is monotonically decreasing at a low Reynolds number, whereas it has a local maximum at a higher Reynolds number. The explanation for this bifurcation is found by looking at the evolution equation of the vorticity component variances using DNS data. Neglecting the temperature-induced fluctuations of the viscosity, the evolution equation for the fluctuating vorticity variances $\overline{\Omega'_\alpha \Omega'_\alpha}$ is

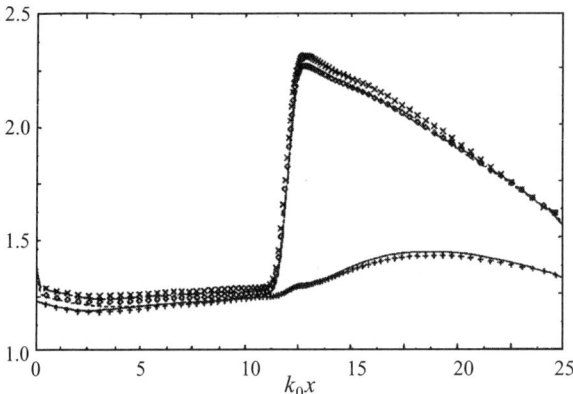

Figure 11.5. Vorticity amplification across shock (DNS data on two computational grids, $M_1 = 1.2$, $Re_t = 238$). Solid line and "+": streamwise component $\overline{\Omega'_1 \Omega'_1}$; dashed line and "x": spanwise component $\overline{\Omega'_2 \Omega'_2}$; dash–dot line and diamonds: spanwise component $\overline{\Omega'_3 \Omega'_3}$. From Lee, Lele, and Moin (1993) with permission of CUP.

11.2 Linear Nondestructive Interaction

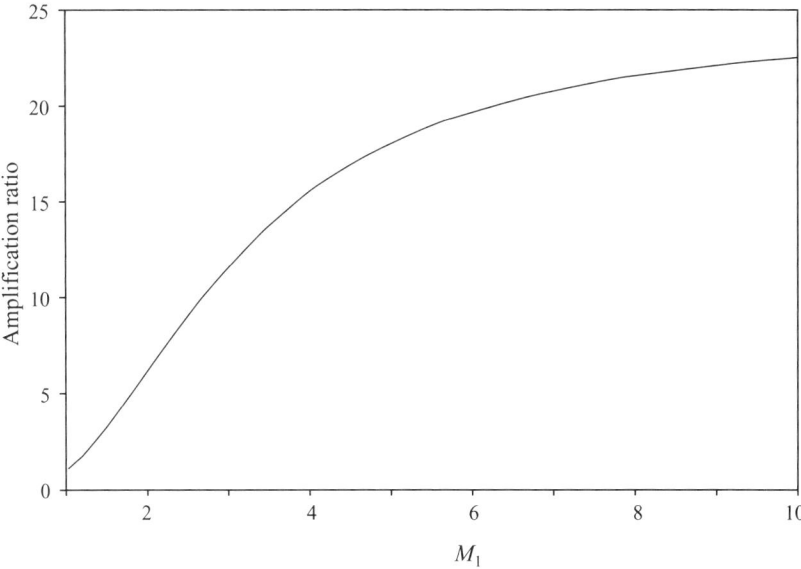

Figure 11.6. Amplification of transverse vorticity components $\overline{\Omega'_2\Omega'_2}$ and $\overline{\Omega'_3\Omega'_3}$ across shock vs. the upstream Mach number M_1: LIA results. From Lee, Lele, and Moin (1997) with permission of CUP.

(without summation over repeated Greek indices)

$$\underbrace{\bar{u}_j \frac{\partial}{\partial x_j}\overline{\Omega'_\alpha \Omega'_\alpha}}_{\text{I}} = \underbrace{2\overline{\Omega'_\alpha \Omega'_j}\bar{S}_{\alpha j}}_{\text{II}} + \underbrace{2\overline{\Omega'_\alpha \Omega'_j S'_{\alpha j}}}_{\text{III}} - \underbrace{2\overline{\Omega'_\alpha \Omega'_\alpha}\bar{S}_{jj}}_{\text{IV}}$$

$$- \underbrace{\overline{\Omega'_\alpha \Omega'_\alpha S'_{jj}}}_{\text{V}} + \underbrace{2\epsilon_{\alpha j k}\overline{\left(\frac{\Omega'_\alpha \frac{\partial \rho}{\partial x_j}\frac{\partial p}{\partial x_k}}{\rho^2}\right)}}_{\text{VI}} - \underbrace{\frac{\partial}{\partial x_k}\overline{\Omega'_\alpha \Omega'_\alpha u'_k}}_{\text{VII}}$$

$$+ \underbrace{2\epsilon_{\alpha j k}\overline{\Omega'_\alpha \frac{\partial}{\partial x_j}\left(\frac{1}{\rho}\frac{\partial \tau_{kl}}{\partial x_l}\right)}}_{\text{VIII}}. \quad (11.19)$$

Inside the shock wave, the transverse component evolution ($\alpha = 2, 3$) that leads to the existence of the jump in the LIA theory is dominated by the vorticity–compression terms (IV + V), the vorticity–mean-compression term IV being the leading term. Downstream of the shock region, the vortex-stretching mechanism (II + III) is balanced by the viscous effects (VIII). In all DNS cases, both the baroclinic production term (VI) and the turbulent transport (VII) are negligible. The dynamics of the streamwise component is different. Inside the shock wave, vortex stretching (II + III) and vorticity compression (IV + V) balance each other, resulting in a negligible influence of the shock, in agreement with inviscid jump relations. Downstream of the shock wave, the streamwise vorticity variance $\overline{\Omega'_1 \Omega'_1}$ is governed by the balance between two dominating mechanisms: turbulent vortex stretching (III) and viscous effects (VIII). In the low-Reynolds-number case, the viscous damping overwhelms the vortex-stretching effects, leading to a monotonous decay. At a higher Reynolds number, the turbulent stretching is large enough to yield the existence of a local downstream maximum.

3. *Turbulence length scales.* Characteristic scales of turbulence are observed to be modified during the interaction in a scale-dependent manner. Let us first discuss the behavior of the 1D spectra $E_\alpha(k_\beta)$, which are defined such that (without summation over Greek indices)

$$\overline{u'_\alpha u'_\alpha} = \int_{k_\beta=0}^{k_\beta=\infty} E_\alpha(k_\beta)\,dk_\beta, \tag{11.20}$$

where u'_α and k_β are the αth component of u' and the βth component of k, respectively. Both LIA and DNS results show that

a. in the longitudinal spectra $E_\alpha(k_1)$, small scales (i.e., large wavenumbers) are more amplified than large scales (i.e., small wavenumbers) and

b. the amplification pattern is more complex for transverse spectra: Higher amplification at small scales is found for $E_1(k_2)$ and $E_2(k_2)$, whereas the large scales are the most amplified for $E_3(k_2)$.

This complex behavior makes it necessary to carry out a specific analysis for each characteristic length scale, as they are spectrum dependent. Defining the integral scale for the dummy variable ϕ as

$$\lambda_\phi(x) = \int_{r=0}^{r=+\infty} C_{\phi\phi}(r, x)\,dr, \tag{11.21}$$

where the transverse two-point correlation $C_{\phi\phi}(r, x)$ is given by (ϕ is assumed to be a centered random variable)

$$C_{\phi\phi}(r, x) = \frac{\overline{\phi(x, y, z, t)\phi(x, y+r, z, t)}}{\overline{\phi(x, y, z, t)\phi(x, y, z, t)}}, \tag{11.22}$$

in which the statistical averaging is carried out over time and homogeneous directions y and z, both DNS and LIA show that (see Fig. 11.7)

a. λ_{u_1}, λ_{u_2}, and λ_ρ exhibit a significant Mach-number-dependent decrease across the shock and

b. λ_{u_3} is largely increased by the interaction.

Now looking at the Taylor microscales (see Fig. 11.8), it is observed that they are all significantly reduced during the interaction, the reduction being more pronounced in the shock normal direction. It is recalled that the Taylor microscale λ_α associated with u'_α and the density microscale λ_ρ are computed here as

$$\lambda_\alpha = \sqrt{\frac{\overline{u'_\alpha u'_\alpha}}{\overline{\frac{\partial u'_\alpha}{\partial x_\alpha}\frac{\partial u'_\alpha}{\partial x_\alpha}}}}, \quad \lambda_\rho = \sqrt{\frac{\overline{\rho'\rho'}}{\overline{\frac{\partial\rho'}{\partial y}\frac{\partial\rho'}{\partial y}}}}. \tag{11.23}$$

4. *Thermodynamic quantities.* The thermodynamic properties of the flow downstream of the shock are also modified by the interaction process. Both DNS and LIA results show that, for an isentropic incident isotropic turbulence, the

11.2 Linear Nondestructive Interaction

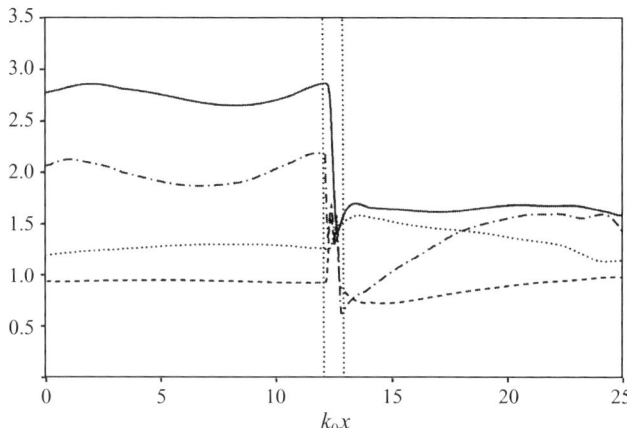

Figure 11.7. Streamwise evolution of turbulence transverse integral scales (DNS, $M_1 = 2$, $M_t = 0.108$, $Re_\lambda = 19$). Dashed line: λ_{u_1}; solid line: λ_{u_2}; dotted line: λ_{u_3}; dashed–dotted line: λ_ρ. From Lee, Lele, and Moin (1997) with permission of CUP.

downstream field remains isentropic for weak shocks such that $M_1 < 1.2$. At higher upstream Mach numbers, the emitted entropy waves have a significant energy because their magnitude becomes comparable to that of acoustic waves. This effect is illustrated by plotting the normalized correlation coefficients (see Fig. 11.9):

$$n_{pp} \equiv \frac{\sqrt{\overline{p'p'}}}{\bar{p}^2} \frac{\bar{\rho}^2}{\sqrt{\overline{\rho'\rho'}}}, \quad n_{\rho T} \equiv 1 + \frac{\sqrt{\overline{T'T'}}}{\bar{T}^2} \frac{\bar{\rho}^2}{\sqrt{\overline{\rho'\rho'}}}, \qquad (11.24)$$

$$C_{\rho T} \equiv 1 + \frac{\bar{\rho}}{\bar{T}} \frac{\sqrt{\overline{\rho'T''}}}{\sqrt{\overline{\rho'\rho'}}}, \qquad (11.25)$$

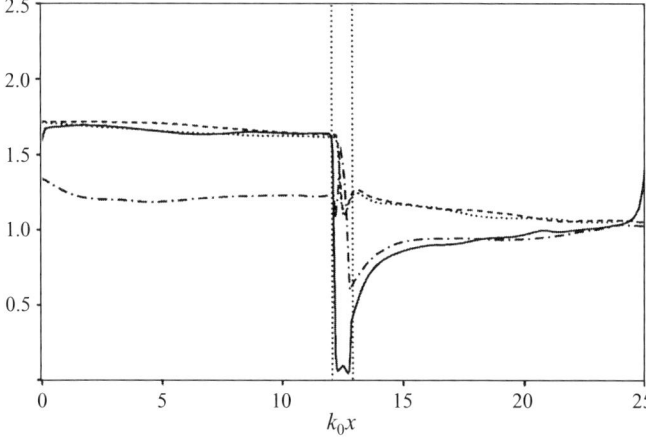

Figure 11.8. Streamwise evolution of turbulence microscales (DNS, $M_1 = 2$, $M_t = 0.108$, $Re_\lambda = 19$). Solid line: λ_{u_1}; dashed line: λ_{u_2}; dotted line: λ_{u_3}; dashed-dotted line: λ_ρ. From Lee, Lele, and Moin (1997) with permission of CUP.

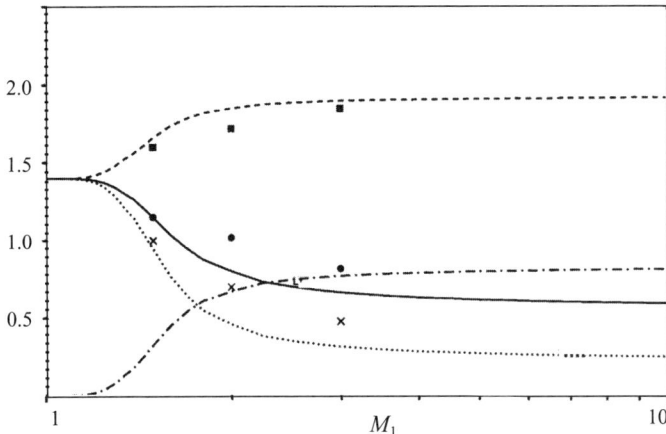

Figure 11.9. Evolution of normalized correlation coefficients in the far-field region downstream of the shock versus the upstream Mach number M_1. $n_{\rho\rho}$: solid line (LIA) and black circles (DNS); $n_{\rho T}$: dashed line (LIA) and black circles (DNS); C_{pT}: dotted line (LIA) and 'x' (DNS); i_s: dashed-dotted line (LIA). From Lee, Lele, and Moin (1997) with permission from CUP.

along with the entropy-fluctuation contribution

$$i_s \equiv \frac{\sqrt{\overline{s's'}}}{c_p^2} \frac{\bar{\rho}^2}{\sqrt{\overline{\rho'\rho'}}}. \tag{11.26}$$

It is observed that the entropy fluctuations are more significant than acoustic fluctuations for $M_1 > 1.65$. But it is worth noting that, downstream of the shock, neither the isentropic hypothesis (which states that the entropy fluctuations are negligible) nor the strong Reynolds analogy proposed by Morkovin for shear flows (which says that the stagnation temperature is constant, which amounts to assuming that acoustic waves have a negligible effect on the density fluctuations) is valid if $M_1 > 1.2$. It is recalled that the latter can be expressed as

$$\frac{\rho'}{\bar{\rho}} = -\frac{T''}{\bar{T}} = (\gamma - 1)M_1^2 \frac{u_1''}{\tilde{u}_1}. \tag{11.27}$$

11.2.4 Acoustic Turbulence–Shock Interaction

The case of an incident purely acoustic isotropic field has been addressed by Mahesh and co-workers (Mahesh, Lele, and Moin, 1995), who used both LIA and DNS. It is observed that this case exhibits very significant differences with respect to the case of an incident purely vortical isotropic turbulence.

The main results are summarized as follows:

1. *Velocity fluctuations.* As in the case of incident vortical turbulence, the interaction yields an increase of the fluctuating kinetic energy just behind the shock wave (see Fig. 11.10). For weak shock waves (LIA analysis in Mahesh, Lele, and Moin, 1995, is done with $M_1 = 1.2$), the kinetic energy decays monotonically

11.2 Linear Nondestructive Interaction

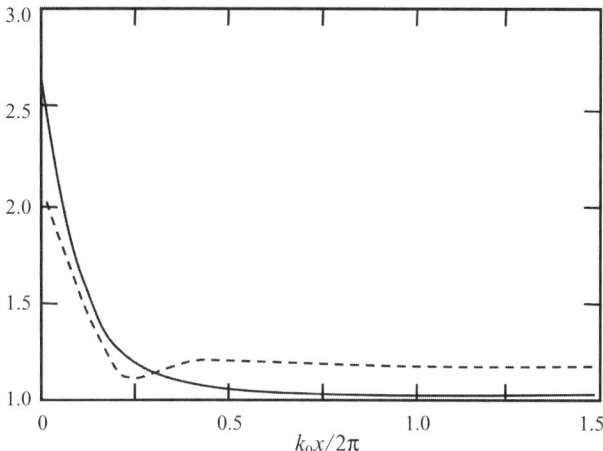

Figure 11.10. LIA prediction of streamwise evolution of turbulent kinetic energy $\mathcal{K}$ downstream of the shock wave. Solid line: $M_1 = 1.2$; dashed line: $M_1 = 2$. From Mahesh, Lele, and Moin (1995) with permission of CUP.

downstream of the shock, whereas for strong shocks ($M_1 = 2$ in Mahesh, Lele, and Moin, 1995) it exhibits local extrema behind the shock. In the former case the evanescent waves are weak and have a negligible impact on the acoustic-energy balance [see Eq. (11.17)], whereas in the latter, the evanescent waves are strong and lead to the existence of a near-field relaxation very similar to the one observed in the case of incident vortical turbulence. But it is worth noting that increasing the Mach number yields a decrease of the amplification immediately behind the shock front.

The behavior of far-field kinetic energy is a bit more complex. Far-field turbulence intensities are plotted versus the upstream Mach number in Fig. 11.11. A first observation is that the shock normal turbulence intensity is higher than the two transverse components. The second, more important, conclusion is that the amplification factor of the far-field kinetic energy does not respond monotonically to an increase in the upstream Mach number: The far-field kinetic energy is lower than the incident kinetic energy for $1.25 \leq M_1 < 1.80$, whereas it is higher for other values. But it is worth noting that the energy of transverse components of velocity decreased over a wider range of upstream Mach numbers. This phenomenon can be understood by decomposing the far-field kinetic energy into a vortical and an acoustic contribution, denoted by $\mathcal{K}_s$ and $\mathcal{K}_d$, respectively (see Fig. 11.12). It is observed that $\mathcal{K}_d$ decays monotonically for $M_1 > 1.2$, whereas $\mathcal{K}_s$ is a strictly increasing function of M_1. The existence of a local minimum is explained by the fact that the solenoidal kinetic energy exceeds the dilatational one for $M_1 > 2$ and the incident kinetic energy for $M_1 > 2.25$. The behavior of the far-field acoustic kinetic energy is governed by the competition between two mechanisms: the amplification of the energy of incident waves, which grows with the Mach number, and the fact that the range of angles corresponding to evanescent emitted waves also increases with M_1, making fewer

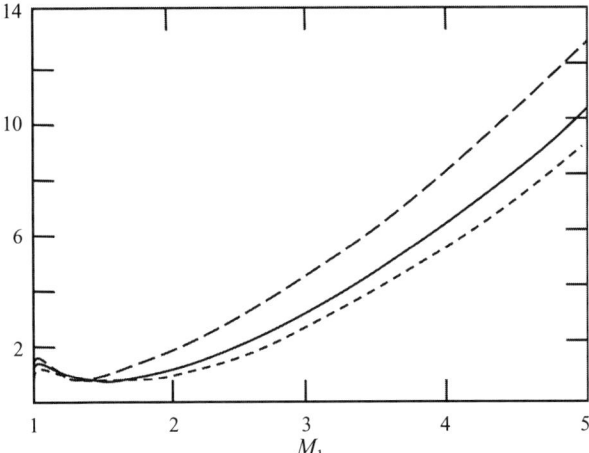

Figure 11.11. LIA analysis of far-field turbulent kinetic energy versus the upstream Mach number M_1. Solid line: turbulent kinetic energy $\mathcal{K}$; long-dashed line: streamwise Reynolds stress $R_{11} = \overline{u'u'}$; short-dashed line: transverse Reynolds stresses $R_{22} = \overline{v'v'}$ and $R_{33} = \overline{w'w'}$. From Mahesh, Lele, and Moin (1995) with permission of CUP.

and fewer emitted waves that contribute to the far field. The LIA analysis shows that, at a high upstream Mach number, the energy of all velocity components scales as M_1^2.

2. *Vorticity*. In this case, the production in the acoustic wave is mainly governed by the baroclinic term [term VI in Eq. (11.19)]. The jump relations for the vorticity components show that transverse components will be the most affected, the streamwise component evolution being governed by other mechanisms, as in the case of incident vortical turbulence.

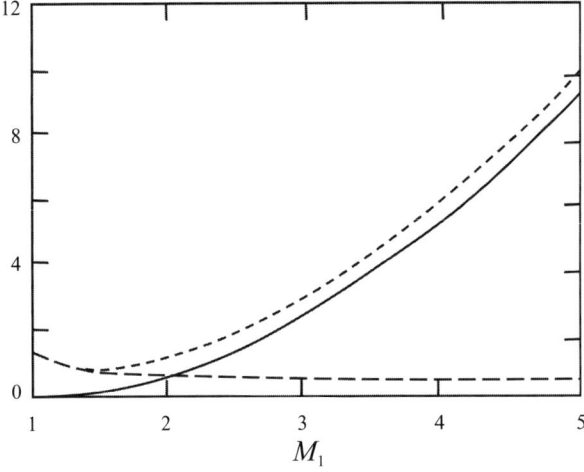

Figure 11.12. LIA analysis of the evolution components of the far-field turbulent kinetic energy versus the upstream Mach number M_1. Long-dashed line: dilatational (acoustic) kinetic energy $\mathcal{K}_d$; solid line: solenoidal (vortical) kinetic energy $\mathcal{K}_s$; short-dashed line: full kinetic energy $\mathcal{K} = \mathcal{K}_s + \mathcal{K}_d$. From Mahesh, Lele, and Moin (1995) with permission of CUP.

11.2 Linear Nondestructive Interaction

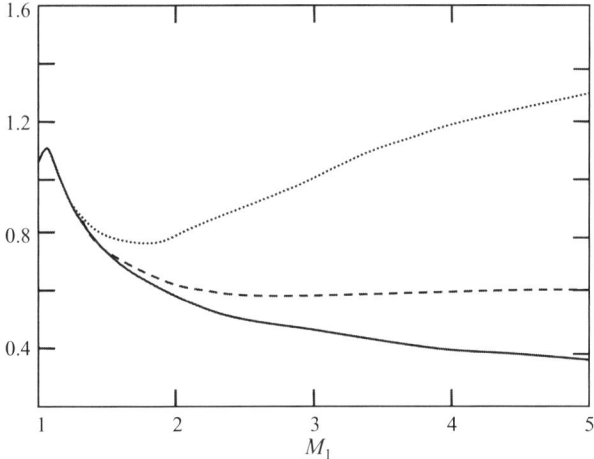

Figure 11.13. LIA prediction of far-field normalized rms fluctuations of thermodynamical quantities. Solid line: $\sqrt{\overline{p'p'}}/\bar{p}$; dashed line: $\gamma\sqrt{\overline{\rho'\rho'}}/\bar{\rho}$; dotted line: $\frac{\gamma}{\gamma-1}(\sqrt{\overline{T'T'}}/\bar{T})$. From Mahesh, Lele, and Moin (1995) with permission of CUP.

3. *Thermodynamic quantities.* The evolution of the thermodynamic quantities in the far field versus the upstream Mach number is displayed in Fig. 11.13. It is observed that the downstream fluctuations are nearly isentropic for $M_1 < 1.5$. At higher Mach numbers the emitted entropy fluctuations are significant relative to the acoustic fluctuations. The dominance of the entropy mode at a high Mach number originates in two phenomena: the emission of stronger and stronger entropy waves and the decrease of acoustic energy in the far field.

11.2.5 Mixed Turbulence–Shock Interaction

We now address the cases in which the incident turbulent field is composed of different types of Kovasznay modes: hybrid vortical–acoustic turbulence (Mahesh, Lele, and Moin, 1995) and vortical–entropic turbulence (Mahesh, Moin, and Lele, 1996 and Mahesh, Lele, and Moin, 1997). These cases are of great interest, as physical turbulence generated in wind tunnels or observed in natural flows is never strictly vortical or acoustic. It is worth recalling here the important conclusion that a Kovasnay mode will generate modes of different natures through nonlinear self-interactions. Therefore the sensitivity of the results previously presented for pure incident fields is of major interest to gain deeper insight into the dynamics of realistic flows. But because the experimental data exhibit a significant dispersion, it can be inferred that their sensitivity must be great (independent of the fact that such experiments are very difficult to perform for technical and technological reasons). Another point is that, in real flows, the distribution of the total energy among the three Kovasznay modes is unknown and cannot usually be controlled. Therefore we hereafter put the emphasis on the theoretical results dealing with the sensitivity of the results rather than give an exhaustive presentation of some realizations.

Let us begin by examining the 2D linearized Euler equation for the vorticity fluctuation about a 1D mean flow. We use it as a simple phenomenological model to describe the amplification of the transverse vorticity components across the shock. The linearized evolution law is

$$\frac{\partial \Omega'}{\partial t} + U \frac{\partial \Omega'}{\partial x} = -\Omega' \frac{\partial U}{\partial x} - \frac{\partial \rho'}{\partial y} \frac{1}{\bar{\rho}^2} \frac{\partial \bar{p}}{\partial x} + \frac{\partial p'}{\partial y} \frac{1}{\bar{\rho}^2} \frac{\partial \bar{\rho}}{\partial x}. \tag{11.28}$$

The usual viscous model (Zel'dovich and Raizer, 2002; Landau and Lifshitz, 1987) for the shock front shows that $(\partial \bar{u}/\partial x) < 0$, $(\partial \bar{p}/\partial x) > 0$, and $(\partial \bar{\rho}/\partial x) > 0$ in the shock region. The first term on the right-hand side of the previous equation corresponds to the compression by the mean-flow gradient. Because $\partial \bar{u}/\partial x$ is negative in the shock region, the net effect of vorticity amplification by the bulk compression is recovered. The two last terms are related to the baroclinic mechanisms. The second term on the right-hand side of Eq. (11.28) involves the fluctuating density and is therefore nonzero for both acoustic and entropy fluctuating modes, whereas the third one is nonzero for acoustic perturbations only. This equation also shows that the baroclinic and the bulk compression contributions can have the same or opposite signs, depending on the respective signs of the vorticity, density, and pressure fluctuations. If the contributions have the same sign, the net amplification of vorticity will be increased by the cooperative interaction, whereas the two mechanisms will tend to cancel in the opposite case, yielding a decrease of the net vorticity fluctuation amplification. One can see that increased amplification is recovered if $\Omega' p' > 0$ or $\Omega' \rho' < 0$. The very important conclusion drawn from that very simple analysis is that the results of the shock–turbulence interaction will be greatly sensitive to the correlation between the Kovasznay modes in the incident field.

11.2.5.1 Influence of the Upstream Entropy Fluctuations

Let us first consider the case of an incident field made of vorticity and entropy modes. We simplify the problem by considering a single-plane entropy wave with amplitude A_e and a single vorticity wave with amplitude A_v with the same wave vector $\mathbf{k}$. According to results presented in Section 12.3, the upstream field is given by

$$\frac{u'}{\bar{u}} = A_v \cos \alpha e^{\iota(\mathbf{k} \cdot \mathbf{x} - \omega t)}, \quad \frac{v'}{\bar{u}} = -A_v \sin \alpha e^{\iota(\mathbf{k} \cdot \mathbf{x} - \omega t)}, \quad \frac{s'}{c_p} = A_e e^{\iota(\mathbf{k} \cdot \mathbf{x} - \omega t)}, \tag{11.29}$$

leading to

$$\frac{\Omega'}{\bar{u}} = -A_v e^{\iota(\mathbf{k} \cdot \mathbf{x} - \omega t)}, \quad \frac{\rho'}{\bar{\rho}} = -\frac{T'}{T} = A_e e^{\iota(\mathbf{k} \cdot \mathbf{x} - \omega t)}, \quad \frac{p'}{\bar{p}} = 0. \tag{11.30}$$

It is seen that Ω' and u' on the one hand and ρ' and T' on the other hand are in phase opposition. Therefore the condition for cooperative interaction $\Omega' \rho' < 0$ is equivalent to $u'T' < 0$. Introducing the complex ratio,

$$\frac{A_e}{A_v} = \varrho e^{\iota \varphi}, \tag{11.31}$$

11.2 Linear Nondestructive Interaction

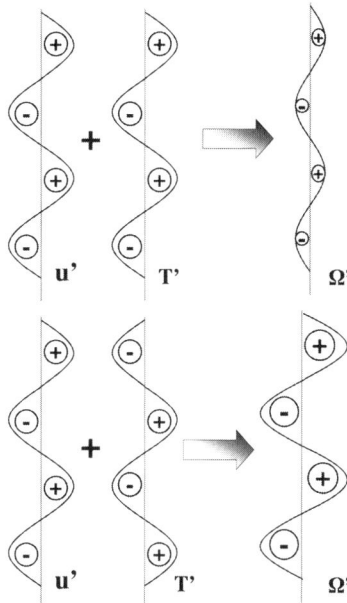

Figure 11.14. Schematic view of the influence of the phase difference between velocity and temperature fluctuations on the emitted vorticity fluctuation. Top: velocity and temperature fluctuations are in phase, leading to a decrease of the vorticity amplification. Bottom: they are in phase opposition, yielding a large increase of the vorticity amplification.

it is seen that cooperative interaction is observed if $\varphi \in]-\pi/2, \pi/2[$, whereas partial cancellation occurs for $\varphi \in]\pi/2, -\pi/2[$. This discussion is illustrated in Fig. 11.14.

DNS and LIA results dealing with the amplification of velocity fluctuations in the case of an incident isotropic turbulent are presented in Fig. 11.15. They show that, in the case in which the streamwise velocity component and temperature fluctuations are strongly anticorrelated ($\overline{u'T'}/u_{\rm rms}T_{\rm rms} \approx -1$), the amplification of all velocity components is greatly enhanced, the effect being more important on the streamwise component. The velocity field still exhibits a near field whose properties are similar to those of the near field generated by a pure vortical incident field. On the opposite, the amplification is reduced when they are correlated, i.e., $\overline{u'T'} > 0$. The evolution of the amplification of the far-field velocity variances with respect to the upstream Mach number is displayed in Fig. 11.16. The LIA predicts that the amplification saturates for $M_1 > 2$, with a remarkable exception: If the upstream fluctuations satisfy Morkovin's hypothesis given in Eq. (11.27), the amplification factor does not saturate and keeps growing with M_1. The main reason is that, if Morkovin's hypothesis is assumed to hold in the upstream region, the relative importance of the entropy modes with respect to the vorticity mode scales like M_1^2.

Similar conclusions hold for the vorticity field: Both DNS and LIA confirm the predictions drawn from the simplified model. The amplification factor of the transverse vorticity components is plotted versus the upstream Mach number in Fig. 11.17. Once again, the amplification is enhanced if $\overline{u_1'T'} < 0$ and exhibits very strong values if the incident turbulent field satisfies Morkovin's hypothesis.

An interesting point is that the interaction with the shock results in a breakdown of Morkovin's hypothesis downstream of the shock wave, even if it holds upstream of the shock. Recalling that the fundamental assumption in Morkovin's

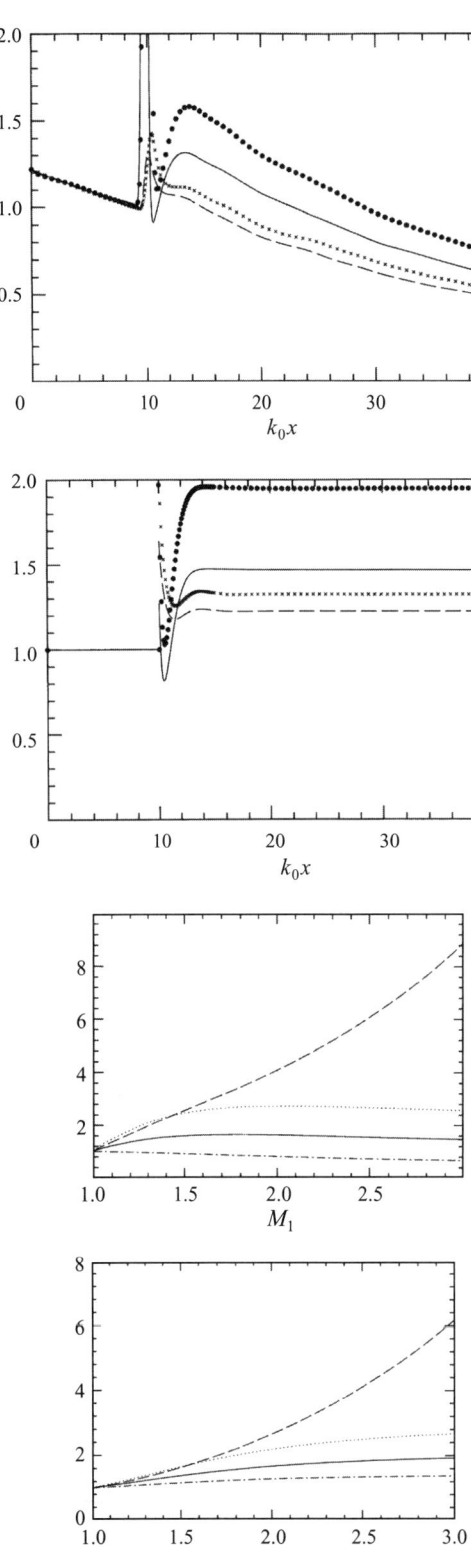

Figure 11.15. Influence of upstream entropy fluctuations on the streamwise evolution of Reynolds stresses at $M_1 = 1.29$. Top: DNS data. Streamwise Reynolds stress $R_{11} = \overline{u'u'}$ for $\overline{u'_1 T'}/u_{\mathrm{rms}} T_{\mathrm{rms}} = -0.06$ (solid line) and -0.84 (black circles); transverse Reynolds stress $R_{22} = \overline{v'v'}$ for $\overline{u'_1 T'}/u_{\mathrm{rms}} T_{\mathrm{rms}} = -0.06$ (dashed line) and -0.84 ("x"). Bottom: LIA analysis, same cases and symbols as for DNS data. From Mahesh, Lele, and Moin (1997) with permission of CUP.

Figure 11.16. LIA analysis of influence of upstream entropy fluctuations on the downstream evolution of Reynolds stresses versus the upstream Mach number M_1. Top: Streamwise Reynolds stress $R_{11} = \overline{u'u'}$. Solid line: pure vortical incident turbulence; dotted line: $\overline{u'_1 T'} < 0$; dashed-dotted line: $\overline{u'_1 T'} > 0$; dashed line: Morkovin's hypothesis satisfied upstream. Bottom: transverse Reynolds stress $R_{22} = \overline{v'v'}$, same symbols as in Fig. 11.15. From Mahesh, Lele, and Moin (1997) with permission of CUP.

11.2 Linear Nondestructive Interaction

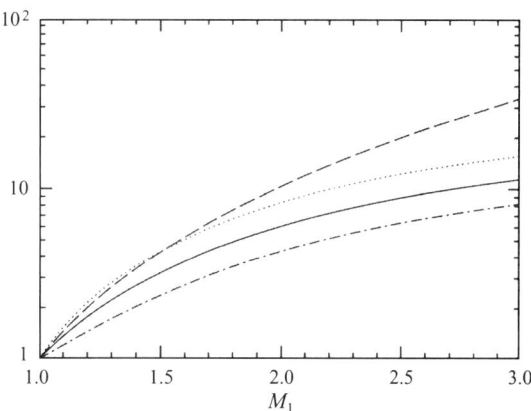

Figure 11.17. LIA analysis of influence of upstream entropy fluctuations on the far-field amplification of transverse vorticity components $\overline{\Omega_2' \Omega_2'} = \overline{\Omega_3' \Omega_3'}$ versus the upstream Mach number M_1. Solid line: pure vortical incident turbulence; dotted line: $\overline{u_1' T'} < 0$; dashed–dotted line: $\overline{u_1' T'} > 0$; dashed line: Morkovin's hypothesis satisfied upstream. From Mahesh, Lele, and Moin (1997) with permission of CUP.

proposal is that the stagnation temperature T^0 is constant in the flow, and decomposing it as follows,

$$T^0 = \bar{T} + T' + \frac{1}{2}\frac{(U + u')^2 + v'^2 + w'^2}{c_p}, \quad (11.32)$$

the Rankine–Hugoniot jump relation for the energy (11.7) yields the continuity of T^0 across the shock wave:

$$\bar{T}_1 + T_1' + \frac{1}{2}\frac{(U_1 + u_1' - u_s)^2 + v_1'^2 + w_1'^2}{c_p}$$
$$= \bar{T}_2 + T_2' + \frac{1}{2}\frac{(U_2 + u_2' - u_s)^2 + v_2'^2 + w_2'^2}{c_p}, \quad (11.33)$$

where u_s is the shock speed associated with the corrugation of the shock front by incident perturbations. Assuming that fluctuations are small enough, one can linearize (11.33), yielding

$$T_1' + \frac{U_1(u_1' - u_s)}{c_p} = T_2' + \frac{U_2(u_2' - u_s)}{c_p}. \quad (11.34)$$

Now assuming that the upstream flow satisfies the Morkovin hypothesis, one obtains the following expression for the linearized fluctuation of the stagnation temperature behind the shock wave:

$$T_2' + \frac{U_2 u_2'}{c_p} = \frac{u_s(U_2 - U_1)}{c_p}. \quad (11.35)$$

Using this expression, one can write

$$\frac{T_2''}{\bar{T}_2} + (\gamma - 1) M_2^2 \frac{u_2''}{U_2} = -(\gamma - 1) M_2 (\mathfrak{C} - 1) \frac{u_s}{\bar{a}_2}, \quad (11.36)$$

where $\mathfrak{C}$ is the compression factor defined by Eq. (12.12). Therefore Morkovin's hypothesis holds downstream of the shock wave if and only if the right-hand side of Eq. (11.36) is zero, which is observed to be false in both LIA and DNS results.

11.2.5.2 Influence of the Upstream Acoustic Fluctuations

The analysis of the influence of upstream acoustic waves is simpler than the one of entropy waves, as these waves cannot be correlated with the vortical fluctuations, their propagation speeds being different.[†] Therefore the emitted far field is obtained by means of a simple superposition of the far fields corresponding to the vortical fluctuations and acoustic fluctuations considered separately.

This is illustrated considering the amplification of the total turbulent kinetic energy $\mathcal{K} = \mathcal{K}_s + \mathcal{K}_d$:

$$\frac{\mathcal{K}_2}{\mathcal{K}_1} = \frac{(\mathcal{K}_s)_2 + (\mathcal{K}_d)_2}{(\mathcal{K}_s)_1 + (\mathcal{K}_d)_1} = (1 - \chi_1)\frac{(\mathcal{K}_s)_2}{(\mathcal{K}_s)_1} + \chi_1 \frac{(\mathcal{K}_d)_2}{(\mathcal{K}_d)_1}, \quad (11.37)$$

where χ_1 is the ratio of dilatational acoustic energy to the total kinetic energy [see Eq. (9.110)] in the upstream state. The amplification factors of the acoustic and vortical components being different, significant differences in the amplification of the total kinetic energy can be observed by varying the value of χ_1.

11.2.6 On the Use of RDT for Linear Nondestructive Interaction Modeling

The LIA theory has been shown to be a very accurate tool to predict and understand the linear nondestructive shock–turbulence interactions. The capability of the RDT theory to account for the same physical effects has also been investigated by many authors, among whom are Jacquin, Cambon, and Blin (1993) (see also the references given therein).

Within the RDT framework, the planar shock is essentially modeled as an unidirectional irrotational compression with a time-varying mean flow already discussed in Subsection 10.2.3.3. Therefore all results presented in this subsection can be used in a straightforward way. Therefore the amplification ratio of turbulent kinetic energy is lower bounded by the solenoidal limit given by Eq. (10.47) and upper bounded by the pressure-released evolution described by relation (10.48). These two evolutions laws are compared with LIA results in Fig. 11.18.

Large discrepancies are observed, the two RDT limits yielding a much larger amplification ratio than both the near- and far-field LIA predictions at almost all upstream Mach numbers. The inaccuracy of RDT for this problem can be understood by looking at Table 11.1, which summarizes the main differences between the RDT and LIA approaches. Of course, because of the isentropy assumption, RDT cannot account for entropy effects discussed in Subsection 11.2.5.1. But by restricting to the case of a purely vortical incoming turbulence described in Subsection 11.2.3, one can observe that the main weaknesses of RDT are as follows:

- The inability to account for the shock corrugation and its feedback on the emitted turbulent field.

[†] It is worth noting here that some interactions exist if the upstream field is composed of a single wave of each type, but these interactions cancel from a statistical viewpoint in fully developed isotropic turbulent flows.

11.3 Nonlinear Nondestructive Interactions

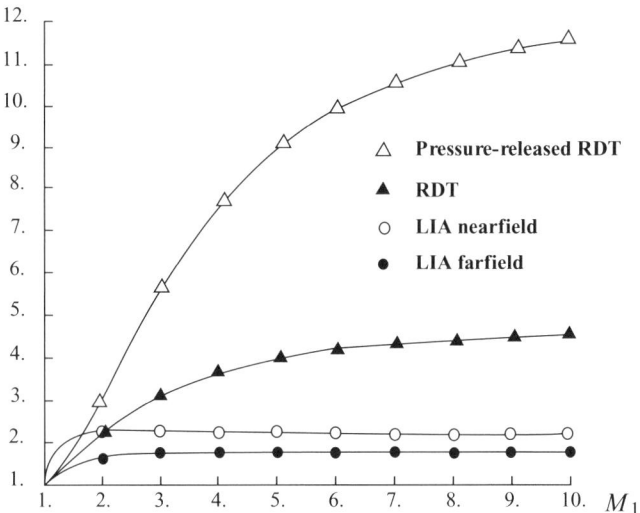

Figure 11.18. Comparison of the turbulent kinetic-energy amplification factor as a function of the upstream Mach number according to different linearized theories for a pure solenoidal upstream disturbance field. Adapted from Jacquin, Cambon, and Blin (1993).

- The inability to predict the existence of evanescent acoustic waves downstream of the shock and the existence of a cutoff incidence angle for incoming disturbances (see details in Subsection 12.4.2).
- The isentropic assumption, because the entropy fluctuations are more significant than the acoustic ones for $M_1 > 1.65$ downstream of the shock. The isentropic assumption for the emitted field is found to be realistic for weak shocks only, with $M_1 < 1.2$.

11.3 Nonlinear Nondestructive Interactions

11.3.1 Turbulent Jump Conditions for the Mean Field

Let us denote the mean and fluctuating velocity components (U_i, V_i, W_i) and (u_i'', v_i'', w_i''), respectively, where the subscripts 1 and 2 refer to the upstream and

Table 11.1. *Main features of shock–turbulence interaction modeling according to linearized theories*

Theory features	LIA	RDT
Intrinsic time scale	none	$1/S$
Intrinsic length scale	none (infinitely small)	none (infinitely large)
Shock viscous effects	yes (linearized Rankine–Hugoniot relations)	no
Shock corrugation effects	yes (linearized response)	no
Vortical disturbances	yes	yes
Acoustic disturbances	yes	yes
Entropy disturbances	yes	no

downstream states, respectively. The mean and fluctuating enthalpies are denoted by $\tilde{h}_i$ and h_i''. Using the same notation as in previous sections, assuming that a frame of reference can be found in which the mean shock wave is stationary and that viscous effects are negligible, Lele (1992) deduced from the Navier–Stokes equations written in conservative form the following jump relations for the mean-flow quantities (the vector normal to the shock wave is chosen to be along the x direction):

$$\bar{\rho}_1 U_1 = \bar{\rho}_2 U_2, \tag{11.38}$$

$$\bar{\rho}_1 U_1^2 + \bar{\rho}_1 \widetilde{u_1'' u_1''} + \bar{p}_1 = \bar{\rho}_2 U_2^2 + \bar{\rho}_2 \widetilde{u_2'' u_2''} + \bar{p}_2, \tag{11.39}$$

$$\bar{\rho}_1 U_1 V_1 + \bar{\rho}_1 \widetilde{u_1'' v_1''} = \bar{\rho}_2 U_2 V_2 + \bar{\rho}_2 \widetilde{u_2'' v_2''}, \tag{11.40}$$

$$\bar{\rho}_1 U_1 W_1 + \bar{\rho}_1 \widetilde{u_1'' w_1''} = \bar{\rho}_2 U_2 W_2 + \bar{\rho}_2 \widetilde{u_2'' w_2''}, \tag{11.41}$$

$$\bar{\rho}_1 U_1 \left(\tilde{h}_1 + \frac{1}{2} K_1 + \frac{1}{2} \mathcal{K}_1 \right) + \bar{\rho}_1 \left(\widetilde{h_1'' u_1''} + U_1 \widetilde{u_1'' u_1''} + V_1 \widetilde{u_1'' v_1''} + W_1 \widetilde{u_1'' w_1''} + \frac{1}{2} \widetilde{\mathcal{K}_1'' \mathcal{K}_1'' u_1''} \right)$$

$$= \bar{\rho}_2 U_2 \left(\tilde{h}_2 + \frac{1}{2} K_2 + \frac{1}{2} \mathcal{K}_2 \right)$$

$$+ \bar{\rho}_2 \left(\widetilde{h_2'' u_2''} + U_2 \widetilde{u_2'' u_2''} + V_2 \widetilde{u_2'' v_2''} + W_2 \widetilde{u_2'' w_2''} + \frac{1}{2} \widetilde{\mathcal{K}_2'' \mathcal{K}_2'' u_2''} \right), \tag{11.42}$$

where the mean-turbulent and mean-flow kinetic energies are defined as

$$\mathcal{K}_i = \widetilde{u_i'' u_i''} + \widetilde{v_i'' v_i''} + \widetilde{w_i'' w_i''}, \quad K_i = U_i^2 + V_i^2 + W_i^2 \tag{11.43}$$

and

$$\widetilde{\mathcal{K}_i'' \mathcal{K}_i'' u_i''} \equiv \frac{1}{\bar{\rho}} \overline{\rho u_i'' (u_i'' u_i'' + v_i'' v_i'' + w_i'' w_i'')}. \tag{11.44}$$

Equations (11.38)–(11.42) show that the mean-flow quantities are directly affected by the jump in the turbulent stresses across the shock and that they cannot be computed separately. Therefore closures for the turbulent terms are required for computing the mean flow downstream of the shock front. Only very few attempts to close the jump conditions are available (Lele, 1992; Zank et al., 2002). Because none of them has been fully assessed, they are not discussed here.

It is worth noting that the mean-flow solutions of the nonlinear jump conditions can be very different from those considered within the LIA framework. A striking example is that if the incident mean flow is normal to the mean shock front (i.e., $V_1 = W_1 = 0$), the mean flow downstream of the shock can deviate from a unidirectional flow (i.e., $V_2 \neq 0$ and/or $W_2 \neq 0$). The observed effects of turbulence are as follows:

1. Turbulent fluctuations increase the mean shock speed.
2. Turbulent fluctuations decrease the efficiency of turbulence amplification across the shock as the amplitude of incident fluctuations is increased.

11.3.2 Jump Conditions for an Incident Isotropic Turbulence

The general jump conditions previously given simplify dramatically in the case of a normal upstream mean flow advecting an isotropic incident field. In this case, the mean flow remains unidirectional and one has

$$V_i = W_i = 0, \quad \widetilde{u_i'' v_i''} = \widetilde{u_i'' w_i''} = 0, \quad i = 1, 2. \tag{11.45}$$

After some algebra, one obtains the following expressions for the mean-flow quantities [to be compared with Eqs. (12.12)–(12.14)]:

$$\frac{\bar{p}_2}{\bar{p}_1} = 1 + \frac{2\gamma}{\gamma+1}\left[\frac{(1-\mathfrak{K})\gamma + (1+\mathfrak{K})}{2}M_1^2 - \mathfrak{K}\right] - \frac{1}{\bar{p}_1}[[\bar{\rho}\widetilde{u''u''}]], \tag{11.46}$$

$$\mathfrak{C} = \frac{\bar{\rho}_2}{\bar{\rho}_1} = \frac{U_1}{U_2} = \frac{1}{\mathfrak{K}}\frac{(\gamma+1)M_1^2}{2+(\gamma-1)M_1^2}, \tag{11.47}$$

$$M_2^2 = \mathfrak{L}^2\left(1 + \frac{\gamma-1}{2}M_1^2\right)$$

$$\times \left[\frac{(1+\mathfrak{L})\gamma - (1-\mathfrak{L})}{2}\frac{(1-\mathfrak{L})\gamma + (1+\mathfrak{L})}{2}M_1^2 - \frac{\gamma-1}{2}\mathfrak{L}^2\right]^{-1}, \tag{11.48}$$

with

$$\mathfrak{L} = \frac{\mathfrak{K}}{\sqrt{1 + \frac{[[H]]}{H_1}}} \tag{11.49}$$

and

$$\mathfrak{K} = \left(1 + \frac{U_1}{U_1 - U_2}\frac{[[H]]}{H_1}\right)\left(1 - \frac{2\gamma}{(\gamma+1)(U_1 - U_2)}\left[\left[\frac{\widetilde{u''u''}}{U}\right]\right]\right)^{-1}, \tag{11.50}$$

where $H = \tilde{h} + U^2/2$ is the mean stagnation enthalpy.

Bibliography

DYAKOV, S. P. (1954). On the stability of shock waves, *Zh. Eksp. Teor. Fiz.* **27**, 288 [Atomic Research Agency Establishment AERE Lib./Trans. 648 (1956)].

FABRE, D., JACQUIN, L., AND SESTERHENN, J. (2001). Linear interaction of a cylindrical entropy spot with a shock wave, *Phys. Fluids* **13**, 2403–2422.

HAYES, W. D. (1957). The vorticity jump across a gasdynamic discontinuity, *J. Fluid Mech.* **2**, 595–600.

JACQUIN, L., CAMBON, C., AND BLIN, E. (1993). Turbulence amplification by a shock wave and rapid distortion theory, *Phys. Fluids A* **5**, 2539–2550.

KONTOROVICH, V. M. (1957). To the question on stability of shock waves, *Sov. Phys. JETP* **6**, 1179 [Atomic Research Agency Establishment AERE Lib./Trans. 648 (1966)].

LANDAU, L. D. AND LIFSHITZ, E. M. (1987). *Fluid Mechanics*, 2nd ed, *Course of Theoretical Physics*, Vol. 6, Butterworth-Heinemann.

LEE, S., LELE, S. K., AND MOIN, P. (1993). Direct numerical simulation of isotropic turbulence interacting with a weak shock wave, *J. Fluid Mech.* **251**, 533–562.

LEE, S., LELE, S. K., AND MOIN, P. (1997). Interaction of isotropic turbulence with shock waves: Effect of shock strength, *J. Fluid Mech.* **340**, 225–247.

LELE, S. K. (1992). Shock-jump relations in a turbulent flow, *Phys. Fluids A* **4**, 2900–2905.

LUBCHICH, A. A. AND PUDOVKIN, M. I. (2004). Interaction of small perturbations with shock waves, *Phys. Fluids* **16**, 4489–4505.

MAHESH, K., LELE, S. K., AND MOIN, P. (1995). The interaction of an isotropic field of acoustic waves with a shock wave, *J. Fluid Mech.* **300**, 383–407.

MAHESH, K., LELE, S. K., AND MOIN, P. (1997). The influence of entropy fluctuations on the interaction of turbulence with a shock wave, *J. Fluid Mech.* **334**, 353–379.

MAHESH, K., MOIN, P. AND LELE, S. K. (1996). The interaction of a shock wave with a turbulent shear flow, *Report No. TF-69*, Department of Mechanical Engineering, Stanford University.

MOORE, F. K. (1954). Unsteady oblique interaction of a shock wave with a plane disturbance, *Tech. Rep. 2879*, NACA.

RIBNER, H. S. (1953). Convection of a pattern of vorticity through a shock wave, *Tech. Rep. 1164*, NACA.

ZANK, P., ZHOU, Y., MATTHAEUS, W. H., AND RICE, W. K. M. (2002). The interaction of turbulence with shock waves: A basic model, *Phys. Fluids* **14**, 3766–3774.

ZEL'DOVICH, Y. B. AND RAIZER, Y. P. (2002). *Physics of Shock Waves and High-Temperature Hydrodynamic Phenomena*, Dover.

12 Linear Interaction Approximation for Shock–Perturbation Interaction

This chapter is devoted to a detailed presentation of the linear interaction approximation (LIA) theory mentioned in Chapter 11. The main assumptions that underlie the LIA are discussed in Subsection 11.2.2 and are not duplicated here. We just recall here that the LIA holds if the following constraints are fulfilled:

1. The fluctuations must be weak in the sense that the distorted shock wave must remain well defined. Numerical experiments led Lee and co-workers (Lee, Lele, and Moin (1993) to propose the following empirical criterion for the linear regime:

$$M_t^2 < \alpha(M_1^2 - 1), \tag{12.1}$$

   where M_t and M_1 are the upstream turbulent and mean Mach numbers, respectively, and $\alpha \approx 0.1$.

2. The time required for turbulent events to cross the shock must be small compared with the turbulence time scale $\mathcal{K}/\varepsilon$ (with $\mathcal{K}$ and ε the turbulent kinetic energy and the turbulent kinetic-energy dissipation rate, respectively), so that nonlinear mechanisms cannot have significant effects.

12.1 Shock Description and Emitted Fluctuating Field

We consider here the interaction of a plane shock with a normal 2D flow in the (x, y) plane. Let the undisturbed shock normal vector and the mean flow be oriented along the x axis. The disturbed shock front is defined as

$$x = x_s(y, t). \tag{12.2}$$

The position of the undisturbed shock is arbitrarily chosen to be $x = 0$. The local instantaneous normal and tangential vectors, $\bm{n}$ and $\bm{t}$, are equal to

$$\bm{n} = \left(1, -\frac{\partial x_s}{\partial y}\right)^T, \quad \bm{t} = \left(\frac{\partial x_s}{\partial y}, 1\right)^T. \tag{12.3}$$

The shock speed in the reference frame associated with the mean shock location is equal to

$$\bm{u}_s = \left(\frac{\partial x_s}{\partial t}, 0\right)^T = (u_s, 0)^T. \tag{12.4}$$

12.1 Shock Description and Emitted Fluctuating Field

The upstream and downstream fields are split as follows:

$$\rho_i(x, y, t) = \bar{\rho}_i + \epsilon \rho_i'(x, y, t), \tag{12.5}$$

$$s_i(x, y, t) = \bar{s}_i + \epsilon s_i'(x, y, t), \tag{12.6}$$

$$T_i(x, y, t) = \bar{T}_i + \epsilon T_i'(x, y, t), \tag{12.7}$$

$$p_i(x, y, t) = \bar{p}_i + \epsilon p_i'(x, y, t), \tag{12.8}$$

$$u_i(x, y, t) = U_i + \epsilon u_i'(x, y, t), \tag{12.9}$$

$$v_i(x, y, t) = 0 + \epsilon v_i'(x, y, t), \tag{12.10}$$

along with

$$x_s(y, t) = 0 + \epsilon \xi(y, t), \tag{12.11}$$

where subscripts 1 and 2 are related to the upstream (incident) and downstream (emitted) fields, respectively. The parameter ϵ is assumed to be a small parameter, i.e., $\epsilon \ll 1$, so that all primed quantities and ξ are of the order of $O(1)$.

Because the mean field obeys the classical Rankine–Hugoniot jump conditions (11.4)–(11.7), which are recovered as the zeroth-order relations by inserting Eqs. (12.5)–(12.10) into jump relations (11.4)–(11.7), the following classical relations hold:

$$\mathfrak{C} \equiv \frac{\bar{\rho}_2}{\bar{\rho}_1} = \frac{U_1}{U_2} = \frac{(\gamma+1)M_1^2}{2+(\gamma-1)M_1^2}, \tag{12.12}$$

$$\frac{\bar{p}_2}{\bar{p}_1} = 1 + \frac{2\gamma}{\gamma+1}(M_1^2 - 1), \tag{12.13}$$

$$M_2 \equiv \frac{U_2}{a_2} = \sqrt{\frac{2+(\gamma-1)M_1^2}{2\gamma M_1^2 - (\gamma-1)}}, \tag{12.14}$$

where M_1 and M_2 are the upstream and downstream Mach numbers, respectively.

The incident field is composed of superimposed plane-propagating waves. Thanks to the linear approximation, one can restrict the analysis to a single incident wave for each Kovasznay mode, with the orientation of the incident wave vector as a free parameter. One can remark that, because the shock is assumed to have no intrinsic length scale, the LIA results will not depend on the wave-vector modulus. Let $\mathbf{k}$ and ω be the wave vector and the frequency of the incident plane wave, respectively. Therefore all fluctuating quantities tied to the incident wave have the following form:

$$\phi(x, y, t) = A_\phi e^{i(\mathbf{k} \cdot \mathbf{x} - \omega t)}, \tag{12.15}$$

where ϕ and A_ϕ are a dummy variable and the corresponding amplitude parameter, respectively. Denoting α as the angle between $\mathbf{k}$ and the x axis, one has $\mathbf{k} \cdot \mathbf{x} = k_x x + k_y y = kr$ with $k_x = k \cos \alpha$, $k_y = k \sin \alpha$, and $r = x \cos \alpha + y \sin \alpha$.

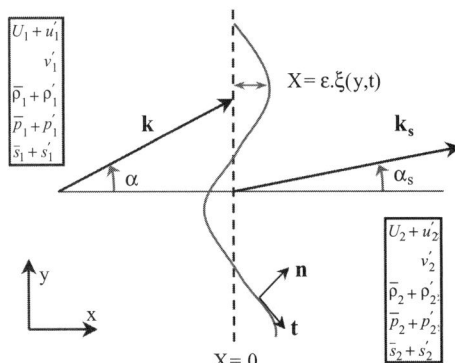

Figure 12.1. Schematic view of the 2D LIA for shock–plane-wave interaction.

Because the shock has no intrinsic scale and is fully governed by incident perturbations, its displacement induced by the perturbation wave just considered is

$$\xi(y,t) = A_\xi e^{i(k_y y - \omega t)} = A_\xi e^{i(k \sin \alpha y - \omega t)}, \tag{12.16}$$

where the amplitude factor A_ξ remains to be computed.

12.2 Calculation of Wave Vectors of Emitted Waves

12.2.1 General

We now address the problem of computing the wave vector of each emitted wave. The problem is schematized in Fig. 12.1.

As previously stated, any incident perturbation triggers the generation of a triad of emitted waves (one wave for each Kovasznay mode) in the downstream region. Continuity at the shock wave requires that the solution in this region have the same transverse wavenumber k_y and frequency as the incident perturbation wave. Therefore all fluctuating variables behind the shock wave have the generic form

$$\phi(x,y,t) = A_\phi F(x) e^{i(k_y y - \omega t)} = A_\phi F(x) e^{i(k \sin \alpha y - \omega t)}, \tag{12.17}$$

where the function $F(x)$ is still unknown and must be such that the emitted fluctuating field obeys the governing equations of the Kovasznay analysis, i.e., the linearized Euler equations. Looking for plane-wave solutions, one can write

$$F(x) = e^{i\tilde{k}_x x}, \tag{12.18}$$

where the wave-vector normal component $\tilde{k}_x$ is such that the dispersion relation associated with the Kovasznay mode under consideration is satisfied.

12.2.2 Incident Entropy and Vorticity Waves

We first address the problem of an incident entropy or vorticity wave, for which $\omega = k \cos \alpha U_1$.

12.2 Calculation of Wave Vectors of Emitted Waves

12.2.2.1 Emitted Entropy and Vorticity Waves

For the emitted entropy and vorticity modes, the linear analysis yields the following single dispersion relation:

$$\mathbf{k} \cdot \bar{\mathbf{u}} = \omega, \tag{12.19}$$

where $\bar{\mathbf{u}}$ is the mean velocity, leading to

$$\omega = k_x U_1 = k \cos \alpha \, U_1 = \tilde{k}_x U_2, \tag{12.20}$$

from which it follows that

$$\tilde{k}_x = \frac{U_1}{U_2} k \cos \alpha = \mathfrak{C} k \cos \alpha, \tag{12.21}$$

where $\mathfrak{C}$ is the compression factor defined by Eq. (12.12). Fluctuating fields associated with the entropy and vorticity modes are therefore of the form

$$\phi(x, y, t) = A_\phi e^{ik(\mathfrak{C} \cos \alpha x + \sin \alpha y - U_1 \cos \alpha t)}, \tag{12.22}$$

where it is important to note that k is the modulus of the wave vector of the *incident* wave. Because the entropy and vorticity modes obey the same dispersion relation, emitted waves associated with these two modes have the same wave vector, i.e., they propagate in the same direction and have the same wavelength. Let $\mathbf{k}_s$ and α_s be the wave vector of the emitted entropy and vorticity waves and the angle between $\mathbf{k}_s$ and the x axis, respectively. Relation (12.22) can be rewritten as follows:

$$\phi(x, y, t) = A_\phi e^{i(\mathbf{k}_s \cdot \mathbf{x} - \omega t)} = A_\phi e^{i(k_s \cos \alpha_s x + k_s \sin \alpha_s y - \omega t)}, \tag{12.23}$$

leading to the two relations

$$\cot \alpha_s = \mathfrak{C} \cot \alpha, \tag{12.24}$$

$$\frac{k_s}{k} = \frac{\sin \alpha}{\sin \alpha_s}. \tag{12.25}$$

12.2.2.2 Emitted Acoustic Waves—Propagative and Nonpropagative Regimes

The dispersion relation for the acoustic mode in the linear approximation is

$$(\omega - \mathbf{k} \cdot \bar{\mathbf{u}})^2 - a^2 k^2 = 0, \tag{12.26}$$

where a is the mean speed of sound. Using this relation in the domain located behind the shock, one obtains

$$(\omega - \tilde{k}_x U_2)^2 - a_2^2 (\tilde{k}_x^2 + k^2 \sin^2 \alpha) = 0. \tag{12.27}$$

Reminding one that $\omega = k U_1 \cos \alpha$, this last relation can be recast as

$$\frac{1}{\mathfrak{C}^2} \left(\frac{1}{M_2^2} - 1 \right) \tilde{k}_x^2 + \frac{2k \cos \alpha}{\mathfrak{C}} \tilde{k}_x - k^2 \left[\cos^2 \alpha - \sin^2 \alpha \frac{1}{(\mathfrak{C} M_2)^2} \right] = 0. \tag{12.28}$$

The discriminant Δ of the preceding quadratic relation determines if $\tilde{k}_x$ is real or imaginary. The discriminant being equal to

$$\Delta = \frac{2k \sin \alpha}{\mathfrak{C} M_2} \sqrt{\left(\frac{\cos \alpha}{\sin \alpha}\right)^2 - \frac{1}{\mathfrak{C}^2}\left(\frac{1}{M_2^2} - 1\right)}, \quad (12.29)$$

it is seen that $\tilde{k}_x$ is real if $0 \leq \alpha \leq \alpha_c$, where the critical angle α_c is such that

$$\cot \alpha_c = \frac{\sqrt{1 - M_2^2}}{\mathfrak{C} M_2} \quad (12.30)$$

and imaginary if $\alpha_c < \alpha \leq \pi/2$. Denoting as $\tilde{k}_i$ and $\tilde{k}_r$ the real and imaginary parts of $\tilde{k}_x$, the emitted acoustic fluctuating field can be expressed as

$$\phi(x, y, t) = A_\phi e^{-\tilde{k}_i x} e^{\iota(\tilde{k}_r x + k \sin \alpha y - \omega t)}, \quad (12.31)$$

showing that the emitted acoustic field decays exponentially behind the shock wave if $\tilde{k}_x$ is not real, i.e., if the angle of incidence of the incident wave is larger than the critical angle α_c. This threshold angle demarcates two regimes for the emitted acoustic waves: the *propagative regime* without damping and the *nonpropagative regime* with damping. The latter is coined as nonpropagative because the emitted wave amplitude is nearly zero at a distance of the order of one wavelength downstream of the shock. Therefore it is possible to identify a *near-field solution* in which the nonpropagative perturbations have a significant contribution and a *far-field solution* in which nonpropagative perturbations are negligible.

Elementary algebra yields the following expression for the roots of Eq. (12.28):

$$\tilde{k}_x^\pm = \frac{-\dfrac{2k \cos \alpha}{\mathfrak{C}} \pm \Delta}{\dfrac{2}{\mathfrak{C}^2}\left(\dfrac{1}{M_2^2} - 1\right)}, \quad (12.32)$$

where Δ is given by (12.29). It is observed that $\tilde{k}_x^+$ is the only physically admissible root, as $\tilde{k}_x^-$ leads to an exponential growth of the solution behind the shock. Therefore $\tilde{k}_x = \tilde{k}_x^+$ hereafter. The corresponding normalized form is

$$\frac{\tilde{k}_x}{k} = \mathfrak{C} \frac{M_2}{1 - M_2^2} \left[-\cos \alpha M_2 + \sin \alpha \sqrt{\cot^2 \alpha - \frac{1}{\mathfrak{C}^2}\left(\frac{1}{M_2^2} - 1\right)} \right]. \quad (12.33)$$

In the nonpropagative regime, the real and imaginary parts of $\tilde{k}_x$ are equal to

$$\frac{\tilde{k}_r}{k} = -\mathfrak{C} \cos \alpha \frac{M_2^2}{1 - M_2^2} \quad (12.34)$$

and

$$\frac{\tilde{k}_i}{k} = \mathfrak{C} \sin \alpha \frac{M_2}{1 - M_2^2} \sqrt{\cot^2 \alpha - \frac{1}{\mathfrak{C}^2}\left(\frac{1}{M_2^2} - 1\right)}, \quad (12.35)$$

12.2 Calculation of Wave Vectors of Emitted Waves

respectively. The emitted acoustic field can be reexpressed by introducing the emitted wave vector $\boldsymbol{k}_a$, which is such that

$$e^{-\tilde{k}_i x} e^{i(\tilde{k}_r x + k \sin \alpha y - \omega t)} = e^{i(k_a(\cos \alpha_a x + \sin \alpha_a y + i\eta x) - \omega t)}, \quad (12.36)$$

where α_a is the angle between $\boldsymbol{k}_a$ and the x axis. The geometrical characteristics of the emitted wave are given by

$$\frac{k_a}{k} = \frac{\sin \alpha}{\sin \alpha_a} \quad (12.37)$$

and

$$\begin{cases} \mathcal{C} \cot \alpha = \cot \alpha_a + \dfrac{1}{M_2 \sin \alpha_a} & \text{(propagative regime)} \\ \dfrac{\cot \alpha_a}{\cot \alpha_a^c} = \dfrac{\cot \alpha}{\cot \alpha_c} & \text{(nonpropagative regime)} \end{cases}, \quad (12.38)$$

and the damping factor is expressed as

$$\eta = \begin{cases} 0 & \text{(propagative regime)} \\ \dfrac{|\cot \alpha_a^c \sin \alpha_a|}{M_2} \sqrt{1 - \left(\dfrac{\cot \alpha}{\cot \alpha_c}\right)^2} & \text{(non propagative regime)} \end{cases}, \quad (12.39)$$

where α_a^c is the angle of the emitted acoustic wave when $\alpha = \alpha_c$. It can be shown that $\cos \alpha_a^c = -M_2$.

12.2.3 Incident Acoustic Waves

12.2.3.1 Fast and Slow Waves

We now consider the case of an incident acoustic wave for which we have $(\omega - k \cos \alpha U_1)^2 = a_1^2 k^2$. The method is similar to the one previously presented for incident entropy and vorticity waves, but the analysis is made a bit more complex because incident waves can be classified into two types: the *fast waves*, which propagate in the direction of the mean flow (i.e., $u_1 U_1 > 0$), and the *slow waves*, which travel in the opposite direction (i.e. $u_1 U_1 < 0$). These two families are demarcated by the stationary Mach waves with an angle of incidence α_M such that $\cos \alpha_M = -1/M_1$. Both fast and slow waves can lead to the generation of vorticity, entropy, and propagating or evanescent acoustic waves.

12.2.3.2 Emitted Entropy and Vorticity Waves

Let us first address the case of emitted entropy and vorticity waves. Using the dispersion relations of incident and emitted waves, one obtains

$$\omega = k \cos \alpha U_1 \pm a_1 k = kU_1 \left(\cos \alpha \pm \frac{1}{M_1} \right) = \tilde{k}_x U_2, \quad (12.40)$$

where $\tilde{k}_x$ can be computed in the same way as in the case of incident vorticity and entropy waves. Signs "+" and "−" in the preceding equation are related to fast and slow waves, respectively. An elegant way to compute the angle of the emitted waves is proposed by Fabre and co-workers (Fabre, Jacquin, and Sesterhenn, 2001), who introduce the angle $\alpha' \in [0, \pi]$ such that

$$\cot \alpha' = \cot \alpha + \frac{1}{M_1 \sin \alpha}. \qquad (12.41)$$

The angle α_s of the emitted entropy and vorticity waves is given by Eq. (12.24) as in the case of an incident entropy–vorticity wave, the angle of incidence α being replaced with α'. The wave-vector modulus, k_s, is still given by Eq. (12.25). The plane-wave operator associated with emitted fluctuating fields is similar to the one found for incident entropy–vorticity waves [Eq. (12.23)], the wave vector k_s being defined as previously stated.

12.2.3.3 Emitted Acoustic Waves

We now turn to the case of emitted acoustic waves. A difference with the case of incident entropy–vorticity waves is that fast and slow waves have different threshold angles, denoted α_c^+ and α_c^-, respectively. These two angles are solutions of

$$\cot \alpha_c^\pm + \frac{1}{M_1 \sin \alpha_c^\pm} = \pm \frac{\sqrt{1 - M_2^2}}{\mathfrak{C} M_2}. \qquad (12.42)$$

For fast waves, the propagative regime is associated with $\alpha \in]0, \alpha_c^+[$ and the nonpropagative regime with $\alpha \in]\alpha_c^+, \alpha_M[$. For slow waves, the propagative and nonpropagative regimes correspond to $\alpha \in]\alpha_c^-, \pi[$ and $\alpha \in]\alpha_M, \alpha_c^-[$, respectively. In both cases, the damping factor η is given by Eq. (12.39), the angle of the incident wave α being replaced with the angle α' defined in Eq. (12.41), and the wavelength k_a is still computed by solving Eq. (12.37). The angle α_a of the emitted wave is defined as follows:

$$\begin{cases} \dfrac{\cot \alpha_a}{\cot \alpha_c^a} = \dfrac{\cot \alpha'}{\cot \alpha_c^+} - \dfrac{1}{M_2}\sqrt{\left(\dfrac{\cot \alpha'}{\cot \alpha_c^+}\right)^2 - 1} & \alpha_a \in]0, \pi[\quad \alpha \in]0, \alpha_c^+[\\[2mm] \dfrac{\cot \alpha_a}{\cot \alpha_c^a} = \dfrac{\cot \alpha'}{\cot \alpha_c^+} & \alpha_a \in]0, \pi[\quad \alpha \in]\alpha_c^+, \alpha_M[\\[2mm] \dfrac{\cot \alpha_a}{\cot \alpha_c^a} = \dfrac{\cot \alpha'}{\cot \alpha_c^-} & \alpha_a \in]\pi, 2\pi[\quad \alpha \in]\alpha_M, \alpha_c^-[\\[2mm] \dfrac{\cot \alpha_a}{\cot \alpha_c^a} = \dfrac{\cot \alpha'}{\cot \alpha_c^+} + \dfrac{1}{M_2}\sqrt{\left(\dfrac{\cot \alpha'}{\cot \alpha_c^-}\right)^2 - 1} & \alpha_a \in]\pi, 2\pi[\quad \alpha \in]\alpha_c^-, \pi[\end{cases} \qquad (12.43)$$

Using these definitions for the damping rate η and the wave vector k_a, the emitted acoustic field is still of the form (12.36).

12.3 Calculation of Amplitude of Emitted Waves

Table 12.1. *First decomposition of the emitted field associated with a single plane incident wave with wavenumber k and frequency* ω

	Acoustic mode	Vorticity mode	Entropy mode
$\frac{u_2'(x,y,t)}{U_2} =$	$F e^{\iota \tilde{k}_x x} e^{\iota(k\sin\alpha y - \omega t)}$ $+$	$G e^{\iota(k\mathfrak{C}\cos\alpha x + k\sin\alpha y - \omega t)}$ $+$	0
$\frac{v_2'(x,y,t)}{U_2} =$	$H e^{\iota \tilde{k}_x x} e^{\iota(k\sin\alpha y - \omega t)}$ $+$	$I e^{\iota(k\mathfrak{C}\cos\alpha x + k\sin\alpha y - \omega t)}$ $+$	0
$\frac{p_2'(x,y,t)}{\bar{p}_2} =$	$K e^{\iota \tilde{k}_x x} e^{\iota(k\sin\alpha y - \omega t)}$ $+$	0 $+$	0
$\frac{\rho_2'(x,y,t)}{\bar{\rho}_2} =$	$\frac{K}{\gamma} e^{\iota \tilde{k}_x x} e^{\iota(k\sin\alpha y - \omega t)}$ $+$	0 $+$	$Q e^{\iota(k\mathfrak{C}\cos\alpha x + k\sin\alpha y - \omega t)}$
$\frac{T_2'(x,y,t)}{\bar{T}_2} =$	$K \frac{\gamma-1}{\gamma} e^{\iota \tilde{k}_x x} e^{\iota(k\sin\alpha y - \omega t)}$ $+$	0 $-$	$Q e^{\iota(k\mathfrak{C}\cos\alpha x + k\sin\alpha y - \omega t)}$
$\frac{s_2'(x,y,t)}{c_p} =$	0 $+$	0 $-$	$Q e^{\iota(k\mathfrak{C}\cos\alpha x + k\sin\alpha y - \omega t)}$

12.3 Calculation of Amplitude of Emitted Waves

12.3.1 General Decompositions of the Perturbation Field

Utilizing either formulation of the exponential operator and the results given in Subsection 9.1.2 dealing with the inviscid Kovasznay decomposition, the perturbation field in the downstream domain associated with a single incident plane wave can therefore be written under the general form* given in Table 12.1, where F, G, H, I, K, and Q are amplitude parameters that will be computed thanks to the boundary conditions, i.e., the linearized Rankine–Hugoniot jump conditions. The fully turbulent emitted field is recovered by summing the emitted perturbations associated with all incident waves, i.e., carrying out the summation over k, α and the wave nature.

It is worth noting that all these parameters are not independent because the fluctuations are solutions of the linearized Euler equations. Substitution into the x-momentum equation yields

$$U_1(-\iota F k \cos\alpha U_1) + \iota U_2 U_1 F \tilde{k}_x = -\iota \frac{\bar{p}_2}{\bar{\rho}_2} K \tilde{k}_x, \tag{12.44}$$

from which it follows that

$$F = \frac{1}{\gamma} \frac{1}{(\mathfrak{C} M_2)^2} \frac{\frac{\tilde{k}_x}{k}}{\cos\alpha - \frac{\tilde{k}_x}{\mathfrak{C} k}} K. \tag{12.45}$$

The y-momentum equation leads to

$$U_1(-H \iota k \cos\alpha U_1) + U_2 U_1 H \iota \tilde{k}_x = -\frac{\bar{p}_2}{\bar{\rho}_2} K \iota k \sin\alpha, \tag{12.46}$$

which can be rearranged like

$$H = \frac{1}{\gamma} \frac{1}{(\mathfrak{C} M_2)^2} \frac{\sin\alpha}{\cos\alpha - \frac{\tilde{k}_x}{\mathfrak{C} k}} K. \tag{12.47}$$

---

* It is chosen here to normalize velocity fluctuations by using U_2, and not U_1. Turning from one formulation to the other one brings in the compression factor $\mathfrak{C}$.

A last constraint is that the vortical velocity field is solenoidal, which is equivalent to

$$U_1 G \imath k \mathfrak{C} \cos \alpha + U_1 I \imath k \cos \alpha = 0, \tag{12.48}$$

leading to

$$I = -\mathfrak{C} \cot \alpha G. \tag{12.49}$$

It is recalled that the entropy, density, pressure, and temperature fluctuations are tied by the two following linearized relations:

$$\frac{\rho'_2}{\bar{\rho}_2} = \frac{1}{\gamma} \frac{p'_2}{\bar{p}_2} - \frac{s'_2}{c_p}, \tag{12.50}$$

$$\frac{T'_2}{\bar{T}_2} = \frac{\gamma - 1}{\gamma} \frac{p'_2}{\bar{p}_2} + \frac{s'_2}{c_p}. \tag{12.51}$$

The preceding system is supplemented by the normalized boundary conditions

$$\frac{1}{U_1} \frac{\partial \xi}{\partial t} = L e^{\imath(k \sin \alpha y - \omega t)}, \quad \frac{\partial \xi}{\partial y} = -\frac{L}{\cos \alpha} e^{\imath(k \sin \alpha y - \omega t)}, \tag{12.52}$$

where L is an amplitude factor for the shock displacement. The downstream perturbation field is therefore parameterized by four independent parameters, namely I, K, L, and Q. The problem is *a priori* well behaved, as there are four unknowns and four jump conditions. The problem can be recast, making the transfer coefficients $Z_F, Z_G, Z_H, Z_I, Z_K, Z_L$, and Z_Q appear, which are defined as

$$F = A Z_F, \quad G = A Z_G, \quad H = A Z_H, \quad I = A Z_I, \quad K = Z_K, \quad L = A Z_L, \quad Q = A Z_Q, \tag{12.53}$$

where A is the complex amplitude of the incident wave (A is therefore identical to the coefficient of the Fourier transform of the incident perturbation field associated with $\boldsymbol{k}$).

The decomposition given in Table 12.1 and Eq. (12.52) can also be rewritten, making the amplitude of each Kovasznay mode explicitly appear. This new expression is given in Table 12.2, where it is chosen here to use the second form of the exponential wave operator to illustrate it.

The shock-front displacement is now expressed as

$$k\xi(y, t) = Z_x e^{\imath(k \sin \alpha y - \omega t)}. \tag{12.54}$$

The coefficient ζ is defined as $\zeta = \sqrt{1 - \eta^2 + 2i\eta \cos \alpha_a}$. The four unknowns are now A_a, A_v, A_s, and A_x, i.e., the normalized amplitudes of the acoustic, vorticity, and entropy modes and shock displacement, respectively. One observes that $\zeta = 1$ in the propagative regime, whereas ζ is imaginary in the nonpropagative regime, showing that the velocity and pressure fluctuations associated with evanescent waves are not in phase.

12.3 Calculation of Amplitude of Emitted Waves

Table 12.2. *Second decomposition of the emitted field associated with a single plane incident wave with wavenumber k and frequency ω*

	Acoustic mode	Vorticity mode	Entropy mode
$\frac{u_2'(x,y,t)}{U_2} =$	$A_a \frac{\cos\alpha_a + \iota\eta}{\gamma M_2 \zeta} e^{-k_a \eta x} e^{\iota(k_a x \cos\alpha_a + k_a y \sin\alpha_a - \omega t)}$	$+ A_v \sin\alpha_s e^{\iota(k_s x \cos\alpha_s + k_s y \sin\alpha_s - \omega t)}$	$+ 0$
$\frac{v_2'(x,y,t)}{U_2} =$	$A_a \frac{\sin\alpha_a}{\gamma M_2 \zeta} e^{-k_a \eta x} e^{\iota(k_a x \cos\alpha_a + k_a y \sin\alpha_a - \omega t)}$	$- A_v \cos\alpha_s e^{\iota(k_s x \cos\alpha_s + k_s y \sin\alpha_s - \omega t)}$	$+ 0$
$\frac{p_2'(x,y,t)}{\bar{p}_2} =$	$A_a e^{-k_a \eta x} e^{\iota(k_a x \cos\alpha_a + k_a y \sin\alpha_a - \omega t)}$	$+ 0$	$+ 0$
$\frac{\rho_2'(x,y,t)}{\bar{\rho}_2} =$	$A_a \frac{1}{\gamma} e^{-k_a \eta x} e^{\iota(k_a x \cos\alpha_a + k_a y \sin\alpha_a - \omega t)}$	$+ 0$	$- A_s e^{\iota(k_s x \cos\alpha_s + k_s y \sin\alpha_s - \omega t)}$
$\frac{T_2'(x,y,t)}{\bar{T}_2} =$	$A_a \frac{\gamma-1}{\gamma} e^{-k_a \eta x} e^{\iota(k_a x \cos\alpha_a + k_a y \sin\alpha_a - \omega t)}$	$+ 0$	$+ A_s e^{\iota(k_s x \cos\alpha_s + k_s y \sin\alpha_s - \omega t)}$
$\frac{s_2'(x,y,t)}{c_p} =$	0	$+ 0$	$+ A_s e^{\iota(k_s x \cos\alpha_s + k_s y \sin\alpha_s - \omega t)}$

For an incident wave with complex amplitude A the unknown amplitudes are given by

$$A_a = A Z_a, \quad A_v = A Z_v, \quad A_s = A Z_s, \quad A_x = A Z_x, \qquad (12.55)$$

where Z_a, Z_v, Z_s, and Z_x are complex transfer functions associated with the second decomposition.

The two decompositions are tied by the following equalities:

$$k_a \eta = \tilde{k}_i, \quad k_a \cos\alpha = \tilde{k}_r, \qquad (12.56)$$

$$Z_F = Z_a \frac{\cos\alpha_a + \iota\eta}{\gamma M_2 \zeta}, \quad Z_G = Z_v \sin\alpha_s, \quad Z_H = Z_a \frac{\sin\alpha_a}{\gamma M_2 \zeta}, \qquad (12.57)$$

$$Z_I = -Z_v \cos\alpha_s, \quad Z_K = Z_a, \quad Z_Q = -Z_s. \qquad (12.58)$$

12.3.2 Calculation of Amplitudes of Emitted Waves

Amplitudes of the emitted Kovasznay modes are related to those of the incident wave through linearized jump conditions. The first step of the procedure consists of substituting Eqs. (12.5)–(12.10) into Rankine–Hugoniot relations (11.4)–(11.7) written in a frame of reference that moves at the local instantaneous speed of the shock wave and to retain terms that are of the order of $O(\epsilon)$. Then, normalizing the fluctuating quantities by using mean-flow variables, one obtains the following equations that are valid at $x = 0$:

$$\frac{1}{\mathcal{C}} \left(\frac{u_2' - \frac{\partial \xi}{\partial t}}{U_2} \right) = \left(\frac{u_2' - \frac{\partial \xi}{\partial t}}{U_1} \right) = \frac{(\gamma-1)M_1^2 - 2}{(\gamma+1)M_1^2} \left(\frac{u_1' - \frac{\partial \xi}{\partial t}}{U_1} \right)$$
$$+ \frac{2}{(\gamma+1)M_1^2} \left(\frac{T_1'}{\bar{T}_1} \right), \qquad (12.59)$$

$$\frac{1}{\mathcal{C}} \left(\frac{v_2'}{U_2} \right) = \left(\frac{v_2'}{U_2} \right) = \left(\frac{v_1'}{U_1} \right) + \frac{2(M_1^2 - 1)}{(\gamma+1)M_1^2} \frac{\partial \xi}{\partial y}, \qquad (12.60)$$

$$\left(\frac{p'_2}{\bar{p}_2}\right) = \frac{4}{(\gamma-1)M_1^2+2}\left(\frac{u'_1-\frac{\partial \xi}{\partial t}}{U_1}\right) - \frac{(\gamma-1)M_1^2+4}{(\gamma-1)M_1^2+2}\left(\frac{T'_1}{\bar{T}_1}\right), \quad (12.61)$$

$$\left(\frac{p'_2}{\bar{p}_2}\right) = \frac{4\gamma M_1^2}{2\gamma M_1^2-(\gamma-1)}\left(\frac{u'_1-\frac{\partial \xi}{\partial t}}{U_1}\right) - \frac{2\gamma M_1^2}{2\gamma M_1^2-(\gamma-1)}\left(\frac{T'_1}{\bar{T}_1}\right), \quad (12.62)$$

where all mean-flow quantities and incident fluctuating perturbations (subscript 1) are known.

The second step of the LIA procedure consists of selecting an incident wave, i.e., choosing its nature and prescribing the incident wave vector $\boldsymbol{k}$ (which is equivalent to prescribing α, k, and ω). One then obtains a set of linear equations for the amplitude of the emitted waves and the shock displacement, expressing both incident and emitted fluctuating fields by using one of the two decompositions previously presented [Table 12.1 and Eq. (12.52) or Table 12.2 and Eq. (12.54)], in which $\boldsymbol{k}_s$, $\boldsymbol{k}_a$, and the damping factor for evanescent waves are computed thanks to the ad hoc relations.

Let us take the second decomposition displayed in Table 12.2 and Eq. (12.54) as an example. This case was exhaustively described by Fabre and co-workers (Fabre, Jacquin, and Sesterhenn, 2001).

In the case of an incident entropy wave, the linearized jump conditions yield the following linear system:

$$\mathcal{A}\begin{pmatrix} Z_v \\ Z_s \\ Z_a \\ Z_x \end{pmatrix} = \begin{pmatrix} -1 \\ -\mathfrak{C} \\ 0 \\ \frac{\mathfrak{C}^2}{(\gamma-1)M_1^2} \end{pmatrix}, \quad (12.63)$$

where the matrix $\mathcal{A}$ is given by

$$\mathcal{A} = \begin{bmatrix} \sin\alpha_s & -1 & \frac{1}{\gamma} + \frac{\cos\alpha_p + i\eta}{\gamma M_2 \zeta} & i(\mathfrak{C}-1)\cos\alpha \\ 2\sin\alpha_s & -1 & \frac{M_2^2+1}{\gamma M_2^2} + 2\frac{\cos\alpha_p + i\eta}{\gamma M_2 \zeta} & 0 \\ -\cos\alpha_s & 0 & \frac{\sin\alpha_p}{\gamma M_2 \zeta} & i(1-\mathfrak{C})\sin\alpha \\ \sin\alpha_s & \frac{1}{(\gamma-1)M_2^2} & \frac{1}{\gamma M_2^2} + \frac{\cos\alpha_p + i\eta}{\gamma M_2 \zeta} & i\mathfrak{C}(1-\mathfrak{C})\cos\alpha \end{bmatrix}. \quad (12.64)$$

For an incident vorticity wave, one obtains

$$\mathcal{A}\begin{pmatrix} Z_v \\ Z_s \\ Z_a \\ Z_x \end{pmatrix} = \begin{pmatrix} \sin\alpha \\ 2\mathfrak{C}\sin\alpha \\ -\mathfrak{C}\cos\alpha \\ \mathfrak{C}^2\sin\alpha \end{pmatrix}, \quad (12.65)$$

where the matrix $\mathcal{A}$ is given by Eq. (12.64) as in the case of an incident entropy wave.

12.3 Calculation of Amplitude of Emitted Waves

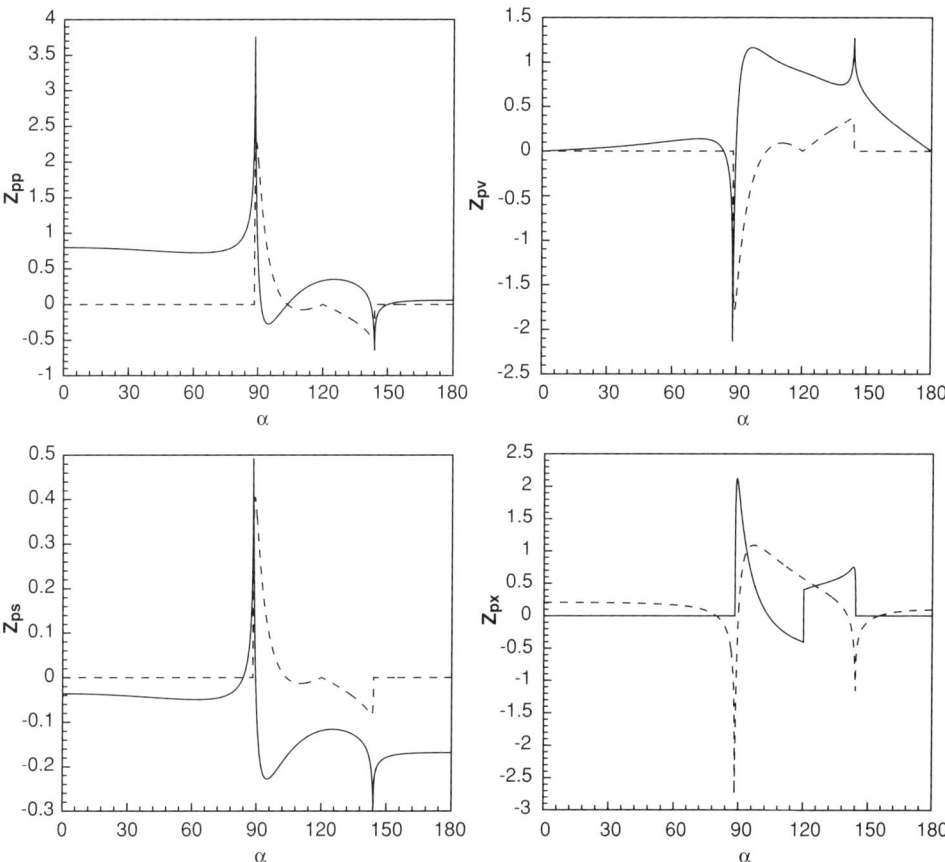

Figure 12.2. LIA transfer functions vs. the angle of the incident wave α in the case of an incident plane acoustic wave. Top: Z_a (left) and Z_v (right); Bottom: Z_s (left) and Z_x (right). Solid line: real part, dashed line: imaginary part. Courtesy of D. Fabre, IMFT, France.

The case of incident acoustic waves leads to

$$\mathcal{A}' \begin{pmatrix} Z_v \\ Z_s \\ Z_a \\ \frac{\sin\alpha}{\sin\alpha'} Z_x \end{pmatrix} = \begin{bmatrix} \frac{1}{\gamma} + \frac{\cos\alpha}{\gamma M_1} \\ \frac{\mathfrak{C}}{\gamma}\left(\frac{M_1^2+1}{M_1^2} + \frac{2\cos\alpha}{M_1}\right) \\ \mathfrak{C}\frac{\sin\alpha}{\gamma M_1} \\ \frac{\mathfrak{C}^2}{\gamma}\left(\frac{1}{M_1^2} + \frac{\cos\alpha}{M_1}\right) \end{bmatrix}, \tag{12.66}$$

where the matrix $\mathcal{A}'$ is given by Eq. (12.64) with α replaced with the angle α' defined in Eq. (12.41) in the last column:

$$\mathcal{A}' = \begin{bmatrix} \sin\alpha_s & -1 & \frac{1}{\gamma} + \frac{\cos\alpha_p + \imath\eta}{\gamma M_2 \zeta} & \imath(\mathfrak{C}-1)\cos\alpha' \\ 2\sin\alpha_s & -1 & \frac{M_2^2+1}{\gamma M_2^2} + 2\frac{\cos\alpha_p + \imath\eta}{\gamma M_2 \zeta} & 0 \\ -\cos\alpha_s & 0 & \frac{\sin\alpha_p}{\gamma M_2 \zeta} & \imath(1-\mathfrak{C})\sin\alpha' \\ \sin\alpha_s & \frac{1}{(\gamma-1)M_2^2} & \frac{1}{\gamma M_2^2} + \frac{\cos\alpha_p + \imath\eta}{\gamma M_2 \zeta} & \imath\mathfrak{C}(1-\mathfrak{C})\cos\alpha' \end{bmatrix}. \tag{12.67}$$

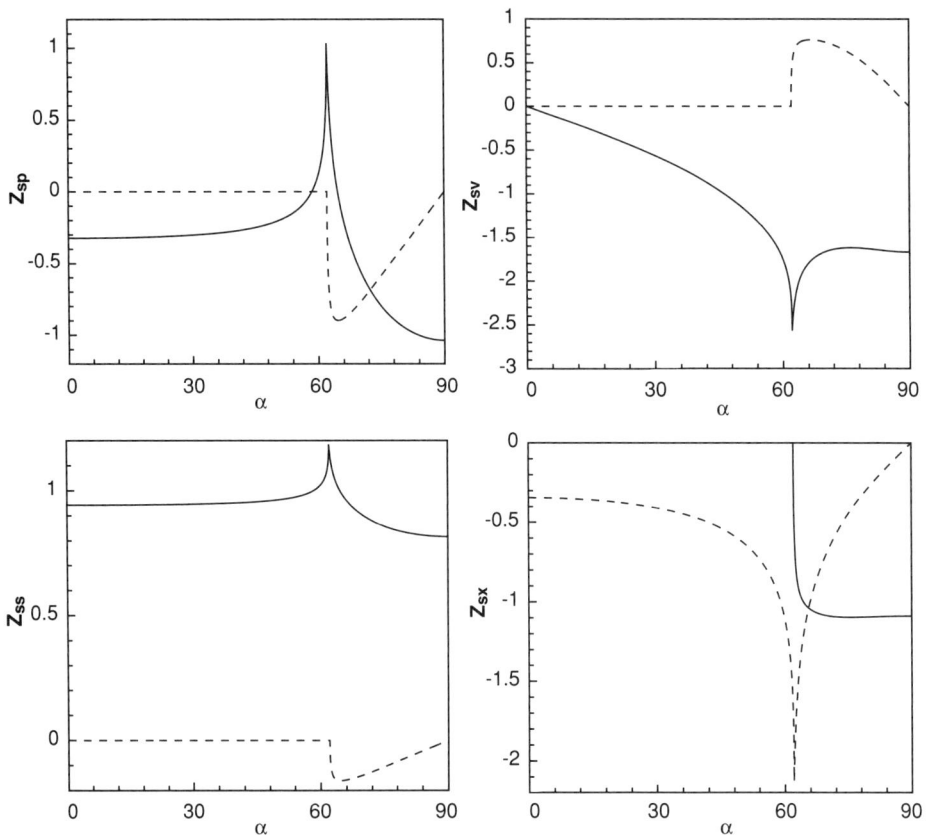

Figure 12.3. LIA transfer functions versus the angle of the incident wave α in the case of an incident plane entropy wave. Top: Z_a (left) and Z_v (right); Bottom: Z_s (left) and Z_x (right). Solid line: real part, dashed line: imaginary part. Courtesy of D. Fabre, IMFT, France.

A careful examination of these results reveals that the transfer functions depend only on γ, M_1 and α, i.e., they are not functions of the wavenumber k. This is coherent with the fact that the shock wave is assumed to have no intrinsic length scale.

The computed transfer functions, angles of emission, and damping factor are displayed in Figs. 12.2–12.5.

12.4 Reconstruction of the Second-Order Moments

The general formulation of the fluctuating field behind the shock wave makes it possible to derive expressions for the second-order statistical moments and to emphasize some fundamental differences between fields generated by propagating and evanescent waves (Mahesh, Lele, and Moin, 1995; Mahesh, Moin, and Lele, 1996). As previously stated, the existence of evanescent waves with significant amplitude just behind the shock yields the existence of a thin region with peculiar behavior, referred to as the near field.

12.4 Reconstruction of the Second-Order Moments

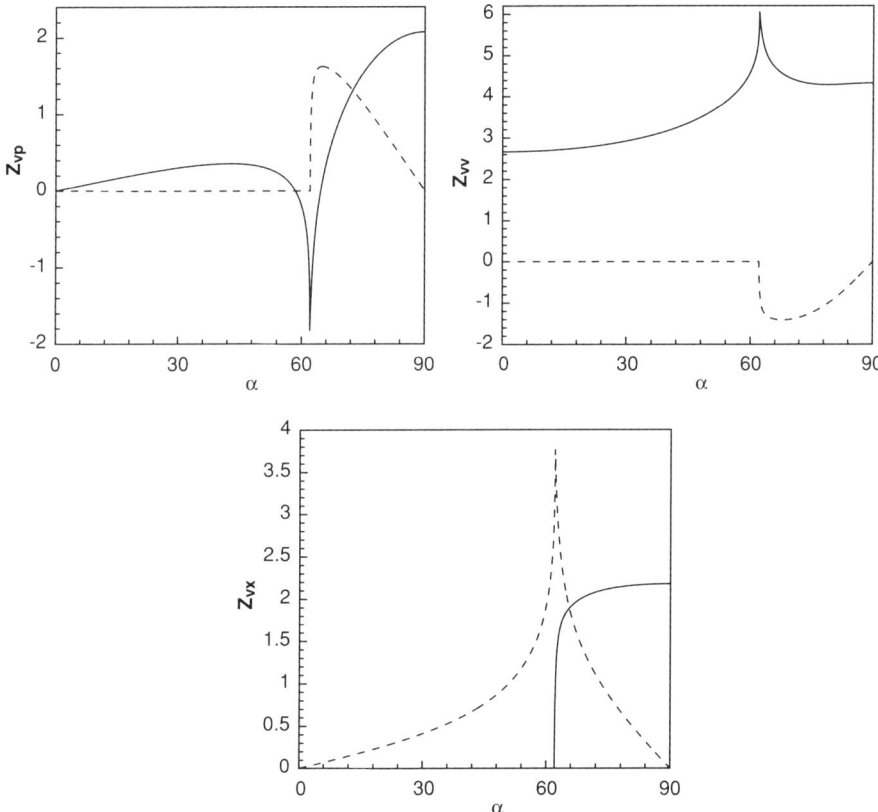

Figure 12.4. LIA transfer functions versus the angle of the incident wave α in the case of an incident plane vorticity wave. Top: Z_a (left) and Z_v (right); Bottom: Z_x. Solid line: real part, dashed line: imaginary part. Courtesy of D. Fabre, IMFT, France.

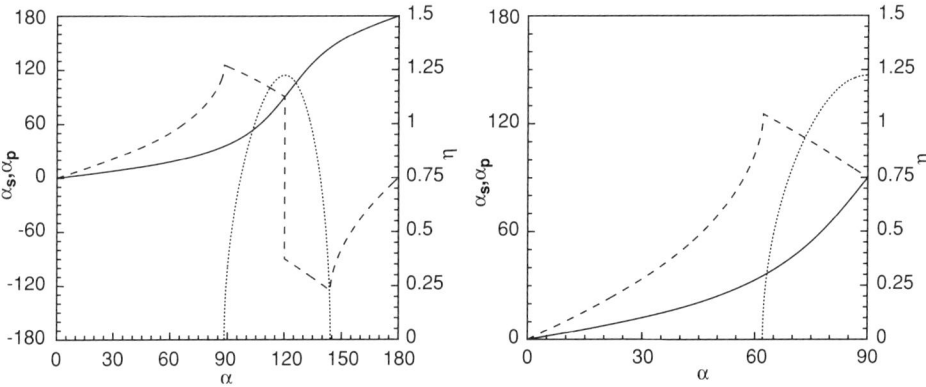

Figure 12.5. Angle of emitted waves and damping parameter η as functions of the angle of the incident wave. Left: incident acoustic wave; Right: incident entropy–vorticity wave. Dotted line: η; solid line: α_s; dashed line: α_a. Courtesy of D. Fabre, IMFT, France.

12.4.1 Case of a Single Incident Wave

Let us first consider the kinetic energy associated with the shock normal velocity component u'_2. The mean fluctuating kinetic energy is defined as

$$\overline{u'^2_2}(x) \equiv \overline{u'_2 u'^*_2}(x), \tag{12.68}$$

where the superscript * and the overbar denote the complex conjugate and averaging over time and the homogeneous direction y, respectively. Using the decomposition given in Table 12.1 along with Eq. (12.53), one obtains

$$\frac{\overline{u'^2_2}(x)}{U_1^2} = \left(|Z_F|^2 e^{\imath(\tilde{k}-\tilde{k}^*)x} + |Z_G|^2 + Z_F Z_G^* e^{\imath(\tilde{k}-k\mathfrak{C}\cos\alpha)x} + Z_F^* Z_G e^{-\imath(\tilde{k}^*-k\mathfrak{C}\cos\alpha)x} \right) |A|^2. \tag{12.69}$$

If $\tilde{k}$ is real, i.e., if the acoustic waves are not damped and the regime is of the propagative type, the preceding expression is identical to

$$\frac{\overline{u'^2_2}(x)}{U_1^2} = |A|^2 \left\{ |Z_F|^2 + |Z_G|^2 \right.$$
$$+ 2[(Z_F)_r(Z_G)_r + (Z_F)_i(Z_G)_i]\cos(\tilde{k} - k\mathfrak{C}\cos\alpha)x$$
$$\left. - 2[(Z_F)_i(Z_G)_r + (Z_F)_r(Z_G)_i]\sin(\tilde{k} - k\mathfrak{C}\cos\alpha)x \right\}, \tag{12.70}$$

where the subscripts i and r are related to the imaginary and real components, respectively; in the case of evanescent waves $\tilde{k}$ is complex, and Eq. (12.69) leads to

$$\frac{\overline{u'^2_2}(x)}{U_1^2} = |A|^2 \left\{ |Z_F|^2 e^{-2\tilde{k}_i x} + |Z_G|^2 \right.$$
$$+ 2[(Z_F)_r(Z_G)_r + (Z_F)_i(Z_G)_i]e^{-2\tilde{k}_i x}\cos(\tilde{k}_r - k\mathfrak{C}\cos\alpha)x$$
$$\left. - 2[(Z_F)_i(Z_G)_r + (Z_F)_r(Z_G)_i]e^{-2\tilde{k}_i x}\sin(\tilde{k}_r - k\mathfrak{C}\cos\alpha)x \right\}. \tag{12.71}$$

A similar expression is derived for the kinetic energy of the transverse component, replacing $\overline{u'^2_2}(x)$ with $\overline{v'^2_2}(x)$, Z_F with Z_H, and Z_G with Z_I in Eqs. (12.69)–(12.71). It is seen that in the propagative regime the kinetic energy has spatially uniform contributions from both the vortical and acoustic modes and an oscillating component whose argument is equal to the phase difference between them. In the nonpropagative regime it has a spatially uniform contribution from the vorticity mode only, an exponentially decaying monotone acoustic component, and exponentially damped components that are due to the correlation between the vorticity and the acoustic mode. Temperature and density variances exhibit a similar behavior, as they are made of a combination of the acoustic and entropy modes.

The vorticity, which depends on the sole vorticity mode, has spatially uniform distribution. The same conclusion holds for the entropy variance, as entropy

12.4 Reconstruction of the Second-Order Moments

fluctuations depend on the sole entropy mode that has no evanescent regime. Quantities that depend on the sole acoustic mode, such as pressure and dilatation, exhibit dramatic changes when switching from one regime to the other one. Let us illustrate this point by using the pressure variance. In the propagative regime, the pressure variance is spatially uniform:

$$\frac{\overline{p_2'^2}(x)}{\bar{p}_2^2} = |Z_K|^2 |A|^2, \tag{12.72}$$

whereas in the nonpropagative regime it experiences an exponential decay,

$$\frac{\overline{p_2'^2}(x)}{\bar{p}_2^2} = e^{-2\tilde{k}_i x} |Z_K|^2 |A|^2. \tag{12.73}$$

Looking at these expressions, one can distinguish between the near-field and the far-field behaviors. In the near-field region, exponentially decaying terms are still important. The width of this region scales as $O(1/k)$, showing that it is very thin. In the far-field region, all decaying terms are negligible.

12.4.2 Case of an Incident Turbulent Isotropic Field

It is important to note that the developments just presented are valid in the case of a single incident wave. We now address the problem of an incident turbulent isotropic field. In the original Cartesian frame of reference, the undisturbed plane shock is assumed to lie in the y–z plane and the normal mean flow is parallel to the x axis.

Let us first consider an isotropic solenoidal velocity field generated by vorticity modes. The energy spectrum of the incident velocity field is assumed to be of the form

$$E_{ij}(\mathbf{k}) = \frac{E(k)}{4\pi k^2} \left(\delta_{ij} - \frac{k_i k_j}{k^2} \right), \tag{12.74}$$

where $E(k)$ is the 3D energy spectrum. Because the velocity field is solenoidal, the velocity vector associated with the wave vector $\mathbf{k}$ is orthogonal to $\mathbf{k}$ and may have a component orthogonal to the plane spanned by $\mathbf{k}$ and the x axis. Therefore the 2D analysis previously presented must be extended by introducing cylindrical coordinates (x, r, ϕ) (see Fig. 12.6). Here, x still refers to the x axis in the original Cartesian frame of reference. Let u_x' and u_r' be the fluctuating-velocity components in the x and r directions, and u_ϕ' be the one in the ϕ direction. The latter is normal to the x–r plane spanned by the x axis and the wave vector $\mathbf{k}$. We also introduce α, the angle between $\mathbf{k}$ and the x axis in the x–r plane, and ϕ is defined as the angle between $\mathbf{k}$ and the y axis in the y–z plane.

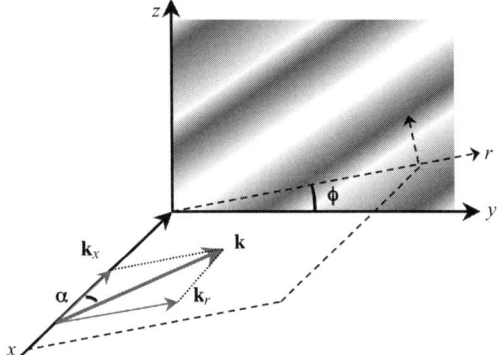

Figure 12.6. Schematic view of the LIA for shock–turbulence interaction: reference frames for the treatment of an isotropic incident field.

The change of frame of reference implies that

$$u' = u'_x, \tag{12.75}$$

$$u'_r = v' \cos \phi + w' \sin \phi, \quad u'_\phi = -v' \sin \phi + w' \cos \phi, \tag{12.76}$$

$$v' = u'_r \cos \phi - u'_\phi \sin \phi, \quad w' = u'_r \sin \phi + u'_\phi \cos \phi, \tag{12.77}$$

and the elemental volumes of integration are tied by the following relation:

$$d\mathbf{k} = k^2 \sin \alpha \, d\alpha \, d\phi \, dk, \tag{12.78}$$

with $k \in [0, +\infty]$, $\alpha \in [-\pi/2, \pi/2]$ and $\phi \in [0, 2\pi]$, along with

$$k_1 = k \cos \alpha, \quad k_2 = k \cos \phi \sin \alpha, \quad k_2 = k \sin \phi \sin \alpha. \tag{12.79}$$

The 2D analysis previously presented obviously holds for the u'_x and u'_r components, as the x–r plane corresponds to the x–y plane in the 2D analysis. The angle α is the same in the two coordinate systems. The method consists therefore of applying the LIA procedure to the u'_x and u'_r components, whereas the u'_ϕ component is left unmodified by the interaction because it is tangential to the shock wave, according to Eq. (11.6). An important point is that the amplitude A of the incident wave that appears in the 2D analysis [see Eqs. (12.53) and (12.55)] is now related to the magnitude of the velocity vector in the x–r plane, leading to

$$|A| = \frac{|u'_1|}{\sin \alpha}, \quad |A|^2 = \frac{E(k)}{4\pi k^2}. \tag{12.80}$$

Now, introducing the spectral components of kinetic energy behind the shock waves in both the Cartesian and cylindrical coordinates systems, Eqs. (12.75)–(12.77) yield the following relations:

$$E_{11} = E_{xx}, \tag{12.81}$$

$$E_{22} + E_{33} = E_{rr} + E_{\phi\phi}, \tag{12.82}$$

$$E_{22} = \cos^2 \phi E_{rr} + \sin^2 \phi E_{\phi\phi} - \sin 2\phi E_{r\phi}, \tag{12.83}$$

$$E_{33} = \sin^2 \phi E_{rr} + \cos^2 \phi E_{\phi\phi} + \sin 2\phi E_{r\phi}. \tag{12.84}$$

12.4 Reconstruction of the Second-Order Moments

The spectral components are linked to the amplitude factor A by

$$|A|^2 = \frac{E^1_{11}}{\sin^2\alpha} = E^1_{11} + E^1_{rr} = E^1_{nn} - E^1_{\phi\phi}, \qquad (12.85)$$

where the superscript "1" refers to the incident perturbation field and E_{nn} is the total kinetic-energy density, from which it follows that

$$E^1_{\phi\phi} = E^1_{nn} - |A|^2 = \frac{E(k)}{4\pi k^2} \qquad (12.86)$$

and

$$E_{22} + E_{33} = E_{rr} + \frac{E(k)}{4\pi k^2}. \qquad (12.87)$$

The streamwise kinetic energy behind the shock wave is defined as

$$\frac{\overline{u_2'^2}(x)}{U_1^2} = \int_{k=0}^{+\infty} \int_{\alpha=-\pi/2}^{+\pi/2} \int_{\phi=0}^{2\pi} E_{11} k^2 \sin\alpha \, d\alpha \, d\phi \, dk, \qquad (12.88)$$

where E_{11} is the amplitude of the emitted wave associated with a single incident wave vector computed with the 2D theory [see Eq. (12.69)]:

$$\frac{E_{11}}{U_1^2} = |A|^2 \Big[|Z_F|^2 e^{\iota(\tilde{k}-\tilde{k}^*)x} + |Z_G|^2$$
$$+ Z_F Z_G^* e^{\iota(\tilde{k}-k\mathcal{C}\cos\alpha)x} + Z_F^* Z_G e^{-\iota(\tilde{k}^*-k\mathcal{C}\cos\alpha)x} \Big]. \qquad (12.89)$$

Remarking that E_{11} is independent of ϕ and that it is symmetric about $\alpha = 0$, one finds that Eq. (12.88) simplifies as

$$\frac{\overline{u_2'^2}(x)}{U_1^2} = 4\pi \int_{k=0}^{+\infty} \int_{\alpha=0}^{+\pi/2} E_{11} k^2 \sin\alpha \, d\phi \, dk. \qquad (12.90)$$

If the transfer functions have the same value at all scales, i.e., if the ratios between the amplitudes of the three Kovasznay modes are scale independent, meaning that kinetic energy, entropy, and pressure are null or exhibit the same spectrum, then the emitted kinetic energy does not depend on the shape of the spectrum of the incident field:

$$\frac{\overline{u_2'^2}(x)}{U_1^2} = \left\{ \int_{\alpha=0}^{+\pi/2} \Big[|Z_F|^2 e^{\iota(\tilde{k}-\tilde{k}^*)x} + |Z_G|^2 + Z_F Z_G^* e^{\iota(\tilde{k}-k\mathcal{C}\cos\alpha)x} \right.$$
$$\left. + Z_F^* Z_G e^{-\iota(\tilde{k}^*-k\mathcal{C}\cos\alpha)x} \Big] \sin\alpha \, d\alpha \right\} \int_{k=0}^{+\infty} E(k) \, dk. \qquad (12.91)$$

An expression for the far field is derived by neglecting evanescent terms and using the fact the correlation terms in the propagative regime integrates to zero:

$$\left(\overline{\frac{u_2'^2}{U_1^2}}\right)_{\text{far-field}} = \left[\int_{\alpha=0}^{\alpha_c}(|Z_F|^2+|Z_G|^2)\sin\alpha\,d\alpha + \int_{\alpha_c}^{+\pi/2}|Z_G|^2\sin\alpha\,d\alpha\right]$$
$$\times \int_{k=0}^{+\infty} E(k)\,dk, \qquad (12.92)$$

where α_c is the threshold angle that demarcates propagating and evanescent emitted acoustic waves (see Section 12.2). Expressions similar to Eq. (12.90) for the transverse components of kinetic energy are

$$\frac{\overline{v_2'^2}(x)}{U_1^2} = \frac{\overline{w_2'^2}(x)}{U_1^2} = \frac{1}{2}4\pi \int_{k=0}^{+\infty}\int_{\alpha=0}^{+\pi/2}\left[E_{rr}+\frac{E(k)}{4\pi k^2}\right]k^2\sin\alpha\,d\alpha\,dk. \quad (12.93)$$

Comparing Eqs. (12.90) and (12.93), one sees that the velocity field is statistically axisymmetric behind the shock wave. Simplified expressions for the far field can also be easily derived.

Expressions for the emitted vorticity components can be derived in the same manner. It is first recalled that, for an incident solenoidal isotropic velocity field, the following relations hold for the Cartesian components of the incident vorticity field:

$$\frac{\overline{\Omega_2'^2}(x)}{U_1^2} = \frac{\overline{\Omega_3'^2}(x)}{U_1^2} = \int_{\alpha=0}^{+\pi/2}(2-\sin^2\alpha)\sin\alpha\,d\alpha \int_{k=0}^{+\infty}\frac{k^2 E(k)}{2}dk. \quad (12.94)$$

It is recalled that the shock normal component of vorticity is not modified during the interaction with the shock wave. The two transverse components of the emitted vorticity field are given by

$$\frac{\overline{\Omega_2'^2}(x)}{U_1^2} = \frac{\overline{\Omega_3'^2}(x)}{U_1^2} = 2\int_{k=0}^{+\infty}\int_{\alpha=-\pi/2}^{+\pi/2}\int_{\phi=0}^{2\pi}\Big(\cos^2\phi\,\{\cos^2\alpha\mathfrak{C}^2|Z_I|^2$$
$$-2\mathfrak{C}\cos\alpha\sin\alpha[(Z_I)_r(Z_G)_r+(Z_I)_i(Z_G)_i]\}$$
$$+\cos^2\alpha\sin^2\phi+\sin^2\alpha|Z_G|^2\Big)k^4|A|^2\sin\alpha\,d\alpha\,d\phi\,dk.$$
$$(12.95)$$

If the transfer functions are the same at all scales, the preceding equation becomes

$$\frac{\overline{\Omega_2'^2}(x)}{U_1^2} = \frac{\overline{\Omega_3'^2}(x)}{U_1^2} = \int_{\alpha=0}^{+\pi/2}\big\{\cos^2\alpha\mathfrak{C}^2|Z_I|^2+\sin^2\alpha|Z_G|^2+\cos^2\alpha\phi$$
$$-2\mathfrak{C}\cos\alpha\sin\alpha[(Z_I)_r(Z_G)_r+(Z_I)_i(Z_G)_i]\big\}\sin\alpha\,d\alpha$$
$$\times \int_{k=0}^{+\infty}\frac{k^2 E(k)}{2}dk. \quad (12.96)$$

It is worth noting that the emitted vorticity field does not depend on the spectrum shape in this case.

12.5 A posteriori Assessment of LIA

The case of an incident isotropic acoustic field is much simpler, as the associated velocity field is purely dilatational and is therefore parallel to the wave vector. Consequently, the velocity vector is entirely contained in the x–r plane in the cylindrical coordinate system previously introduced, i.e., u'_ϕ is identically zero. Therefore this case will not be detailed because it can be directly treated by use of the 2D LIA theory, the 2D x–y plane being taken equal to the x–r plane. It is just recalled that the isotropic spectral tensor is now defined by

$$E^1_{ij}(\mathbf{k}) = \frac{E(k)}{8\pi k^2} \frac{k_i k_j}{k^2} \qquad (12.97)$$

and not by Eq. (12.74).

The last case is the one of scalar quantities, such as pressure and entropy. This case is much simpler because no projection is needed. As an example, the variance of the emitted pressure wave is given by

$$\frac{\overline{p'^2_2}(x)}{\bar{p}^2_2} = 2\pi \int_{k=0}^{+\infty} \int_{\alpha=0}^{+\pi/2} E_{pp} k^2 \sin\alpha \, d\alpha \, dk, \qquad (12.98)$$

where E_{pp} is computed with relations (12.72) and (12.73) for a single incident plane wave, i.e., $E_{pp} = |Z_K|^2 |A|^2$ in the propagative regime and $E_{pp} = e^{-2\bar{k}_i x} |Z_K|^2 |A|^2$ in the nonpropagative regime.

Statistics related to the oscillations of the shock front can also be derived in a similar manner. For the shock speed variance, one obtains

$$\frac{\overline{(\partial \xi/\partial t)^2}}{U_1^2} = 4\pi \int_{k=0}^{+\infty} \int_{\alpha=0}^{+\pi/2} |Z_L|^2 |A|^2 k^2 \sin\alpha \, d\alpha \, dk, \qquad (12.99)$$

which can be rearranged as follows if the modal amplitude ratios are scale independent for an incident isotropic solenoidal velocity field:

$$\frac{\overline{(\partial \xi/\partial t)^2}}{U_1^2} = \int_{\alpha=0}^{+\pi/2} |Z_L|^2 \sin\alpha \, d\alpha \int_{k=0}^{+\infty} E(k) dk. \qquad (12.100)$$

The rms oscillation amplitude is given by

$$\overline{\xi^2} = 4\pi \int_{k=0}^{+\infty} \int_{\alpha=0}^{+\pi/2} \frac{|Z_L|^2}{k^2 \cos^2\alpha} |A|^2 k^2 \sin\alpha \, d\alpha \, dk, \qquad (12.101)$$

from which it follows that

$$\overline{\xi^2} = \int_{\alpha=0}^{+\pi/2} \frac{|Z_L|^2}{\cos^2\alpha} \sin\alpha \, d\alpha \int_{k=0}^{+\infty} \frac{E(k)}{k^2} dk \qquad (12.102)$$

for incident isotropic vortical turbulence if transfer function values are scale independent.

12.5 A posteriori Assessment of LIA

As said at the beginning of this chapter, the LIA is observed to compare well with DNS and experimental data if $M_t^2 < 0.1(M_1^2 - 1)$. But a finer analysis reveals that some discrepancies arise in predicting the emitted field associated with incident waves with an angle of incidence close to the critical angle that demarcates propagating and evanescent waves. The rationale for that is that the energy of the emitted waves is high near the critical angle (the transfer functions exhibit a strong, narrow peak for $\alpha = \alpha_c$), leading to a breakdown of the small-perturbation hypothesis. The amplitude of the fluctuations being too high, the linear approximation should be refined to account for additional nonlinear effects.

Bibliography

FABRE, D., JACQUIN, L., AND SESTERHENN, J. (2001). Linear interaction of a cylindrical entropy spot with a shock wave, *Phys. Fluids* **13**, 2403–2422.

LEE, S., LELE, S. K., AND MOIN, P. (1993). Direct numerical simulation of isotropic turbulence interacting with a weak shock wave, *J. Fluid Mech.* **251**, 533–562.

LEE, S., LELE, S. K., AND MOIN, P. (1997). Interaction of isotropic turbulence with shock waves: Effect of shock strength, *J. Fluid Mech.* **340**, 225–247.

MAHESH, K., LELE, S. K., AND MOIN, P. (1995). The interaction of an isotropic field of acoustic waves with a shock wave, *J. Fluid Mech.* **300**, 383–407.

MAHESH, K., LELE, S. K., AND MOIN, P. (1997). The influence of entropy fluctuations on the interaction of turbulence with a shock wave, *J. Fluid Mech.* **334**, 353–379.

MAHESH, K., MOIN, P., AND LELE, S. K. (1996). The interaction of a shock wave with a turbulent shear flow, *Report No. TF-69*, Department of Mechanical Engineering, Stanford University.

MOORE, F. K. (1954). Unsteady oblique interaction of a shock wave with a plane disturbance, *Tech. Rep. 2879*, NACA.

RIBNER, H. S. (1953). Convection of a pattern of vorticity through a shock wave, *Tech. Rep. 1164*, NACA.

13 Linear Theories. From Rapid Distortion Theory to WKB Variants

13.1 Rapid Distortion Theory for Homogeneous Turbulence

The essentials for RDT were introduced in Chapter 5. More details are given in this chapter for deriving basic equations and calculating statistics. One considers the pure incompressible case in the presence of a mean flow with admissible mean-velocity-gradient matrix $A_{ij}(t)$. In the linear limit, the fluctuating field (u'_i, p') satisfies modified equation (2.29) with the advection–distortion parts written in terms of $A_{ij}(t)$:

$$\underbrace{\frac{\partial u'_i}{\partial t} + A_{jk} x_k \frac{\partial u'_i}{\partial x_j}}_{\text{advection}} + A_{ij} u'_j + \frac{\partial p'}{\partial x_i} = 0. \tag{13.1}$$

13.1.1 Solutions for ODEs in Orthonormal Fixed Frames of Reference

The solution of Eq. (13.1) is most easily obtained by Fourier analysis, with elementary components of the form

$$u'_i(\mathbf{x}, t) = a_i(t) \exp[\imath \mathbf{k}(t) \cdot \mathbf{x}], \tag{13.2}$$

$$p'(\mathbf{x}, t) = b(t) \exp[\imath \mathbf{k}(t) \cdot \mathbf{x}], \tag{13.3}$$

where $\imath^2 = -1$. Evolution equations for the amplitudes are easily obtained from Eq. (13.1). They can be written as follows:

$$\frac{da_i}{dt} + \imath a_i x_j \left(\frac{dk_j}{dt} + A_{nj} k_n\right) + A_{ij} a_j + \imath k_i b = 0. \tag{13.4}$$

Time dependency of the wave vector $\mathbf{k}$ allows one to simplify the advection term by setting

$$\frac{dk_i}{dt} + A_{ji} k_j = 0. \tag{13.5}$$

According to the mathematical treatment of partial derivative equations, this equation also gives the characteristic lines of the operator $\frac{\partial}{\partial t} - A_{nj} k_n \frac{\partial}{\partial k_j}$ in Fourier space. Because the mean trajectories are the characteristic lines of the advection operator $\frac{\partial}{\partial t} + A_{jn} x_n \frac{\partial}{\partial x_j}$, it is clear that eikonal equation (13.5) is the counterpart of

13.1 Rapid Distortion Theory for Homogeneous Turbulence

$\dot{x}_i - A_{ij}x_j$ in Fourier space, changing A_{ij} into $-A_{ji}$ and x into k. A more physical interpretation will be given in the following discussion.

More classically, the pressure contribution b is removed from consideration by the incompressibility constraint, which is equivalent to the orthogonality condition (see Subsection 2.5.4):

$$k_i a_i = 0. \qquad (13.6)$$

Applying the projection operator

$$P_{in} = \delta_{in} - \frac{k_i k_n}{k^2} \qquad (13.7)$$

to Eq. (13.4), one finds

$$\frac{da_i}{dt} - \frac{k_i}{k^2} k_n \frac{da_n}{dt} + P_{in} A_{nj} a_j = 0,$$

with $k_i \frac{da_i}{dt} = -\frac{dk_i}{dt} a_i = A_{ni} k_n a_i$, by using relation (13.5), so that a is found to satisfy the following ODE:

$$\frac{da_i}{dt} = -\underbrace{\left(\delta_{in} - 2\frac{k_i k_n}{k^2} \right) A_{nj}}_{M_{ij}} a_j. \qquad (13.8)$$

ODEs (13.5) and (13.8) are referred to as the *Townsend* or *Kelvin–Townsend equations*. In the matrix **M**, the factor $\frac{k_i k_n}{k^2}$ reflects the contribution from the fluctuating pressure term, with a prefactor 2 that takes into account advection in wave space. As usual, spectral analysis allows for a straightforward treatment of the nonlocal dependence of pressure on velocity. The time dependency of the wave vector represents the convection of the plane wave $\exp[\imath k(t) \cdot x]$ by the base flow. Both the direction and magnitude of k change as wave crests rotate and approach or separate from each other because of mean-velocity gradients.

General solutions that are valid for arbitrary initial data are expressed as follows in terms of linear transfer matrices:

$$k_i(t) = B_{ij}(t, t_0) k_j(t_0), \qquad (13.9)$$

$$a_i(t) = G_{ij}(t, t_0) a_j(t_0), \qquad (13.10)$$

where the universal values for **B** and **G** at $t = t_0$ are subsequently recalled.

In the preceding equations, it is perhaps clearer to specify the wave-vector dependency in a and **G**, especially if we combine elementary solutions of the form given by Eq. (13.2) by Fourier synthesis. As a consequence, the RDT solution can be expressed as follows:

$$\widehat{u}_i(k(t), t) = G_{ij}(k, t, t_0) \widehat{u}_j[k(t_0), t_0], \qquad (13.11)$$

in which the Green's function is eventually determined by the initial conditions*

$$G_{ij}(\mathbf{k}, t_0, t_0) = \delta_{ij} - \frac{K_i K_j}{K^2}, \quad K_i = k_i(t_0). \tag{13.12}$$

Of course, **G** is governed by the same equation as $\mathbf{a}$ or $\hat{\mathbf{u}}$,

$$\dot{G}_{ij} = -M_{in} G_{nj},$$

in which the overdot is a convenient notation to indicate that $\mathbf{k}$ has to be considered as a time-dependent vector in $\mathbf{G}(\mathbf{k}, t, t_0)$.

As for the time dependency of the wave vector, **B** can be directly linked to the Cauchy matrix **F** through

$$B_{ij}(t, t_0) = F_{ji}^{-1}(t, t_0). \tag{13.13}$$

The general definition of the Cauchy matrix for an arbitrary flow has been given previously.

When similar characteristic lines for the advection term are compared in both physical and in Fourier space, i.e., $x_i = F_{ij} X_j$ (trajectory) and $k_i = F_{ji}^{-1} K_j$, their close analogy is obvious. One recovers the conservation of $\mathbf{k} \cdot \mathbf{x}$ (= $\mathbf{K} \cdot \mathbf{X}$) along trajectories and also conservation of the plane wave $\exp[\imath \mathbf{k}(t) \cdot \mathbf{x}(t)]$.

13.1.2 Using Solenoidal Modes for a Green's Function with a Minimal Number of Components

For instance, a reduced Green's function can be used in the Craya–Herring frame of reference, as

$$u^{(\alpha)}[\mathbf{k}(t), t] = g_{\alpha\beta}(\mathbf{k}, t, t') u^{(\beta)}[\mathbf{k}(t'), t']. \tag{13.14}$$

The reduced Green's function $g_{\alpha\beta}$, with only four components instead of nine for G_{ij}, can be generated by solving

$$\dot{u}^{(\alpha)} + m_{\alpha\beta} u^{(\beta)} = 0, \tag{13.15}$$

with

$$m_{\alpha\beta} = e_i^{(\alpha)} A_{ij} e_j^{(\beta)} - \dot{e}_i^{(\alpha)} e_i^{(\beta)}. \tag{13.16}$$

Here, the Einstein convention of summation over repeated indices is used for both Latin (varying from 1 to 3) and Greek (taking only the values 1 and 2) indices. The Craya–Herring frame being orthonormal, it characterizes a solid-body motion when $\mathbf{k}$ is time dependent, so that the entrainment term is simply

$$\dot{e}_i^{(\alpha)} e_i^{(\beta)} = \epsilon_{\alpha 3 \beta} \Omega_E, \quad \Omega_E = -e_i^{(2)} A_{ij} e_j^{(1)} - n_i A_{ij} e_j^{(1)} \frac{k}{k_\perp}, \tag{13.17}$$

---

* A different initialization $G_{ij} = \delta_{ij}$ was prescribed in Townsend (1956, 1976). Equation (13.12) presents some advantages, as $k_i G_{ij} = 0$ can be satisfied at any time, and the RDT Green's function can be more easily related to Kraichnan's response function.

13.1 Rapid Distortion Theory for Homogeneous Turbulence

with $k = |\mathbf{k}|$, $k_\perp = \sqrt{k^2 - (\mathbf{k}\cdot\mathbf{n})^2}$. The last result is simplified if the axial vector $\mathbf{n}$ is chosen along one of the eigendirections of $\mathbf{A}$ (Cambon, Teissèdre, and Jeandel, 1985), so that

$$\Omega_E = -e_i^{(2)} A_{ij} e_j^{(1)}. \tag{13.18}$$

This condition was always fulfilled, and an optimal choice of $\mathbf{n}$, if needed, has been discussed.

Similar equations can be found in terms of helical modes, but they present no additional interest, except if the mean vorticity, or the mean absolute vorticity in the presence of an additional Coriolis force, is completely dominant. A first instance is given in Chapter 4.

13.1.3 Prediction of Statistical Quantities

Throughout this book, we are using an approach that reconciles and simplifies two different approaches:

- The way initiated by Townsend in RDT, who addressed the very definition of a deterministic Green's function, prior to any statistical calculation, without using a local frame of reference. Recent studies (since 1986) in the community of hydrodynamic stability theory have essentially the same starting point, being disconnected from applications to statistics anyway.
- The way initiated by Craya, who put the emphasis on solving statistical equations for second-order and third-order spectral tensors, but with a reduced number of components obtained by projecting these equations and these tensors in the eponymous frame of reference. Craya never considered the fluctuating-velocity field in Fourier space, which Herring did (1974), recovering the local frame of reference, but restricting the statistical approach to axial symmetry only.

Recall that the second way yielded several studies (dealing with single-time two-point and single-point second-order statistics) by J. N. Gence and co-workers (mainly his Ph.D. students), following Courseau and Loiseau (1978) for pure strain and pure shear, namely "pure" rotation, rotating shear, and buoyant flows (without mean stratification). Related publications, mainly written in French, can be obtained from the authors on request.

About the first way, it is perhaps useful to discuss some points dealing with terminology. The time-dependent Fourier modes given in Eq. (13.2), when recovered in the hydrodynamic stability community, were often referred to as "Kelvin modes," ignoring their use in RDT (engineering community) and considering that the first instance of such modes was given by Lord Kelvin. This may be true, but the terminology is misleading, given the huge number of Kelvin modes and Kelvin waves called into play in stability analyses. Let us illustrate this with two examples: For a specialist in aerodynamics, a Kelvin wave is a localized inertial wave confined in the core of a vortex, whereas for an oceanographer, it is a much more complex wave, also dealing with the variation of the Coriolis parameter with latitude, the stable density stratification, and the topology. We hope to have clarified the point here that "mean-Lagrangian–Fourier modes," or "Fourier modes advected by the mean" would be less confusing than Kelvin modes. It is not necessary to recall chronologically all the authors who used a similar approach, from Kelvin to Batchelor and Proudman (1954).

As an example of statistical calculation, the RDT equation for second-order statistics is readily derived from Eq. (13.11) by using Eq. (2.64), leading to

$$\hat{R}_{ij}[\boldsymbol{k}(t),t] = G_{in}(\boldsymbol{k},t,t_0)G_{jm}(\boldsymbol{k},t,t_0)\hat{R}_{nm}[\boldsymbol{k}(t_0),t_0]. \qquad (13.19)$$

Given an initial solution $\hat{R}_{ij}$ at $t = t_0$, one can compute it at later times using relation (13.19), provided that the Green's function $G_{ij}(\boldsymbol{k}, t, t')$ is known. The determination of G_{ij} is thus the main problem in applying homogeneous RDT in practice.

Applications mainly concern second-order, two-point, and one-point correlations, with many results about the history of the RST when the initial data are chosen to be isotropic. Similarly, a "rapid" pressure–strain-rate tensor and dissipation tensor can be calculated.

It is not difficult to reintroduce a laminar viscous effect or an efficiently modeled damping effect, as illustrated by Townsend (1956, 1976) and by some RDT applications reported in Chapters 6 and 8. The viscous factor was calculated in the most general way by Cambon, Teissèdre, and Jeandel (1985) as

$$V_0(\boldsymbol{k},t) = \exp\left(-\nu \int_0^t k^2(t)dt\right),$$

so that

$$V_0(\boldsymbol{k},t) = \exp\left[-k_l k_n \int_{t_0}^t F_{li}(t,t')F_{ni}(t,t')dt'\right]. \qquad (13.20)$$

This equation involves a quadratic form in terms of $\boldsymbol{k}$, using the group relations of $\mathbf{F}$ [such as $\mathbf{F}(t,t').\mathbf{F}(t',t'') = \mathbf{F}(t,t'')$] and can also be given in terms of the "material" wave vector $\boldsymbol{K}$.

Equation (13.19) can also be extended to any order n, thanks to the existence of general solution (13.11), by means of a product of n Green's functions: linear solutions for third-order correlations are considered per se in Chapter 4, for instance, and incorporated in triadic closures for evaluating nonlinear transfer terms.

Complete inviscid RDT equations for the RST and the integral length scales are subsequently given for isotropic initial data:

$$\overline{u_i' u_j'} = \frac{\mathcal{K}(0)}{4\pi} \iint_{|\boldsymbol{K}|=1} e_i^{(\alpha)}(\boldsymbol{k})g_{\alpha\gamma}(\boldsymbol{k},t)g_{\beta\gamma}(\boldsymbol{k},t)e_j^{(\beta)}(\boldsymbol{k})d^2\boldsymbol{K} \qquad (13.21)$$

and

$$\mathcal{E}_{ij}^{(n)} = \overline{u_i' u_j'} L_{ij}^{(n)} \quad \text{(no summation on } i, j\text{)}$$

$$= \frac{1}{2}\int_0^\infty \frac{E(k)}{k}dk \iint_{k_l=0} e_i^{(\alpha)}(\boldsymbol{k})g_{\alpha\gamma}(\boldsymbol{k},t)g_{\beta\gamma}(\boldsymbol{k},t)e_j^{(\beta)}(\boldsymbol{k})d^2\boldsymbol{K}. \qquad (13.22)$$

13.1 Rapid Distortion Theory for Homogeneous Turbulence

The latter equation makes use of the additional condition $K_l = k_l$, as for instance $k_1 = K_1$ and $k_3 = K_3$ for the pure plane shear-flow case in Chapter 6).

13.1.3.1 Initial-Value Problem or Forcing?

Instead of considering the initial-value problem, one may add a forcing term to the linear equation in order to mimic a nonlinear effect and/or a source of noise.

From the general solution for the fluctuating field,

$$\hat{u}(k(t),t) = G_{ij}(k,t,t_0)\hat{u}[k(t_0),t_0] + \int_{t_0}^{t} G_{ij}(k,t,t')f_j[k(t'),t']dt', \quad (13.23)$$

it is possible to derive a related statistical solution. The contribution from the initial value can even be omitted if the Green's function is rapidly decaying. Interesting applications can be found by looking at the RST, or even at its subgrid scale counterpart in LES. Choosing an isotropic white noise for the forcing, with

$$\langle f_i^*(p,t)f_j(k,t')\rangle = \frac{B(k)}{4\pi k^2}\delta^3(k-p)\delta(t-t'), \quad (13.24)$$

the RST obeys the following linear response solution:

$$\overline{u_i'u_j'}(t) = \frac{\int_0^\infty B(k)dk}{4\pi}\int_{t_0}^{t}dt'\left[\iint_{|K|=1} e_i^{(\alpha)}(k)g_{\alpha\gamma}(k,t)g_{\beta\gamma}(k,t)e_j^{(\beta)}(k)d^2K\right]. \quad (13.25)$$

Even if the structure of this equation is similar to its counterpart for the initial-value problem, it may be more interesting to have a steady state at large time, forgetting the intermediate history of the RST. For instance, a bounded steady state $\overline{u_i'u_j'}(\infty)$ is found in the absence of exponential or algebraic growth for $g_{\alpha\beta}$ in the inviscid case. More interesting, a steady state can be obtained even in some cases with exponential growth by reintroducing the viscous factor (13.20). For instance, the convergence of the temporal integral is analyzed in Cambon (1982) for a large class of flows subjected to strain-dominated mean flow, i.e., in the presence of a hyperbolic instability. Because the mean advection is seen in wave space ($K \to k$), this asymptotic analysis depends on only the infrared part of the spectrum $B(k)$, as $B(k) \sim k^x$ when $k \to 0$.

Decomposing Eq. (13.25), or its viscous counterpart, into contributions related or not (in a local way) to the mean-velocity gradient reveals typical coefficients, such as efficient viscosity and *anisotropic kinetic alpha* (AKA) coefficients. For instance, in a quasi-parallel flow dominated by mean shear, one can formally write the Taylor series expansion of $\overline{u_i'u_2'} = R_i$ with respect to the gradient of the large-scale flow,

$$R_i(\infty) = \Lambda_i U_0 - \nu_T S\delta_{i1}, \ldots,$$

where two coefficients Λ_i and ν_T appear (N. Le Provost, private communication). Turbulent viscosity is well known, but RDT can also suggest negative values of ν_T, e.g., in rotating shear flows. Some recent applications of RDT to subgrid scale modeling by B. Dubrulle and co-workers appear to be related to such a calculation of an eddy viscosity.

The first term with Λ_i is known as the Λ effect or AKA effect (Frisch, She, and Sulem, 1987). Surprisingly, this term cannot be removed by Galilean invariance. It can exist only in anisotropic helical turbulence.

13.1.4 RDT for Two-Time Correlations

RDT can be used for evaluating the spectrum of the variance of a passive scalar concentration s', subject to a mean scalar gradient $\overline{s}$:

$$\dot{s}' = \frac{\partial s'}{\partial t} + u'_j \frac{\partial s'}{\partial x_j} = -\frac{\partial \overline{s}}{\partial x_j} u_j,$$

in which the material derivative is replaced with a simple Eulerian time derivative, whereas linear dynamics is used only for the bearer velocity field $\boldsymbol{u}'$. Given the strong analogy of the fluctuating trajectory equation, $\dot{x} = u'_i$, with that of the previously mentioned scalar, this approach can be used for calculating mean-square displacements, or single-particle diffusion, the particle being only a fluid element.

A different approach was initiated by Kaneda and Ischida (2000), who calculated two-time velocity correlations by means of RDT. Homogeneous RDT cannot give access to the Lagrangian two-time correlations, because the "Lagrangian" Fourier mode can afford oversimplified "mean" trajectories, but not the "fluctuating" ones. On the other hand, a simplified Corrsin hypothesis can be advocated for replacing the Lagrangian two-time velocity second-order correlations with their Eulerian counterpart. The two ways, either linearizing both scalar and velocity equations to derive second-order single-time mixed correlations, or applying RDT to two-time second-order velocity correlations by using the simplified Corrsin hypothesis, yield the same final result for single-particle dispersion, but the second is much less demanding about physical assumptions. The simplified Corrsin hypothesis is less stringent than the crude assumption of equating $\dot{s}'$ and $\partial s'/\partial t$ from the very beginning.

The only advantage of the first way is to illustrate what information is gained using the following more sophisticated model: to incorporate the linear operators in a synthetic model of turbulence, usually referred to as kinematic simulation (KS), and to compute individiual random trajectories from the synthetic velocity field that include linear (RDT) dynamics.

Applications to single-particle diffusion by rotating stratified turbulence are performed in Cambon et al. (2004), with comparison of the statistical RDT model, the KS + RDT model, and DNS.

13.2 Zonal RDT and Short-Wave Stability Analysis

The condition of extensional mean flow, having a velocity-gradient matrix **A** uniform in the whole space, is very stringent, as is the statistical homogeneity (spatial invariance of any centered multipoint moment related to the fluctuating flow).

13.2 Zonal RDT and Short-Wave Stability Analysis

It is therefore useful to generalize the linear solutions in the presence of more complex "base" (or mean) flows, either for extending stability analyses, or for modeling nonhomogeneous turbulence.

13.2.1 Irrotational Mean Flows

For irrotational mean flows, for instance, potential flows, a tractable form of inviscid RDT in physical space can be based on the solution of the equation that governs the fluctuating vorticity ($\omega_i = \epsilon_{ijk} u'_{j,k}$), a particular Kelvin equation for the linearized case without mean vorticity:

$$\omega_i(\boldsymbol{x},t) = F_{ji}(\boldsymbol{X},t,t_0)\omega_j(\boldsymbol{X},t_0). \tag{13.26}$$

As we have seen in Chapters 5 and 10, the related Weber equation,

$$u'_i(\boldsymbol{x},t) = F_{ji}^{-1}(\boldsymbol{X},t,t_0)u'_j(\boldsymbol{X},t_0) + \frac{\partial \phi(\boldsymbol{x},t)}{\partial x_i}, \tag{13.27}$$

is particularly useful.

The mean flow may involve complex trajectories, which are defined by

$$x_i = \bar{x}_i(\boldsymbol{X}, t_0, t) \quad \text{with} \quad \dot{\bar{\boldsymbol{x}}} = \bar{u}_i(\boldsymbol{x},t), \tag{13.28}$$

in which Lagrangian coordinates $\boldsymbol{X}$ denote the initial position at time t_0 of a particle, which reaches the position $\boldsymbol{x}$ at time t, and the overdot holds for the related "mean" material derivative. The Cauchy matrix $\boldsymbol{F}$ does not need to be defined again. Of course, the complete solution of type (5.21) requires that the potential term on the right-hand side of (13.27) be expressed in terms of initial data. This can be done by using an incompressibility condition with relevant boundary conditions, and even applications to compressible flows are possible (Goldstein, 1978), as already discussed in Chapter 10. Of course, in the general incompressible case, integral nonlocal dependency, as in (5.21), reappears through the solution for ϕ in Eq. (13.27).

13.2.2 Zonal Stability Analysis With Disturbances Localized Around Base-Flow Trajectories

As soon as the mean flow is rotational, equations such as (13.26) or (13.27) are no longer valid to tackle inhomogeneous RDT. Assuming weak inhomogeneity, considerable progress can be made without the need for irrotational mean flow, although simplifications occur in this case. As discussed earlier, turbulence that is fine-scaled compared with the overall dimensions of the flow can be treated under RDT by following a notional particle advected by the mean flow. Thus the results obtained for strictly homogeneous turbulence can be extended to the weakly inhomogeneous case, but with a mean-velocity-gradient matrix $A_{ij}(t)$ that reflects the $\partial \bar{u}_i / \partial x_j$ seen by the moving particle.

Even if the Green's function related to the canonical base flow (2.46)–(2.47) can give interesting information for linear stability analysis and short-time development of turbulence, this problem is somewhat unphysical in the absence of typical length scales for variation of the base-flow gradients and disturbances. For instance, the Green's function in (13.11) depends on only the orientation, but not on the modulus, of the wave vector. Rather than considering perturbations with an arbitrary wavelength k^{-1} in the presence of the flow (2.46), it is more physical to consider a base flow whose velocity gradients vary over a typical length scale ℓ, and to restrict the validity of the zonal stability analysis to perturbations with much shorter wavelengths, i.e., $k^{-1} \ll \ell$. In so doing, the disturbance field should locally experience advection and distortion effects by the base flow, similar to the effects of an extensional flow with space-uniform gradients. Given *a priori* a length scale separation between base and disturbance flows, one can imagine looking through a mathematical magnifying glass in the vicinity of real base trajectories. This idea has been formalized in the context of flow stability (see the short-wave "geometric optics" of Lifschitz and Hameiri, 1991) using an asymptotic approach based on the WKB method, which is traditionally used to analyze the theoretical ray limit (i.e., short waves) in wave problems. The perturbation solution is written as

$$u'_i(\mathbf{x}, t) = a_i(\mathbf{x}, t) \exp[\imath \Phi(\mathbf{x}, t)/\epsilon], \tag{13.29}$$

with a similar expression for the fluctuating pressure, with amplitude $b(x, t)$, where Φ is a real phase function, ϵ is a small parameter expressing the small scale of the "waves" represented by Eq. (13.29), and $a_i(x, t)$ and $b(x, t)$ are complex amplitudes that are expanded in powers of ϵ according to the WKB technique: $a_i = a_i^{(0)} + \epsilon a_i^{(1)} + \cdots$. Inserting (13.29) into linearized equations (2.29) and (2.30) yields

$$\dot{\Phi} a_i^{(0)} + b^{(0)} \frac{\partial \Phi}{\partial x_i} = 0$$

and $k_i a_i^{(0)} = 0$ at the leading ϵ^{-1} order. Consequently, it is found that $b^{(0)} = 0$ and that

$$\dot{\Phi} = \frac{\partial \Phi}{\partial t} + \bar{u}_j \frac{\partial \Phi}{\partial x_j} = 0, \tag{13.30}$$

i.e., the wave crests of Eq. (13.29) are convected by the mean flow, whose trajectories are given by (13.28). It is then apparent that (13.29) is locally a plane-wave Fourier component of wavenumber

$$k_l(\mathbf{x}, t) = \epsilon^{-1} \frac{\partial \Phi}{\partial x_i}. \tag{13.31}$$

The spatial derivatives of $\dot{\Phi} = 0$ yield an Eikonal equation:

$$\dot{k}_i = -A_{ji}(t) k_j, \tag{13.32}$$

where, as before, $A_{ij} = \partial \bar{u}_i / \partial x_j$ and the dot represents the mean-flow material derivative $\partial/\partial t + \bar{u}_i \partial/\partial x_i$. Finally, at the next ϵ^0 order, one obtains

$$\dot{a}_i^{(0)} = -M_{ij}(t) a_j^{(0)}, \tag{13.33}$$

13.2 Zonal RDT and Short-Wave Stability Analysis

with M_{ij} as in (13.8), after elimination of the pressure by use of the leading-order incompressibility condition $k_i a_i^{(0)} = 0$.

Equations (13.32) and (13.33) have exactly the same form as the basic equations of homogeneous RDT (Townsend's equations) and therefore, together with Eq. (13.28), describe the weakly inhomogeneous case at leading order. The only difference is that, rather than being related to simple time derivatives, the dots represent mean-flow material derivatives, implying that one should follow mean-flow trajectories that differ from one to another. In homogeneous RDT, the different classes of disturbances are only labeled by the direction of the initial wave vector $K = k(t_0)$, and all trajectories, such as ψ = constant in (2.49), are equivalent. In the zonal RDT approach, it is necessary to add the Lagrangian coordinate vector X for labeling different trajectories. In agreement with classic continuum mechanics, one has

$$d\bar{x}_i = F_{ij} dX_j + \bar{u}_i dt \tag{13.34}$$

when differentiating the mean-trajectory equation $x = \bar{x}(X, t_0, t)$, so that $\Phi = 0$ and (13.32) correspond to

$$k \cdot \delta x = K \cdot \delta X, \qquad k_i(X, t) = F_{ji}^{-1}(X, t, t_0) K_j, \tag{13.35}$$

which generalizes (13.9)–(13.13). The latter equations actually correspond to $k \cdot x = K \cdot X$. It is perhaps useful to rewrite the complete system of equations, exhibiting all parameters and dependent variables (Godeferd, Cambon, and Leblanc, 2001):

$$\dot{x}_i = \bar{u}_i(x), \tag{13.36}$$

$$\dot{k}_i = -\frac{\partial \bar{u}_j}{\partial x_i} k_j(X, t), \tag{13.37}$$

$$\dot{a}_i = -\left(\delta_{in} - 2\frac{k_i k_n}{k^2}\right) \frac{\partial \bar{u}_n}{\partial x_j} a_j(X, t), \tag{13.38}$$

with solutions (13.35) for k, and

$$a_i(X, k, t) = G_{ij}(X, K, t, t_0) a_j(X, K, t, t_0).$$

Typical applications are presented in Chapter 8. It is even possible to consider the base flow as unsteady and to directly use the Cauchy matrix for solving (13.38) with (13.35), without the need for numerical solutions of Eqs. (13.36) and (13.37), as illustrated by Guimbard and Leblanc (2006).

13.2.3 Using Characteristic Rays Related to Waves Instead of Trajectories

The WKB method by Lifschitz and Hameiri (1991) is different from those developments that lead to "geometric optics" and "physical optics." Accordingly, the first one (Lifschitz and Hameiri, 1991) is referred to as *short-wave* linear stability analysis, or zonal WKB RDT everywhere in this book, whereas only the second is denoted as "geometric optics" from now on. The starting point of "true geometric optics" is similar to Eq. (13.29), but the spatiotemporal evolution is assumed to be slow, so that x and t in (13.29) ought to be replaced with ϵx and ϵt, respectively. In "true geometric optics," the leading-order approximation is $\epsilon^{(0)}$, and the inhomogeneous dispersion law is exhibited, for instance in injecting Eq. (13.29) with $(x \to \epsilon x, t \to \epsilon t)$ in linearized equations (2.29) and (2.30) so that

$$\dot{\Phi} = \pm \sigma(\nabla \Phi) - \nabla \Phi \cdot \bar{u}, \tag{13.39}$$

in which, for the sake of simplicity of notation, spatial and temporal operators concern "slow" variables. Stressing, as before, that $k = \nabla \Phi$, a Hamiltonian function can be defined as

$$\dot{\Phi} = H(k, x) = \pm \sigma(k) - k \cdot \bar{u}. \tag{13.40}$$

Accordingly, a Hamiltonian dynamical system is derived:

$$\dot{x} = \frac{\partial H}{\partial k}, \tag{13.41}$$

$$\dot{k} = -\frac{\partial H}{\partial x}. \tag{13.42}$$

Because H includes both the dispersion frequency and the Doppler frequency that is due to convection by the mean flow, or $k \cdot \bar{u}$, the right-hand side of Eq. (13.41) is the sum of group and convection velocities, and the related characteristic line is the ray along which energy propagates. Applications of the Hamiltonian dynamical system are used by Galmiche (1999), for instance, in the case of gravity waves propagating in an inhomogeneous medium. Note that the dispersion law cannot appear at the leading order in the Lifschitz–Hameiri WKB method, so that the previous system of Hamiltonian equations reduces to trajectory equation (13.36) and to Eikonal equation (13.32), respectively, with $H = -k \cdot \bar{u}$. The dispersion law is recovered in Lifschitz and Hameiri (1991) at the next order by means of the solution for amplitude equation (13.38), similar to RDT solutions of Chapters 4 and 7. Using the development in terms of slow spatiotemporal variables, in true geometric optics, the next order (ϵ^1) leads to the "physical optics" approximation, which yields conservation of wave action.

In the same context of gravity waves, promising perspectives, with transport of statistical spectra with nonlinear effects and diffusion, are offered by Carnevale and Frederiksen (1983). In the latter work, the Hamiltonian function that appears in Eq. (13.40) is affected by nonlinear dynamics in connection with a simplified version of DIA, and the role of resonant triad interactions is displayed.

Another interesting field of application is aeroacoustics, as a zonal RDT along trajectories can be applied to a weakly compressible flow: Both vortical and entropic modes can be considered, but the acoustic mode is always missed because it is not a short-wavelength one at low Mach numbers. On the other hand, it is possible to extend the zonal analysis along trajectories (linked to velocity u) by a ray method along acoustic rays, which are linked to $u + ak$ and $u + ak$.

13.3 Application to Statistical Modeling of Inhomogeneous Turbulence

The stability analysis framework is no longer discussed in the following discussion for the sake of brevity, and only incompressible turbulence is considered.

Generalization of the Craya equation can be sought in the presence of an arbitrary mean flow, in deriving a complete equation for the two-point velocity-correlation tensor with centered position from Eq. (2.29):

$$R_{ij}(r, x, t) = \overline{u'_i(x - r/2)u'_j(x + r/2)}. \tag{13.43}$$

A Fourier transform can be applied with respect to the separation distance r, so that the equation for the hybrid spectral–physical tensor $\hat{R}(k, x, t)$ can be displayed. Equations for both $\mathbf{R}$ and $\hat{\mathbf{R}}$ are very complicated. Correlations involving the pressure cannot be expressed in terms of velocity only, as in Eq. (2.81), especially if boundary conditions have to be taken into account. Hence it is necessary to add some assumptions or to introduce some multiscale approach. The remaining necessary assumption is the separation of spectral and physical space dependencies of the correlations, for example by treating the statistical inhomogeneity as weak. Even for homogeneous turbulence, going beyond the isotropic case entails a high computational cost for two-point simulations using classical nonlinear closures, a cost that is not insignificant compared with that of DNS. Thus it is currently unattractive to solve the full set of equations resulting from closures such as DIA, TFM, or EDQNM in the inhomogeneous case without simplifications.

An alternative approach can take inhomogeneity into account by means of the basis set of modes used to express the fluctuations, while as far as possible maintaining the structure of equations of the correlation matrix similar to that of the homogeneous case. The modes that substitute for Fourier components may, for instance, be chosen to satisfy the boundary and incompressibility conditions. Accordingly, strong inhomogeneity that is due to solid boundaries can be accommodated by the very definition of the fluctuation modes. This approach is illustrated by the recent work of Turner (1999), who considered the problem of channel flow by using suitably chosen modes whose amplitude equations are analogous to those of Fourier modes in the homogeneous case and that were closed by means of a random phase approximation. The normal modes of the linear problem might well be good candidates in this type of approach.

13.3.1 Transport Models Along Mean Trajectories

Simplified equations for $\hat{R}(k, x, t)$ are suggested by the short-wave analysis of Subsection 13.2.2. In turbulent flows, the fluctuating field is not the single component (13.29), but instead consists of a random superposition of such components. As one might expect, given the behavior of the underlying local Fourier components previously described, it can be shown that, at leading order, weakly inhomogeneous turbulence evolves according to

$$\dot{\hat{R}}_{ij} + M_{ik}\hat{R}_{kj} + M_{jk}\hat{R}_{ik} = 0, \qquad (13.44)$$

where the dot now represents the operator

$$\frac{\partial}{\partial t} + \bar{u}_i \frac{\partial}{\partial x_i} - \frac{\partial \bar{u}_j}{\partial x_i} k_j \frac{\partial}{\partial k_i} \qquad (13.45)$$

and expresses both convection by the mean flow and evolution of the wavenumber of individual Fourier components according to Eq. (13.32). Spectral evolution equation (13.44) corresponds to the RDT limit of its homogeneous equivalent, i.e., the Craya equation, provided that the dot operator is interpreted appropriately. Thus, following the mean flow, the leading-order, local spectral tensor $\hat{R}_{ij}(\mathbf{k}, \mathbf{x}, t)$ behaves as in homogeneous RDT, being given in terms of its initial values and the RDT Green's function. The obvious way to incorporate nonlinearity and viscosity into this description is to use

$$\dot{\hat{R}}_{ij} + M_{ik}\hat{R}_{kj} + M_{jk}\hat{R}_{ik} = T_{ij} - D_{ij} - 2\nu k^2 \hat{R}_{ij} \qquad (13.46)$$

rather than Eq. (13.44) to describe spectral evolution, where T_{ij} could be modeled by a homogeneous spectral closure. For the sake of completeness, the tensor D_{ij} would typically represent inhomogeneous diffusion across the mean streamline.

An interesting alternative, as proposed by Nazarenko, Kevlahan, and Dubrulle (1999), is to derive weakly inhomogeneous RDT by using a Gabor transform and related WKB development. A small parameter like ϵ in Eq. (13.29) appears. It is the ratio of the wavelength of the Fourier mode to the length of its Gaussian envelope. The interest of this method is not to derive the equations for the wave vector and the amplitude of the fluctuating-velocity field (the method previously presented does the job in a simpler and more general way), but to calculate a space-dependent RST by integrating $\hat{R}_{ij}(\mathbf{x}, \mathbf{k}, t)$ as in Eq. (13.21). Consequently, the nonlinear term that expresses the feedback from the RST in (2.27) can be evaluated (it is zero in pure homogeneous RDT).

13.3.2 Semiempirical Transport "Shell" Models

This approach, discussed in Godeferd, Cambon, and Scott (2001), is mainly illustrated by semiempirical transport models, which treat the dependency with respect to the position variable by analogy with one-point modeling. These models cannot incorporate all the information coming from general equation (13.46), but they retain some element of its structure. They are very far from the "shell models" presented by, e.g., Bohr et al. (1998), but they share with them the property that the spectral dependency is retained only through the modulus of the wave vector. Accordingly, it is assumed that primitive equations for $\hat{R}(\mathbf{k}, \mathbf{x}, t)$ are integrated over spherical shells of radius k. Because of spherical averaging, one has to forget the idea of recovering the asymptotic RDT limit, even in the homogeneous case, and one needs to model the "rapid" terms comprising distortion and pressure–strain

13.4 Conclusions, Recent Perspectives

	Euler (fully nonlinear)	Linearized (WKB RDT)	Homogeneous RDT	Pressure-released RDT
Trajectory	Any	Any Coarse grained $\delta\mathbf{x} = F\delta\mathbf{X}$	Linear $\mathbf{x} = F\mathbf{X}$	
Transported scalar (gradient)	$\nabla s(\mathbf{x},t) = (F^t)^{-1}\nabla s(\mathbf{X},0)$	Eikonal solution $\mathbf{k} = (F^t)^{-1}\mathbf{K}$ general case : $\mathbf{k}(\mathbf{X},t)$	Eikonal solution $\mathbf{k} = (F^t)^{-1}\mathbf{K}$	
Kelvin equation	$\omega(\mathbf{x},t) = F\omega(\mathbf{X},0)$	Valid only if $A = A^t$	idem	
Weber equation	$\mathbf{u}(\mathbf{x},t) = (F^t)^{-1}\mathbf{u}(\mathbf{X},0)$ $+\nabla\phi$	Valid only if $A = A^t$	idem	$\mathbf{u}(\mathbf{x},t) = H\mathbf{u}(\mathbf{X},0)$ in irrotational cases : $H = (F^t)^{-1}$

Figure 13.1. Sum of the main useful relations for RDT analysis. It is recalled that **F** is the solution of $\dot{F} = AF$. Transposition is denoted as $\mathbf{A}^t$, instead of $\tilde{\mathbf{A}}$ in the text.

correlations, modeling that is unnecessary in the fully anisotropic theory. Transport models for the joint physical–spectral space energy spectrum $E(k, \mathbf{x})$ have been developed that describe inhomogeneity in a way similar to the diffusive terms in the $k - \varepsilon$ model, but allow a better treatment of dissipation, calculated from the energy spectrum. Examples include the inhomogeneous EDQNM model of Burden (1991), the SCIT (Simplified Closure for Inhomogeneous Turbulence) model developed at Lyon (Touil, Bertoglio, and Parpais, 2000) and the LWN (Local Wave Number) model developed at Los Alamos (Clark and Zemach, 1995). These approaches are extensively discussed in Sagaut, Deck, and Terracol (2006).

As a useful compromise between RSM and subgrid-scale modeling, with seamless transition from RANS to LES, the partially integrated Reynolds stress modeling by Chaouat and Schiestel (2005, 2007) deserves attention. The underlying spectral formalism is not based on closures but on heuristic arguments, whereas spatial Taylor expansions are used for the position coordinates in physical space.

13.4 Conclusions, Recent Perspectives Including Subgrid-Scale Dynamics Modeling

Some analytical relations are summarized in Fig. 13.1, particularly for displaying the role of the Cauchy matrix in both homogeneous and zonal RDT (Eikonal equation) in the general case, and its particular involvement for irrotational mean (base) flows.

Application of RDT to subgrid-scale modeling appears to be attractive, but it is probably premature to report related studies by Leonard or by Dubrulle and coworkers, previously touched on in this chapter, and LES is largely outside the scope of this book. The interested reader is referred to Sagaut (2005) for an exhaustive presentation.

One can just mention a direct use of the Cauchy matrix for deriving pressure-released simplified solutions for the transport of the subgrid-scale stress tensor, very similar to what is done for the RST. This way of improving LES, recently discussed by C. Meneveau (invited talk, ETC11 conference, Porto, June, 2007), is in the line of the seminal study by Crow (1968), and, of course, is in agreement with many instances of pressure-released "solutions" given in this book. We think that this approach is valid for a strain-dominated coarse-grain flow, but is more questionable in the case of a vorticity-dominated flow. In applying the pressure-released approach to a rotational coarse-grain flow, it must be borne in mind that $\mathbf{F}$ must be replaced with $\mathbf{H}$, whose history involves transposed $\mathbf{A}$, not to mention that the full RDT solution can be very different from its pressure-released counterpart (one uses $\dot{\mathbf{F}} = \mathbf{A}\dot{\mathbf{F}}$ and $\dot{\tilde{\mathbf{F}}}^{-1} = -\tilde{\mathbf{A}}\tilde{\mathbf{F}}^{-1}$).

Other applications deal with the transport of the coarse-grain mean-velocity gradient, which is governed by the following equation (as the mean-flow gradient for homogeneous RDT, but with additional diffusive terms, not given explicitly):

$$\dot{A}_{ij} + A_{in}A_{nj} = \frac{\partial^2 P}{\partial x_i \partial x_j} + \text{diffusive and subgrid terms.}$$

A spherical form of the pressure Hessian in terms of Eulerian coordinates gives immediately a closed form for the nondiffusive part of the equation, such as

$$\frac{\partial^2 P}{\partial x_i \partial x_j} = \frac{1}{3}\nabla^2 P \delta_{ij} = \left(\dot{A}_{nn} + A_{mn}A_{nm}\right)\frac{\delta_{ij}}{3},$$

but incorrect dynamical behavior was shown.[†] Recent studies by Chevillard and Meneveau support the proposal that the sphericity of the pressure Hessian is better assessed in Lagrangian coordinates, leading to

$$\frac{\partial^2 P}{\partial x_i \partial x_j} = F_{ni}^{-1} F_{mj}^{-1} \frac{\partial^2 P}{\partial X_n \partial X_m} = \frac{1}{3}\frac{\partial^2 P}{\partial X_n \partial X_n} F_{mi}^{-1} F_{mj}^{-1},$$

so that a particular Cauchy–Green tensor is displayed.

Recall that the additive decomposition in terms of trace, symmetric deviator, and antisymmetric parts of $\mathbf{A}$ can be replaced with a multiplicative decomposition for $\mathbf{F}$, $\mathbf{F} = J^{1/3} \cdot \mathbf{Q}' \cdot \mathbf{S}'$ ($J = 1$ in the incompressible case) or $\mathbf{F} = J^{1/3}\mathbf{S}\mathbf{Q}$, with symmetric ($\mathbf{S}, \mathbf{S}'$) and orthogonal ($\mathbf{Q}, \mathbf{Q}'$) factors being not the same according to the order of the multiplication. A pure symmetric factor is displayed, with $\tilde{\mathbf{F}}^{-1} \cdot \mathbf{F}^{-1} = \mathbf{S}^{-2}$, or not ($\tilde{\mathbf{F}}^{-1} \cdot \mathbf{F}^{-1} = \mathbf{Q}\mathbf{S}^{-2}\tilde{\mathbf{Q}}$).

[†] But it can be seen that even oversimplified admissible mean-velocity gradients used in homogeneous RDT immediately question this assumption for the pressure Hessian.

Bibliography

BATCHELOR, G. K. (1953). *The Theory of Homogeneous Turbulence*, Cambridge University Press.

BATCHELOR, G. K. AND PROUDMAN, I. (1954). The effect of rapid distortion in a fluid in turbulent motion, *Q. J. Mech. Appl. Math.* **7**, 83–103.

BAYLY, B. J. (1986). Three-dimensional instability of elliptical flow, *Phys. Rev. Lett.* **57**, 2160–2163.

BOHR, T., JENSEN, M. H., PALADIN, G., AND VULPIANI, A. (1998). *Dynamical Systems Approach to Turbulence*, Cambridge University Press.

BURDEN, A. D. (1991). Towards an EDQNM closure for inhomogeneous turbulence, in Johansson, A. V. and Alfredson, P. H. eds., *Advances in Turbulence III*, Springer Verlag, p. 387.

CAMBON, C. (1982). Etude spectrale d'un champ turbulent incompressible soumis à des effets couplés de déformation et rotation imposés extérieurement, *Thèse de Doctorat d'Etat*, Université Lyon I, France.

CAMBON, C., GODEFERD, F. S., NICOLLEAU, F., AND VASSILICOS, J. C. (2004). Turbulent diffusion in rapidly rotating flows with and without stable stratification, *J. Fluid Mech.* **499**, 231–255.

CAMBON, C. AND RUBINSTEIN, R. (2006). Anisotropic developments for homogeneous shear flows, *Phys. Fluids* **18**, 085106.

CAMBON, C. AND SCOTT, J. F. (1999). Linear and nonlinear models of anisotropic turbulence, *Annu. Rev. Fluid Mech.* **31**, 1–53.

CAMBON, C., TEISSÈDRE, C., AND JEANDEL, D. (1985). Etude d' effets couplés de rotation et de déformation sur une turbulence homogène, *J. Méc. Théor. App.* **5**, 629–657.

CARNEVALE, G. F. AND FREDERIKSEN, J. S. (1983). A statistical dynamical theory of strongly nonlinear internal gravity waves, *Geophys. Astrophys. Fluid Dyn.* **20**, 131–164.

CHAOUAT AND SCHIESTEL (2005). A new partially integrated transport model for subgrid-scale stresses and dissipation rate for turbulent developing flows, *Phys. Fluids* **17** (6), 065106.

CHAOUAT AND SCHIESTEL (2007). From single-scale turbulence models to multiple-scale and subgrid-scale models by Fourier transform, *Theor. Comput. Fluid Dyn.* **21**, 201–229.

CLARK, T. AND ZEMACH, C. (1995). A spectral model applied to homogeneous turbulence, *Phys. Fluids* **7**, 1674–1694.

COURSEAU, P. A. AND LOISEAU, M. (1978). Contribution à l'analyse de la turbulence homogène anisotrope, *J. Méc.* **17**, 245–297.

CRAIK, A. D. D. AND CRIMINALE, W. O. (1986). Evolution of wavelike disturbances in shear flows: A class of exact solutions of Navier–Stokes equations, *Proc. R. Soc. London Ser. A* **406**, 13–26.

CROW, S. C. (1968). Viscoelastic properties of the fine-grained incompressible turbulence, *J. Fluid Mech.* **33**, 1–20.

FRISCH, U., SHE, Z. S., AND SULEM, P. L. (1987). Large-scale flow driven by the anisotropic kinetic alpha effect, *Physica D* **28**, 382–392.

GALMICHE, M. (1999). Interactions turbulence-champs moyens et ondes de gravité internes dans un fluide stratifié. *Thèse de Doctorat*, Université de Toulouse, France.

GODEFERD, F. S., CAMBON, C., AND LEBLANC, S. (2001). Zonal approach to centrifugal, elliptic and hyperbolic instabilities in Stuart vortices with external rotation, *J. Fluid Mech.* **449**, 1–37.

GODEFERD, F. S., CAMBON, C., AND SCOTT, J. F. (2001). Report on the workshop: Two-point closures and their applications, *J. Fluid Mech.* **346**, 393–407.

GOLDSTEIN M. E. (1978). Unsteady vortical and entropic distortions of potential flows round arbitrary obstacles, *J. Fluid Mech.* **89**, 433–468.

GUIMBARD, D. AND LEBLANC, S. (2006). Local stability of the Abrashkin–Yakubovich family of vortices, *J. Fluid Mech.* **567**, 91–110.

HERRING, J. R. (1974). Approach of axisymmetric turbulence to isotropy, *Phys. Fluids* **17**, 859–872.

HUNT, J. C. R. (1973). A theory of turbulent flow around two-dimensional bluff bodies, *J. Fluid Mech.* **61**, 625–706.

HUNT, J. C. R. AND CARRUTHERS, D. J. (1990). Rapid distortion theory and the 'problems' of turbulence, *J. Fluid Mech.* **212**, 497–532.

KANEDA, Y. AND ISHIDA, T. (2000). Suppression of vertical diffusion in strogly stratified turbulence, *J. Fluid Mech.* **402**, 311–327.

LIFSCHITZ, A. AND HAMEIRI, E. (1991). Local stability conditions in fluid dynamics, *Phys. Fluids A* **3**, 2644–2641.

NAZARENKO, S., KEVLAHAN, N. N., AND DUBRULLE, B. (1999). A WKB theory for rapid distortion of inhomogeneous turbulence, *J. Fluid Mech.* **390**, 325–348.

SAGAUT, P. (2005). *Large-Eddy Simulation for Incompressible Flows*, 3rd ed., Springer.

SAGAUT, P., DECK, S., AND TERRACOL, M. (2006). *Multiscale and Multiresolution Approaches in Turbulence*, Imperial College Press.

TOUIL, H., BERTOGLIO, J. P., AND PARPAIS, S. (2000). A spectral closure applied to inhomogeneous turbulence, in Dopazo, C., ed., *Advances in Turbulence VIII*, CIMNE, Spain, p. 689.

TOWNSEND, A. A. (1956, 1976). *The Structure of Turbulent Shear Flow*, Cambridge University Press.

TURNER, L. (1999). Macroscopic structures of inhomogeneous, Navier–Stokes, turbulence, *Phys. Fluids* **11**, 2367–2380.

14 Anisotropic Nonlinear Triadic Closures

This chapter includes detailed equations that are not given in Chapters 4, 5, and 7. Fundamentals of anisotropic triadic closures are given in Chapter 4. A general discussion is offered on various aspects of these closures. Strong anisotropy is the most original aspect that is emphasized throughout this book, but it is perhaps useful to recall the role of the characteristic time (e.g., eddy damping in EDQNM) for the decorrelation of triple correlations in canonical incompressible HIT, and not only in EDQNM. The closure for compressible quasi-isentropic isotropic turbulence, which is a very interesting case of interaction of "strong" solenoidal turbulence with pseudo-acoustical "weak" wave turbulence, also merits additional discussion. Finally, the theory of "linear response" by Kaneda and co-workers, touched on in Chapter 5, is rediscussed in connection with an approach to weak anisotropy.

14.1 Canonical HIT, Dependence on the Eddy Damping for the Scaling of the Energy Spectrum in the Inertial Range

All technical details about EDQNM for HIT are given in Chapter 3. In this case, and only looking at the power-law slope of the single-time energy spectrum $E(k)$ in the inertial range, all "triadic" theories, including the most sophisticated self-consistent ones, from DIA to TFM, LHDIA and LRA, can be analyzed from the following simple, purely dimensional and local in wave space, argument[*]:

$$F(k) = \int_k^\infty T(k)dk \sim \eta(k)^{(-1)}k^4 E^2 \to \varepsilon,$$

where $\eta(k)$ is the ED term in EDQNM, or a constant external frequency in isotropic or isotropized wave-turbulence theory. A link of the exponent of the power law for $E(k)$, or $E(k) \sim k^{-y}$, to the exponent of the power law for $\eta(k)$, or $\eta(k) \sim k^x$ is immediately derived as a linear law:

$$y = 2 - x/2.$$

---

[*] This analysis can be found in existing literature, e.g., in books by Frisch and Lesieur, but it was suggested by a very concise and pedagogical informal talk, given by W. Bos in the CNRS Summer School in Cargèse (France), August 13–25, 2007.

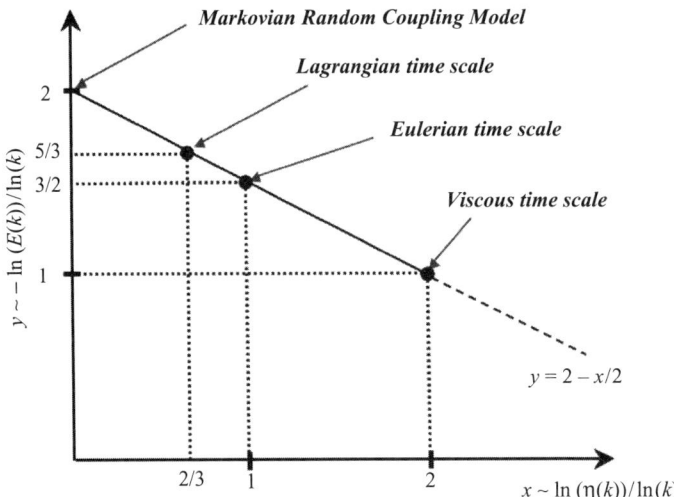

Figure 14.1. Map of a single straight line of the $E(k)$-to-$\eta(k)$ power-law exponents. Suggested by W. Bos.

Some important cases are discussed as follows, and summarized in Fig. 14.1 with $0 \leq x \leq 2$:

- Constant $\eta(k) = f_0 : x = 0$, $y = 2$, $E \sim k^{-2}$. This case is illustrated by some models, such as the Markovian random coupling model, which amount to EDQNM with constant ED. They also illustrate oversimplified cases of wave turbulence, in which $\eta(k)$ is not a nonlinear decorrelation time, but instead the time frequency of the external linear wave operator ($f_0 = 2\Omega$ in rotating turbulence, $f_0 = N$ in stratified turbulence, Alfven frequency in magnetohydrodynamics, etc.).
- "Eulerian" time scale: The sweeping effect, or advection of small scales by the largest ones, seen in the Eulerian framework, suggests $\eta(k) \sim Uk$, yielding $x = 1$, $y = 3/2$, and $E \sim k^{-3/2}$. The wrong exponent of original DIA is immediately derived.
- "Lagrangian" time scale: One of the simplest proposals, by Kraichnan and Orszag, is $\eta(k) \sim \varepsilon^{1/3} k^{2/3}$, yielding $x = 2/3$, $y = 5/3$. More sophisticated proposals, such as $\eta(k) \sim k^{3/2} E^{1/2}$, or $\eta(k) \sim \sqrt{\int_0^k p^2 E(p) dp}$ (see Chapter 3) are also consistent with $x = 2/3$, $y = 5/3$ in the inertial range. The correct Kolmogorov law is recovered, as found in EDQNM and in all self-consistent theories correcting DIA with Lagrangian or semi-Lagrangian approaches, not to mention the very recent self-consistent EDQNM version by Bos and Bertoglio (2006).
- Pure viscous time scale, $\eta(k) \sim \nu k^2$: $x = 2$, $y = 1$, $E(k) \sim k^{-1}$. This corresponds to a late time of decay, or a transient zone between the inertial range and the dissipative one.

It is a bit surprising that a complex spectral-flux term, which results from the difference of two large nonlocal terms {e.g., $pE(q)[k^2 E(p) - p^2 E(q)]$ in Eq. (3.157)}, which are individually high and quasi-balanced, might be so simply evaluated by $k^3 E(k)^2$, by use of pure dimensional analysis, but this works for our simple purpose of deriving power laws.

To speak of "Eulerian" or "Lagrangian" time scales is almost a caricature, without a deep survey of Lagrangian, semi-Lagrangian, and Eulerian theories. We have chosen not to treat in detail these aspects in this book, but let us mention at least two points. On the one hand, the Lagrangian or Eulerian origin of the nonlinear decorrelation time scale of triple correlations can be ignored, as it is done in the local energy transfer theory by McComb (1974), which is also a self-consistent theory (i.e., which involves no adjustable parameter dealing with the Kolmogorov constant) giving the correct point $x = 2/3$, $y = -5/3$ in the diagram. On the other hand, it is important to recall that the parameter $\eta(k)$ comes from Kraichnan's response tensor, which is an essential ingredient of any triadic spectral closure and therefore is revisited in the next section. This response tensor, either random or averaged, is a tangent Green's function to a nonlinear state and therefore is subject to the effects of advection and deformation by the velocity field. In this sense, the RDT Green's function [in fact both $\mathbf{G}(\mathbf{k}, t, t')$ and $\mathbf{F}(t, t')$ in "mean" Eulerian, $\mathbf{G}(\mathbf{K}, t, t')$ in Lagrangian] illustrates in an oversimplified way these advection and deformation effects by the mean flow, even if nonlinear dynamics is essential in the Kraichnan Green's function. Accordingly, we think that a physical discussion taking into account the different advection and deformation aspects cannot ignore their different translations in Lagrangian or Eulerian frameworks.

14.2 Solving the Linear Operator to Account for Strong Anisotropy

14.2.1 Random and Averaged Nonlinear Green's Functions

The concept of *response tensor* is in the heart of all closures inherited from Kraichnan. The most general definition is obtained from writing the perturbation equation for a disturbance field $\delta\hat{u}$ created by an external disturbance $\delta\hat{f}$ (e.g., from a solenoidal stirring force):

$$\delta\hat{u}_i(\mathbf{k}, t) = \iiint\int_{t_0}^{t} \mathcal{G}_{ij}(\mathbf{k}, \mathbf{k}', t, t')\delta\hat{f}_j(\mathbf{k}', t')d^3\mathbf{k}'dt'. \tag{14.1}$$

It is very important to stress that the perturbation is performed around any particular random realization of $\hat{u}$, which is a solution of the fully nonlinear Navier–Stokes equations in Fourier space. The nonlinear term $\widehat{uu}$ leads to the contribution $2\widehat{u\delta u}$ in the $\delta\hat{u}$ equation. Accordingly, the response function is also a random variable, changing from realization to realization of the bearer velocity field $\hat{u}$. Therefore this relation is not only an integral formulation in time, but also in wave-vector space. In this sense, the response tensor before averaging is a tangent Green's function related to a *random and nonlinear* state.

Only the statistical counterpart of $\mathcal{G}$,[†] obtained from statistical ensemble averaging, becomes local in wave space assuming spatial homogeneity, i.e.,

$$\langle \mathcal{G}_{ij} \rangle = G^+_{ij}(\mathbf{k}, t, t') \delta^3(\mathbf{k} - \mathbf{k}'). \tag{14.2}$$

One can just recall that the ED term in the previous section can be related to $\mathbf{G}^+$ by $\eta(k)^{-1} \sim \int G^+(\mathbf{k}, t, t') dt'$.

Considering a homogeneous turbulent velocity field, but subjected to a mean flow that is not itself homogeneous but with uniform velocity gradients, the preceding equation must be modified as follows:

$$\langle \mathcal{G}_{ij} \rangle = G^+_{ij}(\mathbf{k}, t, t') \delta^3[\mathbf{k} - \tilde{\mathbf{F}}^{-1}(t, t') \mathbf{k}']. \tag{14.3}$$

A purely diagonal form is recovered in terms of mean-Lagrangian wave vectors, yielding a term $\delta^3(\mathbf{K} - \mathbf{K}')$.

From these general considerations, we consider that the confusion between $\mathcal{G}$ and $\langle \mathcal{G} \rangle \to G^+$ is highly misleading, even without mean flow. Identification of the random response function with its averaged counterpart is made in Leslie (1973), for instance, and by many other authors. Such oversimplifications do not lead to an incorrect final form of DIA equations, for instance, but this is only because these equations are consistent with a first-loop iterative expansion around a deterministic zeroth-order state for G. The same procedure performed in the presence of a mean flow, using (14.2) instead of (14.3), is nothing but wrong.

Introducing a perturbative expansion in terms of the basic nonlinear term in Navier–Stokes equations, the zeroth-order response function naturally appears as the viscous linear Green's function, which is really deterministic (or "statistically sharp" in Leslie's parlance). In the case of HIT without mean flow, the linear operator reduces to the viscous term, so that the basic Green's function, or zeroth-order response tensor is simply

$$G^{(0)}_{ij}(\mathbf{k}, t, t') = P_{ij}(\mathbf{k}) \exp[-\nu k^2 (t - t')].$$

14.2.2 Homogeneous Anisotropic Turbulence with a Mean Flow

For HAT in the presence of a distorting mean flow, the linear operator inherited from RDT must be accounted for: It is an essential building block for constructing the nonlinear theory, and it can generate the relevant nontrivial zeroth-order response function of any "triadic" closure, including DIA, EDQNM, etc.

In the presence of an additional right-hand-side term $f_i(\mathbf{k}, t)$ in the basic spectral equation,

$$\dot{\hat{u}}_i + M_{ij} \hat{u}_j + \nu k^2 \hat{u}_i = f_i, \tag{14.4}$$

[†] $\mathcal{G}$ must be not confused with the linear Green's function in physical space in Eq. (5.21).

14.2 Solving the Linear Operator to Account for Strong Anisotropy

which represents nonlinearity and possibly random forcing, the basic RDT solution can be generalized as

$$\hat{u}_i(\boldsymbol{k}(t),t) = G^{(0)}_{ij}(\boldsymbol{k},t,t')\hat{u}_j(\boldsymbol{k}(t'),t') + \int_{t'}^{t} G^{(0)}_{ij}(\boldsymbol{k},t,\tau)f_j(\boldsymbol{k}(\tau),\tau)d\tau. \quad (14.5)$$

It is necessary to include the viscous term, so that $\mathbf{G}^{(0)}$ is the viscous Green's function given by

$$G^{(0)}_{ij}(\boldsymbol{k},t,t') = G_{ij}(\boldsymbol{k},t,t')V_0(\boldsymbol{k},t,t'), \quad (14.6)$$

where $\mathbf{G}$ (in pure inviscid RDT) and V_0 are defined in Chapter 13. The latter equation is generic, and similar forms can be found for the equations that govern the statistical moments of $\hat{\boldsymbol{u}}$ at any order.

For instance, the equation for the second-order spectral tensor, or Craya equation (2.81), is formally solved as

$$\hat{R}_{ij}(\boldsymbol{k}(t),t) = G^{(0)}_{im}(\boldsymbol{k},t,t_0)G^{(0)}_{jn}(\boldsymbol{k},t,t_0)\hat{R}_{mn}[\boldsymbol{k}(t_0),t_0]$$
$$+ \int_{t_0}^{t} G^{(0)}_{im}(\boldsymbol{k},t,t')G^{(0)}_{jn}(\boldsymbol{k},t,t')T_{mn}[\boldsymbol{k}(t'),t']dt', \quad (14.7)$$

and similarly for the third order, with a threefold product

$$\mathbf{G}(\boldsymbol{k},t,t') \otimes \mathbf{G}(\boldsymbol{p},t,t') \otimes \mathbf{G}(\boldsymbol{q},t,t'), \quad \boldsymbol{k}+\boldsymbol{p}+\boldsymbol{q}=0,$$

called into play (the detailed equation will be given in the next section).

In a slightly different form, one can introduce a new variable a defined as

$$\hat{u}_i(\boldsymbol{k}(t),t) = G^{(0)}_{ij}(\boldsymbol{k},t,t_0)a_j[\underbrace{\boldsymbol{k}(t_0)}_{K},t], \quad (14.8)$$

which replaces the initial data in the linear solution and can be considered as *slowly varying* in time, where the initial time is fixed at $t' = t_0$. The Green's function can be used for deriving a new equation for the slow variable without any assumption from the exact, Navier–Stokes-type, $\hat{\boldsymbol{u}}$ equation:

$$\dot{a}_i = G^{(0)-1}_{ij}(\boldsymbol{k},t,t_0)P_{jmn}[\boldsymbol{k}(t)]\iiint \underbrace{G^{(0)}_{ms}(\boldsymbol{p},t,t_0)a_s(\boldsymbol{P},t)}_{\hat{u}(p)}\underbrace{G^{(0)}_{nr}(\boldsymbol{q},t,t_0)a_r(\boldsymbol{Q},t)}_{\hat{u}(q)}d^3p,$$
$$(14.9)$$

with $\boldsymbol{q} = \boldsymbol{k} - \boldsymbol{p}$. The latter equation suggests a systematic way to derive a suitable closure. For instance, the idea in applying generalized EDQNM is to transfer the "machinery" of EDQNM procedures/assumptions from the $\hat{\boldsymbol{u}}$ to the slow variables $\boldsymbol{a}$. A cartoon of the optimal procedure, called EDQNM3, and successively applied to the pure rotation case in Chapter 4, can be given as follows:

- The quasi-normal (QN) procedure is the same, working with $\hat{\boldsymbol{u}}$ or with $\boldsymbol{a}$ variables. Fourth-order correlations at three points are expressed in terms of products of second-order correlations.

- The Markovian (M) procedure consists of freezing the time dependency of the slow variables, and of the slow variables only, in the time integral that links third-order to second-order correlations, once the QN assumption used.
- Eddy damping consists of replacing the "bare" viscous RDT Green's function with a possibly renormalized one as

$$G_{ij}^+(\mathbf{k}, t, t') = G_{ij}(\mathbf{k}, t, t')V(\mathbf{k}, t, t'), \qquad (14.10)$$

where the term V, which is substituted into V_0 in Eq. (14.6), will be specified later on.

In view of "exact" equation (14.9), however, specific difficulties linked to the advection term appear, and a unique system of dependent variables has to be chosen.

Time dependence of the wave vectors reflects the advection by the mean flow. The general relation

$$k_i(t) = F_{ji}^{-1}(t, t')k_j(t') \qquad (14.11)$$

is always valid, but for any operator depending on $(\mathbf{k}, t, t')$, one can ask the question of whether a *fixed* wavenumber of reference, such as $\mathbf{K} = \mathbf{k}(t_0)$, is useful or not. In addition to the renormalization of **G** by a scalar term, the related question of renormalizing the Cauchy matrix in Eq. (14.11) can be raised, together with the two-time aspect in general: In spite of some proposals for modeling parallel shear flows with a saturated accumulated mean shear (F_{12} here), e.g., by Maxey and Hunt, we prefer to keep Eq. (14.11) unchanged here. In the presence of solid-body rotation or dominant mean vorticity, for instance, saturating Ωt is meaningless.

To avoid any ambiguity, "mean-Lagrangian" wave vectors will be used, such as $\mathbf{K} = \mathbf{k}(t_0)$, $\mathbf{P} = \mathbf{p}(t_0)$, $\mathbf{Q} = \mathbf{q}(t_0)$, when "slow" variables are concerned, and the time argument, e.g., t or t', will be specified in $\mathbf{k}, \mathbf{p}, \mathbf{q}$.

As far as possible, eigenmode decomposition must be used to diagonalize **G**. At least, a drastic reduction of the number of variables can be obtained in working with the components in the Craya–Herring frame or in similar frames of reference, as used in Cambon (1982), Cambon, Teissèdre, and Jeandel (1985), and in all subsequent papers from the same team.

14.3 A General EDQN Closure. Different Levels of Markovianization

Using the threefold product of Green's functions to express triple correlations in terms of fourth-order ones, the most general EDQN closure for the transfer tensor T_{ij} in the Craya's equation leads to

$$\tau_{ij}(\mathbf{k}(t), t) = P_{jml}[\mathbf{k}(t)] \int_{-\infty}^{t} \iiint_{\mathbf{k}+\mathbf{p}+\mathbf{q}=0}$$
$$\times G_{in}^+(\mathbf{k}, t, t')G_{mr}^+(\mathbf{p}, t, t')G_{ls}^+(\mathbf{q}, t, t')\hat{R}_{vs}[\mathbf{q}(t'), t']$$
$$\times \left\{\frac{1}{2}P_{nvw}[\mathbf{k}(t')]\hat{R}_{wr}(\mathbf{p}(t'), t') + P_{rvw}[\mathbf{p}(t')]\hat{R}_{wn}[\mathbf{k}(t'), t']\right\} d^3\mathbf{p}\,dt'$$

$$(14.12)$$

14.3 A General EDQN Closure

with

$$T_{ij}(\mathbf{k},t) = \tau_{ij}(\mathbf{k},t) + \tau_{ji}^*(\mathbf{k},t). \tag{14.13}$$

If the factor V in Eq. (14.10) is generated only by adding to the "laminar" viscous factor νk^2 a damping term $\eta(\mathbf{k},t)$, the result can be written as

$$V(\mathbf{k},t,t') = \exp\left[-\int_{t'}^{t} \nu k^2(t'') + \eta(\mathbf{k}(t''),t'')dt''\right]. \tag{14.14}$$

The exponential decay factor in Eq. (14.14) is related to the cumulative viscous and eddy damping between t' and t. Notice that, although ED formally appears by means of the revised Green's function G_{ij}^+, unlike viscosity it is really a nonlinear effect, modifying the expression of the third-order moments in terms of the second-order ones.

The time integral in Eq. (14.12) expresses memory of the third-order moments, represented by τ_{ij}, for the fourth-order moments, written as products of $\hat{\mathbf{R}}$. This memory is too long-lasting in the QN model, but ED suppresses memory by progressive attenuation of the Green's function with increasing $t - t'$ by way of the η part of the exponential factor in Eq. (14.14). This will to decrease the importance of third-order memory is taken to its logical conclusion by the Markovianization process. First, the integrand in Eq. (14.14) is approximated by its value at $t'' = t$ to obtain

$$\tilde{V}(\mathbf{k},t,t') = \exp\left[-(\nu k^2(t) + \eta(\mathbf{k}(t),t))(t - t')\right]. \tag{14.15}$$

14.3.1 EDQNM2 Version

Next, the spectral tensors in Eq. (14.12) can be replaced with their values at $t' = t$, and the wave vectors too, leading to the following form (Cambon and Scott, 1999):

$$\tau_{ij}[\mathbf{k}(t),t] = P_{jml}[\mathbf{k}(t)] \iiint_{\mathbf{k}+\mathbf{p}+\mathbf{q}=0} \Psi_{iml;nrs} \hat{R}_{ns}(\mathbf{q},t)$$
$$\times \left[\frac{1}{2}P_{nvw}(\mathbf{k})\hat{R}_{wr}(\mathbf{p},t) + P_{rvw}(\mathbf{p})\hat{R}_{wn}(\mathbf{k},t)\right] d^3\mathbf{p}dt', \tag{14.16}$$

where

$$\Psi_{iml;nrs} = \int_{-\infty}^{t} G_{in}(\mathbf{k},t,t')G_{mr}(\mathbf{p},t,t')G_{ls}(\mathbf{q},t,t')\tilde{V}(\mathbf{k},t,t')\tilde{V}(\mathbf{p},t,t')\tilde{V}(\mathbf{q},t,t')dt'. \tag{14.17}$$

Equation (14.16) yields an EDQNM model, for which the nonlinear transfer term in Eq. (2.67) is determined by $\hat{\mathbf{R}}$ at the current instant of time, rather than by the entire past history of the spectral tensor. This is the essence of Markovianization. This version was a rather logical generalization of the classical approach of Orszag (1970), allowing for mean-flow effects, and was successfully applied to rotating[‡] (Cambon and Jacquin, 1989) and to stably stratified turbulence (Godeferd and Cambon, 1994). On the other hand, it is not the optimal EDQNM version: Depending on the mean-flow features, a simpler or a more sophisticated version can be used.

14.3.2 A Simplified Version: EDQNM1

It is tempting to push the Markovianization one step further and set $t = t'$ in the RDT Green's functions of Eq. (14.17), in which case one obtains equations for the spectral transfer as if there were no mean flow. This amounts to replacing **G** with the identity matrix, so that

$$\Psi_{iml;nrs} = \delta_{in}\delta_{mr}\delta_{ls}\theta_{kpq}(t), \qquad (14.18)$$

with

$$\theta_{kpq} = \nu(k^2 + p^2 + q^2) + \eta(k,t) + \eta(p,t) + \eta(q,t).$$

The only effect of the mean flow on the spectral evolution then appears through the linear operators on the left-hand side of Eq. (2.81). This version is also valid for anisotropic turbulence without mean flow, the isotropic case addressed in Chapter 3 being derived by setting the isotropic form for $\hat{R}$. Finally, a more tractable form of the transfer term is derived from the $e - Z$ decomposition of the anisotropic spectral tensor, in terms of $T^{(e)}$ and $T^{(z)}$ (Cambon, Mansour, and Godeferd, 1997):

$$T^{(e)} = \frac{1}{2}T_{ii}(\mathbf{k})$$
$$= \iiint \theta_{kpq} 2kp \left\{ (e'' + \Re X'')[(xy + z^3)(e' - e) - z(1 - z^2)(\Re X' - \Re X)] \right\} d^3\mathbf{p}$$
$$+ \iint \theta_{kpq} 2kp \left[\Im X'(1 - z^2)(x\Im X - y\Im X') \right] d^3\mathbf{p}, \qquad (14.19)$$

$$T^{(z)} = \frac{1}{2}T_{ij}(\mathbf{k})N_i(-\mathbf{k})N_j(-\mathbf{k})$$
$$= \iiint \theta_{kpq} 2kpe^{-2\imath\lambda} \left\{ (e'' + \Re X'')[(xy + z^3)(\Re X' - X) \right.$$
$$\left. - z(1 - z^2)(e' - e) + \imath(y^2 - z^2)\Im X' \right\} d^3\mathbf{p}$$
$$+ \iint \theta_{kpq} 2kpe^{2\imath\lambda} \left\{ \imath\Im X'(1 - z^2)[x(e + X) - \imath y\Im X'] \right\} d^3\mathbf{p}, \qquad (14.20)$$

with $e = e(\mathbf{k}, t)$, $e' = e(\mathbf{k}, t)$, $e'' = e(\mathbf{q}, t)$, $X = Z(\mathbf{k}, t)e^{2\imath\lambda}$, $X' = Z(\mathbf{p}, t)e^{2\imath\lambda'}$, $X'' = Z(\mathbf{q}, t)e^{2\imath\lambda''}$. Angles λ, λ', and λ'' are defined in Eq. (14.38). With respect to arbitrary

---

[‡] Time dependency of the wave vectors was not considered in the EDQNM2 version for "pure" rotation, because the formalism was developed in the rotating frame, but if the same procedure is applied to equations in the Galilean frame, in the presence of a solid-body "mean" motion, time shifting cannot be neglected, as discussed at the end of this section.

14.3 A General EDQN Closure

anisotropy, only the helicity spectrum and the helicity transfer are omitted, considering that these terms cannot be created and are present only if introduced in initial data. Absence of the helicity spectrum is generally justified in homogeneous turbulence, despite the interest of (random) helical modes for investigating nonlinear interactions. Notice that the latter equations are most easily obtained from the sophisticated EDQNM2 and EDQNM3 versions for rotating turbulence, given in the next section, by setting $\Omega = 0$. This way of deriving equations can appear paradoxical, but, again, it follows from the fact that helical modes give the best basis for both rotating and nonrotating turbulence.

Of course the conventional EDQNM model for 3D isotropic turbulence is recovered using $e = E(k)/(4\pi k^2)$, $Z = 0$, and $\iiint d^3p = 2\pi \iint_{\Delta_k}(pq/k)dpdq$ in Eq. (14.19), whereas the averaging on λ yields $T^z = 0$ in the same conditions.

14.3.2.1 Recovering the Conventional 2D Case With Additional Jetal Mode

Another interesting result is the derivation of an extended isotropic 2D version, setting $k_\perp = k$, $k_\parallel = 0$, and using $\iiint d^3p = \iint_{\Delta_k}(1 - x^2)^{-1/2}dpdq$, $e^{2i\lambda} = e^{2i\lambda'} = e^{2i\lambda''} = -1$. In this case, the expression of $T^e - T^z$ in terms of $e - Z$ is exactly the 2D EDQNM equation used by Leith (1971) and Pouquet et al. (1975). $e - Z$ at $k_\parallel = 0$ is the limit of the toroidal energy spectrum, directly linked to vertical vorticity. In addition, the expression of $T^e + T^z$ in terms of both $e - Z$ and $e + Z$ in this limit is exactly the EDQNM equation of a passive scalar advected by a 2D flow. $e + Z$ is the limit of poloidal energy that represents a purely vertical mode, which is referred to as the jetal mode by Kassinos and Reynolds, and plays the role of the spectrum of the variance of the passive scalar (see also Cambon and Godeferd, 1993). One recovers the fact that a 2D-3C (two-dimensional with three-velocity components) flow, characterized by both $e - Z$ (toroidal = vortical) and $e + Z$ (poloidal = jetal) energy spectra, evolves toward a pure vortical flow, because the energy $e - Z$ is conserved by the inverse cascade, whereas the energy $e + Z$ is rapidly damped.

14.3.3 The Most Sophisticated Version: EDQNM3

In all cases in which the linear RDT effect is shown to be important to the dynamics of triple correlations, it is not possible to use EDQNM1, and EDQNM2 is potentially more relevant. Nevertheless, the Markovianization in EDQNM2 is not completely consistent with the decomposition in terms of slow and rapid terms from the very definition of slow variables in Eqs. (14.8) and (14.9). The most straightforward Markovianization consists of setting $t = t'$ in the slow terms and in the slow terms only. Accordingly, the spectral tensor itself is not to be globally considered as a slow term, but ought to be rewritten in terms of the slow variables as

$$\hat{R}_{vs}(\boldsymbol{q}(t'), t') = G_{vi}^{(0)}(\boldsymbol{q}, t', t_0) G_{sj}^{(0)}(\boldsymbol{q}, t', t_0) A_{ij}(\boldsymbol{Q}, t') \quad (14.21)$$

in Eq. (14.12), so that only A_{ij} is taken at the current instant of time $t' = t$ before time integration is performed. This "optimal" version asks the question of whether the time dependence of the wave vectors has to be considered as rapid or not. On the one hand, truncation of the memory of **F** can be justified: A proposal by Maxey and Hunt for the pure plane shear flow suggests replacing $S(t - t')$, or $\int_{t'}^{t} S(t'')dt''$ for time-evolving mean-shear rate, with a value that saturates toward a constant $S \cdot \tau_{\text{NL}}$ at increasing $t - t'$. On the other hand, in the simplest case of rotating turbulence, but seen in the Galilean frame in the presence of mean solid-body motion (**A** is antisymmetric), there is no physical argument to saturate the phase $\Omega(t - t')$. In this case, **F**, which is a pure orthogonal matrix, and **G**, which has an exact counterpart in the rotating frame, have to be treated in the same way, i.e., as "rapid" terms. Accordingly, we consider that both **F** and **G** in the general case have to be considered as "rapid" for the sake of consistency, and therefore the past history of the wave vectors cannot be neglected.

The resulting closure is rather complicated, involving a lot of **G** factors, together with various **F**-dependent time shifts for wave vectors, with possible simplifications coming only from the group relations verified by **G** and **F** (Cambon, Teissèdre, and Jeandel, 1985). Another technical difficulty comes from the use of the inverse of **G**: This difficulty is avoided, using the reduced $g_{\alpha\beta}$ RDT Green's function, working with Craya–Herring components, with the additional advantage of reducing the number of dependent variables.

The minimum set of equations, in terms of the minimum number of dependent variables, that form the EDQNM3 version is given as follows.

- Definition of "slow" variables a_β, using Lagrangian wave vectors,

$$u^{(\alpha)}(k(t), t) = g_{\alpha\beta}(k, t, t_0) A_\beta(K, t), \qquad (14.22)$$

with easy inversion using $g_{\alpha\beta}^{-1}$.
- Writing the basic Navier–Stokes equations with distorting mean flow in terms of them:

$$\frac{\partial}{\partial t} A^\alpha(K, t) = g_{\alpha\beta}^{-1}(K, t, t_0) \iiint P_{\beta\gamma\delta}(K, P, t) g_{\gamma\mu}(P, t, t_0)$$
$$\times A_\mu(P, t) g_{\delta\nu}(Q, t, t_0) A_\nu(Q, t) d^3 P, \qquad (14.23)$$

with $P_{\beta\gamma\delta}$ given by (2.85).
- To construct EDQN equations, corresponding to (14.12) in terms of A_α correlations tensors, the final closure equation being given for

$$\langle A_\alpha(Q, t) A_\beta(K, t) A_\gamma(P, t) \rangle = A_{\alpha\beta\gamma}(K, P, t) \delta(K + P + Q). \qquad (14.24)$$

- To replace in the EDQN equation for $A_{\alpha\beta\gamma}$ t' with t only in the second-order slow counterparts of $\hat{R}$, such as

$$\langle A_\alpha^*(P, t') A_\beta(K, t') \rangle = A_{\alpha\beta}(K, t') \delta(K - P). \qquad (14.25)$$

14.4 Application of Three Versions to the Rotating Turbulence

The most general EDQNM versions were carried out toward complete achievement for pure rotation only. In this case, the zeroth-order state consists of superimposed oscillating modes of motion, without amplification and interaction: They correspond to neutral dispersive inertial waves. The time integral of a threefold product of Green's functions converges, provided an infinitesimal viscous (or ED) term is added. In the limit of small interactions, two-point closures and theories of wave turbulence share an important background. Even if the latter are developed in the inviscid case, a vanishing damping term is also added, as a mathematical convenience, in order to regularize the resonant operators.

The EDQNM1 version presents no interest because the isotropy is broken by the Green's function only at the level of triple correlations: Started with isotropic initial data, EDQNM1 equations conserve isotropy and are not at all affected by rotation. Equations (14.19) and (14.20), however, remain of interest in some situations, as discussed in Subsection 14.3.2, illustrating the interest of the $e - Z$ decomposition.

Detailed EDQNM2 and EDQNM3 equations are subsequently given in terms of e and Z (without initial helicity).

In EDQNM3 equations subsequently recalled from Cambon, Rubinstein, and Godeferd (2004) [(14.30) and (14.31)], $T^{(e,z,h)}$ are given by volume integrals close to the ones found in the appendix of Cambon, Mansour, and Godeferd (1997) (CMG hereafter). Helicity is ignored here as in CMG, for the sake of brevity. The integrands are completely expressed in terms of (e, Z) through quadratic terms involving triads. The most laborious calculation is for deriving five geometric factors, denoted $A_1(k, p, q), \ldots, A_5(k, p, q)$. Fortunately, these factors were calculated once and for all, and play the same role in EDQNM2 and EDQNM3.

The way to simply move from EDQNM2 to EDQNM3, in the absence of helicity, is found as follows.

The only explicit (in addition to the time dependence of the $e - Z$ variables themselves) time-dependent term in the EDQNM2 integrand of $T^{(e,z)}$ is

$$\exp[-z_{kpq}(t - t')] = \exp[(-\mu_{kpq} - \iota\Omega_{kpq})(t - t')], \quad \Omega_{kpq} = s\sigma_k + s'\sigma_p + s''\sigma_q, \tag{14.26}$$

and its integral gives

$$\int_{-\infty}^{t} e^{-z_{kpq}(t-t')} dt' = \frac{1}{z_{kpq}}. \tag{14.27}$$

The polarization anisotropy is now denoted as ζ in order to avoid confusion with its slow counterpart, which is only relevant here, Z, with the relationship

$$\zeta(s\mathbf{k}, t') = Z(s\mathbf{k}, t')e^{-2\iota s\sigma_k t'}. \tag{14.28}$$

Only Z has to be considered as "slow," so that it has to be frozen to $t' = t$ in the temporal integral over t' resulting from EDQN. Accordingly, the related phase term in ζ will give an additional (versus EDQNM2) contribution to the temporal integrand, with the following modifications:

1. there is no modification for the terms that do not include ζ in $T^{(e)}$,
2. terms containing ζ in T^e are altered in replacing $1/z_{kpq}$ with

$$\int_{-\infty}^{t} e^{-z_{kpq}(t-t') - \imath\Omega_z t'} dt' = \frac{e^{-\imath\Omega_z t}}{z_{kpq} - \imath\Omega_z}, \qquad (14.29)$$

where $\Omega_z = 2s''\sigma_q$ for Z''-type term, $\Omega_z = 2s\sigma_k$ for Z-type term, $\Omega_z = 2s''\sigma_q + 2s\sigma_k$ for ZZ''-type terms, and $\Omega_z = 2s''\sigma_q + 2s'\sigma_p$ for $Z'Z''$-type terms.

Consequently, the EDQNM3 version without helicity of $T^{(e)}$ becomes

$$T^{(e)} = \frac{1}{2^3} \sum_{ss's''} \int C_{kpq}^2 \left[\frac{A_1(sk, s'p, s''q)}{\mu + \imath(s\sigma_k + s'\sigma_p + s''\sigma_q)} e''(e - e') \right] d^3\boldsymbol{p}$$

$$+ \frac{1}{2^3} \sum_{ss's''} \int C_{kpq}^2 \left[\frac{A_2(sk, s'p, s''q)}{\mu + \imath(s\sigma_k + s'\sigma_p - s''\sigma_q)} e^{2\imath s''(\lambda'' - \sigma_q t)} e Z(s''q) \right] d^3\boldsymbol{p}$$

$$+ \frac{1}{2^3} \sum_{ss's''} \int C_{kpq}^2 \left[\frac{A_3(sk, s'p, s''q)}{\mu + \imath(-s\sigma_k + s'\sigma_p + s''\sigma_q)} e^{2\imath s(\lambda - \sigma_k t)} e'' Z(sk) \right] d^3\boldsymbol{p}$$

$$- \frac{1}{2^3} \sum_{ss's''} \int C_{kpq}^2 \left[\frac{A_5(sk, s'p, s''q)}{\mu + \imath(s\sigma_k + s'\sigma_p - s''\sigma_q)} e^{2\imath s''(\lambda'' - \sigma_q t)} e' Z(s''q) \right] d^3\boldsymbol{p}$$

$$+ \frac{1}{2^3} \sum_{ss's''} \int C_{kpq}^2 \left[\frac{A_4(sk, s'p, s''q)}{\mu + \imath(-s\sigma_k + s'\sigma_p - s''\sigma_q)} e^{2\imath s''(\lambda'' - \sigma_q t) + 2\imath s(\lambda - \sigma_k t)} \right.$$

$$\left. \times Z(s''q)Z(sk) \right] d^3\boldsymbol{p}$$

$$- \frac{1}{2^3} \sum_{ss's''} \int C_{kpq}^2 \left[\frac{A_4(sk, s'p, s''q)}{\mu + \imath(s\sigma_k - s'\sigma_p - s''\sigma_q)} e^{2\imath s''(\lambda'' - \sigma_q t) + 2\imath s'(\lambda' - \sigma_p t)} \right.$$

$$\left. \times Z(s''q)Z(s'p) \right] d^3\boldsymbol{p}, \qquad (14.30)$$

where the geometric factors A_1 to A_5 are given in the CMG appendix, and are recalled below. Equations are very symmetric. With respect to EDQNM2, the presence of a Z, or Z', Z'' factor results in changing the corresponding sign in the term $\pm\sigma_k \pm \sigma_p \pm \sigma_q$, and to add the specific time-oscillating phase factor $e^{-2\imath\sigma t}$. $T^{(e)}$ being real, it is possible to retain only $s = 1$ and to replace complex contributions with twice their real part. As in Eqs. (14.19) and (14.20), $e = e(\boldsymbol{k}, t)$, $e' = e(\boldsymbol{p}, t)$, $e'' = e(\boldsymbol{q}, t)$.

14.4 Application of Three Versions to the Rotating Turbulence

The EDQNM3 version of $T^{(z)}$, as follows, is derived from its EDQNM2 counterpart in a similar way, except that the whole term is multiplied, in addition, by the oscillating term $e^{2\imath\sigma_k t}$:

$$T^{(z)} = \frac{1}{2^3}\sum_{s's''}\int C_{kpq}^2 e^{2\imath(\sigma_k t - \lambda)}\left[\frac{A_3(k, -s'p, -s''q)}{\mu + \imath(\sigma_k + s'\sigma_p + s''\sigma_q)}e''(e' - e)\right]d^3\boldsymbol{p}$$

$$+ \frac{1}{2^3}\sum_{s's''}\int C_{kpq}^2 e^{2\imath(\sigma_k t - \lambda)}\left[\frac{A_4(k, -s'p, -s''q)}{\mu + \imath(\sigma_k + s'\sigma_p - s''\sigma_q)}e^{2\imath s''(\lambda'' - \sigma_q t)}eZ(s''\boldsymbol{q})\right]d^3\boldsymbol{p}$$

$$+ \frac{1}{2^3}\sum_{s's''}\int C_{kpq}^2 e^{2\imath(\sigma_k t - \lambda)}\left[\frac{A_1(k, -s'p, -s''q)}{\mu + \imath(-\sigma_k + s'\sigma_p + s''\sigma_q)}e^{2\imath(\lambda - \sigma_k t)}e''Z(\boldsymbol{k})\right]d^3\boldsymbol{p}$$

$$- \frac{1}{2^3}\sum_{s's''}\int C_{kpq}^2 e^{2\imath(\sigma_k t - \lambda)}\left[\frac{A_5(k, -s'p, -s''q)}{\mu + \imath(\sigma_k - s'\sigma_p + s''\sigma_q)}e^{2\imath s'(\lambda' - \sigma_p t)}e''Z(s'\boldsymbol{p})\right]d^3\boldsymbol{p}$$

$$+ \frac{1}{2^3}\sum_{s's''}\int C_{kpq}^2 e^{2\imath(\sigma_k t - \lambda)}\left[\frac{A_2(k, -s'p, -s''q)}{\mu + \imath(-\sigma_k + s'\sigma_p - s''\sigma_q)}e^{2\imath s''(\lambda'' - \sigma_q t) + 2\imath(\lambda - \sigma_k t)}\right.$$

$$\left.\times Z(s''\boldsymbol{q})Z(\boldsymbol{k})\right]d^3\boldsymbol{p}$$

$$- \frac{1}{2^3}\sum_{s's''}\int C_{kpq}^2 e^{2\imath(\sigma_k t - \lambda)}\left[\frac{A_2(k, -s'p, -s''q)}{\mu + \imath(\sigma_k - s'\sigma_p - s''\sigma_q)}e^{2\imath s''(\lambda'' - \sigma_q t) + 2\imath s'(\lambda' - \sigma_p t)}\right.$$

$$\left.\times Z(s''\boldsymbol{q})Z(s'\boldsymbol{p})\right]d^3\boldsymbol{p}. \tag{14.31}$$

Accordingly, all explicit time-dependent oscillating terms cancel for the $T^{(z)}$ term that depends on the third one, $Z(\boldsymbol{k})$.

Let us recall the definition of geometric coefficients[§]:

$$C_{kpq} = \frac{\sin(p, q)}{k} = \frac{\sin(k, q)}{p} = \frac{\sin(k, p)}{q}, \tag{14.32}$$

and

$$A_1(k, p, q) = -(p - q)(k - q)(k + p + q)^2, \tag{14.33}$$

$$A_2(k, p, q) = -(p - q)(k + q)(k + p + q)(k + p - q), \tag{14.34}$$

$$A_3(k, p, q) = (p - q)(k + q)(k + p + q)(-k + p + q), \tag{14.35}$$

$$A_4(k, p, q) = (p - q)(k - q)(k + p + q)(k - p + q), \tag{14.36}$$

$$A_5(k, p, q) = -(p - q)(p + q)(k + p + q)(k + p - q). \tag{14.37}$$

The other geometric coefficients that depend on not only the triad geometry (via moduli k, p, q), but also on the orientation of its plane, are only $\lambda, \lambda', \lambda''$ terms. Following Cambon (1982), Cambon and Jacquin (1989), and Waleffe (1993), they

---

[§] The additional factor $2p/k$ was a mistake in the CMG appendix.

are displayed by substituting into the local frames related to the helical (or complex Craya–Herring) decomposition [$N(sk)$, $N(s'p)$, $N(s''q)$] alternative ones that have their polar axis normal to the plane of the triad rather than to the plane of rotation, so that

$$N(sk) = e^{s\imath\lambda}\underbrace{(\boldsymbol{\beta} + \imath s\boldsymbol{\gamma})}_{W(s)}, \quad N(s'p) = e^{s'\imath\lambda'}\underbrace{(\boldsymbol{\beta'} + \imath s'\boldsymbol{\gamma})}_{W'(s')}, \quad N(s''q) = e^{s''\imath\lambda''}\underbrace{(\boldsymbol{\beta''} + \imath s''\boldsymbol{\gamma})}_{W''(s'')},$$

(14.38)

in which $\boldsymbol{\gamma}$ is the unit vector normal to the plane of the triad, whereas $\boldsymbol{\beta}, \boldsymbol{\beta'}, \boldsymbol{\beta''}$ are unit vectors all located in the plane of the triad, and normal to $\boldsymbol{k}$, $\boldsymbol{p}$, and $\boldsymbol{q}$, respectively. Accordingly, the scalar products in terms of $\boldsymbol{k}, \boldsymbol{p}, \boldsymbol{q}, W, W'$, and W'' depend on only the moduli k, p, q. These scalar products generate all the A_1–A_5 terms.

The last equations, derived from the previous one, which are used in general EDQNM equations [e.g. (14.30) and (14.31)], are

$$\cos\theta_p = p_\parallel/p = -z\cos\theta_k + \sqrt{1-z^2}\sin\lambda, \tag{14.39}$$

$$\cos\theta_q = q_\parallel/q = -y\cos\theta_k - \sqrt{1-y^2}\sin\lambda, \tag{14.40}$$

with $y = \cos(k,q), z = \cos(k,p), \sin(k,q) = C_{kpq}p, \sin(k,p) = C_{kpq}q$. Accordingly,

$$p\cos\theta_p = -q\cos\theta_q = pqC_{kpq}\sin\lambda \tag{14.41}$$

at $k_\parallel = 0$.

The asymptotic limit of wave turbulence in terms of e, Z, h is (Bellet et al., 2006)

$$T^{(e)} = \frac{\pi}{4}\sum_{s',s''} C_{kpq}^2 \frac{A_1(k, s''q, s'p)}{s'C_g(\boldsymbol{p}) - s''C_g(\boldsymbol{q})}\left[e'(e'' - e) + s'h'(s''h'' - h)\right]dS, \tag{14.42}$$

$$T^{(h)} = \frac{\pi}{4}\sum_{s',s''} C_{kpq}^2 \frac{A_1(k, s''q, s'p)}{s'C_g(\boldsymbol{p}) - s''C_g(\boldsymbol{q})}\left[s'H'(e'' - e) + e'(s''h'' - h)\right]dS, \tag{14.43}$$

and

$$T^{(z)} = -Z\frac{\pi}{4}\left[\sum_{s',s''} C_{kpq}^2 \frac{A_1(k, s''q, s'p)}{s'C_g(\boldsymbol{p}) - s''C_g(\boldsymbol{q})}e'dS + \imath \iiint C_{kpq}^2 \frac{A_1(k, s''q, s'p)}{s'C_g(\boldsymbol{p}) - s''C_g(\boldsymbol{q})}e'd^3\boldsymbol{p}\right].$$

(14.44)

Equations (14.42) and (14.43) are essentially the same as in Galtier (2003). The last equation, and Z in general, is ignored in conventional wave-turbulence theory (Waleffe, 1993; Galtier, 2003). The transfer term $T^{(z)}$ is linear in Z, and it is the only term that does not reduce to a surfacic integral ($\iint dS$) over surfaces of resonant triads: The integral $\iiint d^3\boldsymbol{p}$ in Eq. (14.44) denotes a principal-value integral in the vicinity of the resonant surface. Much more complex quadratic interaction terms that involve Z in volumic EDQNM3 [Eqs. (14.30) and (14.31)] are discarded in AQNM when removing rapidly oscillating terms.

14.5 Other Cases of Flows With and Without Production

In the case of magnetohydrodynamic flows, which are not discussed here for the sake of brevity, the use of similar two-point closure/wave-turbulence theories is particularly relevant (see Galtier et al., 2001, for flows dominated by Alfvén waves).

14.5 Other Cases of Flows With and Without Production

Throughout this book, we have distinguished between flows dominated by production and flows dominated by waves. The first class is illustrated by classical shear flows, in which a nonzero production term is displayed in the equations governing the RST. This production is often related to growth of instabilities, when stability analysis is addressed. The second class is illustrated in Chapters 4, 7, and 8 as being the most relevant area to apply spectral closures. Note that the dynamics can be dominated by dispersive waves, which are neutral but for a small part of the configuration space, in which exponential amplification occurs. In the latter case, e.g., for flows with weak ellipticity (i.e., $S \ll \Omega$), the production of energy is nonzero, but classic single-point closure models are of poor relevance, as only particular orientations in wave space are subjected to parametric instability.

14.5.1 Effects of the Distorting Mean Flow

14.5.1.1 Hyperbolic and Elliptic Cases

In the hyperbolic and elliptic cases, with $0 \neq S \neq \Omega$ in Eq. (5.7), the RDT Green's function can display exponential growth at least for particular angles of k ($k_3/k \sim 1/2$ in the case $S \ll \Omega_0$). If the bare zeroth-order response function is modified only by ED, with exponential decorrelation as in Eq. (14.15), convergence is not ensured for the time integral of the threefold product **GGG** in the generic closure relationship. Another type of nonlinear decorrelation operator, e.g., a Gaussian one, could be used.

14.5.1.2 Pure Shear

A less critical situation occurs when $S = \Omega$ (pure plane shear), as the RDT Green's function yields only algebraic growth, so that the viscous term ensures convergence of the time integral involved in the closure. Nevertheless, it is very cumbersome to develop, and especially to solve numerically with enough accuracy, a complete anisotropic EDQNM model in this case. Recall that even calculation of single-point correlations resulting from viscous RDT at high St is not easy (Beronov and Kaneda, private communication). DNSs suggest that fully nonlinear effects yield exponential growth for the turbulent kinetic energy, but computations are very sensitive to cumulated errors (remeshing, low angular resolution at small k, etc.). Such a transition from algebraic growth (linear, small time) to exponential growth (nonlinear) is mimicked by simple models but not really explained. Interesting scaling laws, however, for possible exponential growth follow from self-similarity arguments, as discussed in Chapter 6.

Case	Relevant linear eigenmodes	Conservation laws (inviscid theory)		TKE cascade(s)	Nonlinear effects
		Detailed (isolated triad)	Global (sum over triads)		
Isotropic (3D)	none	➔ Kinetic energy ➔ Helicity	➔ Kinetic energy ➔ Helicity (zero net helicity)	➔ Total (poloidal + toroidal) cascade ➔ existence of both forward and reverse cascades	➔ net forward cascade at high wave numbers ➔ net backward cascade at very low wave numbers
Isotropic (2D)	none	➔ Kinetic energy ➔ Enstrophy	➔ Kinetic energy ➔ enstrophy	➔ existence of reverse cascade only	➔ net forward enstrophy cascade at high wave numbers ➔ net backward energy cascade at high wave numbers
Pure rotation	Inertial waves	➔ Kinetic energy ➔ Helicity	➔ Kinetic energy, ➔ Helicity	Cascade associated to resonant triads	TKE concentration on equatorial plane
Pure stable stratification	➔ Toroidal velocity mode (constant) ➔ Gravity waves	➔ (At very low Froude number only): vertical enstrophy & toroidal kinetic energy	➔ Toroidal kinetic energy ➔ Wave total energy (kinetic = AG + potential)	➔ Toroidal cascade excluding "weak" wave-turbulence ➔ poloidal and potential modes affected by wave-turbulence	Layering due to toroidal cascade, i.e. concentration of TKE on the polar direction
Rotation + stable stratification	➔ Quasi-geostrophic mode (constant) ➔ Inertia-gravity waves	➔ (neglecting wave mode): QG energy & linearized potential enstrophy	➔ QG energy ➔ Wave total energy (kinetic = poloidal + potential)	QG cascade	Multiform, depending on f/N

Figure 14.2. Table of different properties, including eigenmodes (if relevant), conservation law, and cascade processes.

14.5.2 Flows Without Production Combining Strong and Weak Turbulence

A particular class of flows "without production" involves both wavy and nonpropagating modes, the latter being constant in the linear limit. Their dynamics can mix strong and weak turbulence. On the one hand, strong turbulence is concerned only when nonlinear interactions in terms of the nonpropagating modes are considered: That includes the "toroidal turbulence" for pure stratification, the QG turbulence for the stratified rotating case, and the pure solenoidal case for the weakly compressible flow case.

Principal features of flows without production, that are addressed in this book are collected in Fig. 14.2.

Investigation of interactions with waves is a second step in the study of such flows: We are first faced with the problem of interacting acoustic waves in the latter case only, because the solenoidal problem is essentially solved (e.g., using conventional isotropic EDQNM consistent with a Kolmogorov energy spectrum).

14.5.2.1 Buoyant Flows in a Stably Stratified Fluid

In the purely stratified case, gravity-wave turbulence is crucial only if the nonpropagating mode, i.e., the toroidal part of the velocity field, is *a priori* discarded. This

14.5 Other Cases of Flows With and Without Production

removal is generally unphysical, and oversimplified wave-turbulence studies, such as the one by Caillol and Zeitlin (2000), are only marginally relevant. The claim of explaining horizontal layering in the latter paper is highly misleading. In contrast, emphasis on rather strong turbulence is much more relevant, at least at moderate times, looking at the "toroidal cascade," and the transition from a 3D isotropic unstructured flow to a strongly anisotropic, horizontally layered, flow can be described by the statistical theory.

14.5.2.2 Weakly Compressible Isotropic Turbulence

The case of weakly compressible turbulence is not present in Fig. 14.2. In this case, the solenoidal mode plays a role similar to the toroidal mode in stratified turbulence and to the QG mode in the rotating and stratified case, but the pseudo-acoustic mode is not necessarily a wave mode: True acoustic waves are observed at very low wavenumbers, whereas the pseudo-sound regime may hold at higher wavenumbers. As for the case of buoyant turbulence in a stratified fluid, the basic equations in terms of "slow" amplitudes are

$$\frac{\partial a_s}{\partial t} = \sum_{s',s''=0,\pm 1} \int_{k+p+q=0} \exp[-\iota (s\sigma_k + s'\sigma_p + s''\sigma_q) t]$$
$$\times N_{ss's''}(k, p) a_{s'}^*(p, t) a_{s''}^*(q, t) d^3 p. \qquad (14.45)$$

Diffusive terms can be neglected for a preliminary discussion of couplings. Of course, the coupling coefficients $N_{ss's''}$ completely differ from their counterparts in the solenoidal buoyant case subjected to stable stratification, and a_0 is two-component (solenoidal mode) in the compressible case. A similar cartoon, however, can be discussed in both flow cases, depending on the signs (s, s', s''), or triad polarities, as follows:

1. Nonpropagating slow mode, $s = 0$. It is clear that the nonlinear dynamics is dominated by interactions between slow modes only, so that the leading terms may correspond to $s' = s'' = 0$: One recovers the "toroidal turbulence" for the stratified flow case and pure incompressible dynamics for the weakly compressible flow case. The main difference is that incompressible isotropic turbulence is well understood, at least regarding energy spectrum and energy transfer, whereas toroidal turbulence is still under investigation. Consequently, a large-Reynolds-number Kolmogorov energy spectrum can be specified and fixed for the solenoidal mode, as in Fig. 9.5.

2. Our main interest in this subsection is the mode related to $s = \pm 1$, generating "dilatational velocity" and "pressure" contributions, which are closely connected together, or not, by means of a possible acoustic equilibrium. It is very difficult to rank *a priori* the three kinds of interactions $(\pm 1, 0, 0)$, $(\pm 1, \pm 1, 0)$, and $(\pm 1, \pm 1, \pm 1)$ for (s, s', s''). The first one is never resonant, but cannot be completely removed from consideration of whether the order of magnitude of a_0 is much larger than the one of $a_{\pm 1}$. The second one will select resonant "dyads," like $k \pm p = 0$. Only the third one will select resonant triads, such as $k \pm p \pm q = 0$.

It is clear that wave turbulence is only a part of the whole story, which is even irrelevant in some cases, like the toroidal turbulence in the stably stratified case. A simplified EDQNM3 closure strategy is applicable, but the study by Fauchet (1998) has shown the importance of a nonconventional ED term, denoted by V in this chapter, or, more generally of the nonlinear part of Kraichnan's response function.

Finally, we have illustrated closure theories for weakly compressible flows in Chapter 9 by a peculiar study. This viewpoint, which may appear idiosyncratic, is mainly motivated by the existence of detailed asymptotic laws that were derived, with practical interest. There exist important works on this topic, from wave turbulence for nonlinear sound (Zakharov, L'vov, and Falkowich, 1992), to absolute equilibrium in truncated Euler equations. For instance, a generalized $k^{-5/3}$ law can be inferred from Kraichnan (1955) for both solenoidal and acoustic modes, even in the viscous case, but radiation to infinity is excluded in this study.

14.5.3 Role of the Nonlinear Decorrelation Time Scale

When comparing strong turbulence without production and wave turbulence, it is important to stress significant differences:

- Conventional isotropic EDQNM works well, at least for predicting energy spectra and transfers, but we do not know really why! The role of ED is crucial, and even the QN structure results from only a heuristic closure strategy. In any case of strong turbulence less documented, as the toroidal turbulence in stably stratified flows, the conventional isotropic ED probably needs refinement, especially in the spectral region where energy concentrates (quasi-VSHF 1D modes). Recourse to more sophisticated self-consistent closure theories may be useful.
- The "pure" wave-turbulence theory, which appears as a limiting case of QNM closure, also works well, but we know why! A QN structure can be supported by mathematical analysis (Benney and Newell, 1969) or by a physically relevant random-phase approximation. Markovianization results from a rigorous rapid–slow time-scale decomposition, and ED is unimportant in the asymptotic limit. In this limit, there is no need for a significant nonlinear renormalization of the bare dispersion frequency, too.
- The role of ED appears to be very subtle in the "mixed" case, when wave turbulence coexists with strong turbulence. In stably stratified turbulence, a quasi-perfect agreement was found between DNS results and EDQNM2 results (Godeferd and Staquet, 2003), keeping the same ED (inherited from HIT) for all interactions, but only the relatively low-Reynolds-number range, which was allowed in DNS, was investigated. In the case of very high Reynolds number and very low Froude number, with large $ReFr^2$ parameter, which is discussed at the end of Chapter 7, a refined analysis will be needed. The case of quasi-isentropic isotropic turbulence offers a very good example: Keeping the same ED for all interactions yields poor results. Very striking results are found by choosing a Gaussian kernel for V in Eq. (14.10) instead of a exponential one. As discussed in Chapter 9, this cannot be obtained by replacing the acoustic-wave frequency with

a renormalized nonlinear one. A relevant explanation, given in this book but not in the original reference (Fauchet, 1998), is to add a random part to the linear dispersion frequency, in agreement with Kraichnan's random oscillator.

A related problem is the possible need for a "renormalized" wave frequency in wave turbulence. For instance, a nonlinear shift in Rossby wave frequency is demonstrated from statistical theory and DNS (Kaneda and Holloway, 1994; Ishihara and Kaneda, 2001), whereas such a shift seems to be useless in 3D rotating turbulence (inertial wave turbulence) and in MHD turbulence (Alfvén wave turbulence). The study of Galtier et al. (2001) had the merit of showing that even nondispersive waves can generate weak wave turbulence, against a well-established prejudice. Of course, phase mixing results from dispersivity, and naturally damps nonlinearity. The "prejudice," however, is possibly linked to a confusion between a pure advection term by a velocity, V, yielding $\exp(\imath \mathbf{k} \cdot \mathbf{V})$ in Fourier space, and the phase term $\pm \imath \mathbf{k} \cdot \mathbf{V}$ of nondispersive waves,...,forgetting the sign ± 1! Because of the sign, which allows propagation in opposite directions (generally coming from second order in time, Dalembertian-type, operator), wave operators affect the triple correlations – even in the absence of dispersive effects – whereas pure advection terms do not. Rossby waves are characterized by a first order in time operator, and therefore propagate only in one direction. We do not question here the wave terminology, even if Rossby waves could be called "westward advected oscillations" instead of "waves." This is a question of definition of waves, but one can point out the very different nature of inertial, Alfvén, and Rossby waves. The Rossby waves could be less efficient in damping nonlinearity, so that a nonlinear shift would reveal not too weak a turbulence.

14.6 Connection with Self-Consistent Theories: Single Time or Two Time?

The Kraichnan's DIA, in spite of some drawbacks, played a crucial role in the long history and progressive generation of "triadic closure" theory. The two-time aspect is essential in the first version, as well as in its Lagrangian or semi-Lagrangian more sophisticated subsequent variants (see Kaneda, 2007, for a very recent review). The aim of such a two-time and two-point (or even three-point) statistical theory is to derive a consistent set of close equations for both the response tensor, corresponding to $\mathbf{G}^+(\mathbf{k}, t, t')$ in this chapter, and to the two-time spectral tensor, which generalizes our $\hat{R}$ as $\hat{R}(\mathbf{k}, t, t')$.

A theory formulated in terms of two-time statistical tensors can be converted into its single-time counterpart by use of a so-called fluctuation–dissipation theorem. For instance,

$$\hat{R}_{ij}(\mathbf{k}, t, t') = G^+_{in}(\mathbf{k}, t, t') \hat{R}_{nj}(\mathbf{k}, t', t'), \qquad (14.46)$$

ignoring the possible time dependency of the wave vector for the sake of simplicity. In conventional applications, the two-time dependency of the response tensor is *a priori* specified [for instance, exponential $(t - t')$ decorrelation]. EDQNM can be

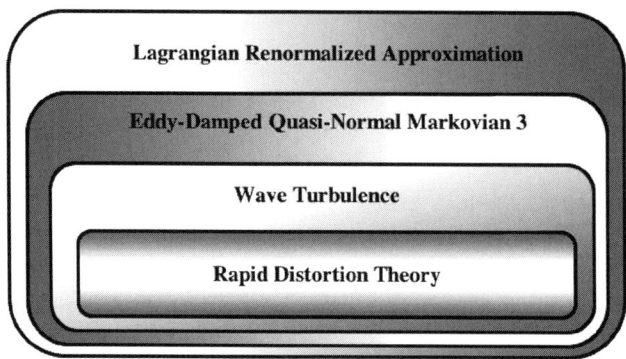

Figure 14.3. Embedded theories.

presented as a by-product of DIA in this way, but we think that it is a rather complicated and indirect way to proceed. Few applications of DIA or EDQNM were made in the context of HAT. One can mention the return to isotropy from a (weak) anisotropic (axisymmetric) case by Herring (1974), and the weakly axisymmetric QG EDQNM model by the same author mentioned in Chapter 8. More sophisticated anisotropic models were developed by Sanderson, Hill, and Herring (1986), using a small number of spherical harmonics. None of these studies were able to incorporate as a building block the RDT Green's function as a natural zeroth-order response tensor, if we exclude wave turbulence, of course.

A general formulation of two-time DIA yields a single Green's function in the nonlinear closure of the equation for the two-time correlation spectral tensor, and the two other factors then appear by means of the preceding fluctuation–dissipation relationship, leading to essentially the same form as Eq. (14.12) with a threefold product of response tensors. Of course, solving (inverting?) the modified response tensor operator for third-order single-time correlations is simpler and more direct, the threefold product even appearing in the basic equation for velocity fluctuation, rewritten in Eqs. (14.9) and (14.23). The final DIA-type evolution equation for the two-time spectral tensor therefore contains an integral whose structure is much the same as EDQN expression (14.12), with terms such as $G_{lq}(\bm{q},t,t')\hat{R}_{qn}(\bm{q}',t')$ replaced with the two-time spectral tensor $\hat{R}_{ln}(\bm{q}',t,t')$, leaving one remaining Green's function from the threefold product, which is replaced with the response tensor.

A more streamlined procedure could be based on EDQNM3, using DIA and subsequent self-consistent theories for improving the ED factor only, in agreement with the strict hierarchy of embedded models/theories displayed in the scheme shown in Fig. 14.3, in which the LRA (Kaneda, 2007) is chosen at the end of the list, as a representation of self-consistent theories with recourse to Lagrangian approach, without excluding other variants. This strategy can be used for deriving complete two-time statistics, as illustrated in RDT by Kaneda and Ishida (2000), in order to have access to $\hat{\mathbf{R}}(\bm{k},t,t')$. The way of solving operators linked to any product of response tensors is applicable. The only difficulty could result from a very complex equation for the response tensor $\mathbf{G}^{+}(\bm{k},t,t')$, with no explicit simplified solution in terms of $\hat{\mathbf{R}}(\bm{k},t,t')$ [such as $\mathbf{G}^{(0)}(\bm{k},t,t')V(\bm{k},t,t')$].

14.7 Applications to Weak Anisotropy

Applications to weakly anisotropic flows have been mentioned previously (Herring, 1974; Sanderson, Hill, and Herring, 1986). In addition, some recent applications of LRA to the response of turbulence to a weak linear operator in the presence of strong nonlinearity deserves attention.

14.7.1 A Self-Consistent Representation of the Spectral Tensor for Weak Anisotropy

The most general decomposition of the spectral tensor (e.g., the $e - Z - H$ decomposition introduced and discussed in Chapter 2), which holds for arbitrary flow anisotropy, involves never more than four real scalars. In addition to the very existence of the polarization anisotropy Z, anisotropy is reflected by the angle dependence of these basic scalars. Looking at the trace of the spectral tensor, it is clear that $E(k)/(4\pi k^2)$ is only the zeroth-degree angular harmonic of $e(\mathbf{k})$ and gives no information on its angular distribution in wave space. Nevertheless, some information about this angular distribution can be obtained by spherically averaging all the components of the spectral tensor $\hat{\mathbf{R}}$, because some weighting factors, such as the projector P_{ij} (for e) or the polarization deviator $N_i N_j$ (for Z) generate angular harmonics until the degree 2. As a result, the following self-consistent decomposition is found:

$$\hat{R}_{ij}(\mathbf{k},t) = \frac{E(k)}{4\pi k^2}\left[\left(1 - 15 H^{(e)}_{pq}\frac{k_p k_q}{k^2}\right)P_{ij} + 5\left(P_{in}P_{jm}H^{(z)}_{nm} + \frac{1}{2}P_{ij}H^{(z)}_{pq}\frac{k_p k_q}{k^2}\right)\right],$$
(14.47)

in which one can identify the contribution from the directional anisotropy as

$$\hat{R}^{(\text{dir})}_{ij} = \left(e - \frac{E}{4\pi k^2}\right)P_{ij} = -15 H^{(e)}_{pq}(k)\frac{k_p k_q}{k^2}\frac{E(k)}{4\pi k^2}P_{ij}(\mathbf{k}),$$
(14.48)

and the contribution of polarization anisotropy¶ as

$$\hat{R}^{(\text{pol})}_{ij} = 5\frac{E(k)}{4\pi k^2}\left[P_{in}(\mathbf{k})P_{jm}(\mathbf{k})H^{(z)}_{nm}(k) + \frac{1}{2}P_{ij}(\mathbf{k})H^{(z)}_{pq}(k)\frac{k_p k_q}{k^2}\right],$$
(14.49)

in addition to the purely isotropic part

$$\hat{R}^{(\text{iso})}_{ij} = \frac{E(k)}{4\pi k^2}P_{ij}(\mathbf{k}).$$

This decomposition, introduced by Cambon and Rubinstein (2006), can generalize many other similar tensorial expansions. It is self-consistent in the sense that it does not involve any adjustable parameter. Given an arbitrary anisotropic $\hat{\mathbf{R}}$, it is possible to derive from it the spherically averaged spectra $H^{(e)}_{ij}(k)$ and $H^{(z)}_{ij}(k)$

¶This form is strictly equivalent to its counterpart in terms of Z, $\Re(Z N_i N_j)$, with $Z = (5/2)E/(4\pi k^2)H^{(z)}_{ij}N_i^* N_j^*$, without using the helical-mode vector $\mathbf{N}$.

defined in Chapter 2, and then to reconstruct its angle-dependent form, up to a given degree of angular harmonics, using Eq. (14.47). The difference between the original, arbitrarily anisotropic, $\hat{\mathbf{R}}$, and its weakly anisotropic approximation generated by $H_{ij}^{(e)}(k)$ and $H_{ij}^{(z)}(k)$ is the contribution from higher-degree harmonics that cannot be reconstructed from $H_{ij}^{(e)}$ and $H_{ij}^{(z)}$ alone. Let us discuss only the case of directional dependence, as a simple example, for illustrating the building of angular dependence. A systematic way to express the angular dependence of a "true" scalar such as e is using scalar spherical harmonics:

$$e = \frac{1}{2}\hat{R}_{ii} = \frac{E}{4\pi k^2} + \sum_{n=1}^{N_0} \sum_{m=-2n}^{2n} e_{2n}^m(k) \underbrace{P_{2n}^m(\cos\theta_k)\exp(\imath m \phi_k)}_{Y_{2n}^m(\theta_k, \phi_k)},$$

whereas the expression derived from (14.47) is

$$e = \frac{E(k)}{4\pi k^2}\left(1 - 15 H_{pq}^{(e)} \frac{k_p k_q}{k^2}\right),$$

in exact agreement with the decomposition in terms of scalar spherical harmonics for $2N_0 = 2$. The symmetric trace-free tensor $E(k)H_{ij}^{(e)}(k)$ includes five independent components, which correspond exactly to the five independent coefficients $e_{2n}^m(k)$ with $n = 1$. Accordingly, this decomposition is consistent with an expansion of the order of $2N_0 = 2$ of the trace of $\hat{\mathbf{R}}$, which is the first nontrivial degree, given the symmetries. A similar link of angular harmonics to $H_{ij}^{(z)}$ components holds for the polarization anisotropy, but is more complicated because tensorial harmonics are called into play.

Among many particular forms and uses of Eq. (14.47), one can mention its application in Cambon, Jeandel, and Mathieu (1981) to the closure of a spherically averaged version of the Craya equation by EDQNM, in which both "rapid" and "slow" spectra of pressure–strain correlations require a partial reconstruction of the angle dependency in Fourier space.

Going back to the approach of linear response, Eq. (14.47) is only used for translating the main results in the presence of a weak shear, as already touched on in Chapter 5. This approach is very different from the one dedicated to strongly anisotropic turbulence, as the linear response is sought with respect to a weak perturbation (the linear RDT operator) to a nonlinear state in statistical equilibrium. In this sense, the tangent response function can be also weakly anisotropic and therefore far from the RDT linear limit $\mathbf{G}^{(0)}$, which is generally very anisotropic for large $t - t'$. A decomposition of $\mathbf{G}^+$ in terms of a pure isotropic factor and a weakly anisotropic one is found consistently.

The linear response has some analogies with the general laws that connect fluxes and forces in statistical theory for continuum media, with similar symmetry properties as the ones prescribed by Onsager. The main equation for the response to a weak mean flow with A_{ij} velocity gradients is

$$\hat{R}_{ij}^{(\text{aniso})} = Q_{ijnm}(\mathbf{k}) A_{nm},$$

14.7 Applications to Weak Anisotropy

and because of the symmetry of the Q_{ijnm} tensor, only the symmetric part, S_{nm}, of A_{nm} is eventually displayed. Translated into our own formalism, this yields $\hat{R}^{(\text{aniso})} = \hat{R}^{(\text{dir})} + \hat{R}^{(\text{pol})}$, and, using Eqs. (14.48) and (14.49),

$$H_{ij}^{(e)}(k) = \frac{1}{15}(B - A)\left(\frac{k}{k_0}\right)^{-2/3} \varepsilon^{-1/3} k_0^{-2/3} S_{ij}, \tag{14.50}$$

$$H_{ij}^{(z)}(k) = \frac{2}{15} A \left(\frac{k}{k_0}\right)^{-2/3} \varepsilon^{-1/3} k_0^{-2/3} S_{ij}. \tag{14.51}$$

For the basic state, a classical Kolmogorov inertial range is recovered, with $E(k) = C_k \varepsilon^{2/3} k^{-5/3}$, so that the dimensional spectra of deviatoric tensors, $E(k)H_{ij}^{(e)}$ and $E(k)H_{ij}^{(z)}$, exhibit a classical scaling like $\varepsilon^{1/3} k^{-7/3}$, as suggested by Lumley. In the preceding equations, k_0 is identified with the wavenumber at which the inertial range can be considered to begin. Accordingly, $k_0^{-2/3}\varepsilon^{2/3}$ is the typical time scale, and $k_0^{-2/3}\varepsilon^{2/3}S_{ij}$ is the relevant nondimensional strain tensor. Finally, A and B are universal constants, obtained in a satisfactory agreement both by DNS for homogeneous pure plane shear and LRA theory (Ishihara, Yoshida, and Kaneda, 2002; Yoshida, Ishihara, and Kaneda, 2003).

14.7.2 Brief Discussion of Concepts, Results, and Open Issues

Given the strong constraints given by weak anisotropy, with a spectral tensor that is necessarily of the form (14.47), and dimensional analysis "à la Lumley," there are very few degrees of freedom, and the main results can be obtained by much simpler, even wrong, ways. The merit of LRA in this case, is to find the result in a rigorous and self-consistent way, avoiding useless oversimplifications. Even if the specific shear-advection term inherited from the linear operator has no significant effect on the tensor $H_{ij}^{(e)}$, which expresses the linear response as in the short-time RDT limit (see Chapter 5), this shear-advection effect is correctly accounted for in the intermediate theoretical steps, so that the confusion between relations (14.2) and (14.3) is avoided from the beginning.

One could expect, at least in the case of pure plane shear, to reconcile an approach to strong anisotropy, more restricted to large scales, and the linear response theory, limited to very small scales.

Not even mentioning the case of combined effects of irrotational strain and vorticity, which leads to elliptical or hyperbolical instabilities with exponential growth in the linear limit, the case of solid-body rotation deserves some attention. It appears that the Coriolis force has no impact on the linear response. This is consistent with an objectivity principle satisfied in continuum mechanics. Nevertheless, it is well known that Chapman–Enskog-type developments for Boltzmann equations can question such objective laws if they are carried out at a sufficient order. The constitutive laws, or fluxes-to-forces relationship, could become explicitly Coriolis dependent in this situation. In the same way, the effect of solid-body rotation can be recovered at a further order (quadratic dependency on Ω?) using LRA.

Another point is the fact that strong anisotropy induced by the Coriolis force at a sufficiently low Rossby number is found to be dominant at small scale, as shown by both wave-turbulence theory and DNS results. The classical picture of strong anisotropy restricted to largest scales is radically questioned. In this situation, it seems to be difficult to match both low-Rossby and high-Rossby limits.

14.8 Open Numerical Problems

The numerical cost of solving EDQNM equations, as well as those issued from similar single-time or even two-time "triadic" theories, is very low in the isotropic case. This cost, and the complexity of the numerical procedure, can blow up, not only in an inhomogeneous configuration, as it is often said, but even in the case of strong anisotropy. The numerical solution of the equations of wave turbulence was demanding in terms of numerical resources, with a particular care for accurately capturing the resonant surfaces with complex shapes.

Reaching very high Reynolds numbers and even asymptotic limits, e.g., vanishing Rossby numbers, is not a problem in solving these statistical model equations, in contrast to DNS. This is the number of angular variables in interaction that is responsible for the high cost, especially because the classical pseudo-spectral scheme is difficult to apply: a factorization like $A(\mathbf{k}) \sum \hat{u}(\mathbf{p})\hat{u}(\mathbf{q})$, which is very simple for basic Navier–Stokes equations, yielding $A(\mathbf{k})\widehat{uu}$ is very cumbersome when one is looking at typical equations in terms of spectral tensors and response functions. A higher accuracy, however, can be obtained in statistical closures, for accounting for typical triads, such as the resonant ones but also the quasi-exact cancellation between some of them. Even very simple quantities affected by phase mixing, e.g., in Chapter 7, whose history consists of damped oscillations with a smooth envelope, are found to exhibit chaotic wrong envelopes after a finite integration time, in any classical pseudo-spectral DNS, because of limited accuracy in terms of $\Delta k/k$, $k_{\|}/k$, etc. It is therefore pertinent to try to solve costly statistical models. Attempts to reduce this cost, using – despite the cumbersome factorization – pseudo-spectral techniques, or Monte Carlo methods, do exist but are outside our scope.

Bibliography

BELLET, F., GODEFERD, F. S., SCOTT, J. F., AND CAMBON, C. (2006). Wave-turbulence in rapidly rotating flows, *J. Fluid Mech.* **552**, 83–121.

BENNEY, D. J. AND NEWELL, A. C. (1969). Random wave closure, *Stud. Appl. Math.* **48**, 29–53.

BOS, W. J. T. AND BERTOGLIO, J.-P. (2006). A single-time two-point closure based on fluid particle displacements, *Phys. Fluids* **18**, 031706.

CAILLOL, P. AND ZEITLIN, W. (2000). Kinetic equations and stationary energy spectra of weakly nonlinear internal gravity waves, *Dyn. Atmos. Oceans* **32**, 81–112.

CAMBON, C. (1982). Etude spectrale d'un champ turbulent incompressible soumis à des effets couplés de déformation et rotation imposés extérieurement, *Thèse de Doctorat d'Etat*, Université Lyon I, France.

CAMBON, C. AND GODEFERD, F. S. (1993). Inertial transfers in freely decaying rotating, stably-stratified, and MPD turbulence, in *Progress in Astronautics and Aeronautics*, ed. H. Branover and Y. Unger, AIAA.

CAMBON C. AND JACQUIN, L. (1989). Spectral approach to non-isotropic turbulence subjected to rotation, *J. Fluid Mech.* **202**, 295–317.

CAMBON, C., JEANDEL, D., AND MATHIEU, J. (1981). Spectral modelling of homogeneous anisotropic turbulence, *J. Fluid Mech.* **104**, 247–262.

CAMBON, C., MANSOUR, N. N., AND GODEFERD, F. S. (1997). Energy transfer in rotating turbulence, *J. Fluid Mech.* **337**, 303–332.

CAMBON, C. AND RUBINSTEIN, R. (2006). Anisotropic developments for homogeneous shear flows, *Phys. Fluids* **18**, 085106.

CAMBON, C., RUBINSTEIN, R., AND GODEFERD, F. S. (2004). Advances in wave-turbulence: rapidly rotating flows, *New J. Phys.* **6**, 73, 1–29.

CAMBON, C. AND SCOTT, J. F. (1999). Linear and nonlinear models of anisotropic turbulence, *Annu. Rev. Fluid Mech.* **31**, 1–53.

CAMBON, C., TEISSÈDRE, C., AND JEANDEL, D. (1985). Etude d' effets couplés de rotation et de déformation sur une turbulence homogène, *J. Méc. Théor. Appl.* **5**, 629–657.

FAUCHET, G. (1998). Modélisation en deux points de la turbulence isotrope compressible et validation à l'aide de simulations numériques, *Thèse de Doctorat, Ecole Centrale de Lyon* (in French).

GALTIER (2003). A weak inertial wave turbulence theory, *Phys. Rev. E* **68**, O15301-1-4.

GALTIER, S., NAZARENKO, S., NEWELL, A. C., AND POUQUET, A. (2001). A weak turbulence theory for incompressible MHD, *J. Plasma Phys.* **63**, 447–488.

GODEFERD, F. S. AND CAMBON, C. (1994). Detailed investigation of energy transfers in homogeneous stratified turbulence, *Phys. Fluids* **6**, 284–2100.

GODEFERD, F. S. AND STAQUET, C. (2003). Statistical modelling and direct numerical simulations of decaying stably stratified turbulence. Part 2. Large-scale and small-scale anisotropy, *J. Fluid Mech.*, **486**, 115–159.

HERRING, J. R. (1974). Approach of axisymmetric turbulence to isotropy, *Phys. Fluids* **17**, 859–872.

ISHIHARA, T. AND KANEDA, Y. (2001). Energy spectrum in the enstrophy transfer range of two-dimensional forced turbulence, *Phys. Fluids* **13**(2), 544–547.

ISHIHARA, T., YOSHIDA, K., AND KANEDA, Y. (2002). Anisotropic velocity correlation spectrum at small scale in a homogeneous turbulent shear flow, *Phys. Rev. Lett.* **88**(15), 154501.

KANEDA, Y. (2007). Lagrangian renormalized approximation of turbulence, *Fluid Dyn. Res.* **39**, 526–551.

KANEDA, Y. AND HOLLOWAY, G. (1994). Frequency shifts of Rossby waves in Geostrophic Turbulence, *J. Phys. Soc. Jpn.* **63**, 2974.

KANEDA, Y. AND ISHIDA, T. (2000). Suppression of vertical diffusion in strongly stratified turbulence, *J. Fluid Mech.* **402**, 311–327.

KRAICHNAN, R. H. (1955). Statistical mechanics of an adiabatically compressible fluid, *J. Acoust. Soc. Am.* **27**, 527–530.

LEITH, C. E. (1971). Atmospheric predictability and two-dimensional turbulence, *J. Atmos. Sci.* **28**, 145–161.

LESLIE, D. C. (1973). *Developments in the Theory of Turbulence*, Clarendon.

McCOMB, W. D. (1974). A local energy transfer theory of isotropic turbulence, *Phys. Fluid A* **7** (5), 632.

ORSZAG, S. A. (1970). Analytical theories of turbulence, *J. Fluid Mech.* **41**, 363–386.

POUQUET, A., LESIEUR, M., ANDRÉ, J.-C., AND BASDEVANT, C. (1975). Evolution of high Reynolds number two-dimensional turbulence, *J. Fluid Mech.* **75**, 305–319.

SANDERSON, R. C., HILL, J. C., AND HERRING, J. R. (1986). Transient behavior of a stably stratified homogeneous turbulent flow. In *Advances in Turbulence* (Ed. G. Comte-Bellot and J. Mathieu), Springer, pp. 184–190.

WALEFFE, F. (1993). Inertial transfers in the helical decomposition, *Phys. Fluids A* **5**, 677–685.

YOSHIDA, K., ISHIHARA, T., AND KANEDA, Y. (2003). Anisotropic spectrum of homogeneous turbulent shear flow in Lagrangian renormalized approximation, *Phys. Fluids* **15**(8), 2385–2397.

ZAKHAROV, V. E., L'VOV, V. S., AND FALKOWICH, G. (1992). Kolmogorov spectra of turbulence. 1. Wave turbulence. In *Springer Series in Nonlinear Dynamics*, Springer.

15 Conclusions and Perspectives

Description and knowledge of turbulent flows is advancing well, particularly with the increasing development of numerical resources (Moore's law) and detailed measurements using more and more particle image velocimetry (PIV), stereoscopic particle image velocimetry (SPIV), and particle tracking velocimetry (PTV). Well-documented databases are created that can support techniques of data compression using a dramatically reduced number of modes (POD, wavelet coefficients, master modes, etc.).

Behind this attractive show window, however, the advance of our conceptual understanding of turbulent flows is much less satisfactory. Advances in numerics, experiments, data-compression schemes, are first beneficial to applied studies, for instance those using a smart combination of techniques (often referred to as *multiphysics*, with hybrid RANS–LES methods, and many others). Turbulent flows are well reproduced in the vicinity of a well-documented "design-point," but this modeling is questioned far from it ("far" in the parameter's space, or simply in elapsed time for unsteady processes). Efficiency of data-compression schemes, for instance, is ellusive because a low-dimension set of modes, identified and validated near the design-point, can lose its relevance far from it.

We hope that this book will contribute to an honest and up-to-date survey of turbulence theory, with the special purpose of reconciling different angles of attack. In this sense, the atomization of the community into competing, and/or too (deliberately) self-isolating, chapels, is perhaps one of the main impediments for advancing theory. The difference of parlances or jargons is a related aspect, despite the universality of the mathematical formalism.

15.1 Homogenization of Turbulence. Local or Global Homogeneity? Physical Space or Fourier Space?

One may go back to the theory of homogeneous turbulence (or "homogenization of turbulence") by George Batchelor (1920–2000), following a very interesting recent essay by Moffatt (2002). It is usefully recalled that Batchelor was aware from the very beginning of the importance of Kolmogorov's approach, including the celebrated 4/5 law. He published a deep analysis of the theory as early as 1946 and 1947, having read the four-page seminal article in *Doklady* (Comptes Rendus of USSR Academy of Sciences). It is therefore irrelevant to oppose Batchelor's approach to turbulence to Kolmogorov's. One may evoke the meeting held in Marseille (1961), which is often mentioned as the "Solvay meeting" of turbulence, quoted as a "watershed for turbulence" by K. Moffat:

> Kolmogorov was there, together with Obukhov, Yaglom, and Millionshchikov; von Karman and G. I. Taylor were both there – the great father figures of prewar research in turbulence – and the place was humming with all the current stars of the subject – Stan Corrsin, John Lumley, Phillip Saffman, Les Kovasznay, Bob Kraichnan, Ian Proudman, and George Batchelor himself, among many others.

Finally, it is recalled how Kolmogorov himself questioned the validity of his K41 theory, opening Pandora's box with a scale-dependent, intermittent distribution of $\varepsilon(r)$. This resulted in both a large interest for internal intermittency, and a frustration that afflicted Batchelor and many others from 1960 onward.

We think, however, that very important progress in the theory was made following Kraichnan's approach, even before the early sixties, not to mention linear theory such as RDT, and that it is a pity to underestimate related studies, as is often done in the "intermittency and scaling" community, especially after the publication of Frisch's book. In addition, the development of practical models, mainly based on single-point closures, in RANS and (more recently) in LES, was very useful for turbulence in engineering and environment, with almost no impact of new developments of theory of internal intermittency, but often a strong connection with the spectral approach. Unjustified* reluctance to look at a formalism in Fourier space, and strong (justified) interest for a statistical approach in terms of velocity (or vorticity, pressure, etc.) *increments* can explain partly such an underestimation.

As a first example, it can be shown (e.g., Chapter 3) that the Kolmogorov law $\langle \delta u_\parallel^3 \rangle(r)/r = -(4/5)\varepsilon$ is as "exact" as its counterpart in Fourier space $\int_k^\infty T(k)dk = \varepsilon$ is. In the same way, more general (not only valid at very high Reynolds number) laws were given by von Karman and Howarth in 1938 in physical space and in Fourier space by Lin and von Karman in 1949.

The concept of local homogeneity raises very important questions. On the one hand, the use of increments (e.g., velocity increments) for defining two-point statistics allows for a better approach to local homogeneity, even if "local homogeneity" is almost an oxymoron, because homogeneity means translational invariance. In addition, structure functions of the order of 2 and 3 can be obtained from measurements more easily than second-order spectra and transfer spectra,† and the spectral approach holds little interest for higher-order statistical moments (higher than 4). From this viewpoint, local homogeneity (and very often isotropy) is assumed at relatively small scales in exploiting physical and numerical experiments in rather complex flows; the assessment that "the flow is considered as homogeneous and isotropic in the center of a von Karman flow, in the centerline of the plane channel, near the

---

* Even the more that studies about scaling and intermittency are often supported by conventional pseudo-spectral DNS!
† Some measurements, however, give direct access to spectral information, and even to anisotropic one, such as scattering of ultrasound waves (C. Baudet, S. Fauve) or light (D. Grésillon).

15.2 Linear Theory, "Homogeneous" RDT, WKB Variants, and LIA

centerline of a jet, etc." can be found in many recent papers, whereas the same assessment would have been considered ridiculous 20–30 years ago. This viewpoint, getting rid of inhomogeneous–anisotropic large scales in rather complex flows and focusing on small scales, considered as homogeneous–isotropic–intermittent, is not wrong, partly thanks to the use of incremental statistics. This is questionable, however, from a dynamical viewpoint: Apparent local isotropy can result from quasi-balanced inhomogeneous flux terms that are present in the transport equations. More generally, we have shown from a dynamical approach that the universality of small scales, independent of the way of injecting energy at large scales, is really questioned in many cases, even at very high Reynolds numbers. Despite significant advantages, the conventional viewpoint has some negative results:

- Not to encourage the building of smart experimental facilities, in which homogeneity can be really assumed in a very large spatial domain, following the ones presented in Chapters 4–8.
- To consider as marginally relevant the theoretical approach to flows, such as those studied in Chapters 4, 7, and 8, which can be really considered as homogeneous, but strongly anisotropic, at almost any scale.

Other arguments, which illustrate the interest of considering Fourier space (modal decomposition related to the Helmholtz decomposition, treatment once and for all of pressure fluctuations, giving the minimal number of dynamical modes), are presented in Chapters 1 and 2. Regarding the description of the cascade, it is important to stress that triadic spectral description, not even mentioning "closure," carries on much more information than third-order structure functions do. It accounts for triple correlations at three points, and not only at two points, and allows us to identify exact operators that underlie detailed conservation laws, such as Eqs. (3.122) and (3.124) for detailed conservation of both energy and helicity, Eqs. (3.214)–(3.216) for detailed conservation of energy and enstrophy, Eqs. (7.16)–(7.18) for detailed conservation of toroidal energy and vertical enstrophy, and Eqs. (8.6) to (8.8) for detailed conservation of QG energy and potential vorticity. Another point that deserves to be emphasized is the power of Waleffe's instability hypothesis that, starting from the exact detailed conservation laws and the stability analysis of a low-dimensional system, leads to *accurate predictions* dealing with triadic tranfers and induced cascades, even in nonhomogeneous cases.

15.2 Linear Theory, "Homogeneous" RDT, WKB Variants, and LIA

It is usually said that the "problem(s) of turbulence" come(s) from the nonlinearity of basic Navier–Stokes-type equations. This is only partially true, as "burgulence" (i.e., pseudo-turbulent behavior exhibited by the solution of the Burgers' equations), not to mention its 3D generalization to the cosmological gas, is essentially solved and understood. The quadratic advection term is probably always involved in the problem, but the projection onto a solenoidal subspace, in connection with the pressure term, is another important ingredient, at least in the nearly incompressible flow case. As a slightly different illustration (from basic dynamical equations, again), the advection term is completely removed in a pure Lagrangian alternative to Navier–Stokes equations, but nonlinear complicated operators rear their ugly heads through pressure and diffusive terms.

As another trivial remark, the validity of a linear approach depends on the state in which one performs the linearization.

Linear theories addressed here retain at least a part from exact dynamical equations, and include a straightforward treatment of the pressure term, together with the Helmholtz decomposition for purely incompressible and weakly compressible fluctuatings flows.

Homogeneous RDT offers interesting possibilities for reconciling the stability analysis and statistical approach, when it is consistent with exponential instability (hyperbolical instability in Chapter 5, barotropic instability for rotating shear, its baroclinic extension, and elliptical-flow instability in Chapter 8). Related destabilizing effects are mimicked by much simpler single-point RSM models, such as for the shear flow rotating around the spanwise direction, but only RDT or more sophisticated linear stability analyses really explain why, in connection with dominant pressure-released modes. In other cases, in which the destabilizing effect comes from a narrow band of angular modes in wave space, with the "rapid" fluctuating pressure allowing a resonant amplification to periodic "production," the "rapid" response of any RSM is poor (e.g. elliptical flow instability and periodic compression with swirl).

Even when RDT gives very few results about the evolution of statistics, it could suggest a good choice of eigenmodes for improving fully nonlinear theories, as illustrated in Chapter 4–8. Identification of a deterministic Green's function, possibly expressed in terms of a minimal number of solenoidal modes, from the basic linearized equation that governs the fluctuating field, is shown to be the best way for using linear theory: It is possible to predict the impact of the linear operator on any statistical moment, showing for instance a poor relevance of RDT dealing with single-time second-order moments, in contrast with interesting information given for two-time second-order statistics and third-order statistics (see Chapters 4, 7, 8, and 13).

WKB variants allow us to relax the assumption of homogeneity in the linear theory, or to suggest at least a nice illustration of what could be "local homogeneity" from a dynamical viewpoint. In contrast to homogeneous RDT, it is possible to identify localized instable zones in a base flow, which is smooth but more realistic than the admissible mean flows of homogeneous RDT and to quantify their contribution: An example of competing centrifugal, elliptical and hyperbolical instabilities is given in Chapter 8 for simple nonparallel flows with adjacent eddies.

It is important to point out some limitations. A generic instability such as the Kelvin–Helmholtz one cannot be afforded, even if homogeneity is relaxed. Despite a very promising extension of RDT to stratified flows with shearing effects, and possible prediction of baroclinic mechanisms, and to compressible shear (Chapter 10), this drawback cannot be ignored. For compressible shear flows, a possible depletion

15.3 Multipoint Closures for Weak and Strong Turbulence

of nonlinearity (with respect to the incompressible case) can explain an unexpected relevance of linear theory, at least in the homogeneous case.

Among the canonical-flow cases addressed in this book, only the case of the incompressible shear flow is a bit disappointing, restricting the approach to RDT: the important mechanism of redistribution of energy between RST components by nonlinear pressure terms, which is of course discarded, can be mimicked by very simple single-point closures.

WKB RDT can be applied to compressible flows, but its implicit ingredient of short-wave disturbance yields discarding the acoustic mode. Some extensions can be found in replacing the base-flow trajectories with the acoustic rays, as is touched on in Chapter 13. On the other hand, LIA has much in common with a purely homogeneous linear theory. Because there is no length scale given by the base flow, there is no restriction of the wavenumber range for the disturbance flows. One can say that the typical length scale of the mean flow is infinite in homogeneous RDT (or equivalently for the extensional base flow in stability analysis), whereas it is zero (the shock-wave thickness) in LIA. As in RDT, a transfer matrix can link upstream and downstream modal amplitudes of the disturbance field, but an entropic disturbance mode can be accounted for. Some wavelike response of the shock wave and its linkage to the full linear transfer matrix for the disturbance field is another useful feature, with no equivalent in RDT. A very striking result of LIA, beyond statistical results, is the possibility of advecting a temperature spot across the shock wave and giving rise to a pair of corotating vortices. In addition to a mathematical transfer term from the upstream entropy mode to downstream vortical mode via the baroclinic torque, a nice formation of structure is found!

15.3 Multipoint Closures for Weak and Strong Turbulence

An assessment of multipoint closures can be proposed. It appears that their use for "production-dominated" flow is probably a too complicated task, given the "return of investment" that one can expect. On the other hand, application to flows "without production," which consist of only nonpropagating neutral modes and wave modes in the linear eigenmode decomposition, is very promising.

The latter case includes incompressible HIT as the simplest, the whole velocity field being a trivial neutral mode in the linear inviscid limit.

In the particular case of turbulence subjected to pure rotation, the complex structural anisotropy is created by the nonlinear cascade, with the angular dependence of energy in wave space reflecting the loss of dimensionality. Such behavior occurs in other flow configurations in which the presence of dispersive waves is more important than the classic "production" mechanisms. Even without additional mean strain (such as the elliptical-flow unstable case), pure rotation induces complex "rapid" and "slow" effects, for which even the basic principles of single-point closures are questionable. Single-point closures look particularly poor because there is no production by the Coriolis force, whereas the dynamics is dominated by waves whose anisotropic dispersivity is induced by fluctuating pressure.

This suggests discriminating "turbulence dominated by production effects" from "turbulence dominated by wavy effects." In short, single-point closures are well adapted to simple turbulent flow patterns of the first class in rather complex geometry, whereas multipoint closures are more convenient for complex turbulent flows in simplified geometry, as illustrated by the second class.

15.3.1 The Wave-Turbulence Limit

Mathematical developments in the area of wave-turbulence theory (WT) have recently renewed interest in flows that consist of superimposed dispersive waves, in which nonlinear interactions drive the long-time behavior. Individual modes are of the kind

$$u'_i(\mathbf{x}, t) = a_i(t) \exp[\iota(\mathbf{k} \cdot \mathbf{x} - \sigma_\kappa t)], \tag{15.1}$$

with a known analytical dispersion law for $\sigma_k = \sigma(\mathbf{k})$. Similar averaged nonlinear amplitude equations can be found using either WT or multipoint closures (MPC), the advantages and drawbacks of which are briefly discussed below.

In the case of wave turbulence, statistical homogeneity and quasi-normal assumptions have equivalent counterparts, obtained by assuming *a priori* Gaussian random phases for the wave fields. In addition, *isotropic* dispersion laws such as $\sigma_k = |\mathbf{k}|^\alpha$ in Eq. (15.1) are almost exclusively treated in wave turbulence for deriving Kolmogorov spectra, with the key hypothesis of constant and isotropic energy fluxes across different scales associated with a wavenumber $|\mathbf{k}|$. By contrast, in geophysical flows, dispersion laws are anisotropic, with, for instance, $\sigma = \beta k_x/k^2$ in the case of Rossby waves, $\sigma = \pm 2\Omega k_\parallel/k$ for inertial waves, and $\sigma = \pm N k_\perp/k$ for gravity waves (k_x, $k_\parallel$, and $k_\perp$ are the components of the associated wave vector respectively, in the zonal direction, and the directions parallel or perpendicular to the rotation–gravity vectors). In the latter two 3D cases, this anisotropy is reflected by the strange conical – St. Andrew's cross – shape of isophase surfaces in typical experiments with a localized point forcing (see views of this type in Figure 4.10) and by angular-dependent energy drains when looking at nonlinear interactions, as illustrated in Chapters 4, 7, and 8.

At least if Eulerian correlations are considered, the MPC and wave-turbulence theories share in general an important background. Kinetic equations for mean spectral-energy densities of waves are found in wave turbulence, similar to homogeneous MPC. Their slow evolution is governed by similar energy transfer terms, which are cubic in terms of wave amplitudes (triads). There is also a possibility that these transfers involve fourth-order interactions (quartets) in wave turbulence when triple resonances are forbidden by the dispersion laws and/or by geometric constraints (e.g., shallow waters). Resonant quartets seem to be particularly relevant when resonances are seen in a Lagrangian description. When triple resonances are allowed, for instance in cases of rotating turbulence, stably stratified turbulence and magnetohydrodynamic turbulence, wave-turbulence kinetic equations have exactly the same structure as their counterpart in elaborated MPC. Hence, wave turbulence

15.3 Multipoint Closures for Weak and Strong Turbulence

and MPC have a common limit at very small interaction parameters (e.g., Rossby number, Froude number, magnetic number in magnetohydrodynamics). Of course, interactions between neutral modes, if they are present, and wave modes cannot be investigated by the pure theory of wave turbulence.

15.3.2 Coexistence of Weak and Strong Turbulence, With Interactions

When eigenmodes consist of nonpropagating, neutral, and wavy modes, one can expect very complex cascade processes. Wave turbulence, dominated by resonant triads, is the only modality in the absence of the nonpropagating mode. Accordingly, inertial wave turbulence in 3D rotating turbulence is really relevant if the Rossby number is sufficiently small.‡ When strong nonlinearity mediated by interactions that involve only the nonpropagating mode and weak nonlinearity involving at least a wave mode are face to face, the former can be considered as dominant. Both the toroidal cascade and QG cascade are therefore of interest in stratified and rotating turbulence. Note that the emergence of toroidal (idem QG) cascade is found in neglecting wavy modes in triadic interactions; this does not mean that waves have no effect; in contrast, this is because gravity waves (idem inertia-gravity) waves severely damp nonlinear contributions other than the pure toroidal (idem QG) ones by angle-dependent phase mixing that toroidal (idem QG) cascade emerges. For weakly compressible flows addressed in Chapter 9, the solenoidal turbulence is already well known, so that a pseudo-acoustical cascade appears as a relevant theme.

15.3.3 Revisiting Basic Assumptions in MPC

The derivation of statistical equations of MPC is often a very formal skill, so that these theories can be considered opaque and complicated. Let us mention (Moffatt, 2002) again:

> and the new approaches, particularly Kraichnan's (1959) DIA, were of such mathematical complexity that it was really difficult to retain that essential link between mathematical description and physical understanding, which is so essential for real progress.

In the same vein, A. Craya, in the early 1960s, evoked about DIA the Mona Lisa's (La Joconde) smile, having his strange beauty but some ambiguity. As a very interesting survey, Y. Kaneda proposed no fewer than seven different ways to derive DIA equations. The essential ingredient is a formal development around a Gaussian field, but the effective second-order spectral tensor and response tensor are only eventually defined by the final set of coupled equations that govern them, so that they cannot be specified *a priori*, and they can significantly differ from

---

‡ The 2D manifold appears as the limit of the wavy inertial mode at vanishing dispersion frequency, it is therefore a low-dimension slow mode, but not at all a 3D nonpropagating mode, filling all the space, as the toroidal mode is in stably stratified turbulence.

their zeroth-order counterpart. The derivation of EDQNM is less subtle, but the conventional presentation is often too close to a cooking recipe, with heuristic procedures (ED, Markovian) called into play in order to correct an initially too crude QN model. To derive EDQNM from DIA, using a specified form for the response tensor, and a so-called fluctuation–dissipation theorem to translate two-time correlations into single-time ones, may give information about ED and Markovianization; but this is really a too complicated and indirect way. Comparing EDQNM with WT theory, especially using the deep analysis of zero cumulant assumption by Benney and Newell (1969) quoted in Chapter 4, is really enlightening. It is first possible to understand why QN closure can be an *intrinsic (exact?) closure* in wave turbulence, getting rid of ED because the damping by phase mixing of dispersive waves is a very efficient and physical process, whereas Markovianization is bound by the natural separation into rapid phase terms and slowly evolving amplitudes of waves. It is perhaps necessary to think in a more physical way of the use of cumulants, and of a more convincing link between the fourth-order cumulants and the third-order ones, yielding the basic concept of ED for strong turbulence. Nth-order cumulants at N points represent the difference between statistical moments of the order of N and their factorized expression in terms of products of moments of smaller order. In this sense, a convergence to zero is ensured, which is not valid for the moments themselves, as soon as the points in the configuration space are sufficiently separated. Instead of speaking of a quasi-Gaussian distribution, which is often questioned in turbulent flow, one may address a pdf at four points, which reduces to almost a product of pdf's for sufficient separation lengths. The QN assumption, or more generally the EDQN one, could be more physically funded by an argument of maximum factorization of four-point pdf's, or maximum decorrelation between the different points, which is less constraining and does not use the word "Gaussian." Of course, this is a very preliminary proposal: We have in mind a four-point distribution without specifying more the configuration space (physical, Fourier, other?). As a simple illustration, factorization would be achieved for any tetrad including at least a long leg: very large tetrads with more than one long leg, flat tetrads with only one long leg.

15.4 Structure Formation, Structuring Effects, and Individual Coherent Structures

The two-point anisotropic description is more powerful, even if homogeneity is assumed, than is generally recognized. In rotating and stratified turbulence the anisotropic spectral description, with angular dependence of spectra and cospectra in Fourier space, allows quantification of columnar or pancake structuring in physical space. Among various indicators of the thickness and width of pancakes, which can be readily derived from anisotropic spectra, integral length scales $L_{ij}^{(n)}$ related to different components and orientations are the most useful. As another illustration (see Fig. 6.5), the streaklike tendency in shear flows can be easily found in calculating by RDT both the $L_{11}^{(1)}$ component, which gives the streamwise length of the streaks, and $L_{11}^{(3)}$, which gives the spanwise separation length of the streaks (as usual, 1 and 3 refer to streamwise and spanwise coordinates, respectively). In pure homogeneous RDT at constant shear rate, both length scales can be calculated analytically and their ratio (elongation parameter) is found to increase as $(St)^2$, $S = \partial U_1/\partial x_2$ being the shear rate. Of course, more realistic quantitative aspects of the true streaky structures found in the near-wall region are not captured, as discussed in Chapter 6.

It is often said that phase information is lost in homogeneous turbulence, but this is true only for single-time second-order statistics, and even does not exclude dynamical phase mixing, as illustrated by damped oscillations toward equipartition (equipartition in terms of poloidal and toroidal energy components for rapid rotation, with nontrivial transient evolution from initial imbalance if initial data are anisotropic, equipartition in term of poloidal and potential energy components for strong stratification, and similar evolution from initial imbalance). More informative and surprising phase mixing is found for two-time second-order statistics, even in the pure linear regime: This illustrates that dispersive waves can drive the Lagrangian diffusion (passive tracers, single-particle displacement), a role that is often attributed to purely spatial structures, such as coherent vortices, in the turbulence community. Finally, nonlinear formation of structures in rotating and stratified flows, which is emphasized in this book, means formation of vortex structures, – waves *are structures too* but are spatiotemporal (delocalized in space) coherent events. In fact, a subtle interplay of linear and nonlinear effects is called into play.

As a final remark, statistical indicators in homogeneous anisotropic turbulence can quantify some average characteristics of structures (e.g., aspect ratios of cigar-shaped and pancake-shaped structures, vorticity skewness for quantifying asymmetry in terms of cyclonic and anticyclonic vorticity for cigar-shaped structures), whereas information on their dynamics can be given by statistical equations. In addition, some individual coherent structures, localized in space (and in time?), if not really obtained in statistical model equations, are found in snapshopts from DNS, as realizations of homogeneous turbulence.

15.5 Anisotropy Including Dimensionality, a Main Theme

This is emphasized throughout this book, except in Chapters 3 and 9. It appears as a multifold and rich property of turbulent flows, even those without production, and affects both the multiscale energy distribution and the cascade process, possibly until smallest scales. Our viewpoint contrasts with what is currently admitted in the turbulence community. In the engineering community, anisotropy is considered as characterized only by the deviatoric part of the RST (b_{ij} is "THE" anisotropy tensor), despite the more general investigation introduced by Reynolds and Kassinos in their structure-based modeling approach. In the physicist community, inhomogeneity–anisotropy is considered only for the largest scales, generally out of investigation, whereas scales that merit attention are seen as homogeneous–isotropic–intermittent. If attention is paid to anisotropy, with recent studies using the SO(3) symmetry group, this concerns only the small anisotropy identified by a very small number of angular harmonics.

Finally, this book includes the material to revisit a general theory of axisymmetric turbulence. Axial symmetry with and without mirror symmetry is the simplest symmetry for an exhaustive statistical and dynamical approach to strong anisotropy, in both spectral and physical space, using all the theoretical tools used herein, including the most sophisticated ones. Application to magnetohydrodynamic flows with external strong magnetic fields could be the next step. This step can be useful for a collaboration between specialists of turbulence in fluid and specialists of turbulence in plasmas. The existence of the International Thermonuclear Experimental Reactor (ITER) worldwide project critically needs such a collaboration. Problems of turbulence, such as the "anomalous" heat and mass (for ions) transfer in the radial direction, is expected to be a severe problem in a future huge Tokamak. Geodynamo and astrophysical turbulence are other instances.

15.6 Deriving Practical Models

Finally, one may anticipate some criticism against this book: too many equations, too few practical results! A striking feature of several homogeneous flows discussed in this book is that they escape turbulence models used in engineering applications. The test is fair, as we have considered the best adapted mathematical formalism to deal with the subtleties of the problem, from $\mathcal{K} - \varepsilon$ models to anisotropic MPC, with a lot of intermediate links.

The way to derive more practical applications, from useful simple scaling laws to once-and-for-all calculation of parameters (eddy diffusivity, anisotropic ratios, etc.), must be discussed.

The terms appearing in the rate equations for Reynolds stress models in homogeneous turbulence can be exactly expressed as integrals over Fourier space of spectral contributions derived from the second-order spectral tensor $\hat{R}_{ij}$, which is the Fourier transform of double correlations at two points, and from the third-order "transfer" spectral tensor T_{ij}. All one-point quantities in the equation that governs $\overline{u'_i u'_j}$ can be expressed as integrals over wavenumber space, as for Eq. (2.63). The equation for the dissipation rate $\varepsilon = \nu \overline{\omega_i \omega_i}$ (in quasi-homogeneous and quasi-incompressible turbulence) can be derived from the exact equation that governs the fluctuating-vorticity field ω_i. Recall that the practical procedure for deriving the ε-equation hardly uses the latter exact equation and consists of basing the equation for $\dot{\varepsilon}/\varepsilon$ on the equation for $\dot{\mathcal{K}}/\mathcal{K}$ with adjustable constants.

About single-point closures, one may recall that the knowledge of the mean (Reynolds averaged) flow together with the RST in every point (with a possible limited time dependence[¶]) would have been the Holy Grail in turbulence modeling

---

[¶] Let us recall that there is no conceptual obstacle against unsteady RANS, and that seminal studies about 1975 dealt with the time development of Reynolds stresses in homogeneous turbulence subject to a given mean flow.

15.6 Deriving Practical Models

20 or 30 years ago. More information can be required now. In this sense the criticism against single-point closure techniques deals less and less with their incorrect closure assumptions, and more and more with the unsufficient information carried out by them. More information about low-probability events, dramatic unsteadiness, coherent structures, for predicting hazards, is needed in engineering and in environmental flows. Two-point two-time statistics can be useful for predicting dispersion processes and radiated noise (e.g., applying acoustic analogies to quasi-incompressible vortical flows). Looking at passive and reactive scalar fields, information on pdf's is needed too.

Bibliography

Moffat, H. K. (2002). G. K. Batchelor and the homogenization of turbulence, *Annu. Rev. Fluid Mech.* **34**, 19–35.

Index

Absolute potential vorticity, 247
Absolute vorticity, 256
Acoustic equilibrium
    strong form, 294, 300, 308, 349
    weak form, 294–296, 308, 315
Acoustic power, 305
Acoustic production spectrum, 305
Anisotropy
    anisotropy tensor b_{ij}, 27–28
    asymptotics (homogeneous shear), 196
    circulicity tensor, 27
    dimensionality tensor, 27, 41
    directional, 40
    Fourier space description, 38–42
    local isotropy breakdown, 214–216
    polarization, 40
    stropholysis tensor, 27, 42

Baroclinic instability, 258
Barotropic instability, 243
Boussinesq approximation, 223
Bradshaw number, 253
Brunt–Väisälä frequency, 223
Buoyancy efficiency parameter, 240
Burgers' number, 248

Coherent structures, 98–102
    dynamics, 159–163, 187–188, 204–213, 319–321, 355
    Vortex
        Q-criterion, 100
        Δ-criterion, 101
        λ_2-criterion, 101
        Burgers', 98
        Horiuti criterion, 101
        Lund–Rogers criterion, 100
        Swirling-length criterion, 101
    Vortex sheet
        Burgers', 99
        curved sheet, 102
        flat sheet, 102
        Horiuti criterion, 102
    Vortex-tube dynamics, 102–109
Conservation laws
    Detailed
        helicity, 80–81
        kinetic energy, 80–82
        potential enstrophy, 246
        Quasi-geostrophic energy, 246
        toroidal energy, 230, 232
        vertical toroidal enstrophy, 230, 232
    Global
        compressible case, 290
        helicity, 80–81
        kinetic energy, 80–81
Coriolis force, 17, 131, 137, 148, 267
Correlation function
    longitudinal $f(r)$, 70
    transverse $g(r)$, 70

Dissipation
    dilatational, 287, 290, 301, 307, 316–317
    evolution equation, 75
    solenoidal, 23, 287, 290, 301, 316–317

Eddy-damping approximation, 88, 90, 298–303, 423–425
EDQNM, 3, 86–94, 149–153, 186–187, 206, 233, 296, 298–300
Eikonal equation, 182, 416
Ekman number, 132
Elliptic instability, 243
Energy cascade, 75, 83–85, 108, 111–112, 133–135, 229–231, 245–250
Energy spectrum, 51
    Batchelor spectrum, 63
    dilatational part, 290, 307, 319
    dissipative range, 54
    inertial range, 55

Energy spectrum (*cont.*)
  Saffman spectrum, 64
  self-similar solution, 61
  solenoidal part, 289, 319
  total acoustic energy, 291
Ensemble average, 19
Entropy (definition), 274
Ertel theorem, 220, 248

Favre average, 282
Froude number, 220

Geometric optics, 415
Gravity waves, 225–226
Green's function, 21, 183, 227, 298, 347, 406–409, 425–426

Helicity spectrum, 39
Hyperviscosity, 122

Ince equation, 254
Inertial waves, 139–143, 226
Intermittency, 6, 72

Kármán–Howarth equation, 71
Kelvin theorem, 16
Kelvin waves, 106
Kelvin–Helmholtz instability, 107, 239, 319, 351
Kinetic energy
  acoustic, 307, 364
  dilatational, 290, 307, 371, 378
  evolution equation, 24
  Lin's equation, 74
  negative production, 167
  solenoidal, 290, 371, 378
Kinetic-energy decay
  Birkhoff–Saffman invariant, 63
  decay exponent, 62–69
  incompressible regimes, 57–58
  Loitsyansky invariant, 63
  Oberlack's invariants, 64
  PLE hypothesis, 60, 65–66, 118
  self-similarity, 59–69
  self-similarity breakdown, 66–67
Kolmogorov constant, 54, 94
Kolmogorov's 4/5 law, 73, 92, 250
Kolmogorov's hypothesis (1941), 54, 72
Kovasznay mode
  acoustic mode, 276
  entropy mode, 276
  vorticity mode, 275

Lagrangian description
  Fourier space, 181–183
  physical space, 10–12, 16–19
LIA, 8, 361–363, 384–404
Lin's equation, 74, 151, 231, 249, 290

Mach number
  convective, 327
  distortion, 329
  gradient, 329, 348
  turbulent, 291, 300, 308, 314, 316, 329, 348
Manifold
  2D, 143
  slow, 143, 155
Markovianization, 88, 428–437
Mean-flow admissibility
  compressible case, 333–336
  incompressible case, 24–26
Mesoscales, 247
Modal decomposition, 5
  Chu–Kovasznay, 6, 274–279, 386
  helical, 32–33, 36–38, 76
  Helmholtz, 5, 220, 279–281
  poloidal–toroidal, 5, 31–32, 36–38, 228, 280
  quasi-geostrophic/ageostrophic, 243–246
  wave-vortex, 5, 220, 243

Navier–Stokes symmetries, 59

POD, 42
Polytropic coefficient, 355
Potential enstrophy, 247
Prandtl number, 224, 275
Pressure
  equation, 347
  equation (compressible RDT), 339
  Poisson equation, 14, 34, 278, 280, 339
  rapid term, 21–23
  role in turbulence dynamics, 120–122
  slow term, 21–23, 79
  solenoidal case, 14
  spectrum (compressible case), 290, 314
  spectrum (incompressible part), 296, 304
  splitting (compressible case), 280
  variance, 290
Pressure-released
  rotating stratified turbulence, 253
  strained turbulence, 179, 330–332
  turbulence, 14
Probability density function, 45, 56
Proudman theorem, 132, 154
Pseudo-sound, 300, 306

Quasi-geostrophic energy, 246
Quasi-normal hypothesis, 45, 56, 86–90

Rankine–Hugoniot jump conditions
  basic, 360
  linearized, 393
  mean flow, 385
  mean turbulent flow, 379–381

Index

vorticity, 361
RDT, 3, 426
    baroclinic instability, 256–259
    elliptical instability, 259–265
    fundamentals, 180–184
    inhomogeneous, 418
    pressure-released, 179, 253, 330–332
    pure shear case, 199–206
    rotating stratified turbulence, 243–245
    shear and rotation–buoyancy, 251–253
    stable stratification case, 232–233
    vs. LIA, 378–379
    zonal, 415
Reynolds decomposition
    definition, 20
    fluctuating flow equations, 20
    mean-flow equations, 20
    Reynolds stress tensor, 22–27
Reynolds stress equations
    compressible flows, 281–286
    Craya's equations, 36, 74
    general case, 22
    homogeneous shear, 197, 328
    irrotational strain, 174–178
    pure rotation, 135–139
    stable stratification, 224–225
Richardson number, 255
Richardson number (rotational), 253
Rossby number, 131, 220

Scale invariance, 343
Self-sustaining turbulent cycle, 209–213
Shear rapidity, 195, 334, 348
Shell models, 6, 418
Shock instabilities, 358
Shocklets, 315–318, 321–323, 355

Spectral-energy transfers
    helical modes, 76
    phase scrambling, 133, 143–144, 225, 294
    scalar term $T(k)$, 74
Statistical moments
    high-order, 43
    multipoint, 19
    single-point, 19
    two-point correlation, 70
    velocity-correlation tensor, 20
Strong Reynolds analogy, 370
Synoptic scales, 243

Taylor's frozen-turbulence hypothesis, 50
Tilting vorticity, 264
Townsend's equations, 415
Triadic interactions
    distant, 81
    forward (F-type), 84
    generals, 76–80
    local, 81
    nonlocal, 81
    resonant, 143, 145–148, 298
    reverse (R-type), 84
    Waleffe's instability hypothesis, 81–86, 119, 146, 230
Turbulence stream-function vector, 27

Vortex wrapping, 206
Vorticity, 15–16, 282, 289
VSHF, 31, 221, 228

Wave turbulence, 5, 45, 145–148, 150, 245–246, 297–298
Weber equation, 15, 178
WKB method, 415

Zig-zag instability, 220, 238